Preface

Why write yet another book on organic chemistry? Well, I have been ⌐_____ long time, about 40 years. I taught the entire year organic course and I used my own book for most of that time. I wrote the book the way I like to teach organic. Over the years, many of my students encouraged me to publish it. I thought about it, but I could never bring myself to take on such a huge project.

In 2012 I retired to half-time teaching. No longer was I assigned to teach the year-long course. Instead, I occasionally get to teach the one quarter, survey organic course. Survey courses are often taught as "memorization" courses, full of vocabulary terms, but short on substance. The logic of organic chemistry is so beautiful (and understandable) that I could not bring myself to teach organic chemistry that way. I took my longer book and cut out large amounts of material that is usually included in a longer course. I mostly included what I felt were the essential "logic" points to acquire an appreciation of organic chemistry.

I want this book to be accessible (inexpensive), so I self-published it at Amazon. Of course, the downside is that I do not have an editor or reviewers to help me proof my book. I hereby designate you to be my editors. Let me know what works and what doesn't. I tried not to water down the essential logic of organic chemistry (too much), but I strove for the "simpler" explanations over more complex explanations.

I can't tell you that this is an easy book, but I will tell you that if you work your way through the book, you will have good insight into how organic chemistry works. Also, biochemistry is just organic chemistry on complicated molecules, and you will have a good foundation for learning how biochemistry works too.

You should be able to analyze and digest new material not covered in the book with the approach provided here. Part of the fun of learning organic chemistry is to use your newfound knowledge to speculate about how something happens, and it's really neat when you turn out to be correct. Have fun with it, but try to use the guiding principles in your arguments. It is the give and take in your thinking that leads you to a more complete understanding of how Nature works. Teaching carried to its ultimate goal would make you independent of your teacher, possibly to progress beyond those who teach you now. This book is another step along that path.

You already know it, but I will repeat the obvious. You must read and reread the material. You must work and rework problems, and then do some more problems. As much as possible, use a pencil and paper to write out your thoughts, questions and ideas as you go along. Some students have told me that "white" boards work better for them (write, erase, write, erase, to the n power). Write your questions down in your book or on a separate piece of paper to organize later or to bring up with your professor or TA at an office hour. Work through the frustrations we all encounter in any challenging endeavor. Use your professor, teaching assistant, notes, book and fellow students to clear up confusing points. Others can't do the work of learning for you, but they can often direct you toward an easier path or bring you back to the right path.

Just how many times do you have to reread the material and redo problems? There is no single answer that works for every individual. It depends on how much time you have available to study, your discipline, your persistence, your innate ability to learn new things and your prior knowledge in related areas. If your foundation is weak, you will have to work harder than someone with a good foundation. But you will be building a more solid foundation for your future learning in subsequent courses, such as biochemistry, or for use in your career.

If you put in the work to understand the ideas in this book, your pursuit can become an energizing experience, and you might not even be able to stop yourself as you discover the answers to some of the mysteries of life. Is organic a struggle to survive or an adventure to enjoy? It greatly depends on what goes on inside your own head. What goes on there far surpasses in importance what your instructor says in class or what is written on these pages. Learning organic chemistry is an incredibly rich intellectual endeavor that will provide you with endless hours of mind-expanding enjoyment, if you let it.

There is an old Chinese saying about leaders, which I have slightly modified to fit my philosophy about teaching and learning.

A teacher is best when students barely know he exists,
Not so good when students obey and acclaim him,
Worse when they despise him.
But of a good teacher who talks little,
When his work is done, his aim fulfilled,
They will say "We did it ourselves".

Preface

The Essential Logic of Organic Chemistry

CreateSpace

First Edition (December, 2016)

ISBN-13: 978-1541060319
ISBN-10: 1541060318

Nonfiction > Reference > Organic Chemistry

THE ESSENTIAL LOGIC OF ORGANIC CHEMISTRY
(aka, How to Cure the Benzene Blues)

Phil Beauchamp

Table of Contents

Solutions for problems are available separately.   217 problems

CHAPTER 1 – ATOMIC STRUCTURE AND SHAPES OF ATOMS IN MOLECULES

Atomic Structure Since molecules are constructed from atoms, we will briefly review the structure and properties of atoms.

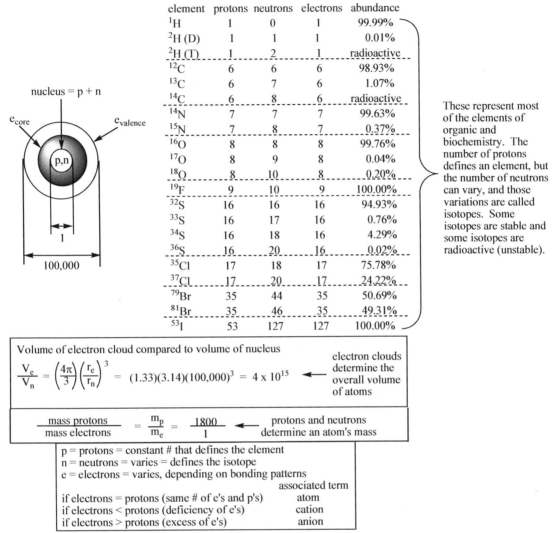

element	protons	neutrons	electrons	abundance
^{1}H	1	0	1	99.99%
^{2}H (D)	1	1	1	0.01%
^{2}H (T)	1	2	1	radioactive
^{12}C	6	6	6	98.93%
^{13}C	6	7	6	1.07%
^{14}C	6	8	6	radioactive
^{14}N	7	7	7	99.63%
^{15}N	7	8	7	0.37%
^{16}O	8	8	8	99.76%
^{17}O	8	9	8	0.04%
^{18}O	8	10	8	0.20%
^{19}F	9	10	9	100.00%
^{32}S	16	16	16	94.93%
^{33}S	16	17	16	0.76%
^{34}S	16	18	16	4.29%
^{36}S	16	20	16	0.02%
^{35}Cl	17	18	17	75.78%
^{37}Cl	17	20	17	24.22%
^{79}Br	35	44	35	50.69%
^{81}Br	35	46	35	49.31%
^{53}I	53	127	127	100.00%

nucleus = p + n

e_{core} $e_{valence}$

p,n

1

100,000

These represent most of the elements of organic and biochemistry. The number of protons defines an element, but the number of neutrons can vary, and those variations are called isotopes. Some isotopes are stable and some isotopes are radioactive (unstable).

Volume of electron cloud compared to volume of nucleus

$$\frac{V_e}{V_n} = \left(\frac{4\pi}{3}\right)\left(\frac{r_e}{r_n}\right)^3 = (1.33)(3.14)(100,000)^3 = 4 \times 10^{15}$$ ← electron clouds determine the overall volume of atoms

$$\frac{\text{mass protons}}{\text{mass electrons}} = \frac{m_p}{m_e} = \frac{1800}{1}$$ ← protons and neutrons determine an atom's mass

p = protons = constant # that defines the element
n = neutrons = varies = defines the isotope
e = electrons = varies, depending on bonding patterns

	associated term
if electrons = protons (same # of e's and p's)	atom
if electrons < protons (deficiency of e's)	cation
if electrons > protons (excess of e's)	anion

What if you were the nucleus of an atom?

Let's assume you are 5 feet tall and weigh 112 pounds. The diameter of the electron cloud would be approximately 500,000 feet. Since there are about 5,000 feet in a mile the diameter would be about 100 miles (approximately the distance of Los Angeles to San Diego). The volume would be $(4/3)(\pi)r^3$, or about 500,000 miles^3. Amazingly, the entire electron cloud would only have a mass of $(1/1800) \times (112)$ pounds $\approx$ 1 ounce! (Real air would weigh about 10^{16} pounds = 10 quadrillion pounds!)

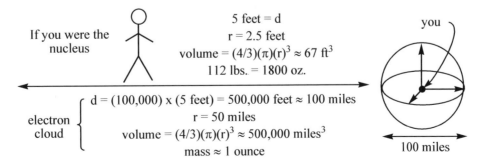

If you were the nucleus

5 feet = d
r = 2.5 feet
volume = $(4/3)(\pi)(r)^3 \approx 67$ ft^3
112 lbs. = 1800 oz.

you

electron cloud
d = (100,000) × (5 feet) = 500,000 feet ≈ 100 miles
r = 50 miles
volume = $(4/3)(\pi)(r)^3 \approx 500,000$ miles^3
mass ≈ 1 ounce

100 miles

What do atomic orbitals look like?

Atomic Orbital - a region in space about an atom where there is a high probability of finding up to two electrons (2 e-). The shape of an orbital is predicted by the mathematics of quantum mechanics (we just need to know what the pictures look like). Where the electrons are is where the bonds and lone pairs will be. The bonds give rise to the shape and polarity of the molecule which in turn affects the chemistry we observe and study.

a. **"s orbitals"** are spherical in shape. Electrons have both particle (mass) and wave (phase) nature. The wave nature of electrons (phase) may change as the distance increases from the nucleus but at any fixed distance from the nucleus the phase nature of s electrons will be the same (think of layers on an onion). Nodes are regions in space where the probability of finding an electron is zero (the phase is zero).

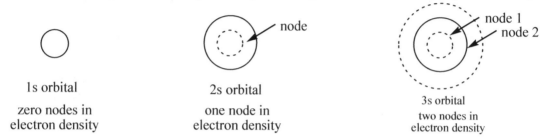

1s orbital
zero nodes in
electron density

2s orbital
one node in
electron density

3s orbital
two nodes in
electron density

b. **"p orbitals"** have a dumbbell shape with two lobes at 180° to one another with opposite phase nature. All p lobes intersect at the nucleus with zero probability of finding any electron density (= node).

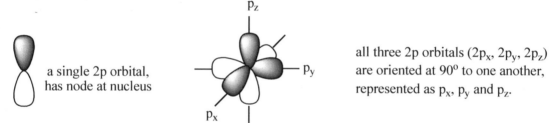

a single 2p orbital,
has node at nucleus

all three 2p orbitals ($2p_x$, $2p_y$, $2p_z$) are oriented at 90° to one another, represented as p_x, p_y and p_z.

c. **"d and f orbitals"** have more complicated shapes and are not important for beginning organic chemistry. However they are important for understanding the inorganic world and many biochemical molecules involve the chemistry of d orbitals (like iron in hemoglobin, cobalt in vitamin b_{12}, and zinc, copper, manganese and others in proteins.). However, we will not discuss these in our course.

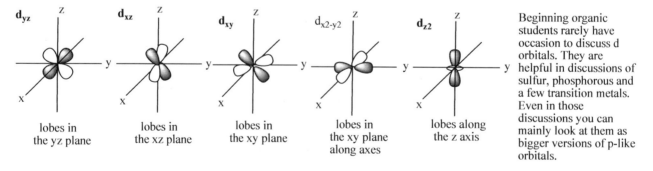

d_{yz}
lobes in
the yz plane

d_{xz}
lobes in
the xz plane

d_{xy}
lobes in
the xy plane

d_{x2-y2}
lobes in
the xy plane
along axes

d_{z2}
lobes along
the z axis

Beginning organic students rarely have occasion to discuss d orbitals. They are helpful in discussions of sulfur, phosphorous and a few transition metals. Even in those discussions you can mainly look at them as bigger versions of p-like orbitals.

Part of the periodic table is shown below. Many of these elements are used in organic chemistry and biochemistry. We will use some of these in our discussions, though often only briefly. The main elements in our course will be C, H, O and N, with Cl, Br and I contributing smaller roles. S and P play important roles in biochemical examples.

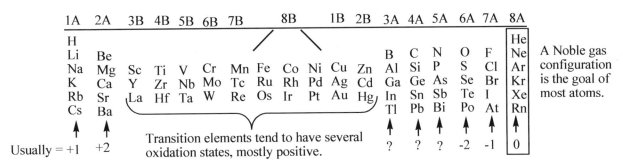

Elements on the left side and in the middle tend to lose electrons forming cations, while elements on the right side tend to gain electrons forming anions or to share electrons forming covalent bonds. Carbon, in the middle right, usually shares electrons in covalent bonds (single, double and triple).

Periodic Table (another view) – Atomic orbitals (and electrons) are found in the following locations of the periodic table. [$Z_{effective}$ = (# protons) – (core electrons)]

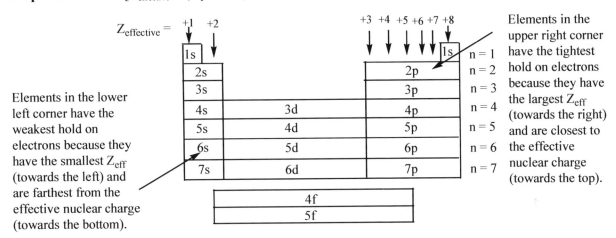

Atomic Configuration provides a representation of the electrons about a single isolated atom. Isolated atoms look different than atoms in molecules. Isolated atoms use *atomic* orbitals (s,p,d,f) to fill in electrons according to a few simple rules. The electrons closest to the nucleus are held the tightest and called the core electrons (full shells), while those in the incompletely filled outermost shell are called valence electrons. The valence electrons largely determine the bonding patterns and chemistry of the atom in order to gain a Noble gas configuration.

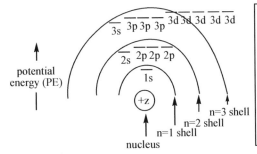

Electrons fill into atomic orbitals according to the following rules.

1. **Pauli Exclusion Principle** - only two electrons may occupy any orbital and those electrons must have opposite spins
2. **Hund's Rule** - electrons entering a subshell containing more than one orbital (p,d,f) will spread themselves out over all of the available orbitals with their spins in the same direction until the final subshell is over half filled, when they begin to pair up.
3. **Aufbau Principle** - orbitals are filled with electrons in order of increasing energy (lowest PE to highest PE)

Problem 1 – Write an atomic configuration for H, He, Li, Be, B, C, N, O, F, Ne, Na, Mg, S, Cl, Br, I.

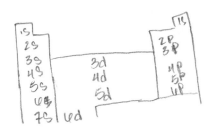

Chapter 1

Are electrons all the same?

(e_{val}) = valence electrons = The outermost layer of electrons determines the bonding patterns of atoms. The usual goal is to attain a noble gas configuration. This is accomplished by losing e's (forming cations) or gaining e's (forming anions) or sharing e's (covalent bonds)

(e_{core}) = core electrons = The innermost layer(s) of electrons (usually full shells or subshells) are held too tightly for bonding (sharing) and not usually considered in the bonding picture. These e's cancel a portion of the nuclear charge (called shielding) so that the valence e's only see part of the nuclear charge, called $Z_{effective}$.

$Z_{effective}$ = (# protons) - (core e's) = the effective nuclear charge. This is the net positive charge felt by the valence e's (bonding and lone pairs). $Z_{effective}$ = same # as the column of the main group elements.

Effective nuclear charge - $Z_{effective}$ is the net positive charge felt by the valence electrons (bonding and lone pair electrons). It can be estimated by subtracting the number of core electrons from the total nuclear charge.

$$Z_{effective} = (Z_{total}) - (\text{# core electrons}) = \text{approximate positive charge felt by the valence electrons}$$

$Z_{effective}$ is an important parameter in determining how tightly an atom pulls electrons to itself and helps determine the polarity of its bonds. Polarity is one of our most important concepts.

Problem 2 – What is the total nuclear charge and effective nuclear charge for each of the atoms below? How does this affect the electron attracting ability of an atom?

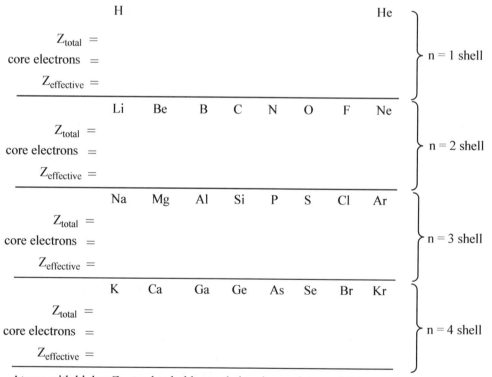

Atoms with higher Z_{eff} tend to hold onto their valence electrons tighter.
Atoms with the same Z_{eff}, but lower valence shells (higher in a column) also hold valence electrons tighter because the valence electrons are closer to the nucleus having the same Z_{eff}.

Chapter 1

Ionization potential as a measure of an atom's electron attracting power.

Ionization potential is the amount of energy needed to remove an electron from an atom, molecule or ion and provides an experimental measure of an atom's electron attracting power. An atom's electron attracting power is a key idea in establishing the concept of polarity.

The curved arrow shown in the equation below may be new to you. In organic chemistry, we use curved arrows to show how key electrons change in a chemical reaction (it's called *arrow pushing*). As electrons leave one location and move to a new location, curved arrows keep track of this movement. When you learn how to use curved arrows, you will understand, pretty well, how organic chemistry and biochemistry work. We will include curved arrows at every opportunity to maximize your exposure to these new tools. The half-headed arrow below in the following equation indicates single electron movement (less common for us). Later we will show two electron movement with full-headed arrows (more common for us).

Experimental evidence for an atom's attracting power for electrons can be seen in a number of properties, such as ionization potential, atomic radius, ionic radius and others. A few examples are provided in the tables below.

Table 1 – First ionization potentials of atoms (energy to remove an electron from a neutral atom)

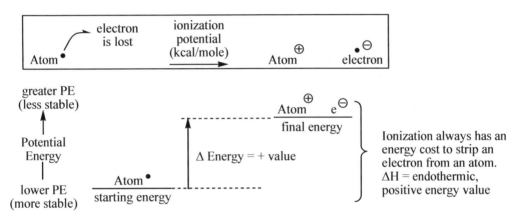

Energy to ionize an electron from a neutral atom = IP1 (units are kcal/mole). Compare rows and columns.							
Group 1A	Group 2A	Group 3A	Group 4A	Group 5A	Group 6A	Group 7A	Group 8A
H +314							He +568
Li + 124	Be +215	B + 192	C + 261	N + 335	O + 315	F + 402	Ne + 499
Na +118	Mg +177	Al +138	Si +189	P +242	S +239	Cl +300	Ar +363
K +99	Ca +141	Ga +138	Ge +182	As 226	Se +225	Br +273	Kr +323
$Z_{eff} = +1$	$Z_{eff} = +2$	$Z_{eff} = +3$	$Z_{eff} = +4$	$Z_{eff} = +5$	$Z_{eff} = +6$	$Z_{eff} = +7$	$Z_{eff} = +8$

The general trend, across a row, is that the ionization potential gets larger and the hold on electrons is stronger. Why? The answer is found in the size of $Z_{effective}$. As we move from lithium (IP$_1$ = +124) to carbon (IP$_1$ = +261) to fluorine (IP$_1$ = +402), the valence shell stays the same (n = 2), but the effective nuclear charge holding those electrons keeps increasing, from +1 to +4 to +7 (second row elements are all shielded by the n=1 shell with 2 electrons). Because each extra electron goes into the same shell (n = 2), there is essentially no shielding of the nucleus by any of the additional electrons (the electrons are doing their best to avoid one another). It's a lot harder to pull an electron away from a +7 charge than a +1 charge. It seems reasonable that fluorine has a stronger attraction for electrons than the other second row elements, even when those electrons are shared in a chemical bond. There are a few deviations that we will ignore.

In a column, all of the atoms have the same $Z_{effective}$. What changes in a column is how far away the valence electrons are from that same effective nuclear charge (Z_{eff}). Valence electrons in inner quantum shells are held tighter because they are closer to same Z_{eff}. This is evident by the larger energy required to ionize elements at the top of each column.

Problem 3 - Which atom in each pair below probably requires more energy to steal away an electron (also called ionization potential, IP)? Why? Are any of the comparisons ambiguous? Why? Check your answers with the data in the I.P. table on the previous page.

a. C vs N   b. N vs O   c. O vs F   d. F vs Ne

Problem 4 - Explain the atomic trends in ionization potential in a row (Na vs Si vs Cl) and in a column (F vs Cl vs Br).

Atomic radii and ionic radii also support these periodic table trends in electron attracting power. Smaller radii, below, indicate 1. in a row (shell) there is a stronger contraction of the electron clouds due to higher $Z_{effective}$ attraction for the electrons and 2. in a column, closer distance of electrons to a particular $Z_{effective}$ nuclear charge, due to electrons being in a lower principle quantum shell.

When an atom acquires a negative charge (becomes an anion), it expands its size due to the greater electron-electron repulsion. When an atom loses electron density, the remaining electron cloud contracts. Remember, most of an atom's mass is in the nucleus, but most of its size is due to the electron cloud.

Table 2

Neutral atomic radii in picometers (pm) = 10^{-12} m [100 pm = 1 angstrom]							
H = 53							He = 31
Li = 167	Be = 112		B = 87	C = 67	N = 56	O = 48	F = 42	Ne = 38
Na = 190	Mg = 145		Al = 118	Si = 111	P = 98	S = 88	Cl = 79	Ar = 71
K = 243	Ca = 194	3d....	Ga = 136	Ge = 125	As = 114	Se = 103	Br = 94	Kr = 88
Rb = 265	Sr = 219	4d....	In = 156	Sn = 145	Sb = 133	Te = 123	I = 115	Xe = 108

Problem 5 - Explain the atomic trends in atomic radii in a row (Li vs C vs F) and in a column (C vs Si vs Ge).

Table 3

Cations (on the left) and anions (on the right) radii in picometers (pm) = 10^{-12} m   [100 pm = 1 angstrom]

Li^{+1} = 90	Be^{+2} = 59		B^{+3} = 41	C =	N^{-3} = 132	O^{-2} = 126	F^{-1} = 119
Na^{+1} = 116	Mg^{+2} = 86		Al^{+3} = 68	Si =	P =	S^{-2} = 170	Cl^{-1} = 167
K^{+1} = 152	Ca^{+2} = 114	3d....	Ga^{+3} = 76	Ge =	As =	Se^{-2} = 184	Br^{-1} = 182
Rb^{+1} = 166	Sr^{+2} = 132	4d....	In^{+3} = 94	Sn =	Sb =	Te^{-2} = 207	I^{-1} = 206

cations   anions

Problem 6 - In the tables above:

a.   Explain the cation distances compared to the atomic distances (Li, Be, B). (Tables 2 and 3)

b.   Explain the anion distances compared to the atomic distances (N, O,F). (Tables 2 and 3)

c.   Explain the cation distances in a row (Li, Be, B). (Table 3)

d.   Explain the anion distances in a row (N, O, F). (Table 3)

Chapter 1

Atoms in Molecules Behave Differently

Atoms behave differently in bonding situations with other atoms. The goal is to acquire a Noble gas configuration by ionic, polar covalent or pure covalent bonding. This is often described as the "octet rule," (8 valence electrons) for main group elements (1A through 8A). The octet rule is almost always true for second row elements (C, N, O and F). It is mostly true for elements in higher rows, but there are many exceptions.

Bonds represent an overlap of orbitals, which are a region where negative electron density is present. Electrons allow positive nuclei to remain in close proximity (come together). In our course covalent bonds will represent two shared electrons between two bonded atoms. We propose two types of covalent bonds: The first are sigma bonds, where electron density is directly between the two bonded atoms. The second type are pi bonds, where electron density is above and below (or in front and in back) of a line connecting the two bonded atoms. Sigma bonds are always the first bond and pi bonds are the second (and third) bonds overlapping a sigma bond. Pi bonds are represented by sideways overlap of two p orbitals.

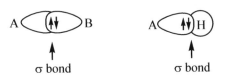

possible sigma bonds - the electron density is directly between the two bonded atoms, sigma bonds are always the first bond

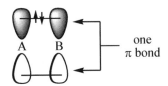

pi bonds - the electron density is above and below the two bonded atoms (or in front and in back), none is directly between, pi bonds are always second or third bonds

Electronegativity measures the attraction an atom has for electrons in chemical bonds. There are many proposed scales to do this, we will use the Pauling scale. The larger the number the greater is the attracting power of an atom for the electrons in chemical bonds. Electronegativity will determine nonpolar, polar and ionic characteristics of bonds, and when shapes are included it determines the same attributes in molecules. Compare rows and compare columns in the following table. The largest numbers are found in the upper right corner (strongest attraction for electrons) and the smallest numbers are found in the lower left corner (weakest attraction for electrons). Group 8A elements (Noble gases) towards the top don't have a value listed because the ones toward the top don't make bonds.

X (chi) is the symbol used for electronegativity, which is the property the indicates an atoms attraction for electrons in chemical bonds with other atoms.

Approximate electronegativity values for some main group elements.

Group 1A $Z_{eff}=+1$	Group 2A $Z_{eff}=+2$		Group 3A $Z_{eff}=+3$	Group 4A $Z_{eff}=+4$	Group 5A $Z_{eff}=+5$	Group 6A $Z_{eff}=+6$	Group 7A $Z_{eff}=+7$	Group 8A $Z_{eff}=+8$
H = 2.2								He = none
Li = 1.0	Be = 1.5		B = 2.0	C = 2.5	N = 3.0	O = 3.5	F = 4.0	Ne = none
Na = 0.9	Mg = 1.2		Al = 1.5	Si = 1.9	P = 2.2	S = 2.6	Cl = 3.2	Ar = none
K = 0.8	Ca = 1.0	3d elements	Ga = 1.6	Ge = 1.9	As = 2.2	Se = 2.5	Br = 3.0	Kr = 3.0
Rb = 0.8	Sr = 0.9	4d elements					I = 2.7	Xe = 2.7

Simplistic estimate of bond polarities using differences in electronegativity between two bonded atoms.

A—••—B   bond polarity based on $\Delta X = \left| X_A - X_B \right|$

$\Delta X \leq 0.4$ considered to be a pure covalent bond (non-polar)
$0.4 < \Delta X < (1.4 - 2.0)$ considered to be a polar covalent bond (permanent charge imbalance)
$(1.4 - 2.0) < \Delta X$ considered to be an ionic bond (cations and anions)

Depending on the magnitude of the difference in electronegativity between the bonded atoms, we can classify the polarity or charge separation, *qualitatively*, as nonpolar bonds (electrons are almost equally shared and there at most a slight difference in electronegativity, that we ignore), as polar bonds (electrons are shared and there is a moderate difference in electronegativity) or as ionic bonds (electrons are completely transferred from the less electronegative element to the more electronegative element when there is a large difference in electronegativity). One arbitrary calculation used for making these distinctions is shown below. We will use this calculation to qualitatively classify bond polarity, even though many exceptions are known (see d and e).

Example: Classify each type of bond below as ionic, polar covalent or pure covalent.

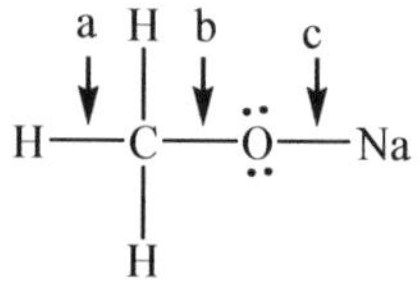

$\Delta X_a = |X_H - X_C| = |2.2 - 2.5| = 0.3 =$ pure covalent bond

$\Delta X_b = |X_C - X_O| = |2.5 - 3.5| = 1.0 =$ polar covalent bond

$\Delta X_c = |X_O - X_{Na}| = |3.5 - 0.9| = 2.6 =$ ionic bond

rewritten as ionic

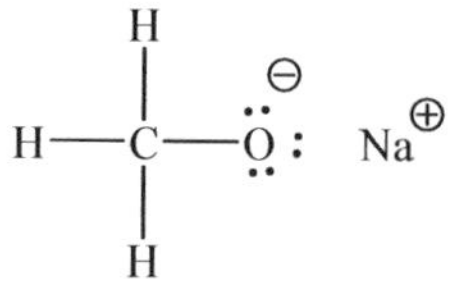

Problem 7 Classify each bond type below as pure covalent, polar covalent or ionic according to our simplistic guidelines above. If a bond is ionic, rewrite it showing correct charges.

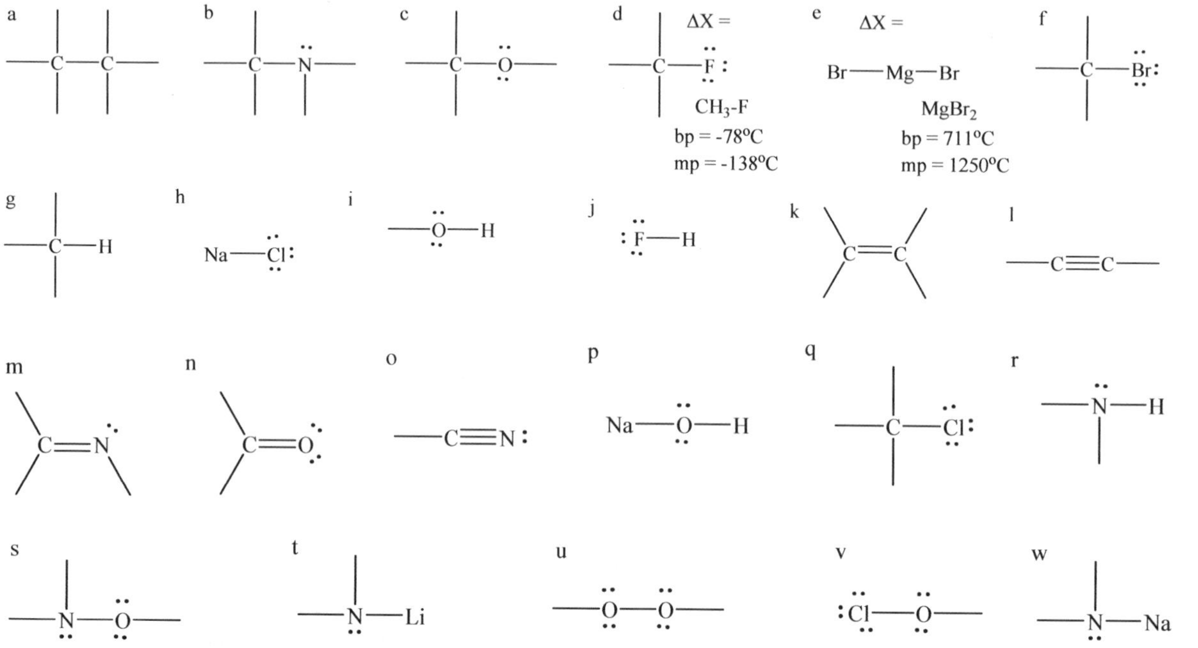

Elemental hydrogen, fluorine, oxygen and nitrogen are examples of atoms sharing electrons evenly because the two atoms competing for the bonded electrons are the same. Each line drawn between two atoms symbolizes a two electron bond. These examples also illustrate that pure covalent bonding can occur with single, double and triple bonds and shows that some elements have lone pairs of electrons.

H—H
hydrogen molecules with a single bond, each H has a duet of electrons (like helium)

:F̈—F̈:
fluorine molecules with a single bond, each F has a octet of electrons (like neon)

:Ö=Ö:
oxygen molecules with a double bond, each O has an octet of electrons (like neon)

:N≡N:
nitrogen molecules with a triple bond, each N has an octet of electrons (like neon)

If the two bonded atoms are not the same, then there will be different attractions for the electrons. One atom will have a greater pull for the electrons and will claim a greater portion of the shared electron density. This will make that atom polarized partially negative, while the atom on the other side of the bond will be polarized partially positive by a similar amount. Because there are two opposite charges separated along a bond, the term "dipole moment" (μ) is used to indicate the magnitude of charge separation. Bond dipoles depend, not just on the amount of charge separated, but also the distance by which the charges are separated, as indicated by their bond lengths. Due to the limited time available in our course we will not explore this relationship beyond using the *dipole moment* (μ) as a qualitative measure of bond or molecule polarity (when

3D shape is also considered). Bond dipoles are simple because they are just the straight line distance between two atoms. Molecules are more complicated because they can have many bonds that are at all possible angles and depend on the shape of the entire molecule.

The symbols + and - represent qualitative charge separation forming a bond dipole. Alternatively, an arrow can be drawn pointing towards the negative end of the dipole and a positive charge written at the positive end of the dipole.

Two qualitative pictures of a bond dipole. B is assumed to be more electronegative than A.

$$ \overset{\mu}{+\longrightarrow} $$

$$ \text{A}\underset{|\text{--------}|}{\overline{\quad\quad}}\text{B} \qquad \text{or} \qquad \overset{\delta+ \qquad \delta-}{\text{A}\underset{|\text{--------}|}{\overline{\quad\quad}}\text{B}} \qquad X_B > X_A $$
$$ \quad d \qquad\qquad\qquad\qquad\qquad\qquad d $$

$$ \text{bond } \mu = \left(\begin{array}{c} \text{amount of} \\ \text{charge} \\ \text{separated} \end{array} \right) \times \left(\begin{array}{c} \text{distance} \\ \text{between} \\ \text{charges in cm} \end{array} \right) = \text{\# with units of Debye} $$
$$ (D = 10^{-18} \text{ esu-cm}) $$

$$ \text{bond } \mu = (\quad e \quad) \times (\quad d \quad) = \text{bond dipole moment} $$

	$\overline{\mu}$
H-H	0.00 D
Cl-Cl	0.00 D
H-Cl	1.05 D

e = electrostatic charge (sometimes written as q)

The absolute value of a unit charge on a proton or an electron is $\pm 4.8 \times 10^{-10}$ esu

d = distance between the opposite charges

This is often given in angstrums (A = angstrum), but converted to cm for use in calculations ($1A = 10^{-8}$ cm)

	ΔX	$d_{X\text{-}H}$	$\mu_{molecule}$	mp (°C)	bp (°C)	density
H-F	$\Delta X_{F\text{-}H} = 1.8$	92 pm	1.86 D	-84	+20	1.15 g/dm³ (gas)
H-Cl	$\Delta X_{Cl\text{-}H} = 1.0$	127 pm	1.05 D	-114	-85	1.49 g/dm³ (gas)
H-Br	$\Delta X_{Br\text{-}H} = 0.8$	141 pm	0.82 D	-87	-67	3.31 g/dm³ (gas)
H-I	$\Delta X_{I\text{-}H} = 0.5$	161 pm	0.38 D	-51	-35	2.85 g/mL (liquid)*

* Notice there is a different phase.

Because there are two factors that make up bond dipole moments (charge and distance), bonds that appear to be more polar based on electronegativity differences might have similar dipole moments if they are shorter than less polar bonds. The methyl halides provide an example of this aspect. Fluoromethane has a more polar bond, but shorter bond length, while iodomethane has a less polar bond, but longer bond length. The opposing trends of the halomethanes produce similar dipole moments in all of the molecules. Polarity, polarizability and shape have a large influence on physical properties (mp, bp, solubility, etc). We will briefly review these in Chapter 2.

Chapter 1

electronegativity

$X_H = 2.2$	$X_C = 2.5$	$X_N = 3.0$	$X_O = 3.5$
$X_F = 4.0$	$X_{Cl} = 3.2$	$X_{Br} = 3.0$	$X_I = 2.7$

μ = dipole mement (measure of polarity)

d = bond distance in picometers (10^{-12} m)

	ΔX	d_{C-X}	$\mu_{molecule}$	mp (°C)	bp (°C)	density
CH_3-F	$\Delta X_{C-F} = 1.5$	139 pm	1.86 D	-142	-78	0.53 g/dm^3 (gas)
CH_3-Cl	$\Delta X_{C-Cl} = 0.7$	178 pm	1.87 D	-97	-24	0.92 g/dm^3 (gas)
CH_3-Br	$\Delta X_{C-Br} = 0.5$	193 pm	1.81 D	-94	+3	3.97 g/dm^3 (gas)
CH_3-I	$\Delta X_{C-I} = 0.2$	214 pm	1.62 D	-66	+42	2.28 g/mL (liquid)*

* Notice there is a different phase.

	ΔX	d_{X-H}	$\mu_{molecule}$	mp (°C)	bp (°C)	density
CH_4	$\Delta X_{CH} = 0.3$	109 pm	0.00 D	-182	-164	0.66 g/dm^3 (gas)
NH_3	$\Delta X_{NH} = 0.8$	102 pm	1.42 D	-78	-33	0.73 g/dm^3 (gas)
H_2O	$\Delta X_{OH} = 1.3$	96 pm	1.85 D	0	+100	1.00 g/mL (liquid)*
H-F	$\Delta X_{FH} = 1.8$	92 pm	1.86 D	-84	+20	0.99 g/mL (liquid)*

* Notice there is a different phase.

Hybridization

One theory of bonding mixes atomic orbitals to make hybrid orbitals (sp, sp^2 and sp^3 in our course). Hybrid creations are mixtures that retain some of the properties of the components mixed together (roses, cattle, atomic orbitals). Linus Pauling developed the hybridization theory in the 1930s and it has proven so useful that it is still taught in all organic text books of today because it can explain the shapes and properties of molecules that we observe in a simple way. A qualitative model, minus all of the complicated mathematics is pretty easy for beginning students to follow and it is the approach that we will take here.

The Hybridization Model for Atoms in Molecules - These are the shapes we have to explain and the discussion that follows will show how we do it.

ethane
tetrahedral carbon atoms

HCH bond angles ≈ 109°
HCC bond angles ≈ 109°

ethene
trigonal planar carbon atoms

HCH bond angles ≈ 120° (116°)
CCH bond angles ≈ 120° (122°)

ethyne
linear carbon atoms

HCC bond angles = 180°

allene
trigonal planar carbon atoms
at the ends and a linear
carbon atom in the middle

HC_aH bond angles ≈ 120°
HC_aC_b bond angles ≈ 120°
$C_aC_bC_a$ bond angles = 180°

Hybridization is how we are going to explain the shapes above.

Chapter 1

1. sp hybridization – carbon and other atoms of organic chemistry and biochemistry

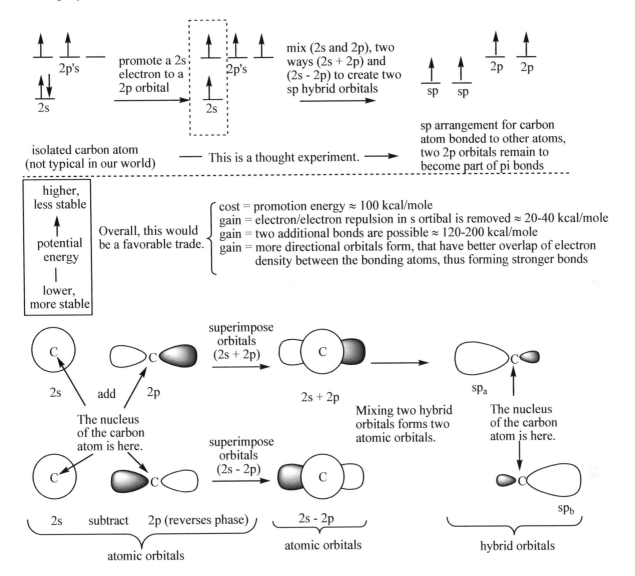

mix (2s and 2p), two ways (2s + 2p) and (2s - 2p) to create two sp hybrid orbitals

promote a 2s electron to a 2p orbital

sp arrangement for carbon atom bonded to other atoms, two 2p orbitals remain to become part of pi bonds

isolated carbon atom (not typical in our world) —— This is a thought experiment. ——

higher, less stable

↑

potential energy

|

lower, more stable

Overall, this would be a favorable trade.

{
cost = promotion energy ≈ 100 kcal/mole
gain = electron/electron repulsion in s ortibal is removed ≈ 20-40 kcal/mole
gain = two additional bonds are possible ≈ 120-200 kcal/mole
gain = more directional orbitals form, that have better overlap of electron density between the bonding atoms, thus forming stronger bonds
}

The nucleus of the carbon atom is here.

superimpose orbitals (2s + 2p)

Mixing two hybrid orbitals forms two atomic orbitals.

The nucleus of the carbon atom is here.

superimpose orbitals (2s - 2p)

2s   add   2p   2s + 2p sp_a

2s   subtract   2p (reverses phase)   2s - 2p sp_b

atomic orbitals   atomic orbitals   hybrid orbitals

The Complete Picture of an sp Hybridized Carbon Atom

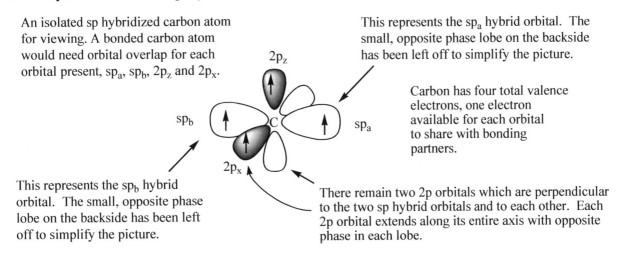

An isolated sp hybridized carbon atom for viewing. A bonded carbon atom would need orbital overlap for each orbital present, sp_a, sp_b, $2p_z$ and $2p_x$.

This represents the sp_a hybrid orbital. The small, opposite phase lobe on the backside has been left off to simplify the picture.

Carbon has four total valence electrons, one electron available for each orbital to share with bonding partners.

$2p_z$

sp_b C sp_a

$2p_x$

This represents the sp_b hybrid orbital. The small, opposite phase lobe on the backside has been left off to simplify the picture.

There remain two 2p orbitals which are perpendicular to the two sp hybrid orbitals and to each other. Each 2p orbital extends along its entire axis with opposite phase in each lobe.

Chapter 1

Two sp carbon atoms bonded in a molecule of ethyne (…its common name is acetylene)

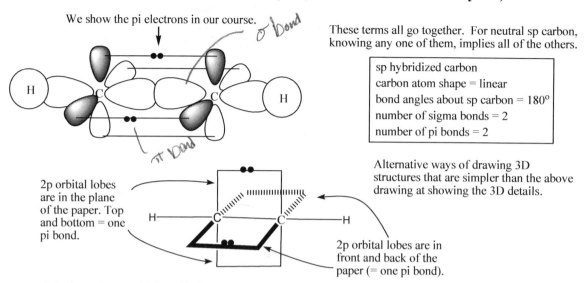

We show the pi electrons in our course.

These terms all go together. For neutral sp carbon, knowing any one of them, implies all of the others.

sp hybridized carbon
carbon atom shape = linear
bond angles about sp carbon = 180°
number of sigma bonds = 2
number of pi bonds = 2

2p orbital lobes are in the plane of the paper. Top and bottom = one pi bond.

Alternative ways of drawing 3D structures that are simpler than the above drawing at showing the 3D details.

2p orbital lobes are in front and back of the paper (= one pi bond).

3D ethyne drawn with 2p orbitals as lines and pi electrons explicitly drawn in, in a manner similar to showing lone pair electrons. In this book we will usually draw pi bonds this way in 3D structures.

Ethyne is composed of five bonds (three sigma = the first bonds and two pi = the second and third bonds).

H—C≡C—H
    Each line represents a bond. While the three simple lines of the triple bond appear equivalent, we know that the first bond formed is a sigma bond of overlapping sp hybrid orbitals. The second and third bonds are overlapping 2p orbitals, above and below and in front and in back. Since the C-H bonds are single bonds, we know that they are sigma bonds too, using hybrid orbitals. This is how you will determine the hybridization of any atom in a structure. Knowing how many pi bonds are present will tell you how many 2p orbitals are being used in those pi bonds. The remaining s and 2p orbitals must be mixed together in hybrid orbitals (in this example, only an s and a 2p remain to form two sp hybrid orbitals).

- -

HCCH
    The connections of the atoms are implied by the linear way the formula is drawn. You have to fill in the details about the number of bonds and where they are from your understanding of each atom's bonding patterns. A C-H bond can only be a single bond so there must be three bonds between the carbon atoms to total carbon's normal number of four bonds. This means, of course, that the second and third bonds are pi bonds, using 2p orbitals, leaving an s and 2p orbitals to mix, forming two sp hybrid orbitals.

- -

≡
    A bond line formula only shows lines connecting the carbon atoms and leaves off the hydrogen atoms. Every end of a line is a carbon (two in this drawing) and every bend in a line is a carbon (none in this drawing). You have to figure out how many hydrogen atoms are present by substracting the number of lines shown (bonds to non-hydrogen atoms) from four, the total number of bonds of a neutral carbon (4 - 3 = 1H in this drawing). The shape of the carbon atoms must be linear, because we know the hybridization is sp.

- -

C_2H_2
    This is the ultimate in condensing a structure. Merely writing the atoms that are present and how many of them there are provides no details about the connectivity of the atoms. It only works for extremely simple molecules that have only one way that they can be drawn. Ethyne is an example of such molecule. Other formulas may have several, hundreds, thousands, millions, or more ways for drawing structures ($C_{20}H_{42}$ has about 366,000 ways). Formulas written in this manner are usually not very helpful.

- -

sp hybridized carbon

carbon atom shape = linear
hybridization = sp
bond angles about sp carbon = 180°
number of sigma bonds = 2
number of pi bonds = 2

All of the details in this group go together. If you have any one of them, you should be able to fill in the remaining details.

Problem 8 – Draw a 3D representation of hydrogen cyanide, HCN. Show lines for the sigma bond skeleton and the lone pair of electrons in its proper location. Show two dots for the lone pair. Also show pi bonds represented in a manner similar to above. What is different about this structure compared with ethyne above (# bonds, lone pairs, polarity, all of them)? Would such a drawing work if oxygen is switched in for nitrogen? What would be different?

2. sp² hybridization

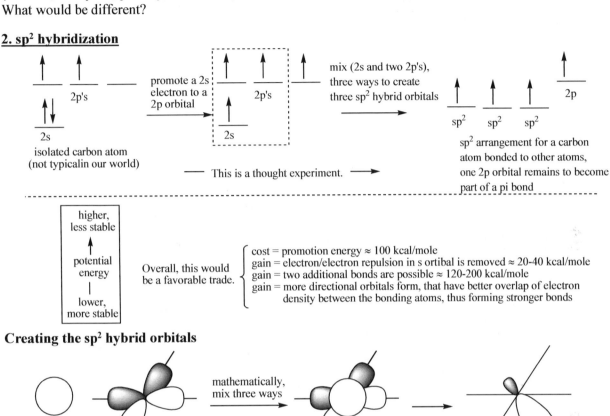

higher,
less stable
↑
potential
energy
↓
lower,
more stable

Overall, this would be a favorable trade.

$\begin{cases}$ cost = promotion energy ≈ 100 kcal/mole
gain = electron/electron repulsion in s ortibal is removed ≈ 20-40 kcal/mole
gain = two additional bonds are possible ≈ 120-200 kcal/mole
gain = more directional orbitals form, that have better overlap of electron
density between the bonding atoms, thus forming stronger bonds $\end{cases}$

Creating the sp² hybrid orbitals

2s

2pₓ and 2p_y
(2 lines define a plane)

mathematically,
mix three ways

one example of
mixing 2s+2p+2p

Similar phase mixes
constructively in the
right front quadrant

sp²ₐ

The "mixing" process symbolized
here is repeated two additional ways,
creating three sp₂ hybrid orbitals, all
in one plane.

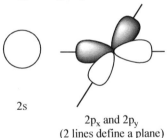

top-down
view
120°
sp²c
120°
sp²b
120°
sp²a

All three sp² hybrid orbitals lie in a plane and divide a circle into three equal wedges of 120°. The descriptive term for the shape is trigonal planar. This picture shows the sp² hybrid orbitals without their small backside lobes and no 2p orbital is shown. These hybrid orbitals will form sigma bonds, which are always first bonds.

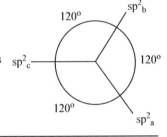

An isolated sp² hybridized carbon atom for viewing. A bonded carbon atom would need orbital overlap for each orbital present, sp²ₐ, sp²b, sp²c and 2pz.

Carbon has four total valence electrons, one electron available for each orbital to share with bonding partners.

side-on
view

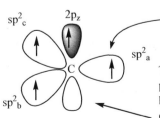

These represent sp² hybrid orbitals. The small, opposite phase lobe on the backside has been left off to simplify the picture.

There remains one 2p orbital perpendicular to the three sp² hybrid orbitals. The 2p orbital extends along the entire axis with opposite phase in each lobe.

Chapter 1

Alternative ways to draw sp^2 carbon (or other atoms). Dash lines indicate in back of the page, wedge lines, indicate in front of the page, simple lines are in the page.

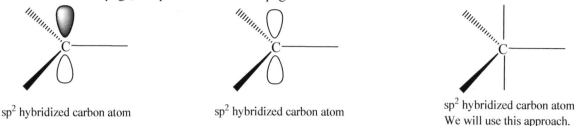

sp^2 hybridized carbon atom   sp^2 hybridized carbon atom   sp^2 hybridized carbon atom
                                                                              We will use this approach.

Two sp^2 carbon atoms bonded in a molecule of ethene (its common name is ethylene). Ethene is composed of six bonds (five sigma and one pi).

We show the pi electrons in our course.

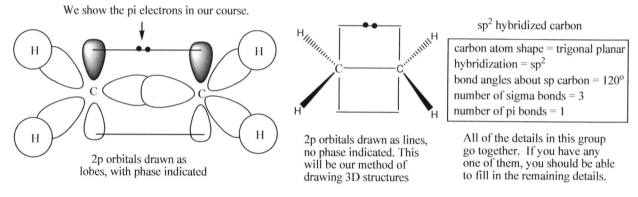

2p orbitals drawn as lobes, with phase indicated

2p orbitals drawn as lines, no phase indicated. This will be our method of drawing 3D structures

sp^2 hybridized carbon

| carbon atom shape = trigonal planar |
| hybridization = sp^2 |
| bond angles about sp carbon = 120° |
| number of sigma bonds = 3 |
| number of pi bonds = 1 |

All of the details in this group go together. If you have any one of them, you should be able to fill in the remaining details.

Two sp^2 carbon atoms bonded in a molecule of ethene (...its common name is ethylene). Ethene is composed of six bonds (five sigma and one pi). A dash indicates a bond behind the page and a wedge indicates a bond in front of the page.

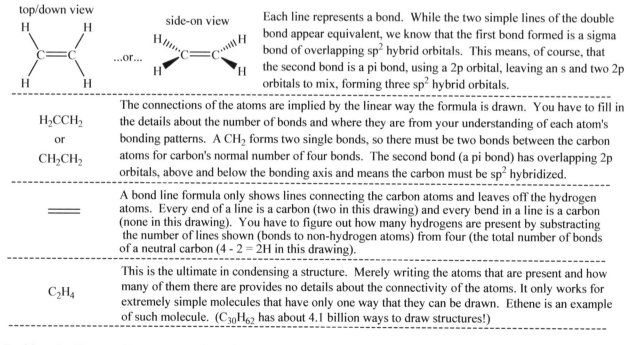

top/down view

side-on view

...or...

Each line represents a bond. While the two simple lines of the double bond appear equivalent, we know that the first bond formed is a sigma bond of overlapping sp^2 hybrid orbitals. This means, of course, that the second bond is a pi bond, using a 2p orbital, leaving an s and two 2p orbitals to mix, forming three sp^2 hybrid orbitals.

H$_2$CCH$_2$
or
CH$_2$CH$_2$

The connections of the atoms are implied by the linear way the formula is drawn. You have to fill in the details about the number of bonds and where they are from your understanding of each atom's bonding patterns. A CH$_2$ forms two single bonds, so there must be two bonds between the carbon atoms for carbon's normal number of four bonds. The second bond (a pi bond) has overlapping 2p orbitals, above and below the bonding axis and means the carbon must be sp^2 hybridized.

A bond line formula only shows lines connecting the carbon atoms and leaves off the hydrogen atoms. Every end of a line is a carbon (two in this drawing) and every bend in a line is a carbon (none in this drawing). You have to figure out how many hydrogens are present by substracting the number of lines shown (bonds to non-hydrogen atoms) from four (the total number of bonds of a neutral carbon (4 - 2 = 2H in this drawing).

C$_2$H$_4$

This is the ultimate in condensing a structure. Merely writing the atoms that are present and how many of them there are provides no details about the connectivity of the atoms. It only works for extremely simple molecules that have only one way that they can be drawn. Ethene is an example of such molecule. (C$_{30}$H$_{62}$ has about 4.1 billion ways to draw structures!)

Problem 9 – Draw a 3D representation of methanal (common name = formaldehyde), H$_2$C=O. Show lines for the sigma bond skeleton and the lone pairs of electrons with two dots for each lone pair. Also show pi bonds represented in a manner similar to above. What is different about this structure compared with ethene above (# bonds, lone pairs, polarity, all of these)? Draw a 3D representation of H$_2$C=NH. Can fluorine be switched in for oxygen? What would change? Formal charge is discussed on page 26.

3. sp³ hybridization

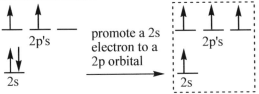 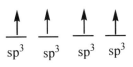

promote a 2s electron to a 2p orbital

mix (2s and three 2p's), four ways to create four sp³ hybrid orbitals

sp³ arrangement for carbon atom bonded to other atoms, no 2p orbitals remain so no pi bonds can form

isolated carbon atom (not typical in our world) —— This is a thought experiment. ——➤

higher, less stable

↑

potential energy

|

lower, more stable

Overall, this would be a favorable trade.

cost = promotion energy ≈ 100 kcal/mole
gain = electron/electron repulsion in s ortibal is removed ≈ 20-40 kcal/mole
gain = two additional bonds are possible ≈ 120-200 kcal/mole
gain = more directional orbitals form, that have better overlap of electron density between the bonding atoms, thus forming stronger bonds

Creating the sp³ hybrid orbitals

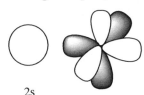

2s

$2p_x$, $2p_y$, $2p_z$

mathematically, mix four ways

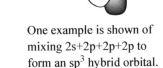

One example is shown of mixing 2s+2p+2p+2p to form an sp³ hybrid orbital.

Similar phases interact constructively in the front, right, upper octant where the large lobe will be located.

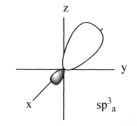

The "mixing" process symbolized here is repeated three additional ways, creating four sp³ hybrid orbitals.

Carbon has four total valence electrons, one electron available for each orbital to share with bonding partners. Remember, a wedge indicates a bond in front of the page and a dash indicates a bond in back of the page, for 3D perspective.

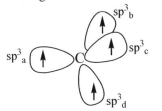

=

=

sp³ orbitals minus the small backsidelobes, our intuition about the bond angles fails us here. The angles are 109.5°, but vary in real compounds so we will just use 109° in our course.

sp³ orbitals drawn using our 3D conventions

A tetrahedron has four equivalent triangular sides. The atoms at the ends of the bonds with carbon define the vertices of the tetrahedron. The carbon is sitting in the middle of the tetrahedron.

One sp³ carbon atom bonded in methane and two sp³ carbon atoms bonded in ethane

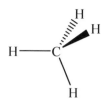

> carbon atom shape = tetrahedral
> hybridization = sp³
> bond angles about sp carbon = 109°
> number of sigma bonds = 4
> number of pi bonds = 0

All of the details in this group go together. If you have any one of them, you should be able to fill in the remaining details.

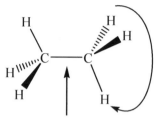

A single bond allows rotation to occur about the carbon-carbon bond, which alters the shape and the energy of the molecule. There are no second or third bonds between atoms, so no pi bonds using 2p orbitals. Only hybrid orbitals and sigma bonds (seven total bonds).

C-C single bond rotation
$\longrightarrow$

dashes - behind the page
wedges - in front of the page
simple lines - in the page

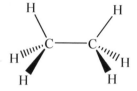

This is a different shape than the structure at the left. We say the are different conformations (discussed in a later chapter).

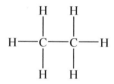

Each line represents a bond. Since there are only single bonds, we know that they must be sigma bonds. There cannot be any pi bonds becasue there are no second or third bonds between the same two atoms. The 2s and all three 2p orbitals must all be mixed, meaning that the hybridization has to be sp³ and all of the terms that go along with sp³ hybridization. Remember, the actual bond angles are approximately 109° in tetrahedral atoms.

H_3CCH_3
or
CH_3CH_3

The connections of the atoms are implied by the linear way the formula is drawn. You have to fill in the details about the number of bonds and where they are located from your understanding of each atom's bonding patterns. A CH_3 has three single bonds between carbon and hydrogen, so there can only be one additional bond between the carbon atoms to total carbon's normal number of four bonds. This means, of course, that there is no pi bond, using a 2p orbital, leaving the 2s and all three 2p orbitals to mix, forming four sp³ hybrid orbitals.

———

A bond line formula only shows lines connecting the carbon atoms and leaves off the hydrogen atoms. Every end of a line is a carbon (two in this drawing) and every bend in a line is a carbon (none in this drawing). You have to figure out how many hydrogens are present by substracting the number of lines shown (bonds to non-hydrogen atoms) from four (the total number of bonds of a neutral carbon (4 - 1 = 3H on each carbon atom in this drawing).

C_2H_6

This is the ultimate in condensing a structure. Merely writing the atoms that are present and how many of them there are provides no details about the connectivity of the atoms. A structure can be generated only for extremely simple molecules that have only one way that they can be drawn. Ethane is an example of such molecule. ($C_{40}H_{82}$ has about 62 trillion ways it can be drawn.)

Problem 10 – Draw a 3D representation of methanol, H_3COH or CH_3OH, methanamine, H_3CNH_2 or CH_3NH_2 and fluoromethane CH_3F or H_3CF or FCH_3. Show lines for the sigma bond skeleton and a line with two dots for lone pairs, in a manner similar to above. What is different about this structure compared with ethene above (# bonds, lone pairs, polarity, all of these)?

Chapter 1

Summary – Key features to determine hybridization of carbon atoms in organic chemistry.

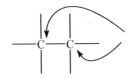

There are no second or third bonds between the same two atoms so no 2p orbitals are used to make any pi bonds. The hybridization must be sp^3 ($2s + 2p + 2p + 2p = 4sp^3$) and all four atomic orbitals are mixed to form four sp^3 hybrid orbitals.

2s 2p 2p 2p

No pi bonds, so all atomic orbitals are used in hybridization = sp^3

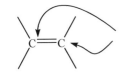

There is a second bond between the two carbon atoms. This must be a pi bond and uses 2p orbitals. The hybridization must be sp^2 ($2s + 2p + 2p = 3sp^2$) and three atomic orbitals are mixed to form three sp^2 hybrid orbitals.

2s 2p 2p ~~2p~~

One pi bond, so only the 2s and two of the 2p's are used in hybridization = sp^2

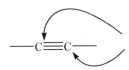

There is a second and third bond between the two carbon atoms. These must be two pi bonds and using two 2p orbitals. The hybridization must be sp ($2s + 2p = 2sp$) and two atomic orbitals are mixed to form two sp hybrid orbitals.

2s 2p ~~2p~~ ~~2p~~

Two pi bonds, so only the 2s and one of the 2p's are used in hybridization = sp

---

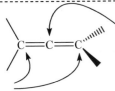

sp^2, see above

There is a second bond with the atom on the left and again with the atom on the right. These must be two pi bonds and using two 2p orbitals. The hybridization must be sp ($2s + 2p = 2sp$) and two atomic orbitals are mixed to form two sp hybrid orbitals. The hybridization of the end carbons is sp^2 (see the second example above). The planar shapes of the atoms of the two end carbons are twisted 90° relative to one another because the 2p orbitals on the middle carbon making the pi bonds with them are angled at 90° relative to one another.

2s 2p ~~2p~~ ~~2p~~

Two pi bonds on the center carbon, so only the 2s and one of the 2p's are used in hybridization = sp

Chapter 1

Bond line formulas do not show carbon and hydrogen atoms. A carbon atom is implied at every bend in a line and at every end of a line (and occasionally as a dot). They are easy and fast to draw but make larger demands on you for interpretation because they leave out lots of detail. Heteroatoms (N, O, S, halogens) are drawn in, as is a hydrogen atom on an oxygen or nitrogen and a hydrogen atom in an aldehyde.

Two structures showing all of the atoms. It's a lot of work to draw structures this way.

Number of hydrogen atoms on a carbon = (4) - (number of bonds shown)
(you don't see these)

Same molecule using a bond-line structure showing only non-hydrogen bonds. A carbon atom is implied at every bend and every end of a line and every dot. This one is a lot easier and faster to draw.

Carbon #	hybridization	bond angles	shape	number of hydrogen atoms	σ bond	π bonds	lone pairs
1	sp	180°	linear	1	2	2	0
2	sp	180°	linear	0	2	2	0
3	sp^3	109°	tetrahedral	2	4	0	0
4	sp^3	109°	tetrahedral	2	4	0	0
5	sp^2	120°	trigonal planar	1	3	1	0
6	sp^2	120°	trigonal planar	1	3	1	0
7	sp^2	120°	trigonal planar	0	3	1	0
8	sp	180°	linear	0	2	2	0
9	sp^2	120°	trigonal planar	2	3	1	0
10	sp^3	109°	tetrahedral	1	4	0	0
11	sp^3	109°	tetrahedral	2	4	0	0
12	sp^3	109°	tetrahedral	0	4	0	0
13	sp^3	109°	tetrahedral	3	4	0	0
14	sp^3	109°	tetrahedral	3	4	0	0

Chapter 1

Problem 11 - What is the hybridization of all carbon atoms in the structure below? What are the bond angles, shapes, number of sigma bonds, number of pi bonds and number's of attached hydrogen atoms? Bond line formulas are shorthand, symbolic representations of organic structures. Each bend represents a carbon, each end of a line represents a carbon and each dot represents a carbon. All carbon/carbon bonds are shown. The number of hydrogen atoms on a carbon is determined by the difference between four and the number of bonds shown.

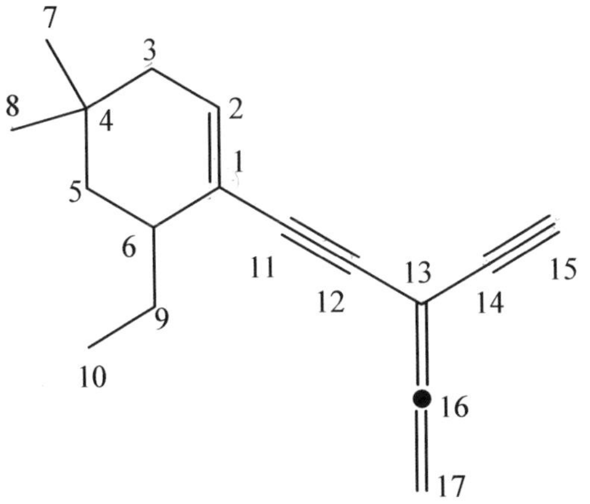

Carbon #	hybridization	bond angles	shape	number of hydrogen atoms	σ bond	π bonds	lone pairs
1							
2							
3							
4							
5							
6							
7							
8							
9							
10							
11							
12							
13							
14							
15							
16							
17							

Chapter 1

Other Atoms in Organic Chemistry

How do other atoms of organic chemistry fit into this picture? Except for hydrogen, which only has 1s electrons and no hybrid orbitals, all other atoms are compatible with the same hybridization model. Nitrogen adds another proton for a $Z_{eff} = +5$ and has 5 valence electrons. Oxygen (sulfur is similar) adds yet another proton for a $Z_{eff} = +6$ and has 6 valence electrons. Fluorine (Cl, Br, I are similar) has a $Z_{eff} = +7$ and 7 valence electrons.

Fully bonded neutral carbon, nitrogen, oxygen and fluorine all have an octet of valence electrons. However, this is accomplished in different arrangements of bonds and lone pairs of electrons (nonbonded electron pairs). The increase in Z_{eff} makes each subsequent atom hold on more tightly to its valence electrons (higher electronegativity). All atoms shown below have achieved a Noble Gas configuration. Stable octets for the second row (period) elements are present, like Ne, and a doublet for hydrogen (first row/period) which becomes like He. Atoms from the third, fourth and greater rows sometimes deviate from the octet rule due to d orbital availability (but only rarely in our course).

Other Atoms in Organic Chemistry – nitrogen, oxygen and halogens

Valence Possibilities for Neutral Atoms

Atom	sp^3	sp^2	sp	sp	# bonds	# lone pairs	Z_{eff}	X
carbon val. e- = 4 needs 4e- bonds = 4	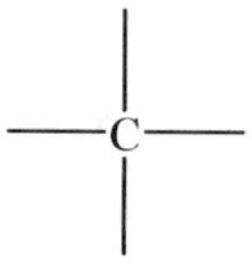 —C—	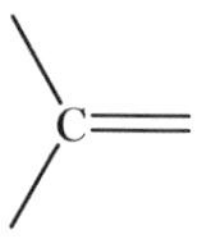 C=	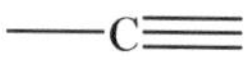 —C≡	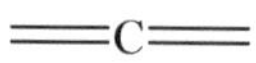 =C=	4	0	+4	2.5
nitrogen val. e- = 5 needs 3e- bonds = 3	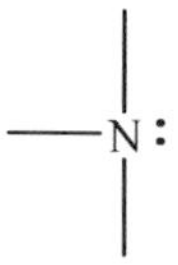 —N:	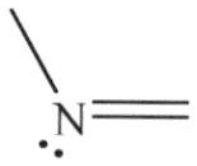 N=	 :N≡	too many bonds	3	1	+5	3.0
oxygen val. e- = 6 needs 2e- bonds = 2	 —O:	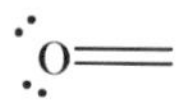 O=	too many bonds		2	2	+6	3.5
fluorine val. e- = 7 needs 1e- bonds = 1	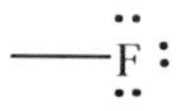 —F:	too many bonds			1	3	+7	4.0
hydrogen val. e- = 1 needs 1e- bonds = 1	 —H (1s orbital)	too many bonds			1	0	+1	2.2

Problem 12 - What types of orbitals do the lone pair electrons occupy in each example above, according to our hybridization model? Hint: What is the hybridization of the atom? It has to use the same kind of hybrid orbital to hold the lone pair electrons...unless it is part of a resonant system (discussed later).

Problem 13 – Can a nitrogen atom bond in the second sp pattern above for carbon (four bonds)? Can an oxygen atom bond in the first sp pattern above for carbon (three bonds)? How about the second sp pattern of carbon (four bonds)? Can a fluorine atom bond in the sp^2 pattern for carbon (two bonds)? Are any of these reasonable possibilities? Have you ever seen a structure with four bonds to a nitrogen atom (NH_4^+) or three bonds to an oxygen atom (H_3O^+) or two bonds to a fluorine atom (H_2F^+)? Would there be any necessary changes for the atoms in such arrangements (notice the positive charge)? We'll discuss the answers to these questions soon. (Formal charge is discussed on page 26.)

Chapter 1

Problem 14 - What is the hybridization of all carbon atoms in the structure below? What are the bond angles, shapes, number of sigma bonds, number of pi bonds, number of lone pairs and number's of attached hydrogen atoms? What is the hybridization of the lone pair orbitals? Bond line formulas are shorthand, symbolic representations of organic structures. Each bend represents a carbon, each end of a line represents a carbon and each dot represents a carbon. All carbon/carbon bonds are shown. The number of hydrogen atoms on a carbon is determined by the difference between four and the number of bonds shown.

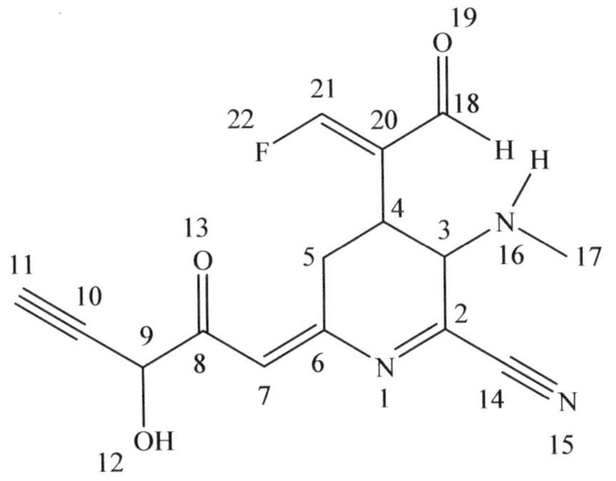

Atom#	hybridization	bond angles	shape	number of hydrogen atoms	σ bond	π bonds	lone pairs	hybridization of lone pairs
1								
2								
3								
4								
5								
6								
7								
8								
9								
10								
11								
12								
13								
14								
15								
16								
17								
18								
19								
20								
21								
22								

When an atom does not have its normal number of bonds and lone pairs, it is usually less stable (more reactive). Situations include cations, anions and free radicals. You need to be able to easily classify an atom's status in such situations. Formal charge helps us do that.

Formal Charge – a convention designed to indicate an excess or deficiency of electron density compared to an atom's neutral allocation of electrons.

$$\text{Formal Charge} = \left(Z_{\text{effective}}\right) - \left(\begin{array}{c}\text{electrons in}\\\text{lone pairs}\end{array}\right) - \left(\frac{1}{2}\right)\left(\begin{array}{c}\text{electrons in}\\\text{covalent bonds}\end{array}\right)$$

This number never changes for an atom and represents positive charge not cancelled by core electrons.

Total valence electrons allocated to an atom assuming electrons are shared evenly in bonds. This number varies depending on the bonding arrangement. It is negative because electrons are negative.

Rules of Formal Charge

1. When an atom's total valence electron credit exactly matches its Z_{eff}, there is no formal charge.

2. Each deficiency of an electron credit from an atom's normal number of valence electrons produces an additional positive charge.

3. If the formal charge calculation shows excess electron credit over an atom's normal number of valence electrons, a negative formal charge is added for each extra electron.

4. The total charge on an entire molecule, ion or free radical is the sum of all of the formal charges on the individual atoms.

Common Examples

When an atom has a full octet, each bond missing from its normal number adds an extra -1 charge. A two electron bond with one electron credit is changed to a lone pair with a full two electron credit. Any organic atom has this possibility. (H, C, N, O, F, etc.) Also, all hybridization states are possible. The last two examples show normally trivalent atoms (B and Al) picking up an extra two electron bond and one electron credit, for the minus charge.

When an atom has a full octet, each extra bond over its normal number of bonds in its neutral state adds an extra +1 charge. An atom needs a lone pair of electrons to create positive formal charge in this manner. A lone pair with full electron credit is changed to a bond with only one electron credit. This is not possible for carbon, but is possible for nitrogen, oxygen and fluorine (or any atom with a lone pair). Again, all hybridization states are possible.

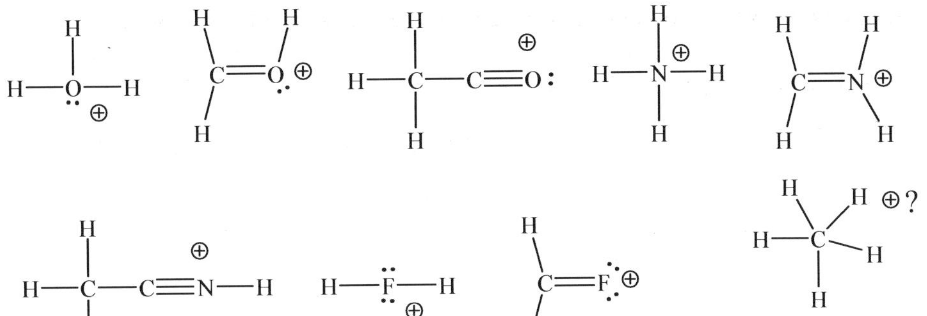

Wrong! Carbon cannot exceed four bonds. (See next)

A second way to create positive formal charge is to remove both of the shared electrons with a bonding partner. This is the most common way to create positive formal charge on carbon and forms a *carbocation*. The hybridization usually changes since there is less electron/electron repulsion. The empty orbital is usually a 2p orbital. Notice the curved arrows showing two electron movement (full head) as X leaves (called the leaving group).

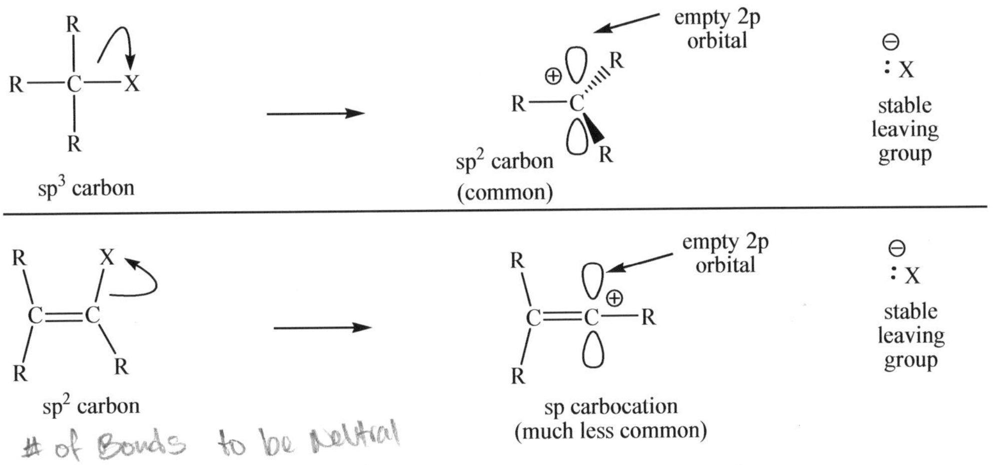

# of Bonds to be neutral

C	4
N	3
O	2
F	1
H	1

If not octeted, then calc. Formal Charge via

FC = #e⁻ − # bond − # elec in lone pair

When an atom has more than their # bonds they get (+1) per bond over

" has less than their # bonds they get (−) per bond under

* Remember
4, 3, 2, 1, 1
C N O F H

Chapter 1

★ More Bonds & Full octets are Better

Concise Rules for Drawing 2D Lewis Structures, including resonance structures

1. You need to begin with a condensed line formula to provide clues about the connectivity of the atoms. From that you can draw a sigma skeletal framework showing all of the sigma bonds from the condensed line formula.

2. Sum all of the valence electrons available for bonding and lone pairs from the atoms present. Reduce this number by 1 for each positive charge and increase this number by 1 for each negative charge.

3. Each sigma bond counts for two electrons. Subtract this number of electrons from the total number of electrons to determine how many electrons are available for lone pairs and pi bonds.

4. Use any remaining electrons as lone pairs to fill in octets on electronegative atoms first (F > O > N > other halogens > C). If the electrons are still not used up, add the additional electrons to carbon atoms at alternate positions to maximize charge separation where possible (minimize electron-electron repulsion) until all electrons have been used. It is helpful for you to write in the formal charge at this point (but not necessary). Formal charge will suggest the best way to group lone pair electrons used to make pi bonds.

5. A lone pair of electrons on an atom next to a neighbor atom with an incomplete octet can be shared to form a second (double) or third (triple) bond with the neighbor atom (these will be pi bonds). If this can be done to a particular electron deficient atom by more than one neighbor atom, then resonance structures are present. The best resonance structure will have the maximum number of bonds. If two structures have the same number of bonds, then a secondary consideration is to minimize charge separation (the formal charge from rule 4 will help here). If residual charge is necessary, it is better to be consistent with electronegativity (e.g. negative formal charge is better on oxygen than on carbon). Your goals, in order, are: 1. to maximize the number of bonds and 2. to minimize formal charge and 3. to keep formal charge consistent with electronegativity, if the other two goals have been met.

6. Resonance structures require a specific use of 3 kinds of arrows.

a. Double headed arrows are placed between resonance structures.

$$\left(\begin{array}{c}\text{resonance}\\\text{structure 1}\end{array}\right) \longleftrightarrow \left(\begin{array}{c}\text{resonance}\\\text{structure 2}\end{array}\right)$$

   A double headed resonance arrow shows that these structures both contribute to an actual structure that cannot be drawn as a simple single Lewis structure.

b. A full headed curved arrow indicates two electron movement.

electron
pair begins
here

electron
pair moves
here

A full headed curved arrow indicates two electron movement to show where an electron pair will be located in the "next" structure, whether used in resonance structures or used in a reaction equation.

c. A half headed curved arrow indicates one electron movement.

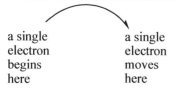

a single
electron
begins
here

a single
electron
moves
here

A half headed curved arrow indicates one electron movement to show where an electron will be located in the "next" structure (as in free radicals), whether used in resonance structures or used in a reaction equation.

Chapter 1

Problem 15 – Several 2D Lewis sigma bond frameworks are provided below, with reasonable resonance possibilities. Use the first structure to add in valence electrons to generate an acceptable Lewis structure. Write in the formal charge wherever present in the atoms shown. Resonance structures have double headed arrows between them. All of the resonance contributors must be considered together to evaluate the true nature of a particular structure. You should include correct curved arrows to show necessary electron movement to form each subsequent resonance structure (no arrows on the last structure). Resonance structures show the electrons at different locations of p orbitals. Resonance is one of your most important concepts in organic and biochemistry.

a. CH_3NO_2

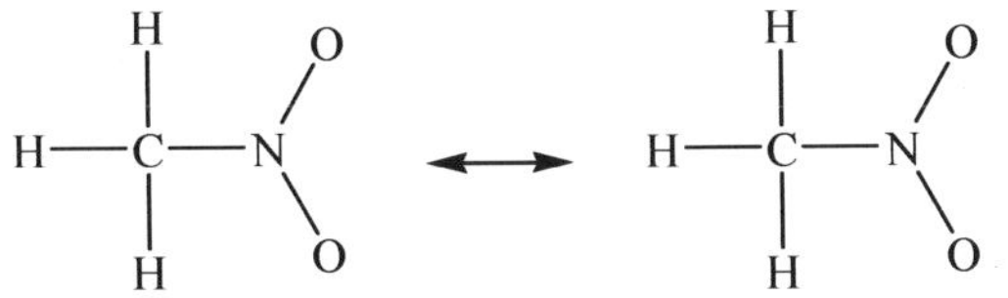

1 x C =
3 x H =
1 x N =
2 x O =
total e- =

b. CH_2N_2

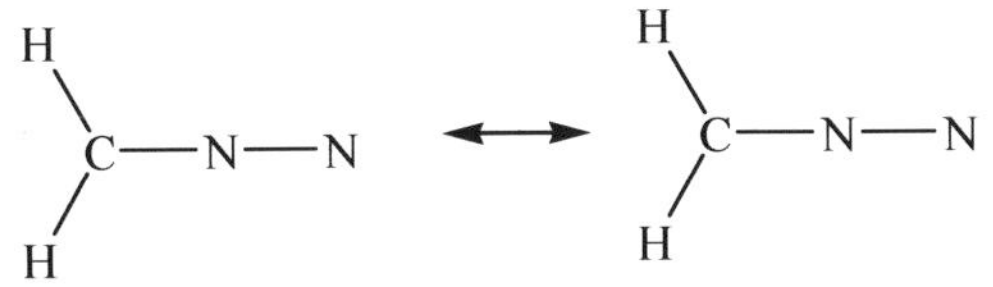

1 x C =
2 x H =
2 x N =
total e- =

c. $NH_3CH_2CO_2$

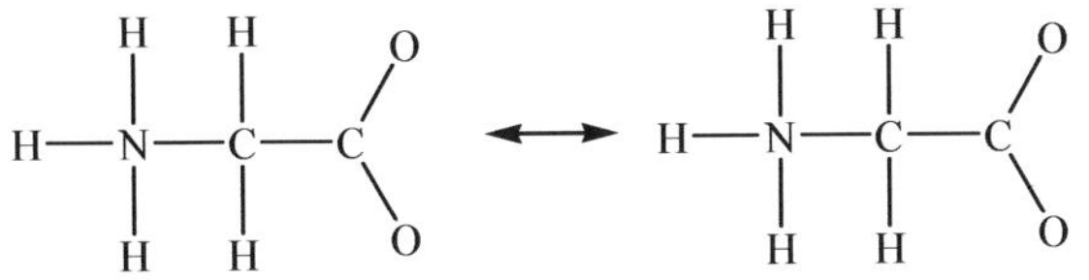

2 x C =
5 x H =
1 x N =
2 x) =
total e- =

d. CH_3NO_2

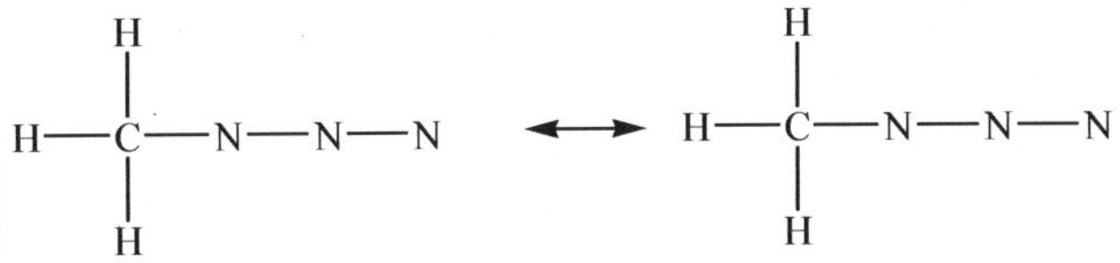

1 x C =
3 x H =
1 x N =
2 x O =
total e- =

e. CO_3^{-2}

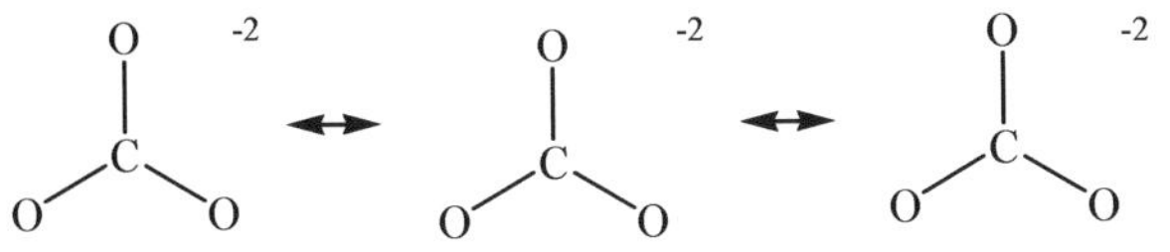

1 x C =
3 x O =
-2   =
total e- =

f. $C(NH_2)_3^{\oplus}$

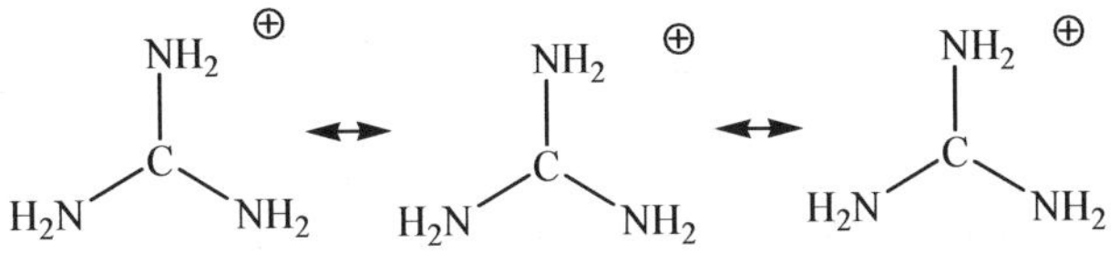

1 x C =
3 x N =
6 x H =
+1   =
total e- =

Other possibilities (various resonance structures are possible)

g. $NO_2^{\oplus}$ h. $NO_2^{\ominus}$ 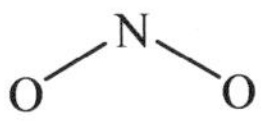   i. $N_3^{\ominus}$    j. CO_2 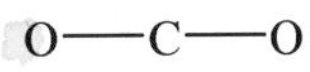   k. SO_2 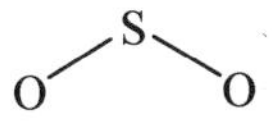   l. $NO_3^{\ominus}$ 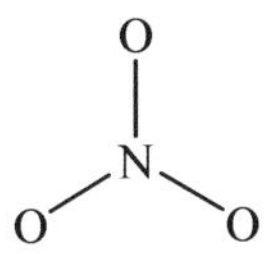   m. SO_3^{-2}

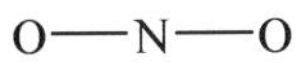

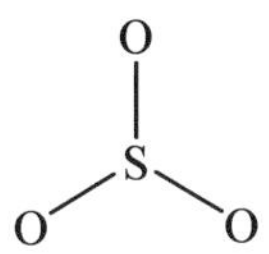

n. SO_4^{-2} o. PO_4^{-3} p. O_3 q. $HCOCH_2^{\ominus}$ r. $CH_3COCH_2^{\ominus}$ s. $CH_3O_2CCH_2^{\ominus}$

t. $HC(OH)CH_3^{\oplus}$ u. $CH_3C(OH)CH_3^{\oplus}$ v. $CH_3O(OH)CCH_3^{\oplus}$ w. $NCCH_2^{\ominus}$ x. $HNCCH_3^{\oplus}$

Add in any necessary formal charge.

methane   methyl free radical   methyl carbocation   methyl carbanion   methyl singlet carbene   methyl doublet carbene

50%   40%   10%
theoretical calculation of relative contributions.

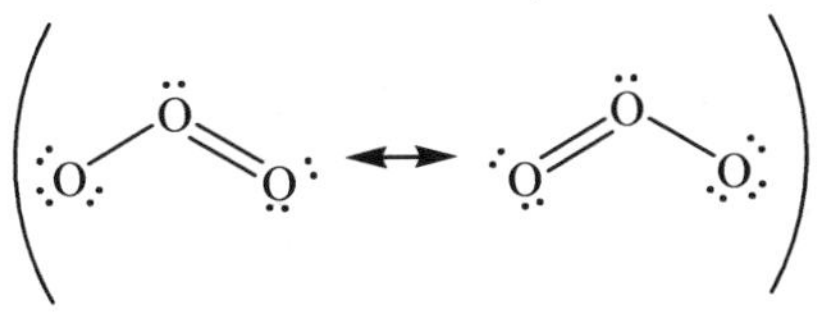

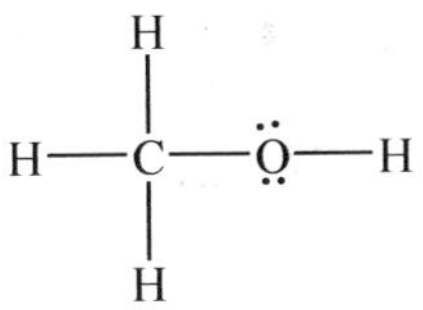

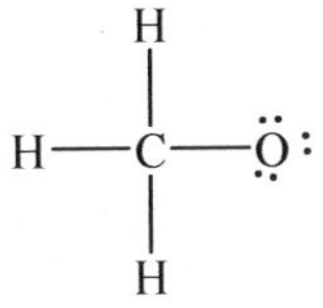

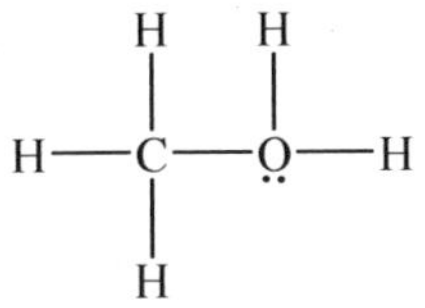

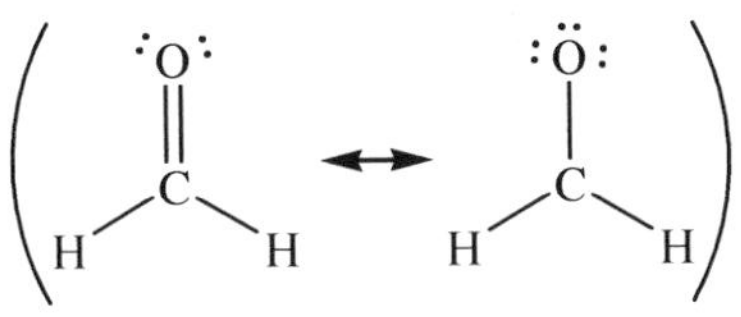

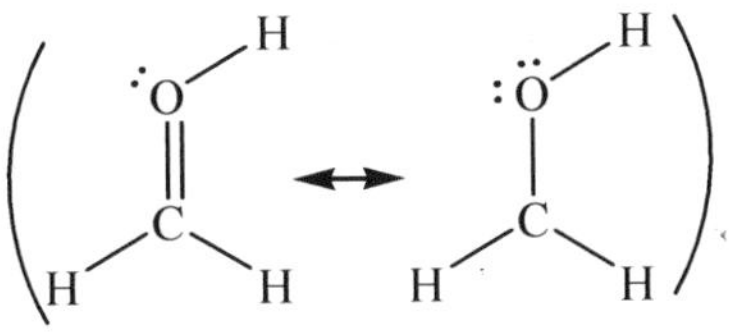

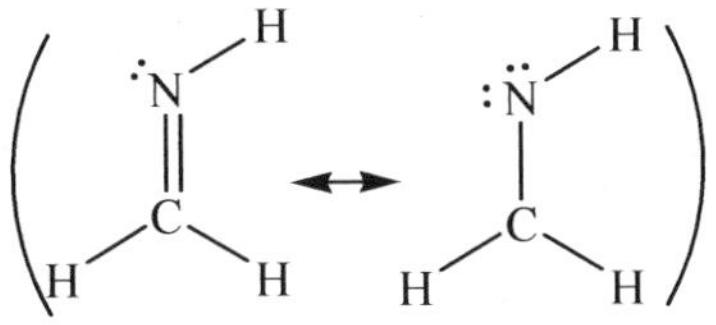

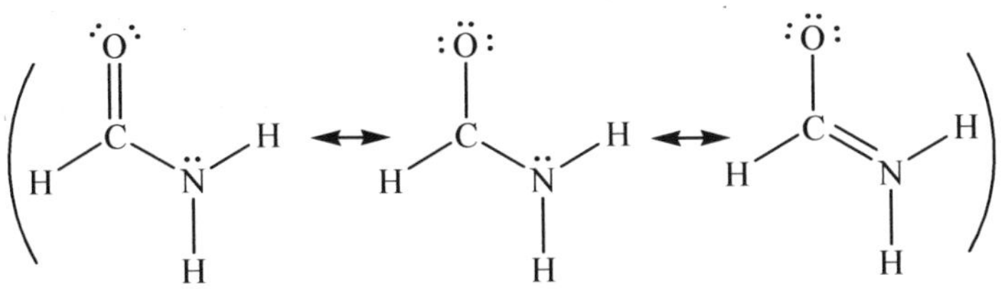

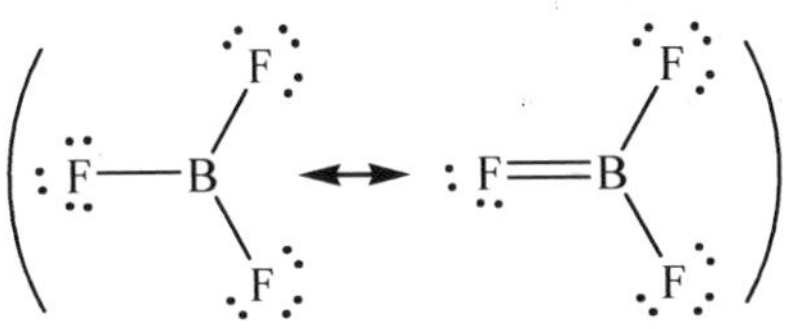

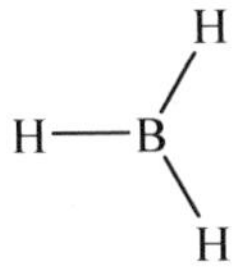

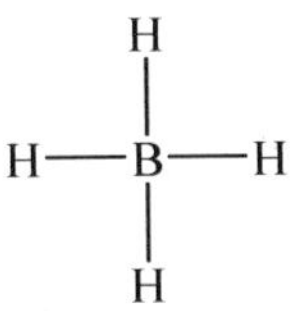

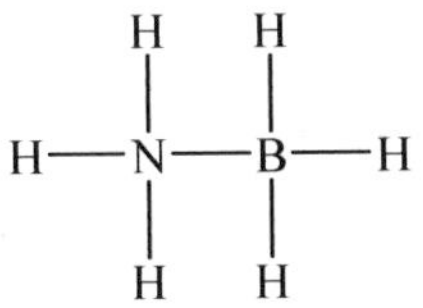

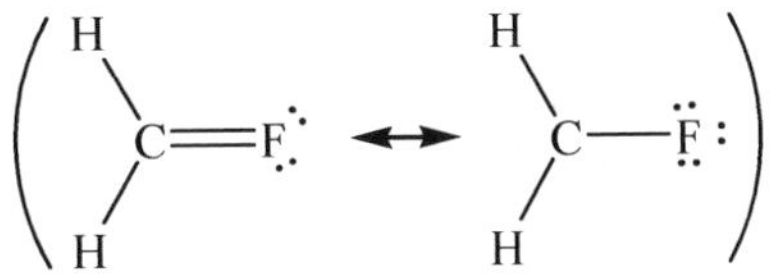

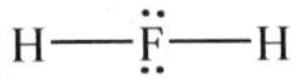

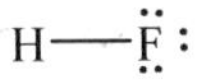

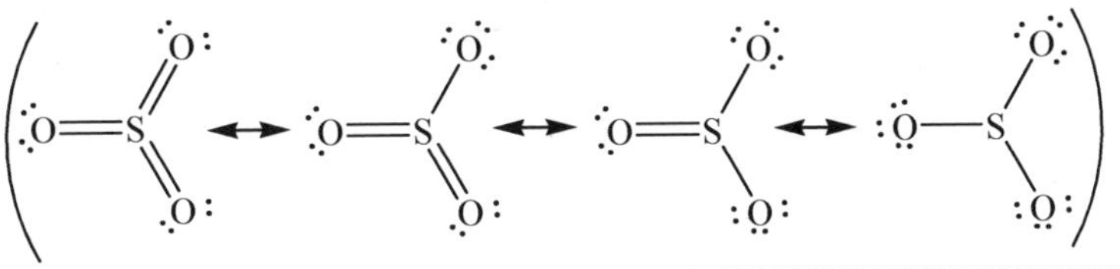

The first two structures violate the octet rule, but are often rationalized based on sulfur's 3d orbitals to accept additional electron density.

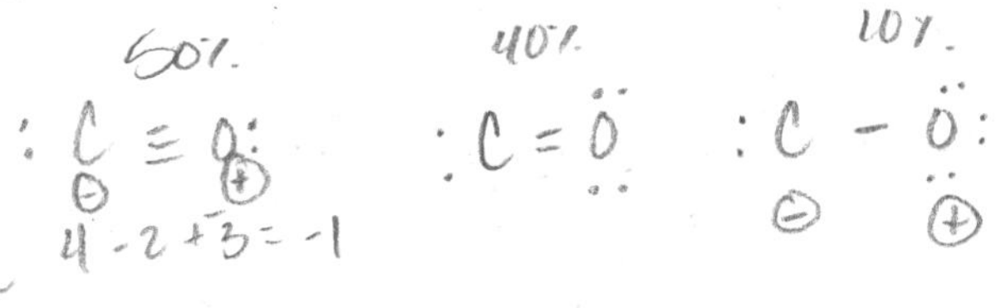

50%   40%   10%

Goals

a. Maximize # of Bonds

b. Minimize Formal Charge

c. Keep formal Charge Consistent w/ electronegativity if other goals were met.

:C≡O:
⊖   ⊕
4 - 2 + 3 = -1

:C=Ö
⊖

:C—Ö:
⊖   ⊕

Chapter 1

Practice Drawing Lewis Structures (2D and 3D).

1. Explicitly draw the connections of the atoms as indicated from the formula. Single lines are two electron bonds. The first bond is always a sigma bond made from hybrid orbitals and second and/or third bonds are always pi bonds made from 2p orbitals.

sigma connections

$CH_3CH_2CH_3$ | CH_2CHCH_3 | $HCCCH_3$ | $CH_2CHCCCH_3$ | H_2CCCH_2 | C_6H_6 6C ring

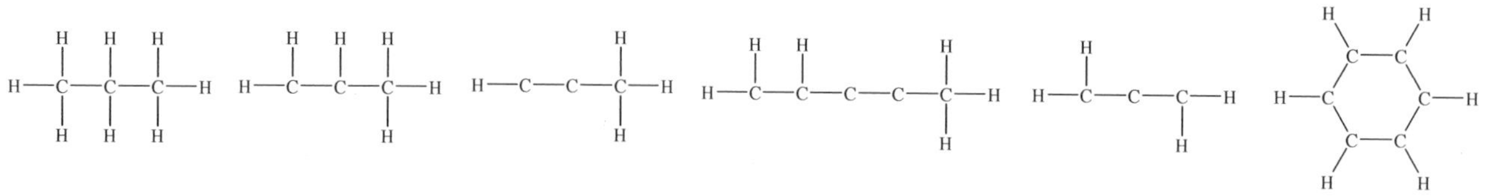

2. Count the total number of valence (bonding) electrons indicated from the bonding atoms and any formal charge present (add an electron for every negative charge and subtract an electron for every positive charge).

3C = 12 e	3C = 12 e	3C = 12 e	5C= 20 e	3C = 12 e	6C = 24 e
8H = 8 e	6H = 6 e	4H = 4 e	6H= 6 e	4C = 4 e	6H = 6 e
total = 20 e	total = 18 e	total = 16 e	total = 26 e	total = 16 e	total = 30 e

3. Add in any extra valence electrons to complete octets for atoms deficient in electrons. Begin first with more electronegative atoms, then continue on carbon atoms, until there are no more electrons to add in. When possible maximize separation of lone pair electrons. Examples with nitrogen, oxygen and halogen atoms are provided after these 'all carbon' examples.

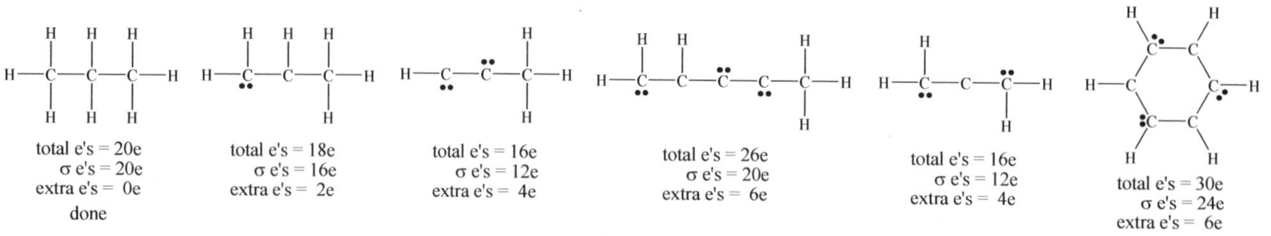

4. Use any lone pair electrons to share with neighbors lacking a full octet to make second and/or third bonds (which are pi bonds). Try to do this in a manner to minimize charge formation and compatible with an atoms neutral bonding patterns (C has 4 bonds, N has 3 bonds and 1 lone pair, O has 2 bonds and 2 lone pairs and F has 1 bond and 3 lone pairs). What is the hybridization, bond angles and descriptive shape for each carbon here?

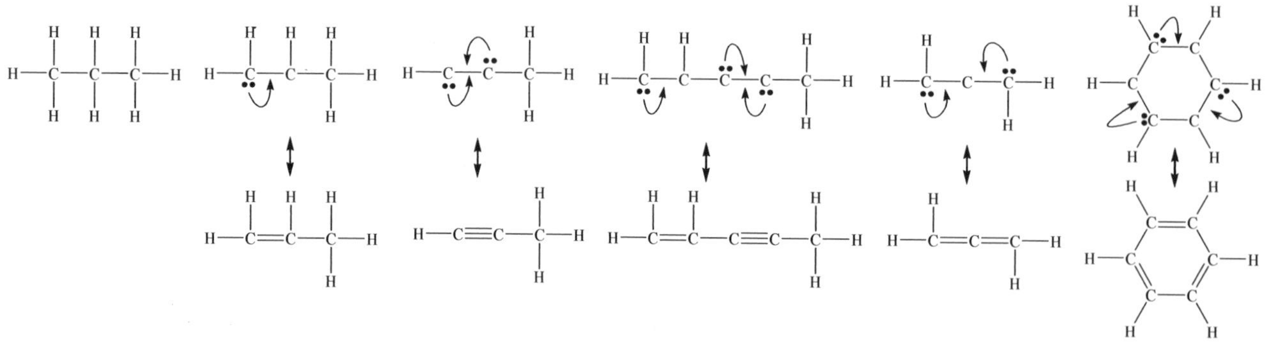

In all of the above structures, carbon has four bonds, zero lone pairs and zero formal charge. Once an acceptable Lewis structure is generated, you can determine the hybridization, shape and bond angles of each atom using the number of pi bonds that each atom has. Remember, pi bonds use 2p orbitals that are not available for hybridization.

Chapter 1

Nitrogen adds an extra proton (positive charge) in the nucleus and an extra valence electron in the bonding valence electrons. Neutral nitrogen has 3 bonds and 1 lone pair for itself. Follow the procedure above to draw 2D Lewis structures for the given molecules. What is the hybridization, bond angles and descriptive shape for each nonhydrogen atom below?

$CH_3CH_2NH_2$
1° amine

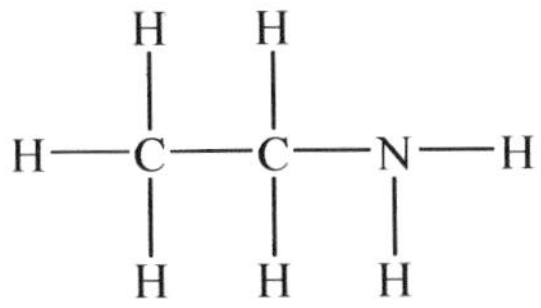

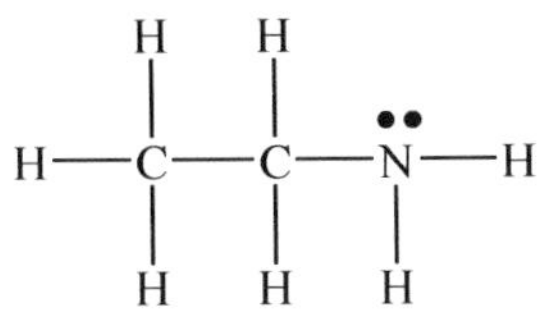

CH_3CHNCH_3
imine
(resonance?)

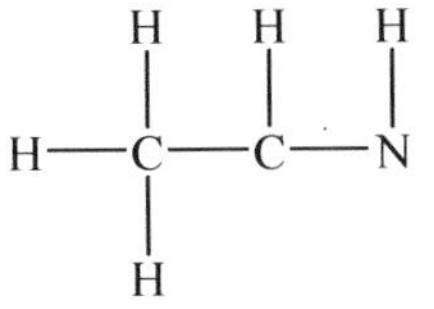

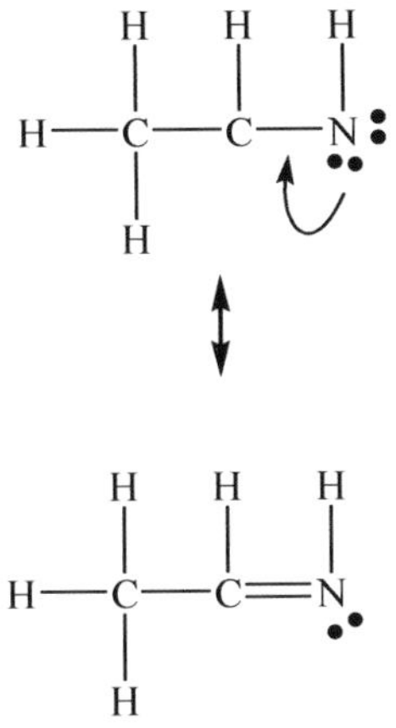

CH_3CN
nitrile
(resonance?)

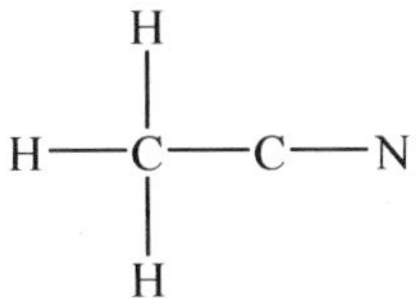

2C = 8 e
1N = 5 e
7H = 7 e
total = 20 e

total e's = 20e
σ e's = 18e
extra e's = 2e

2C = 8 e
1N = 5e
5H = 5 e
total = 18 e

total e's = 20e
σ e's = 16e
extra e's = 4e

2C = 8 e
1N = 5 e
3H = 3 e
total = 16 e

total e's = 20e
σ e's = 14e
extra e's = 6e

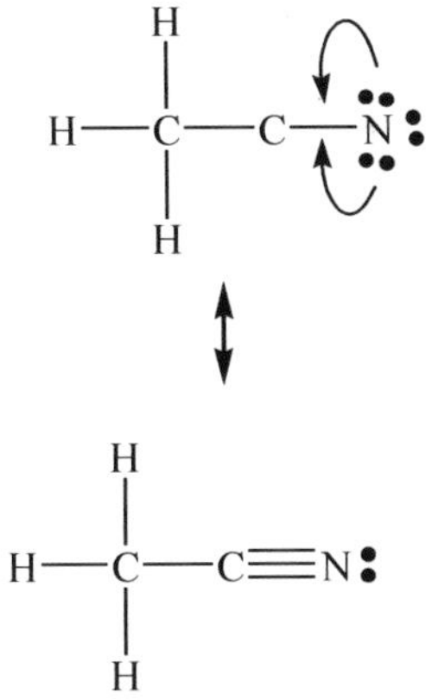

Oxygen adds an extra proton (positive charge) in the nucleus and an extra valence electron in the bonding electrons. Neutral oxygen has 2 bonds and 2 lone pairs for itself. Follow the procedure above to draw 2D Lewis structures for the given molecules. What is the hybridization, bond angles and descriptive shape for each nonhydrogen atom below?

CH_3CH_2OH
1° alcohol

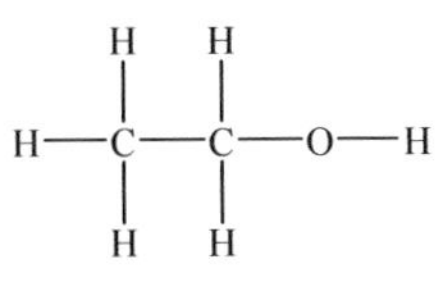

CH_3OCH_3
ether

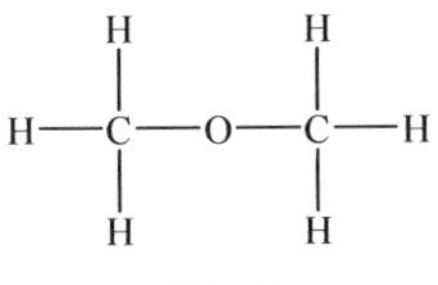

$CH_3CO_2CH_3$
ester

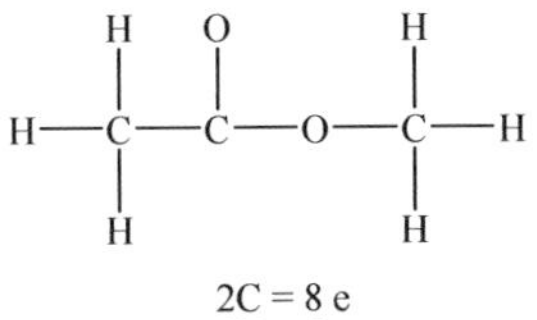

CH_3CONH_2
1° amide

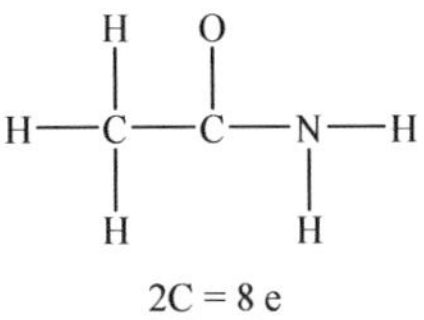

CH_3COCH_3
ketone

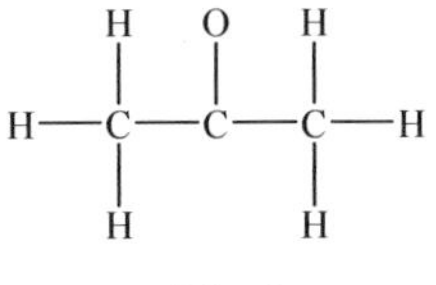

2C = 8 e
1N = 5 e
7H = 7 e
total = 20 e

total e's = 20e
σ e's = 18e
extra e's = 2e

2C = 8 e
1N = 5 e
7H = 7 e
total = 20 e

total e's = 20e
σ e's = 18e
extra e's = 2e

2C = 8 e
1N = 5 e
7H = 7 e
total = 20 e

total e's = 20e
σ e's = 18e
extra e's = 2e

2C = 8 e
1N = 5 e
7H = 7 e
total = 20 e

total e's = 20e
σ e's = 18e
extra e's = 2e

2C = 8 e
1N = 5 e
7H = 7 e
total = 20 e

total e's = 20e
σ e's = 18e
extra e's = 2e

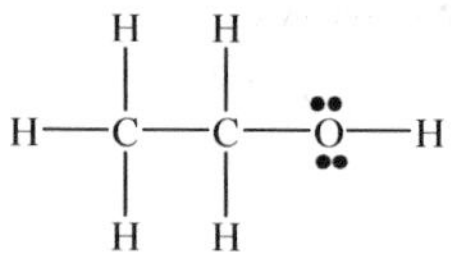

Chapter 1

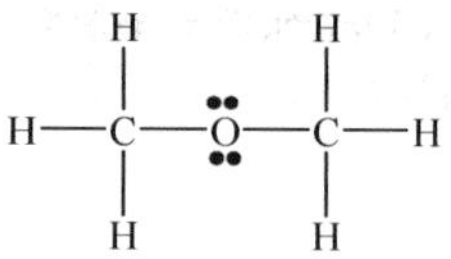

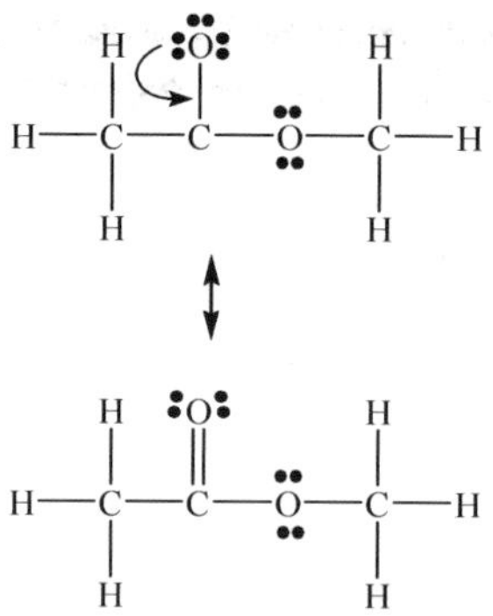

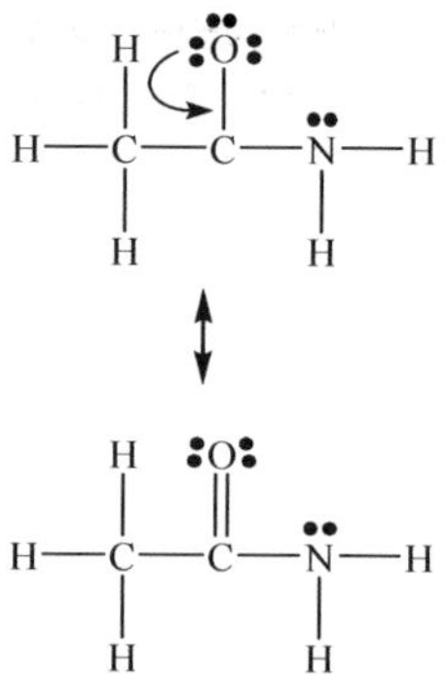

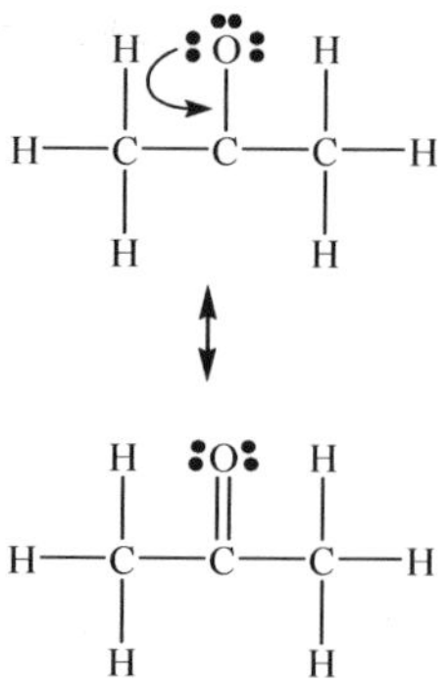

There are actually 3 acceptable ways to draw resonance structures for the ester and the amide (and related functional groups). The neutral structure, with full octets, is best, followed by the third resonance structure because it has maximum bonds and full octets, followed by the second resonance structure, which shows charge consistent with electronegativities, but 1 fewer bond and an incomplete octet on carbon.

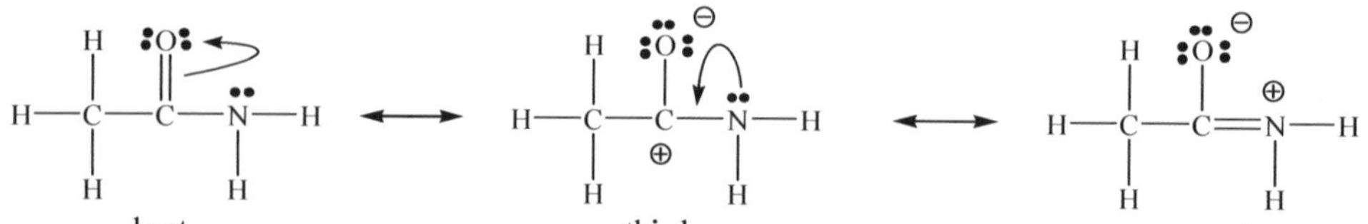

best   third   second

Fluorine adds an extra proton (positive charge) in the nucleus and an extra valence electron in the bonding electrons. Neutral fluorine has 1 bond and 3 lone pairs for itself. Follow the procedure above to draw 2D Lewis structures for the given molecules. What is the hybridization, bond angles and descriptive shape for each nonhydrogen atom below?

CH_3CH_2F
fluoroalkane

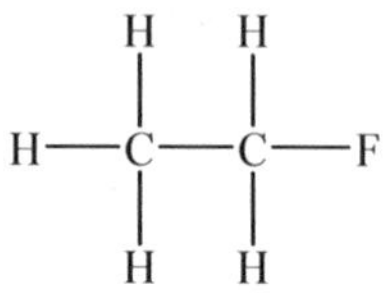

CH_3COF
acid fluoride

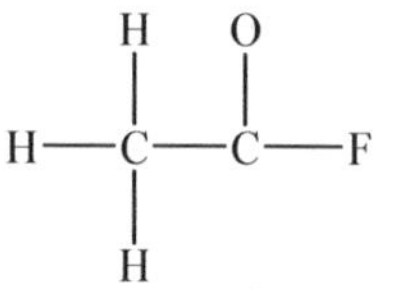

2C = 8 e
1F = 7 e
5H = 5 e
total = 20 e

total e's = 20e
σ e's = 12e
extra e's = 6e

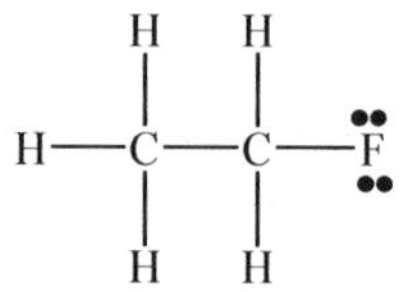

2C = 8 e
1F = 7 e
1O = 6 e
7H = 3 e
total = 24 e

total e's = 24e
σ e's = 12e
extra e's = 12e

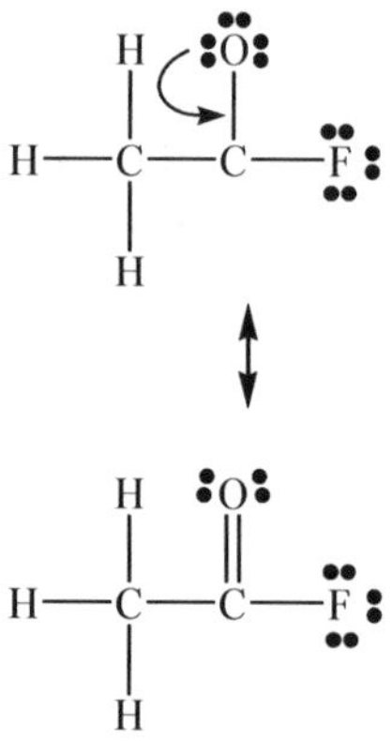

done

Chapter 1

Problem 16 - Draw two dimensional Lewis structures for the following structures. Include two dots for any lone pair electrons. What is the hybridization, bond angles and descriptive shape for each nonhydrogen atom below?

a. $CH_3CH_2(CH_3)_2CH$
alkane

b. $(CH_3)_3CCH_2Br$
1° bromoalkane

c. CH_3COCH_3
ketone

d. $H_2CCHCONH_2$
alkene 1° amide

e. $CH_3CHCHCH_3$
alkene

f. $(CH_3)_2CHNHCH_3$
2° amine

g. $CH_3CHCHCHO$
aldehyde

h. $(CH_3)_3CCN$
nitrile

i. $(CH_3)_2CHNH_2$
1° amine

j. $CH_2CHN(CH_3)CH_2CH_3$
3° amine

k. k. $CH_3CH_2CO_2H$
carboxylic acid

l. HCO_2H
carboxylic acid

m. $CH_3(CH_2)_3CH_2OH$
1° alcohol

n. $H_2CCHCCH$
alkenyne

o. $(CH_3)_2CHCO_2CH_3$
ester

p. $(CH_3)_2CHCOBr$
acid bromide

q. C_6H_6
aromatic

r. $C_6H_5CH_3$
aromatic

s. $HO_2CCH(Br)CHO$
acid/aldehyde

t. $(CH_3)_2CHCCCH_3$
alkyne (6 C ring)

Problem 17 - Draw two dimensional Lewis structures for the following structures. Include two dots for any lone pair electrons. Some of these have formal charges. What is the hybridization, bond angles and descriptive shape for each nonhydrogen atom below?

a. $(CH_3)_3NO$
amine oxide
formal charge

b. CH_3NO_2
nitromethane
formal charge

c. $H_3NCH(CH_3)CO_2$
amino acid
formal charge

d. $NCCHCHCH_3$
alkene nitrile
cis or trans

e. $C_6H_5CHCH_2$
styrene (6 C ring)

f. $CH_2CHCOCH(CH_3)_2$
alkene ketone

g. $HCCH_2CHO$
alkyne aldehyde

h. $CH_3C_6H_4OH$
a phenol (6 C ring)

i. $BrCH_2CHClCH_3$
1°/2° halide

j. $(CH_3)_2CCHCO_2H$
alkene carboxylic acid

k. $HCCCH_2NH_2CH_2CH(CH_3)_2^+$
alkyne ammonium ion

l. H_2CCO
ketene

m. $OHCCHC(CH_3)CO_2^{\ominus}$
aldehyde alkene carboxylate ion

n. $CH_3CH_2C(CH_3)_2CONH_2$
amide

o. $(CH_3)_2CHCOCl$
acid chloride

p. $CCH(CH_3)CONH_2$
nitrile amide

q. $(CH_3)_2CHCO_2CH_3$
ester

r. $CH_3COC_6H_5$
aromatic ketone (6 C ring)

s. $CH_3CCCONHCH_3$
alkyne amide

t. $CH_3OCH_2CH_2COCH_3$
ether ketone

Chapter 1

Guidelines for Three Dimensional Lewis Structures (build molecular models to better visualize the structures)

Guideline 1 – Draw a 2D Lewis structure as described above. This will allow you to determine the connections of the atoms, the hybridization, the shape, the bond angles, the number of sigma and pi bonds and the nature of any lone pairs of electrons.

Guideline 2 – Draw all reasonable 2D resonance structures to allow a determination of each atom's hybridization. Look for more than one way to share electrons (lone pairs next to an empty 2p orbital or next to a pi bond). The hybridization of any atom is determined by the resonance structure in which that atom has its maximum number of bonds. This is a very important idea that you will probably overlook on occasion. If I make a big deal about it here, maybe you won't forget about it…but you probably will!

Guideline 3 – Place as many atoms as possible in the plane of the paper when you begin to draw your 3D structure. You only need to know how to draw three different shapes (sp, sp² and sp³), but you really need to know how to draw them well. If there is a triple bond or an aromatic ring present, it is often helpful to begin drawing your 3D structure from that feature first.

a. If a triple bond is present in your 2D Lewis structure, draw it first. (RCCR, RCN, RCO⁺)

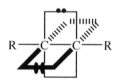

3D triple bond, in plane of page

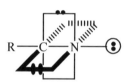

The lone pair is in an sp hybrid orbital.

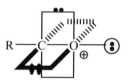

The lone pair is in an sp hybrid orbital and oxygen has a positive formal charge.

b. There are many types of double bonds possible. These pi bonds are drawn in the plane of the page, but they can be drawn in and out of the page too (next page). Additional pi bonds, not included below, are -N=O and -N=N-. (condensed line formulas of structures, just below: R₂CCR₂, R₂CNR, R₂CO, R₂CF⁺)

2D pi bond, in plane of page

The lone pair is in an sp² hybrid orbital.

The lone pairs are in sp² hybrid orbitals.

The lone pairs are in sp² hybrid orbitals and fluorine has a positive formal charge.

There are only two possible shapes for an sp² hybridized atom, if the pi bond is drawn in the plane of the paper. Look for the parallel bond relationships to decide if lines are dashes, wedges or simple and which way they point.

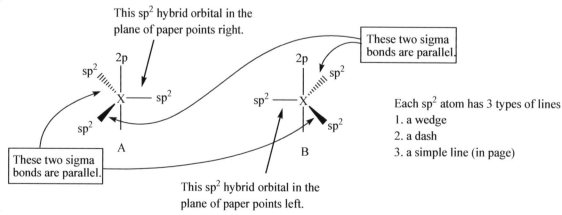

Each sp² atom has 3 types of lines
1. a wedge
2. a dash
3. a simple line (in page)

Additional ways an sp^2 pi bond can be drawn when the pi bond is in or out of the plane of the paper. Look for parallel bond relationships (arrows point at parallel bonds).

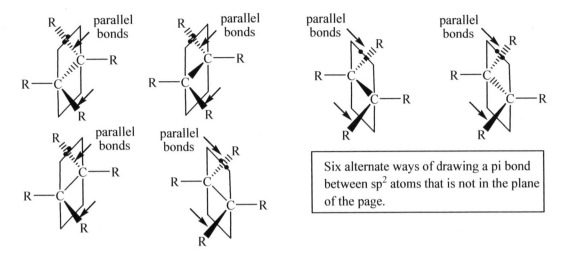

Six alternate ways of drawing a pi bond between sp^2 atoms that is not in the plane of the page.

c. Pi bonds can be joined together in an almost endless number of possible ways. (two triples, triple plus a double, two doubles, cumulated, aromatic…etc.).

2D Lewis structure

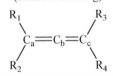

3D Lewis structure

Notice that several bonds are parallel.
R$_1$-C$_a$ is parallel with C$_b$-C$_c$ and C$_d$-R$_6$.
R$_2$-C$_a$ is parllel with C$_b$-R$_3$, C$_c$-R$_4$ and C$_d$-R$_5$.
C$_a$-C$_b$ is parallel with C$_c$-C$_d$

In the left structure, the back pi bond was drawn first so the sigma bond between the two pi bonds is a wedge line. In the right structure, the front pi bond was drawn first so the sigma bond is a dashed line. Use the parallel relationship of the sigma bonds to help you draw the correct perspective in these structures.

2D Lewis structure cumulene (allene)
(a little misleading)

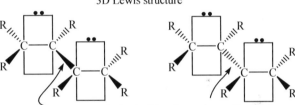

The 2D structure of allene looks like the C$_a$-R$_1$ and C$_c$-R$_4$ bonds are parallel, but they are NOT.

3D Lewis structure

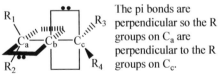

The pi bonds are perpendicular so the R groups on C$_a$ are perpendicular to the R groups on C$_c$.

2D Lewis structure aromatic (benzene). Add resonance arrows and draw a second equivalent resonance structure.

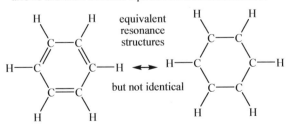

equivalent resonance structures

but not identical

3D Lewis structure

Chapter 1

Problem 18 – Draw the following structures using 3D representations. (If you need help, use the keys.)

a R—C≡C—C≡C—R   b R—C≡C—C≡N: c

d

e

d. Drawing sp³ hybridization may be the most difficult to represent in three dimensions. Simple single bonds allow unlimited rotations that can be in the page, in front of the page or in back of the page. To make your drawing easier, assume that tetrahedral atoms are in a zig-zag conformation (shape) in the plane of the paper. Usually, we will draw two simple lines, in the plane of the paper and one wedged line (in front) and one dashed line (behind). The two simple lines will have to be drawn close to one another, as will the dashed and wedged lines. These can be difficult to draw accurately and we will allow shapes to be drawn that are *pretty close*".

These lines are drawn correctly.

Simple lines can be drawn close together four different ways, pointing up, pointing right, pointing down and pointing left.

Wedged and dashed lines can be drawn close together four different ways, pointing down, pointing left, pointing up and pointing right.

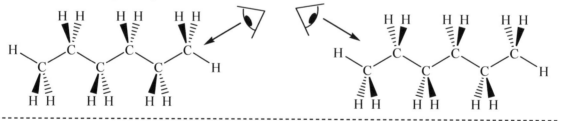

These lines are drawn **incorrectly**. They imply 180° bond angles. Don't draw them this way!

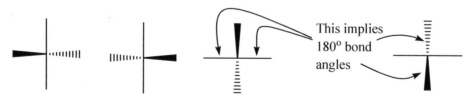

This implies 180° bond angles

Drawing the simple lines or the wedged and dashed lines opposite one another implies that there are 180° bond angles...but we know that sp³ atoms have approximately 109° bond angles.

Chapter 1

Point of view provides yet another complication in drawing tetrahedral 3D atoms. Stay consistent.

The perspective of these two atoms is consistent when viewed from the left. The dashes are parallel and the wedges are parallel.

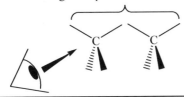

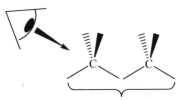

The perspective of these two atoms is consistent when viewed from the left. The dashes are parallel and the wedges are parallel.

The perspective of these two atoms is consistent when viewed from the right. The dashes are parallel and the wedges are parallel.

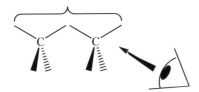

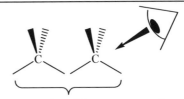

The perspective of these two atoms is consistent when viewed from the right. The dashes are parallel and the wedges are parallel.

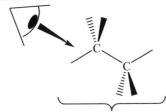

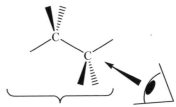

The perspective of these two atoms is consistent when viewed from the left. The wedges pointing up are parallel to the dashes pointing down, and vice versa.

The perspective of these two atoms is consistent when viewed from the right. The wedges pointing up are parallel to the dashes pointing down, and vice versa.

None of the perspectives of these pairs of atoms is consistent when viewed from the either side. All of these are drawn incorrectly.

   inconsistent viewing perspectives      **Wrong!**

   **Wrong!**      inconsistent viewing perspectives

As much as possible, draw a sequence of sp³ atoms in a zig-zag shape. Besides there being a logical "energetic" reason for drawing a structure this way (studied later), it is easier to draw and see on a piece of paper. The zig-zag bonds in the plane of the paper all use simple lines and there will be a wedged and dashed line coming off each sp³ atom. At alternate positions they will be pointing up, when the chain is up, and then down, when the chain is down. Try to keep in mind the point of view restrictions shown above.

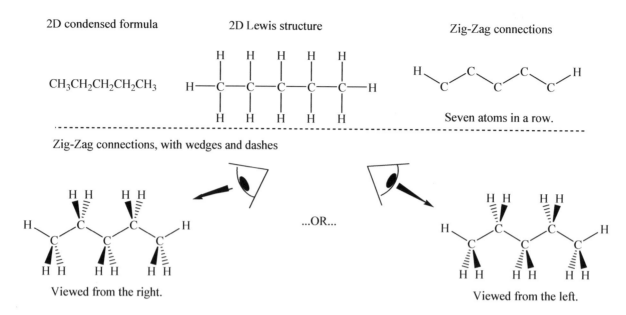

It's easy to switch in an sp³ heteroatom in place of an sp³ carbon atom (N, O or S). Just fill in a lone pair where there was a hydrogen atom. Remember too, single bonds allow for rotations so there are many possible variations that one can use to draw these structures.

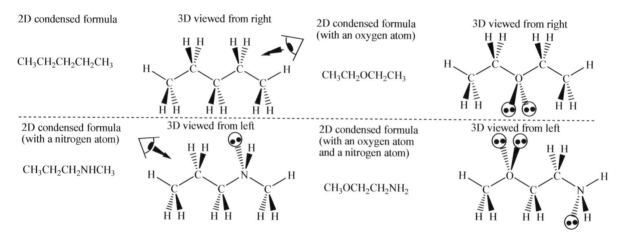

Here are two examples, showing all three possible hybridization states in one molecule, and some heteroatoms thrown in to boot.

Condensed line formula

$CH_3CCCHCHCH_3$

2D Lewis structure

3D Lewis structures

Triple bond in the plane, double bond in and out of the plane, viewed from the left.

Chapter 1

Condensed line formula

$$NCCCC_6H_4COCHCHCH_2OCH_2CH_2NH_3$$

2D Lewis structure

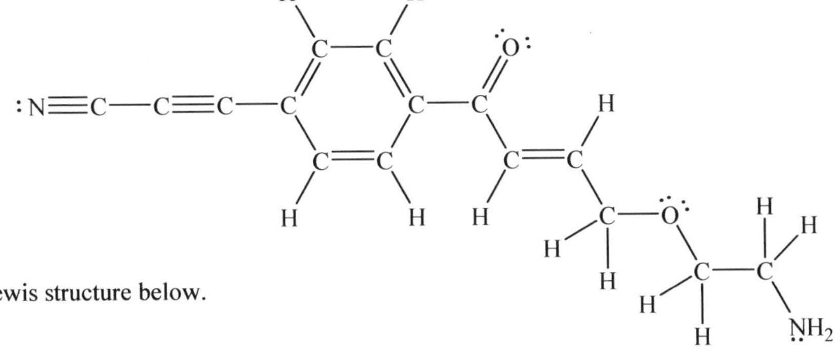

3D Lewis structure below.

This is a difficult 3D Lewis structure to draw. Whatever you draw on the paper is a good approximation of what is in your head. Errors on the page imply errors in your mind. Be careful and precise in what you write down and your thinking will be much clearer.

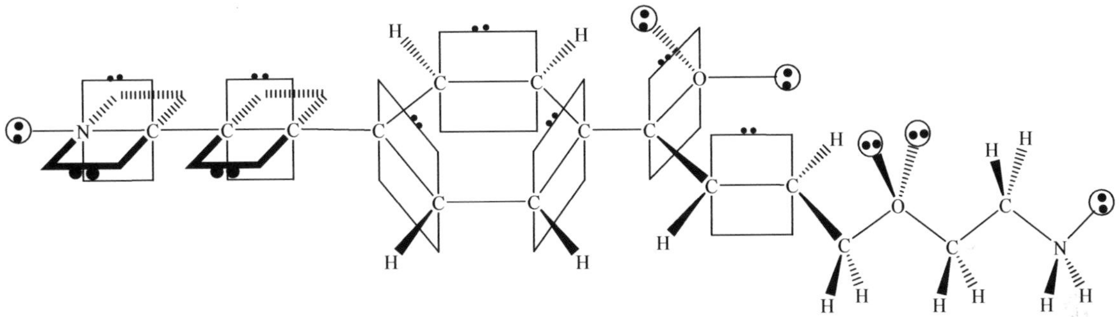

2D and 3D resonance structures – common themes

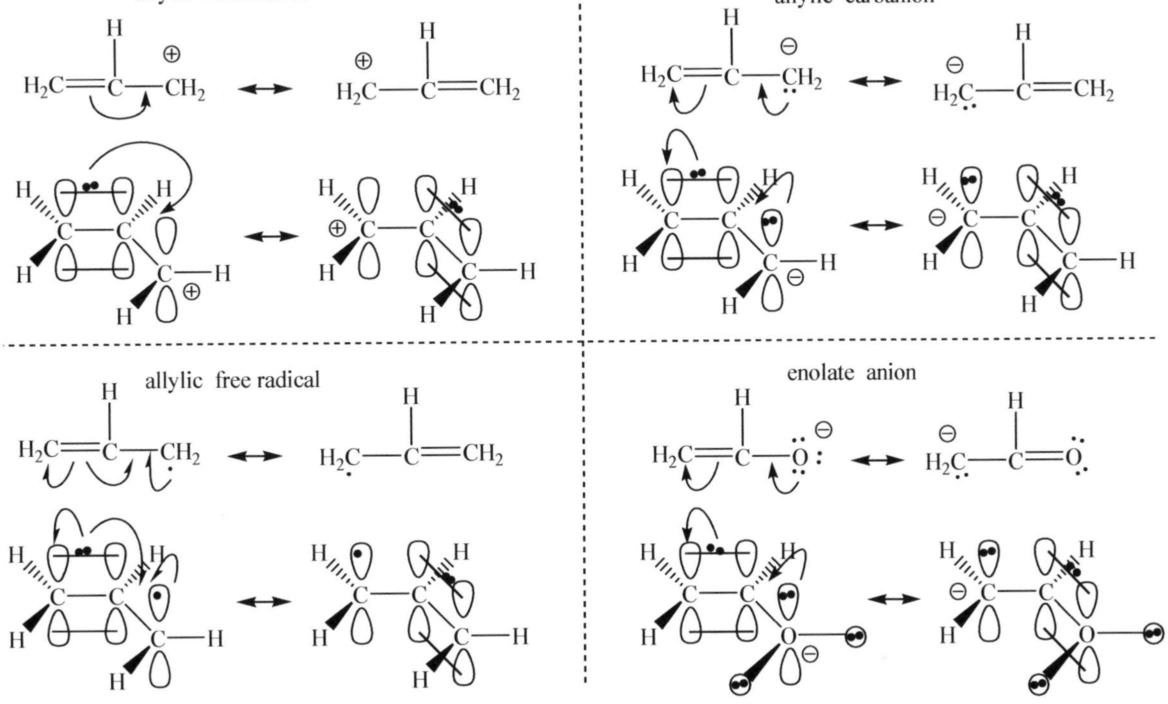

allylic carbocation

allylic carbanion

allylic free radical

enolate anion

Chapter 1

Summary of possible 3D shapes for carbon, nitrogen, oxygen (and sulfur) and fluorine (and Cl, Br, I)

Carbon: Four ways to draw carbon in organic chemistry.

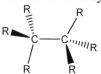 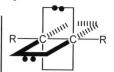 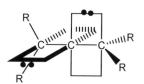

hybridization: sp³
bond angles: 109°
number of sigma bonds: 4
number of pi bonds: 0
number of lone pairs: 0

hybridization: sp²
bond angles: 120°
number of sigma bonds: 3
number of pi bonds: 1
number of lone pairs: 0

hybridization: sp
bond angles: 180°
number of sigma bonds: 2
number of pi bonds: 2
number of lone pairs: 0

hybridization: sp
bond angles: 180°
number of sigma bonds: 2
number of pi bonds: 2
number of lone pairs: 0

Nitrogen: Four ways to draw nitrogen in organic chemistry.

 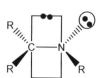 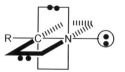 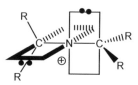

hybridization: sp³
bond angles: 109°
number of sigma bonds: 3
number of pi bonds: 0
number of lone pairs: 1

hybridization: sp²
bond angles: 120°
number of sigma bonds: 2
number of pi bonds: 1
number of lone pairs: 1

hybridization: sp
bond angles: 180°
number of sigma bonds: 1
number of pi bonds: 2
number of lone pairs: 1

hybridization: sp
bond angles: 180°
number of sigma bonds: 2
number of pi bonds: 2
number of lone pairs: 0
positive formal charge

Oxygen: Three ways to draw oxygen in organic chemistry.

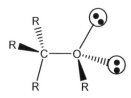 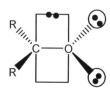 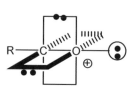

hybridization: sp³
bond angles: 109°
number of sigma bonds: 2
number of pi bonds: 0
number of lone pairs: 2

hybridization: sp²
bond angles: 120°
number of sigma bonds: 1
number of pi bonds: 1
number of lone pairs: 2

hybridization: sp
bond angles: 180°
number of sigma bonds: 1
number of pi bonds: 2
number of lone pairs: 1
positive formal charge

Fluorine: Two ways to draw fluorine in organic chemistry.

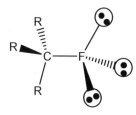 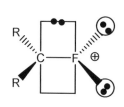

hybridization: sp³
bond angles: 109°
number of sigma bonds: 1
number of pi bonds: 0
number of lone pairs: 3

hybridization: sp²
bond angles: 120°
number of sigma bonds: 1
number of pi bonds: 1
number of lone pairs: 2
positive formal charge

Chapter 1

Problem 19 – First, convert the condensed line formulas of the following hydrocarbons into 2D Lewis structures. Next, draw 3D structures for each of the 2D structures. You should show the bonds in front of the page as wedges and bonds in back of the page with dashed lines and bonds in the plane of the page as simple lines. Show the 2p orbitals for pi bonds along with their electrons. If you cannot figure out how the atoms are connected, there are some clues just below.

Condensed line formulas (If you need hints, partial connections are just below.)

a.
H_2CCH_2

b.
CH_2CHCH_3

c.
H_3CCCH

d.
CH_2CCH_2

e.
$CH_2CHCHCH_2$

f.
$H_2CCCHCH_3$

g.
$CH_3CCCH_2CH_2CH_3$

h.
H_2CCCCH_2

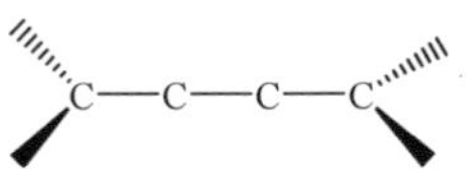

i.
$C_6H_5CCCHCH_2$

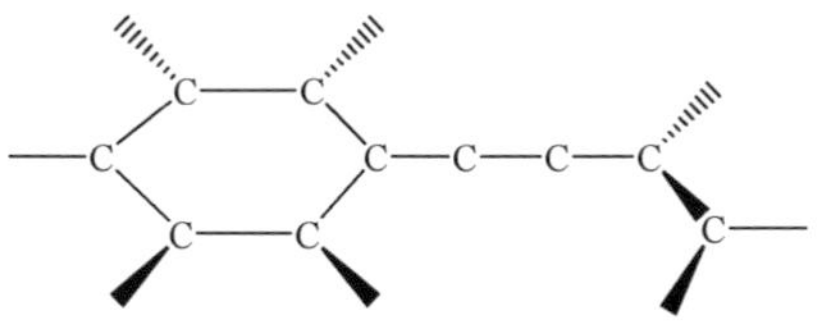

j.
$CH_3CCCCCH_3$

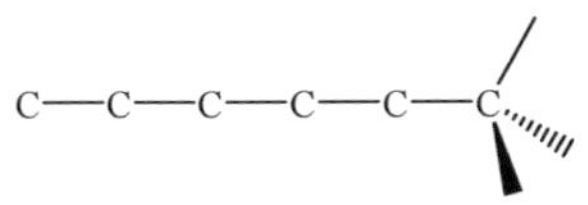

k.
$H_2CCHCCH$

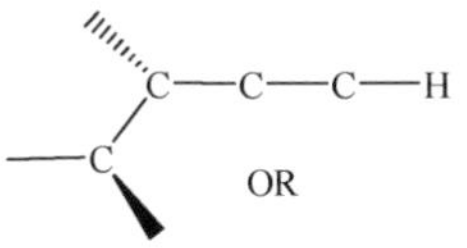

l.
$CH_3CH_2CH_2CCCHCH_2$

Hints: Possible 3D skeletal connections for problem 19.

a.
H_2CCH_2

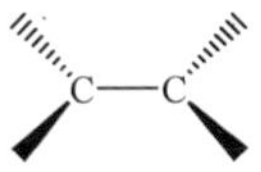

b.
CH_2CHCH_3

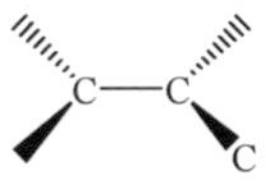

c.
H_3CCCH

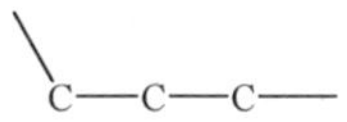

d.
CH_2CCH_2

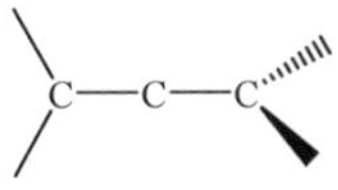

e.
$CH_2CHCHCH_2$

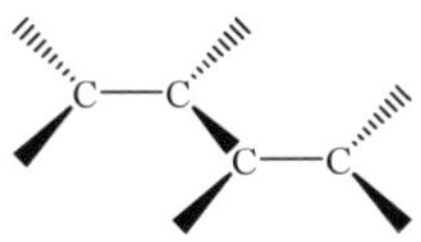

f.
$H_2CCCHCH_3$

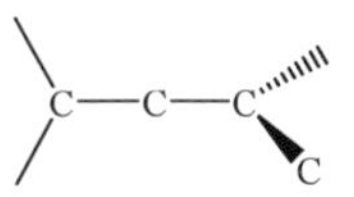

g.
$CH_3CCCH_2CH_2CH_3$

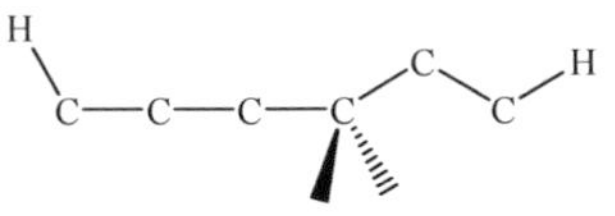

h.
H_2CCCCH_2

i.
$C_6H_5CCCHCH_2$

k.
$H_2CCHCCH$

j.
$CH_3CCCCCH_3$

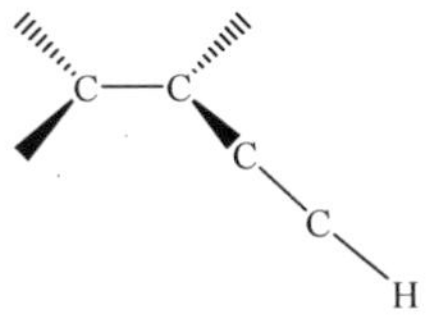

l.
$CH_3CH_2CH_2CCCHCH_2$

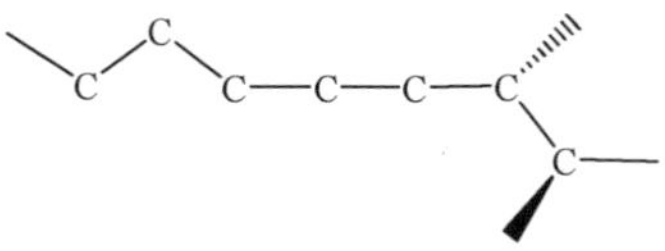

Problem 20 – This problem is very similar to problem 19, except heteroatoms (N, O and F) are substituted in for some of the carbon atoms and some structures have formal charge. If resonance structures are present, decide an atom's hybridization based on the resonance structure where it has its maximum number of bonds. If you cannot figure out how the atoms are connected, there are some clues just below.

Condensed line formulas

a.
CH_2CHCHO
alkene, aldehyde

b.
$HCCCO_2H$
alkyne, acid

c.
H_3CCH_2CN
alkane, nitrile

d.
$CH_3NHCH_2CH_3$
2° amine

e.
CH_3CCCN
alkyne, nitrile

f.
$CH_3CH_2COCH_3$
ketone

g.
$CH_3CHCHCH_2OH$
alkene, alcohol

h.
CH_2CHN_3
has formal charge
alkene, azide

i.
$OHCC_6H_4CN$
aldehyde, aromatic, nitrile

j.
CH_3CHN_2
has formal charge
diazoalkane

k.
$CH_2CHC_6H_4CCCO_2CH_3$
six atom ring
alkene, aromatic, alkyne, ester

l.
$ClCH_2CH_2CONHCH_3$
chloro, 2° amide

m.
$O_2NCHCHCCCH_3$
has formal charge
nitro, alkene, alkyne

n.
$CH_3C_5H_4N$
six atom
heterocyclic ring
pyridine

o.
CH_3CHCO
ketene

p.
$CH_3OCH_2CH_2NHCH_2CH_2OH$
ether, 2° amine, alcohol

Hints: Possible 3D skeletal connections for problem 20.

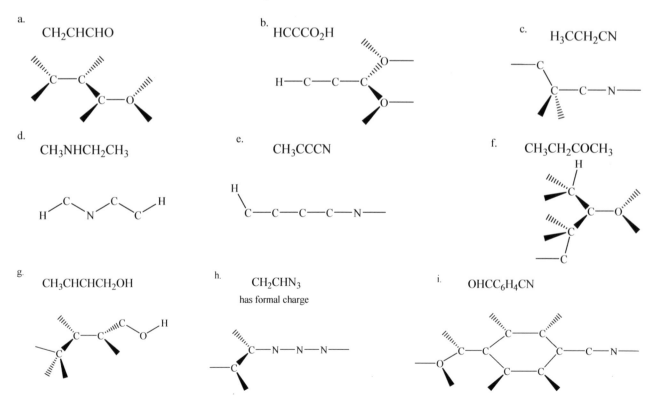

a. CH_2CHCHO

b. $HCCCO_2H$

c. H_3CCH_2CN

d. $CH_3NHCH_2CH_3$

e. CH_3CCCN

f. $CH_3CH_2COCH_3$

g. $CH_3CHCHCH_2OH$

h. CH_2CHN_3
has formal charge

i. $OHCC_6H_4CN$

Chapter 1

g.

CH₃CHCHCH₂OH

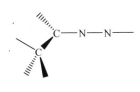

h.

CH₂CHN₃

has formal charge

i.

OHCC₆H₄CN

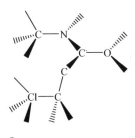

j.

CH₃CHN₂

has formal charge

k.

CH₂CHC₆H₄CCCO₂CH₃

six atom ring

l.

ClCH₂CH₂CONHCH₃

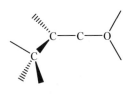

m.

O₂NCHCHCCCH₃

has formal charge

n.

CH₃C₅NH₄

six atom
heterocyclic ring

o.

CH₃CHCO

p.

CH₃OCH₂CH₂NHCH₂CH₂OH

Chapter 1

CHAPTER 2 – TYPES OF BONDING (and how it affects physical properties)

Melting Points (what are they?)

Compact regular shaped molecules pack together more efficiently, typically having closer contacts and forming stronger lattice structures than irregular shaped molecules. Well ordered lattice structures tend to have higher melting points, when other factors are similar. There is not a large difference in the forces of interaction in the solid versus the liquid phases because the molecules are close together in both. Typically, about one out of ten molecules leaves its lattice position open in a liquid. This allows greater mobility for the molecules in a liquid than in a solid. It also expands the volume slightly so that liquids, in general, are less dense than the solid phase. Normally solids sink to the bottom when both liquid and solid are together. We will judge melting points mainly on the symmetry of the lattice structures. More symmetrical molecules often form a more compact lattice structure which allows closer contact and stronger intermolecular forces of interaction, requiring more energy to break down and thus, a higher melting point.

Fortunately for us, water is an exception to the general observation that solids sink. Instead of ice sinking in water, it floats. Because of hydrogen bonding, ice (solid water) expands and is less dense than liquid water. If that weren't the case, the earth would be a giant ice cube. What would happen to all of our flowing water? Every summer a little water would form on the top and the rest would remain a big iceberg on the bottom. That would be a big problem for living creatures and life as we know it.

Boiling Points (what are they?)

Gas molecules, on the other hand, behave independently of one another having extremely weak intermolecular forces. In fact, the ideal gas law assumes (as a first approximation) there are zero forces of interaction in the gas phase ($PV = nRT$). The boiling point is an indication of how much energy it takes to completely separate one molecule from all of the other molecules in the liquid phase. A boiling point is an excellent indicator of the strength of interactions among the molecules. This is NOT the same as the attraction of atoms for the electrons in bonds, which significantly determines the bond energy between two atoms. More polar molecules require more energy to boil (higher bp) because each molecule has a stronger attraction for its neighbor molecules. Also, a greater accumulation of any particular type of force requires more energy to overcome the greater forces of attraction (= higher bp).

More polar = Higher BP

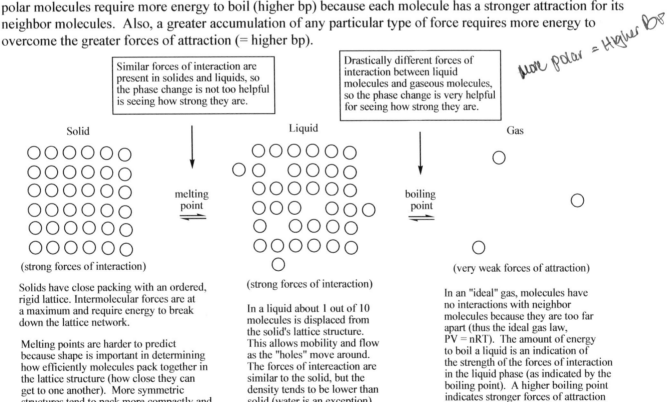

Similar forces of interaction are present in solides and liquids, so the phase change is not too helpful is seeing how strong they are.

Drastically different forces of interaction between liquid molecules and gaseous molecules, so the phase change is very helpful for seeing how strong they are.

Solid — melting point — Liquid — boiling point — Gas

(strong forces of interaction)

(strong forces of interaction)

(very weak forces of attraction)

Solids have close packing with an ordered, rigid lattice. Intermolecular forces are at a maximum and require energy to break down the lattice network.

Melting points are harder to predict because shape is important in determining how efficiently molecules pack together in the lattice structure (how close they can get to one another). More symmetric structures tend to pack more compactly and have, relatively, higher melting points.

In a liquid about 1 out of 10 molecules is displaced from the solid's lattice structure. This allows mobility and flow as the "holes" move around. The forces of intereaction are similar to the solid, but the density tends to be lower than solid (water is an exception).

In an "ideal" gas, molecules have no interactions with neighbor molecules because they are too far apart (thus the ideal gas law, $PV = nRT$). The amount of energy to boil a liquid is an indication of the strength of the forces of interaction in the liquid phase (as indicated by the boiling point). A higher boiling point indicates stronger forces of attraction in the liquid phase.

Chapter 2

Types of Intermolecular Forces of Attraction

A variety of examples is shown in the table below, from completely nonpolar to ionic. The physical properties change greatly, depending on the overall intermolecular forces of attraction (generally, ionic > polar > nonpolar). The boiling points can provide some insight into the strengths of these forces.

Examples	Mol. Wt.	Melting point (^{o}C)	Boiling point (^{o}C)
NaCl (ionic)	58.4	+801	+1465 (ionic salt)
LiF (ionic)	25.9	+845	+1676 (ionic salt)
H-H	2.0	-259	-253
CH_4	16.0	-182	-164
NH_3	17.0	-78	-33
H_2O	18.0	0	+100
HF	20.0	-183	+20
CH_3CH_3	30.0	-183	-89
CH_3NH_2	31.0	-94	-6
CH_3OH	32.0	-98	+65
CH_3F	34.0	-142	-78

$\Delta T = 131^{o}C$
$\Delta T = 133^{o}C$
$\Delta T = 80^{o}C$

molecular substances

Absolute zero = 0 K = -273°C

Often when we see a series of numbers, such as the temperatures in the above table, our eyes glaze over and we move on to the next blurry group of words. But this time, let's do a thought experiment to give those temperatures more meaning. Let's fill two imaginary pots with water. One we merely set in front of us and one we put over our imaginary stove burner and bring to a boil. The pot in front of us, at room temperature, is about 25°C, while the pot boiling on the stove is about 100°C, a mere 75°C higher. Now for the thought experiment: stick your imaginary hand into each of the pots of water. What? You say I'm crazy? You know that 75°C is a huge difference in temperature. The imaginary cold water does nothing to your imaginary hand, while the imaginary boiling water cooks it. If you want to try this as a real experiment, substitute a real hot dog for your imaginary hand. Now, think about what the much larger differences in temperature in the table above are telling us while you are eating your real hot dog.

Ionic bonding - Ion-Ion interactions are very strong forces of attraction between oppositely charged ions.

The elements at the edges of the periodic table tend to lose or gain valence electrons, depending on which side they are. A complete transfer of electrons forms ions (cations lose electrons and anions gain electrons). Neutral salts require charge balance: (total negative charge) = (total positive charge), so that the net charge is zero.

If one were to mix diatomic chlorine gas, Cl_2, with metallic sodium, Na, quite likely an explosion would occur in a violent transfer of electrons. Both elements are dangerously reactive. The metallic sodium atoms would give up their electrons to the chlorine diatomic molecules, breaking the covalent bond between the chlorine atoms. In so doing, each element would attain the desired Noble gas configuration as oppositely charged ions and a large amount of energy would be released (the lattice energy of the NaCl salt), as the reaction is very exothermic. The metallic lattice structure of sodium atoms would disintegrate and the gaseous chlorine would disappear. The ions formed would surround themselves with opposite charge and avoid similar charge, forming an ionic lattice of table salt. There is a dramatic change in physical and chemical properties when this reaction occurs (e.g. mp, bp, water solubility, etc.).

Chapter 2

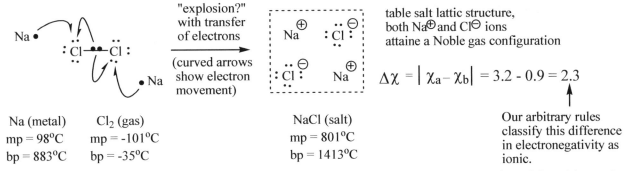

Na (metal) Cl_2 (gas)

mp = 98°C   mp = -101°C

bp = 883°C   bp = -35°C

NaCl (salt)

mp = 801°C

bp = 1413°C

table salt lattic structure, both $Na^{\oplus}$ and $Cl^{\ominus}$ ions attaine a Noble gas configuration

$$\Delta\chi = |\chi_a - \chi_b| = 3.2 - 0.9 = 2.3$$

Our arbitrary rules classify this difference in electronegativity as ionic.

    The solubility of many salts in water is often high because water has a concentrated partial positive end (the protons) that can interact strongly with anions and a concentrated partial negative end (the oxygen) that can interact strongly with cations, leading to very high overall solvation energies. However, solvation energy has to be larger than the very large lattice energy (energy of the cations interacting with the anions) or the salt will not dissolve.

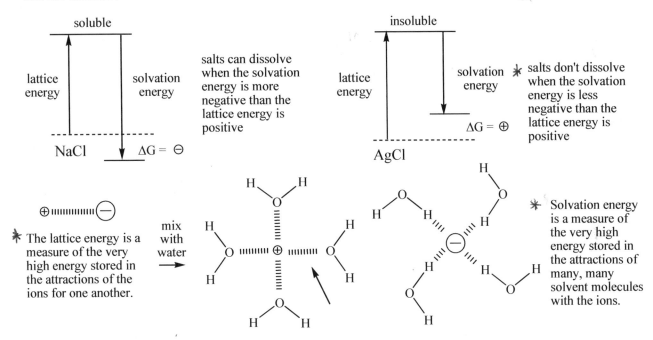

    Ionic substances, in general, tend to have "omni directional bonding" (omni = "all"), which tends to produce strongly bonded lattice structures that are difficult to break down, leading to very high melting and boiling points. Sometimes the temperatures are too high and the salt decomposes instead of melting or boiling. There are many types of lattice structures, depending on the size of the charges and the size of the ions. A salt's lattice energy (the energy holding the salt together) is typically very large, explaining their very high melting points. We will simplistically represent all ionic substances with the figure below, cations surrounded by anions and anions surrounded by cations.

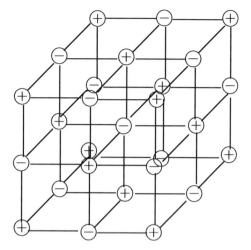

Lattic structure - depends on
the size and charge of the ions.

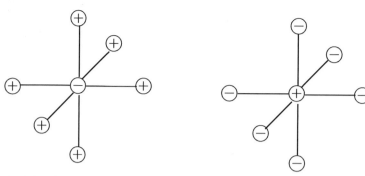

Each ion is surrounded on many sides by oppositely charged ions. To introduce the disorder of a liquid (melt) or a gas (boil) requires a very large input of energy (mp indicates the amount of energy required to break down the ordered lattice structure and boiling point indicates the amount of energy required to remove an ion pair from the influence of all neighbors. Ionic bonds (ionic attractions on all sides) can only be bronke at great expense in energy.

The first number is the melting point and the second number is the boiling point (in °C).

The BP & MP of Ionic molecules are very high

	$F^{\ominus}$	$Cl^{\ominus}$	$Br^{\ominus}$	$I^{\ominus}$	O^{-2}	N^{-3}
$Li^{\oplus}$	846 / 1717 sol. = 1.3g ΔX = 3.0	610 / 1383 sol. = 550g ΔX = 2.2	550 / 1289 sol. = 1670g ΔX=2.0	467 / 1178 sol. = 1510g ΔX=1.7	1570 / 2563 sol. = reacts ΔX=2.5	813 / ? sol. = reacts ΔX=2.5
$Na^{\oplus}$	996 / 1787 sol. = 404g ΔX = 3.1	801 / 1465 sol. = 359g ΔX = 2.3	747 / 1447 sol. = 905g ΔX = 2.1	660 / 1304 sol. = 1840g ΔX=1.8	1132 / 1950^{dec} sol. = reacts ΔX=2.6	unstable
$K^{\oplus}$	858 / 1517 sol. = 920g ΔX = 3.2	996 / 1787 sol. = 344g ΔX = 2.4	734 / 1435 sol. = 678g ΔX = 2.2	681 / 1345 sol. = 1400g ΔX=1.9	>300^{dec} / - sol. = reacts ΔX=2.7	not found
Be^{+2}	554 / 1169 sol. = v.sol. ΔX = 2.5	399 / 482 sol. = 151 ΔX = 1.7	508^{sub} / - sol. = v.sol. ΔX = 1.5	480 / 590 sol. = explodes ΔX = 1.2	2907 / 3500 sol. = ? ΔX = 2.0	2200 / 2240^{dec} sol. = soluble ΔX=1.5
Mg^{+2}	1263 / 2260 sol. = 0.13 ΔX = 2.8	714 / 1412 sol. = 543 ΔX = 2.0	711^{dec} / - sol. = 1020 ΔX = 1.8	637^{dec} / - sol. = 1480 ΔX = 1.5	2852 / 3600 sol. = 0.01 ΔX = 2.3	1500 / ? sol. = ? ΔX=1.8
Ca^{+2}	1418 / 2533 sol. = 0.01 ΔX = 3.0	772 / 1935 sol. = 745 ΔX = 2.2	730 / 1935 sol. = soluble ΔX = 2.0	779 / 1100 sol. = 660 ΔX = 1.7	2572 / 2850 sol. = 1 ΔX = 2.5	1195 / ? sol. = reacts ΔX=2.0
Al^{+3}	1291 / ? sol. = 7 ΔX = 2.5	192 / ? sol. = 460 ΔX = 1.7	98 / 265 sol. = reacts ΔX = 1.5	189 / 360^{sub} sol. = reacts ΔX = 1.2	2072 / 2977 sol. = 0 ΔX = 2.0	2200 / 2517 sol. = reacts ΔX = 1.5

$\Delta X = |X_a - X_b|$

Chapter 2

Problem 1 - Predict the formula for the combination of the following pairs of ions. What kinds of melting points would be expected for these salts?

	$F^{\ominus}$	$NO_2^{\ominus}$	$NO_3^{\ominus}$	$HCO_3^{\ominus}$	PO_4^{-3}	CO_3^{-2}
$K^{\oplus}$						
Ba^{+2}						
Zn^{+2}						
Al^{+3}						

Covalent Bonding (Molecules)

A single neutral hydrogen atom would have a single valence electron. This is not a common occurrence in our world because such a hydrogen atom would be too reactive. However, if one were able to generate a source of hydrogen atoms, they would quickly join together in simple diatomic molecules having a single covalent bond with the two hydrogen atoms sharing the two electrons and attaining the helium Noble gas configuration (duet rule). Such a reaction would be very exothermic.

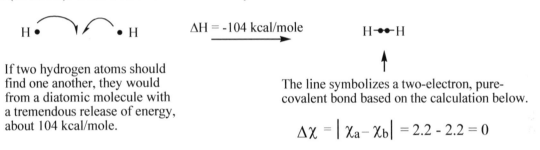

H •   • H $\Delta H = -104$ kcal/mole   H—••—H

If two hydrogen atoms should find one another, they would from a diatomic molecule with a tremendous release of energy, about 104 kcal/mole.

The line symbolizes a two-electron, pure-covalent bond based on the calculation below.

$$\Delta\chi = |\chi_a - \chi_b| = 2.2 - 2.2 = 0$$

The atoms of organic chemistry, H, C, N, O, S and halogens, tend to attain a Noble gas configuration by sharing electrons in covalent bonds of molecules. Simple formulas often have only one choice for joining the atoms in a molecule (CH_4, NH_3, H_2O, HF). As the number of atoms increases, however, there are many more possibilities, especially for carbon structures (sometimes incredible numbers of possibilities, $C_{40}H_{82} > 62$ trillion isomers!). These possibilities may require single, double and/or triple bonds in chains or rings of atoms. Carbon, nitrogen, oxygen and fluorine can be bonded in all combinations, according to their valencies. We study many of the bond types shown below.

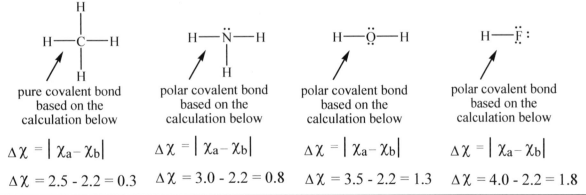

pure covalent bond based on the calculation below	polar covalent bond based on the calculation below	polar covalent bond based on the calculation below	polar covalent bond based on the calculation below

$\Delta\chi = |\chi_a - \chi_b|$ $\Delta\chi = |\chi_a - \chi_b|$ $\Delta\chi = |\chi_a - \chi_b|$ $\Delta\chi = |\chi_a - \chi_b|$

$\Delta\chi = 2.5 - 2.2 = 0.3$ $\Delta\chi = 3.0 - 2.2 = 0.8$ $\Delta\chi = 3.5 - 2.2 = 1.3$ $\Delta\chi = 4.0 - 2.2 = 1.8$

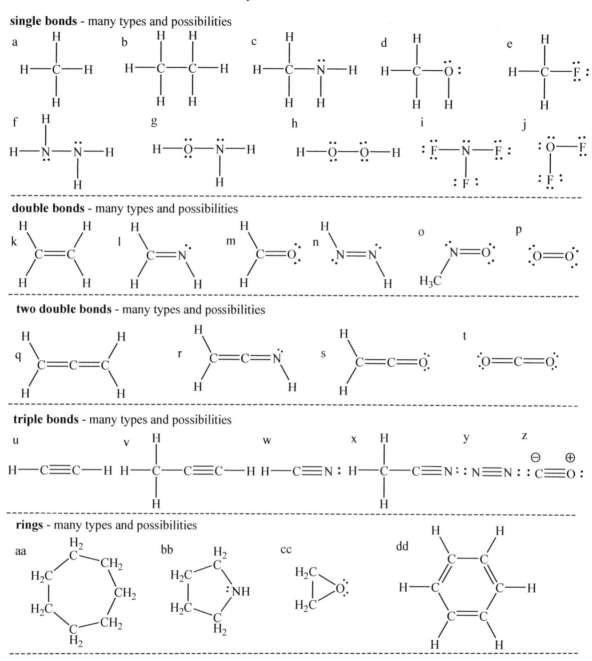

single bonds - many types and possibilities

double bonds - many types and possibilities

two double bonds - many types and possibilities

triple bonds - many types and possibilities

rings - many types and possibilities

Problem 2 – Supply bond dipole arrows to any polar bonds above (according to our arbitrary rules). Make sure they point in the right direction.

Carbon tends to share electron density with 2-4 atoms to form its octet. Bond energies to carbon also tend to be strong, which leads to an infinite variety of possible stable chains and stable rings with itself and other atoms. Hydrogen and the halogen atoms only form one bond with carbon, when they are present, and are found covering the surface of the chains and rings (they are like the skin of a molecule). Oxygen atoms and nitrogen atoms, on the other hand, form two and three bonds, respectively. They can be found in the interior of chains and rings or on the surface of chains and rings. The physical properties (mp, bp, solubilities, etc.) of molecular substances are very different from ionic substances. The strong attractions in covalent bonds do not change when a substance melts or boils since none of those bonds break. In contrast to ionic salts, there are much weaker forces that attract one molecule to another in covalent substances, and this shows in differences in their physical properties (mp, bp, etc.).

Chapter 2

2. Dispersion forces / van der Waals interactions / London forces (nonpolar attractions)

Most of the examples in the table are gases at room temperature, an indication that there are no strong forces of interaction between the molecules. However, clearly there is some attraction between the molecules, because at room temperature bromine is a liquid and iodine is a solid, and ultimately each gaseous substance in the table does condense to a liquid and solidify to a solid. Also there is quite a range of differences in melting and boiling points among the different compounds.

Substance	MW (g/mol)	mp (^{o}C)	bp (^{o}C)	Dipole Moment (μ)	Phase at room temperature	
H——H	2	-259	-253	0.0	gas	
N≡N	28	-210	-196	0.0	gas	absolute zero
O=O	32	-218	-183	0.0	gas	
F——F	38	-219	-188	0.0	gas	$0 K = -273^oC = -460^oF$
Cl——Cl	71	-101	-35	0.0	gas	
Br——Br	160	-7	+ 59	0.0	liquid	} puzzle?
I——I	254	+114	+184	0.0	solid	

A plot of the trends in melting points and boiling points in the halogen family and Nobel Gas family helps us to see the differences more clearly (below). The difference melting and boiling points show there are differences in the forces of attraction between the halogen molecules even though all of the molecules are nonpolar. Bromine is even a liquid and iodine a solid at room temperature. The higher boiling points ($I_2 > Br_2 > Cl_2 > F_2$) are due to greater polarizability of the larger atoms, where the electrons are held less tightly.

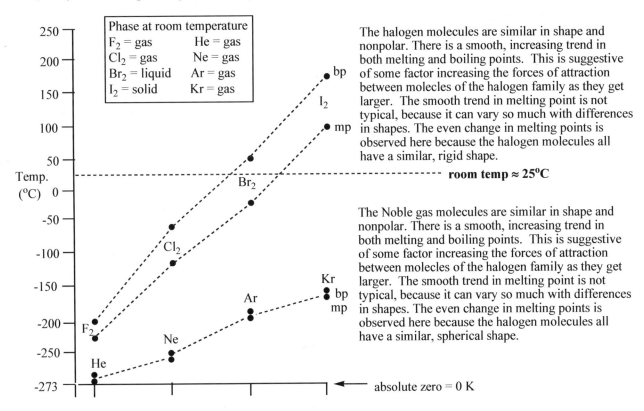

The halogen molecules are similar in shape and nonpolar. There is a smooth, increasing trend in both melting and boiling points. This is suggestive of some factor increasing the forces of attraction between molecles of the halogen family as they get larger. The smooth trend in melting point is not typical, because it can vary so much with differences in shapes. The even change in melting points is observed here because the halogen molecules all have a similar, rigid shape.

The Noble gas molecules are similar in shape and nonpolar. There is a smooth, increasing trend in both melting and boiling points. This is suggestive of some factor increasing the forces of attraction between molecles of the halogen family as they get larger. The smooth trend in melting point is not typical, because it can vary so much with differences in shapes. The even change in melting points is observed here because the halogen molecules all have a similar, spherical shape.

Dispersion forces are temporary fluctuations of negative electron clouds from one direction to another, relative to the less mobile and more massive positive nuclear charge. These fluctuations of electron density induce fleeting, weak dipole moments. ***Polarizability*** is the property that indicates how well this fluctuation of electron density can occur about an atom. Within a column (same Z_{eff}), larger atoms are more polarizable, because they do not hold as tightly to their valence electrons as smaller atoms, since the electrons are farther away from Z_{eff}. Thus, atoms lower in a column are more polarizable than atoms higher up. In a row it's harder to predict. Atoms to the right have a larger Z_{eff} which should make them less polarizable, but it might seem that lone pairs are held less tightly by only one atom, instead of two atoms in a bond. However, fluorine is not very polarizable (it holds on very hard to even its lone pairs) and appears to indicate that larger Z_{eff} is more important. Picture a cotton ball (polarizable electron clouds = sticky) and a marble (nonpolarizable electron cloud = not sticky).

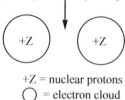

In a nonpolar molecule the $\oplus$ and $\ominus$ are centered, on average. This would seem to indicate that in nonpolar molecules there is no polarity or attraction between molecules. So why do such substances liquify and solidify? Why aren't they always gases?

Dispersion Forces

Fast moving electrons shift position relative to slow moving nuclei, creating a temporary imbalance of charge,

which induces a similar distortion of the electron clouds in neighbor structures and a weak attraction for neighbor molecules.

$\oplus$ and $\ominus$ are not centered creating temporary polarity.

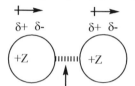

Weak, fluctuating polar forces of attraction between molecules.

+Z = nuclear protons
◯ = electron cloud

The simplest molecules we can study are diatomics (two atom molecules). While the examples in the table above are not organic molecules, they are simple and we can use this simplicity to learn important ideas that apply to organic molecules. Simple diatomic molecules must have a linear geometry, since two points determine a straight line. When both atoms are the same, the bonding electrons must be shared evenly. There can be no permanent distortion of the electron clouds toward either atom, so there are no polar bonds.

Periodic trends in polarizability, α.

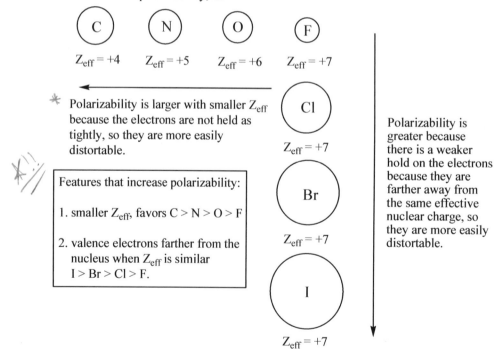

Polarizability is larger with smaller Z_{eff} because the electrons are not held as tightly, so they are more easily distortable.

Features that increase polarizability:

1. smaller Z_{eff}, favors C > N > O > F

2. valence electrons farther from the nucleus when Z_{eff} is similar I > Br > Cl > F.

Polarizability is greater because there is a weaker hold on the electrons because they are farther away from the same effective nuclear charge, so they are more easily distortable.

Dispersion forces are cumulative, so when the contact surface area is larger, the interactions are stronger (because there are more of them). Higher molecular weight alkanes have more carbon atoms to interact than lower molecular weight alkanes (even though only similar weak dispersion forces are present in both).

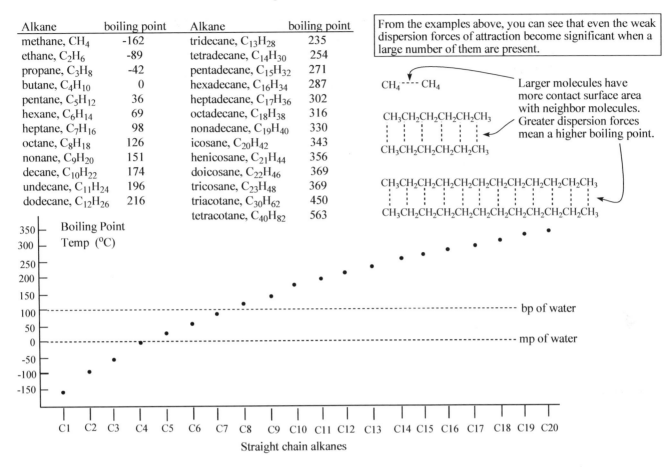

Alkane	boiling point	Alkane	boiling point
methane, CH_4	-162	tridecane, $C_{13}H_{28}$	235
ethane, C_2H_6	-89	tetradecane, $C_{14}H_{30}$	254
propane, C_3H_8	-42	pentadecane, $C_{15}H_{32}$	271
butane, C_4H_{10}	0	hexadecane, $C_{16}H_{34}$	287
pentane, C_5H_{12}	36	heptadecane, $C_{17}H_{36}$	302
hexane, C_6H_{14}	69	octadecane, $C_{18}H_{38}$	316
heptane, C_7H_{16}	98	nonadecane, $C_{19}H_{40}$	330
octane, C_8H_{18}	126	icosane, $C_{20}H_{42}$	343
nonane, C_9H_{20}	151	henicosane, $C_{21}H_{44}$	356
decane, $C_{10}H_{22}$	174	doicosane, $C_{22}H_{46}$	369
undecane, $C_{11}H_{24}$	196	tricosane, $C_{23}H_{48}$	369
dodecane, $C_{12}H_{26}$	216	triacotane, $C_{30}H_{62}$	450
		tetracotane, $C_{40}H_{82}$	563

From the examples above, you can see that even the weak dispersion forces of attraction become significant when a large number of them are present.

$CH_4 ---- CH_4$

Larger molecules have more contact surface area with neighbor molecules. Greater dispersion forces mean a higher boiling point.

In alkane isomers (having the same number of atoms, C_nH_{2n+2}), more branching reduces contact with neighbor molecules and weakens the intermolecular forces of attraction. Linear alkanes have stronger forces of attraction than their branched isomers because they have a greater contact surface area with their neighbor molecules. Branches tend to push neighbor molecules away. This is very evident in boiling points, where all of the forces of attraction are completely overcome and the linear alkane isomers have higher boiling points (stronger attractions) than the branched isomers. The strength of these interactions falls off as the 6th power of distance. A structure twice as far away will only have 1/64 the attraction for its neighbor. *Linear – usually higher BP than Branched, i.e.*

$\Delta BP = 42°C$

bp = -42°C

bp = -0.5°C

More atoms increase the contact surface area with neighbor molecules (not isomers).

$\Delta BP = 12°C$

bp = -0.5°C

bp = -12°C

$\left(\frac{1}{1}\right)^6 = 1$

$\left(\frac{1}{2}\right)^6 = \frac{1}{64}$

Less branching increases contact surface area with neighbor molecules in these isomers.

Melting points are not so predictable because there is not much difference in forces of interaction in solids and liquids, because molecules are close to one another in both. The strength of the lattice structure is the major factor in determining melting points because it determines how close the molecules can get to one another. Highly symmetrical, rigid shapes tend to pack together more efficiently (more closely) than unsymmetrical, flexible molecules and have higher melting points.

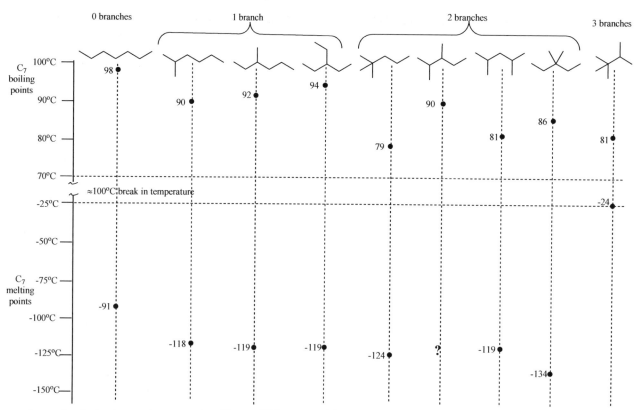

A particularly dramatic example of how different melting points can be, in very similar structures, is found in two isomers of C_8H_{18}. Both of these hydrocarbons are nonpolar molecules. Only C-C and C-H bonds are present. Dispersion forces are the primary means of interaction with neighbor structures. The first example is octane (bp = 126°C) and the second example is highly substituted 2,2,3,3-tetramethylbutane (bp = 106°C).

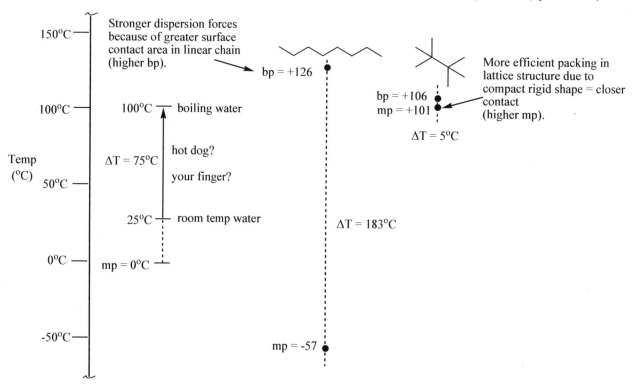

Both molecules look symmetrical, as drawn. However, octane is like a wet piece of spaghetti and can assume all sorts of irregular shapes, while 2,2,3,3-tetramethylbutane has, essentially, only one shape. Its highly compact and regular cylinder shape produces a very efficient packing arrangement that is hard to alter. The symmetrically branched isomer melts at a much higher temperature. Possible shapes are shown below for each of the structures.

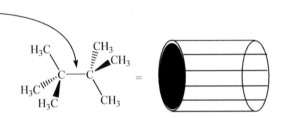

Many different shapes are possible for octane. The structures do not pack together efficiently and it takes very little energy to disrupt the solid lattice structure, and thus the very low melting point.

On the other hand, rotation about the central C-C bond does not change the overall shape of 2,2,3,3-tetramethylbutane. The molecule always has the same shape of a compact cylinder. A regular consistent packing arrangement in the lattice structure results. If the melting point was only 5°C higher, it would sublime (solid → gas).

Problem 3 - The following pairs of molecules have the same formula; they are isomers. Yet, they have very different melting points. Match the melting points with the correct structure and provide an explanation for the difference. Hint: single bonds can rotate and pi bonds tend to be rigid, fixed shapes.

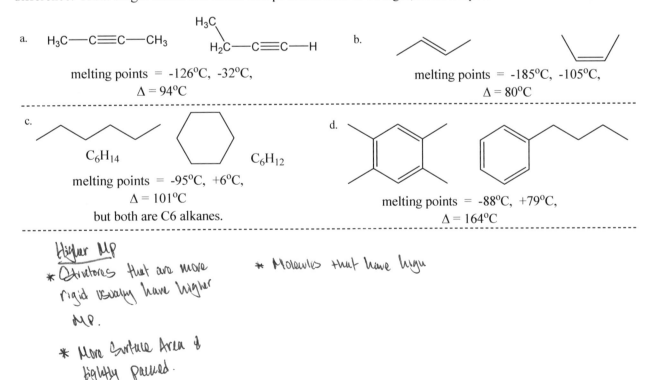

a. $H_3C-C\equiv C-CH_3$

H_3C
$H_2C-C\equiv C-H$

melting points = -126°C, -32°C,
Δ = 94°C

b.

melting points = -185°C, -105°C,
Δ = 80°C

c.

C_6H_{14} C_6H_{12}

melting points = -95°C, +6°C,
Δ = 101°C
but both are C6 alkanes.

d.

melting points = -88°C, +79°C,
Δ = 164°C

Higher MP
* Structures that are more rigid usually have higher MP.

* Molecules that have high

* More Surface Area & tightly packed.

Ions with covalent bonds

Sometimes there is a mix of covalent and ionic bonds. Anions and/or cations can be connected with covalent bonds and still have a net overall negative or positive charge, and both charges are even possible in the same molecule (e.g. amino acids). The following ions provide additional examples of ions with covalent bonds from your prior chemistry courses.

nitrite   nitrate   sulfite   sulfate   phosphate

perchlorate   chlorate   chlorite   hypochlorite   bicarbonate   carbonate

cyanide   amide anion   hydroxide   hydronium ion   ammonium ion   glycine (amino acid)

2a. Dipole-dipole interactions (in between polarity) – Since the partial charges present in molecules having dipole moments are less than full charges and the bonds are very directional (not "omni"), attractions for neighbor molecules are weaker than is found in ionic salts. However, polar molecules usually have stronger attractions than nonpolar molecules of similar size and shape. Many polar molecules below are compared with a similar size and shape nonpolar molecules to show how polarity can affect boiling points (a better indication of the strength of attractions among neighbor molecules when other factors are similar). A higher boiling point indicates stronger attractions. Dipole-dipole interactions represent moderate forces of attraction between partially polarized molecules. The molecular dipole moments are indicators of charge imbalance due to a difference in electronegativity, bond length and molecular shape. Fluorine is unusual in that it makes very polar bonds but is not very polarizable and holds on very tightly to its lone pairs, so it sometimes surprises us with lower than expected boiling points. Sometimes atoms that hold onto their electrons very tightly are called 'hard' (think 'marbles') and atoms that have polarizable electrons are called 'soft' (think 'cotton balls').

Melting points are better indicators of packing efficiency and boiling points are better indicators of forces of attraction between the molecules. Remember, resonance structure are not different molecules, just different arrangements of the electrons. Miscellaneous compounds are compiled below, representing a variety of factors discussed in this topic.

Useful Comparisons

nonpolar	polar	more polar		nonpolar	polar	more polar
$\mu = 0.0$ D	$\mu = 2.98$ D			$\mu = 0.0$ D	$\mu = 2.3$ D	
bp = -81°C	bp = +26°C			bp = -104°C	bp = -20°C	
mp = -84°C	mp = -13°C			mp = -169°C	mp = -92°C	
H_2O sol. = insoluble	H_2O sol. = miscible			H_2O sol. = 2.9mg/L	H_2O sol. = 400g/L	
$pK_a = 25$	$pK_a = 9.2$			$pK_a = 44$	$pK_a = NA$	

$\Delta T_{bp} = 107$°C $\Delta T_{bp} = 84$°C

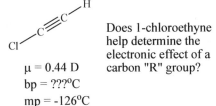

Chapter 2

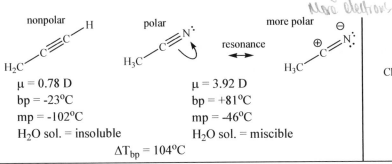

nonpolar

$\mu = 0.78$ D
bp = -23°C
mp = -102°C
H₂O sol. = insoluble

polar — resonance — more polar

$\mu = 3.92$ D
bp = +81°C
mp = -46°C
H₂O sol. = miscible

$\Delta T_{bp} = 104$°C

$\mu = 0.44$ D
bp = ???°C
mp = -126°C

Does 1-chloroethyne help determine the electronic effect of a carbon "R" group?

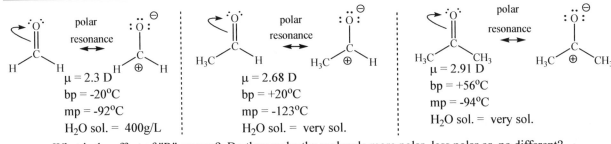

polar resonance

$\mu = 2.3$ D
bp = -20°C
mp = -92°C
H₂O sol. = 400g/L

polar resonance

$\mu = 2.68$ D
bp = +20°C
mp = -123°C
H₂O sol. = very sol.

polar resonance

$\mu = 2.91$ D
bp = +56°C
mp = -94°C
H₂O sol. = very sol.

What is the effect of "R" groups? Do they make the molecule more polar, less polar or no different?
Electron donation or electron withdrawal through sigma bonds is called an inductive effect.

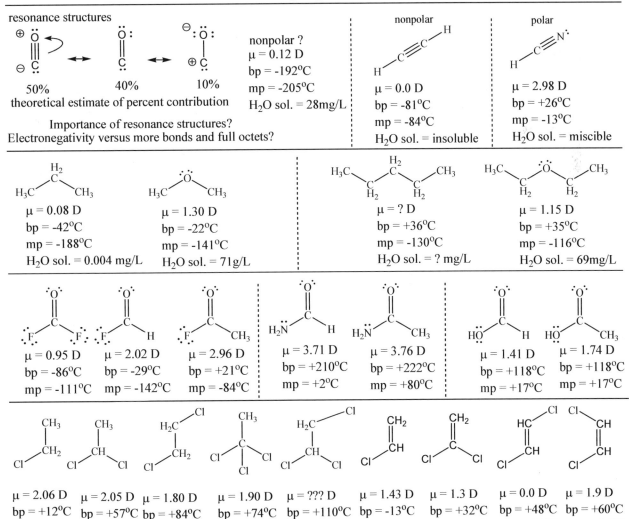

resonance structures

50%   40%   10%
theoretical estimate of percent contribution

Importance of resonance structures?
Electronegativity versus more bonds and full octets?

nonpolar ?
$\mu = 0.12$ D
bp = -192°C
mp = -205°C
H₂O sol. = 28mg/L

nonpolar

$\mu = 0.0$ D
bp = -81°C
mp = -84°C
H₂O sol. = insoluble

polar

$\mu = 2.98$ D
bp = +26°C
mp = -13°C
H₂O sol. = miscible

$\mu = 0.08$ D
bp = -42°C
mp = -188°C
H₂O sol. = 0.004 mg/L

$\mu = 1.30$ D
bp = -22°C
mp = -141°C
H₂O sol. = 71g/L

$\mu = ?$ D
bp = +36°C
mp = -130°C
H₂O sol. = ? mg/L

$\mu = 1.15$ D
bp = +35°C
mp = -116°C
H₂O sol. = 69mg/L

$\mu = 0.95$ D
bp = -86°C
mp = -111°C

$\mu = 2.02$ D
bp = -29°C
mp = -142°C

$\mu = 2.96$ D
bp = +21°C
mp = -84°C

$\mu = 3.71$ D
bp = +210°C
mp = +2°C

$\mu = 3.76$ D
bp = +222°C
mp = +80°C

$\mu = 1.41$ D
bp = +118°C
mp = +17°C

$\mu = 1.74$ D
bp = +118°C
mp = +17°C

$\mu = 2.06$ D
bp = +12°C
mp = -139°C

$\mu = 2.05$ D
bp = +57°C
mp = -97°C

$\mu = 1.80$ D
bp = +84°C
mp = -35°C

$\mu = 1.90$ D
bp = +74°C
mp = -33°C

$\mu = ???$ D
bp = +110°C
mp = -37°C

$\mu = 1.43$ D
bp = -13°C
mp = -154°C

$\mu = 1.3$ D
bp = +32°C
mp = -122°C

$\mu = 0.0$ D
bp = +48°C
mp = -81°C

$\mu = 1.9$ D
bp = +60°C
mp = -81°C

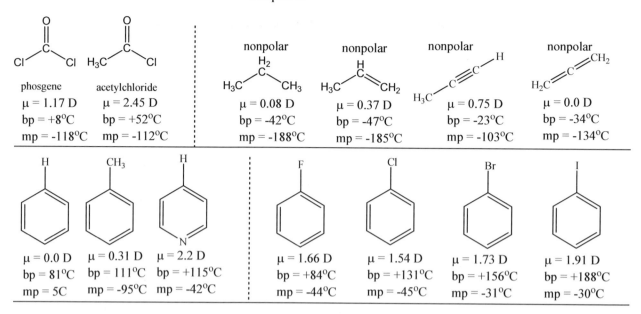

phosgene
$\mu = 1.17$ D
bp = +8°C
mp = -118°C

acetylchloride
$\mu = 2.45$ D
bp = +52°C
mp = -112°C

nonpolar
$\mu = 0.08$ D
bp = -42°C
mp = -188°C

nonpolar
$\mu = 0.37$ D
bp = -47°C
mp = -185°C

nonpolar
$\mu = 0.75$ D
bp = -23°C
mp = -103°C

nonpolar
$\mu = 0.0$ D
bp = -34°C
mp = -134°C

$\mu = 0.0$ D
bp = 81°C
mp = 5C

$\mu = 0.31$ D
bp = 111°C
mp = -95°C

$\mu = 2.2$ D
bp = +115°C
mp = -42°C

$\mu = 1.66$ D
bp = +84°C
mp = -44°C

$\mu = 1.54$ D
bp = +131°C
mp = -45°C

$\mu = 1.73$ D
bp = +156°C
mp = -31°C

$\mu = 1.91$ D
bp = +188°C
mp = -30°C

2b. Hydrogen bonds – Hydrogen bonds represent a very special dipole-dipole interaction. Molecules that have this feature have even stronger attractions for neighbor molecules than normal polar bonds would suggest. Solvents that have an O-H or an N-H bond are called "protic solvents" and can both donate and accept hydrogen bonds (because they also have lone pairs of electrons). They generally have higher boiling points than similar sized structures without any "polarized hydrogen atoms". The reasons for a polarized hydrogen atom's strong attraction for electron density is that there are no other layers of electrons around a hydrogen atom (hydrogen is the only atom to use the n=1 shell in bonding). When a hydrogen atom's electron cloud is polarized away from the hydrogen atom in a bond with an electronegative atom, usually oxygen or nitrogen in organic chemistry, an especially strong polarization results. In a sense the hydrogen atom is "desperate" for additional electron density and strongly attracted to any source that can supply this, such as a lone pair on a neighbor molecule, or even a pi bond. We call such interactions "hydrogen bonds". A molecule that has such a polarized hydrogen is classified as a hydrogen bond donor. A molecule that has a partial negatively charged region that can associate with such a hydrogen is classified as a hydrogen bond acceptor. Quite often the hydrogen bond acceptor is a lone pair of electrons on another oxygen or nitrogen atom, but fluorine, chlorine bromine or sulfur may provide lone pair acceptor sites as well. Weak hydrogen bonds can even form with C=C pi bonds.

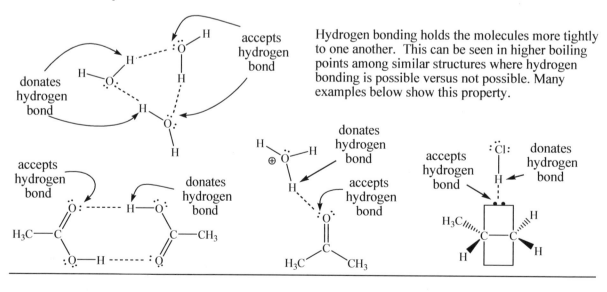

Hydrogen bonding holds the molecules more tightly to one another. This can be seen in higher boiling points among similar structures where hydrogen bonding is possible versus not possible. Many examples below show this property.

Chapter 2

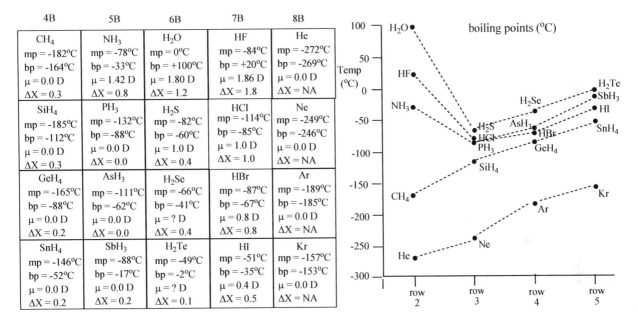

three hydrogen bonds in G-C base pair

guanine   cytosine

two hydrogen bonds in A-T base pair

adenine   thymine

Which base pair binds more tightly, GC, AT or are they about the same?

Problem 4 – Provide an explanation for the different boiling points in each column.

4B	5B	6B	7B	8B
CH₄ $\mu = 0.0$ D mp = -182°C bp = -164°C ΔX = 0.3	NH₃ mp = -78°C bp = -33°C μ = 1.42 D ΔX = 0.8	H₂O mp = 0°C bp = +100°C μ = 1.80 D ΔX = 1.2	HF mp = -84°C bp = +20°C μ = 1.86 D ΔX = 1.8	He mp = -272°C bp = -269°C μ = 0.0 D ΔX = NA
SiH₄ mp = -185°C bp = -112°C μ = 0.0 D ΔX = 0.3	PH₃ mp = -132°C bp = -88°C μ = 0.0 D ΔX = 0.0	H₂S mp = -82°C bp = -60°C μ = 1.0 D ΔX = 0.4	HCl mp = -114°C bp = -85°C μ = 1.0 D ΔX = 1.0	Ne mp = -249°C bp = -246°C μ = 0.0 D ΔX = NA
GeH₄ mp = -165°C bp = -88°C μ = 0.0 D ΔX = 0.2	AsH₃ mp = -111°C bp = -62°C μ = 0.0 D ΔX = 0.0	H₂Se mp = -66°C bp = -41°C μ = ? D ΔX = 0.4	HBr mp = -87°C bp = -67°C μ = 0.8 D ΔX = 0.8	Ar mp = -189°C bp = -185°C μ = 0.0 D ΔX = NA
SnH₄ mp = -146°C bp = -52°C μ = 0.0 D ΔX = 0.2	SbH₃ mp = -88°C bp = -17°C μ = 0.0 D ΔX = 0.2	H₂Te mp = -49°C bp = -2°C μ = ? D ΔX = 0.1	HI mp = -51°C bp = -35°C μ = 0.4 D ΔX = 0.5	Kr mp = -157°C bp = -153°C μ = 0.0 D ΔX = NA

Answer: Each dotted line represents the boiling points in the hydrides in a column in the periodic table, plus the Noble gases. In the Noble gases there is a continuing trend towards higher boiling point as the main atom gets larger with more and more polarizable electron clouds. Where bonds to hydrogen become polar, hydrogen bonding becomes important, leading to stronger attractions for neighbor molecules and higher boiling points deviate from the trends observed due to dispersion forces. Deviations are clearly seen with NH_3, H_2O and HF. Melting points are not plotted.

$\mu = 0.08$ D
bp = -42°C
mp = -188°C
H_2O sol. = 40mg/L

$\mu = 1.30$ D
bp = -22°C
mp = -141°C
H_2O sol. = 71g/L

$\mu = 1.69$ D
bp = +65°C
mp = -98°C
H_2O sol. = miscible

$\mu = 1.69$ D
bp = +78°C
mp = -114°C
H_2O sol. = miscible

$\mu = 1.68$ D
bp = +98°C
mp = -126°C
H_2O sol. = miscible

$\mu = 1.66$ D
bp = +118°C
mp = -90°C
H_2O sol. = 73g/L

Problem 5 – Provide an explanation for the different boiling points in each group. In part b, an H-F bond is expected to be more polar than an H-O bond, so why does HF boil lower than H_2O? Think of an "H bond" donor site as a hand that can grab and a lone pair as a handle that can be held onto. If you only had one hand, it really doesn't make any difference how many handles are available to grab hold of and if you only have one handle, it doesn't make any difference how many hands you have. But, if you have two hands and two handles a much stronger grip is possible with twice the handholds.

Chapter 2

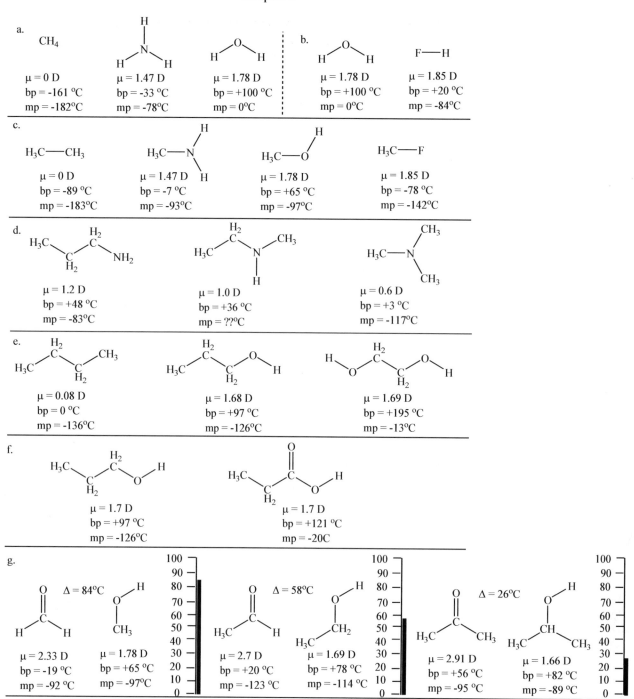

a.

CH₄

μ = 0 D
bp = -161 °C
mp = -182°C

H
|
N
/ | \
H   H

μ = 1.47 D
bp = -33 °C
mp = -78°C

H
\
O
/
H

μ = 1.78 D
bp = +100 °C
mp = 0°C

b.

H
\
O
/
H

μ = 1.78 D
bp = +100 °C
mp = 0°C

F—H

μ = 1.85 D
bp = +20 °C
mp = -84°C

c.

H₃C—CH₃

μ = 0 D
bp = -89 °C
mp = -183°C

H₃C—N

μ = 1.47 D
bp = -7 °C
mp = -93°C

H₃C—O

μ = 1.78 D
bp = +65 °C
mp = -97°C

H₃C—F

μ = 1.85 D
bp = -78 °C
mp = -142°C

d.

μ = 1.2 D
bp = +48 °C
mp = -83°C

μ = 1.0 D
bp = +36 °C
mp = ??°C

μ = 0.6 D
bp = +3 °C
mp = -117°C

e.

μ = 0.08 D
bp = 0 °C
mp = -136°C

μ = 1.68 D
bp = +97 °C
mp = -126°C

μ = 1.69 D
bp = +195 °C
mp = -13°C

f.

μ = 1.7 D
bp = +97 °C
mp = -126°C

μ = 1.7 D
bp = +121 °C
mp = -20C

g.

Δ = 84°C

μ = 2.33 D
bp = -19 °C
mp = -92 °C

μ = 1.78 D
bp = +65 °C
mp = -97°C

Δ = 58°C

μ = 2.7 D
bp = +20 °C
mp = -123 °C

μ = 1.69 D
bp = +78 °C
mp = -114 °C

Δ = 26°C

μ = 2.91 D
bp = +56 °C
mp = -95 °C

μ = 1.66 D
bp = +82 °C
mp = -89 °C

* Branched
Mole. tends to
push Neighboring
molecules away.

Chapter 2

The effect of larger size atoms having similar Z_{eff} is to allow more distortion of the electron cloud and larger fluctuating, transient dipoles. Stronger attractions for neighbor molecules leads to higher boiling points. The halomethanes show a similar trend, even though their dipole moments are very similar.

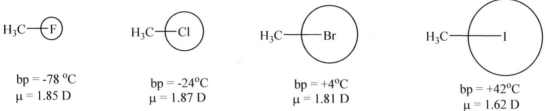

bp = -78 °C	bp = -24°C	bp = +4°C	bp = +42°C
μ = 1.85 D	μ = 1.87 D	μ = 1.81 D	μ = 1.62 D

Larger, less tightly held valence shell of iodine is more polarizable.

Problem 6 – Provide an explanation for the different boiling points in each series.

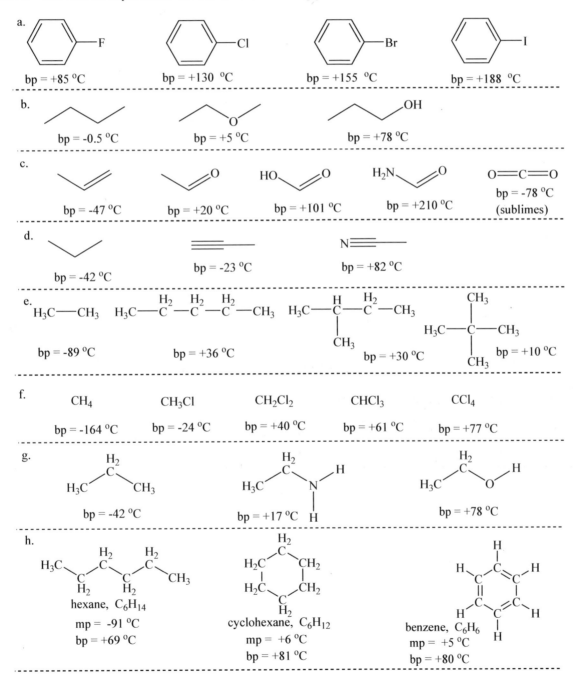

a.

bp = +85 °C   bp = +130 °C   bp = +155 °C   bp = +188 °C

b.

bp = -0.5 °C   bp = +5 °C   bp = +78 °C

c.

bp = -47 °C   bp = +20 °C   bp = +101 °C   bp = +210 °C   bp = -78 °C (sublimes)

d.

bp = -42 °C   bp = -23 °C   bp = +82 °C

e.

bp = -89 °C   bp = +36 °C   bp = +30 °C   bp = +10 °C

f.

CH₄   CH₃Cl   CH₂Cl₂   CHCl₃   CCl₄

bp = -164 °C   bp = -24 °C   bp = +40 °C   bp = +61 °C   bp = +77 °C

g.

bp = -42 °C   bp = +17 °C   bp = +78 °C

h.

hexane, C₆H₁₄   cyclohexane, C₆H₁₂   benzene, C₆H₆
mp = -91 °C   mp = +6 °C   mp = +5 °C
bp = +69 °C   bp = +81 °C   bp = +80 °C

Chapter 2

Problem 7 – Match the given boiling points with the structures below and give a short reason for your answers. (-7°C, +31 °C, +80 °C, +141°C, 1420°C)

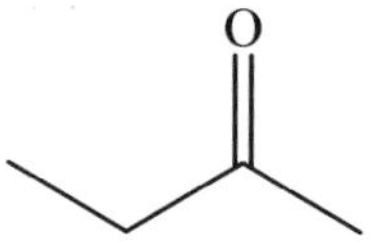

2-butanone	2-methyl-1-butene	propanoic acid	potassium chloride	2-methylpropene
MW = 72 g/mol	MW = 70 g/mol	MW = 74 g/mol	MW = 74.5 g/mol	MW = 56 g/mol

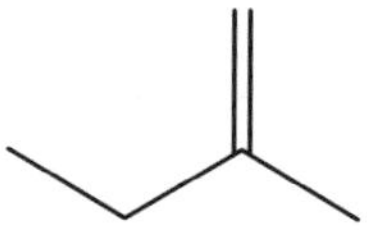

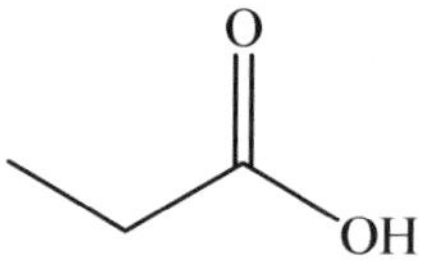

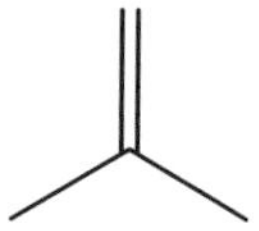

KCl

Solutes, Solvents and Solutions

Mixing occurs easily when different substances, having similar forces of interaction, are combined. This observation is summarized in the general rule of "Like Dissolves Like". To discuss the solution process, there are a few common terms that we need be familiar with.

Solutes are substances that are dissolved in a solvent. There may be as few as 1 solute (ethanol in water) or as many as 1000's of dissolved solutes (blood).

The *solvent* is the liquid in which the solute(s) is(are) dissolved. The solvent is usually the major component of the mixture. Sometimes when water is present, it is considered to be the solvent, even when present in only 1% (as is often the case in sulfuric acid/water mixtures).

A *solution* is the combination of the solvent and the solute(s).

To dissolve a solute in a solvent, the forces of interaction among solute molecules must be overcome and the forces of interaction among solvent molecules must be overcome. These are energy expenses. In return, new interactions develop among solute and solvent molecules. These are energy gains. The balance between these energy expenses and energy gains determines whether a solute will dissolve. A solution is formed when the energy gains are greater than the energy expenses. The important interactions are listed below in decreasing order of energy importance.

Possible energy expenses due to (solute / solute) and (solvent / solvent) interactions		Possible energy gains from (solute / solvent) interactions
1. ion-ion (lattice energy)	increasing energy	1. ion-dipole (solvation energy)
2. hydrogen bonds		2. hydrogen bonds
3. dipole-dipole		3. dipole-dipole
4. dispersion forces		4. dispersion forces

The dielectric constant, ε, is a measure of the necessary work to create charge in a medium. It is a useful parameter that indicates a solvent's ability to allow charge separation, such as when ions are dissolved. The reference medium is a "vacuum" (ε = 1), but air is about the same (ε = 1). Think how difficult it is to separate charge in air (lightening is a possible result). On the other hand, solvents that are polar (with a large dipole moment) are able to insert themselves in between opposite charges, allowing the opposite charges to separate more easily. The dielectric constant, ε, essentially indicates the fraction of work needed to separate charge compared to a vacuum (same as air, ε = 1). For example, water (with its large, opposite polarity) has a dielectric constant of 78, indicating that it is 78 times easier to separate charge in water than in air. Consider how easy is it to dissolve NaCl in water (it happens at room temperature) and how hard it is to boil solid NaCl (bp > 1400°C).

Chapter 2

The medium (solvent) influences the work to separate charge by orienting its polar molecules opposite to the charged plates (think of ions in solution). This spreads out the charge throughout the medium. In a sense, the solvent is delocalizing charge in a manner analogous to resonance.

Work to create charge on plates.

$$= \frac{(terms)}{\varepsilon}$$

ε = dielelectric constant (ε_{air} = 1), (ε_{H2O} = 78)

The work to create charge in water is to 1/78 of the work to create charge in air because of polar water's ability to delocalize the charge throughout itself.

battery creates charge on plates

Solvent Classifications of polarity
polar: dielectric constant > 15
nonpolar: dielectric constant < 15

Solvent as protic or aprotic
protic solvent: has a polarized hydrogen atom on an oxygen (O-H) or nitrogen (N-H) atom
aprotic: no polarized hydrogen atoms

Solvents are classified as polar or nonpolar based on their dielectric constant (see table below). If the dielectric constant is greater than 15, then the solvent is considered polar. If less than 15, then the solvent is considered nonpolar. This is example of an arbitrary division, created for our convenience.

Water and alcohols (= "organic water") are in the special class of solvents that have a polarized hydrogen atom in a covalent bond (O-H > N-H). Solvents in this class tend to be especially good at interacting with both types of charge. Such solvents are called "*protic solvents*", because they can both donate and accept hydrogen bonds. In our course, protic solvents have an O-H bond or an N-H bond and tend to be better at solvating (thus, dissolving) polar and ionic solutes than solvents without this feature.

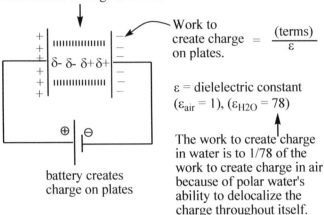

Protic solvents usually have an O-H bond.

Protic Solvents good at solvating both Cations & Anions

Protic solvents are good at solvating both cations and anions. Many partial negative solvent dipoles (lone pairs on oxygen or nitrogen atoms) can take the place of an anion, while many partial positive solvent dipoles (polarized hydrogen atoms) can take the place of the cation.

Solvents that are classified as polar (ε > 15, see below), but do not have a polarized hydrogen are called *polar aprotic solvents*. They tend to be very good at solvating positive charge, but not so good at solvating negative charge. This turns out to be very helpful when anions need to react in chemical reactions, and the use of such solvents is very common in organic chemistry when that is the goal.

Polar Aprotic Solvents - Polar Solvents but don't have a polarized hydrogen
* Don't have OH or N-H bonds.

Aprotic solvents do NOT have an O-H (or N-H) bond.

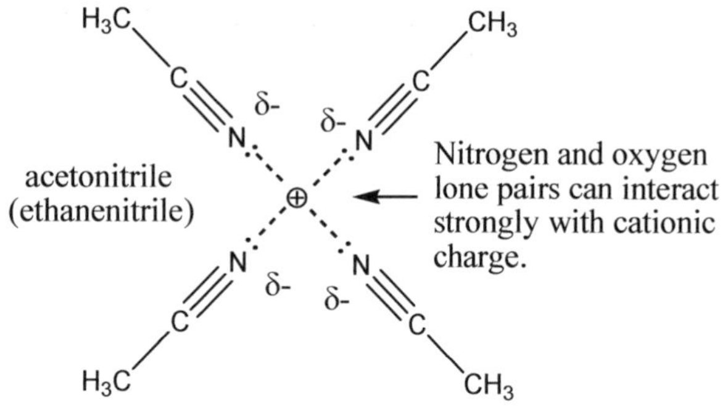

acetonitrile
(ethanenitrile)

Nitrogen and oxygen lone pairs can interact strongly with cationic charge.

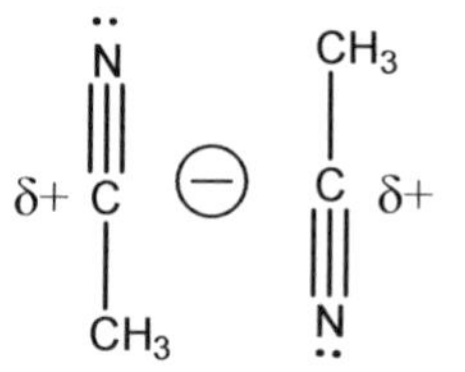

Polar aprotic solvents interact poorly with anionic charge, so the anions tend to be available to react.

Polar aprotic solvents are good at solvating cations and not so good at solvating anions.

Solvents with $\varepsilon < 15$ and no polarized hydrogen atom are called nonpolar solvents and generally do not mix well with polar and ionic substances. In biochemistry such substances are referred to as hydrophobic (don't mix well with water) or lipophilic (mix well with fatty substances). This is a good feature because they can keep different aqueous solutions (blood vs. cytosol) apart and prevent mixing (cell membranes).

Three different kinds of solvents that chemists tend to talk about are: nonpolar solvents, polar aprotic solvents and polar protic solvents. Examples are shown below.

Common Solvents	Formula	dielectric constant (ε)	protic/aprotic
water	H_2O	78	protic
methanol (MeOH)	CH_3OH	33	protic
ethanol (EtOH)	CH_3CH_2OH	24	protic
2-propanol (i-PrOH)	$(CH_3)_2CHOH$	20	protic
methanoic acid (formic acid)	HCO_2H	≈ 50	protic
*ethanoic acid (acetic acid, HOAc)	CH_3CO_2H	6	protic
2-propanone (acetone)	CH_3COCH_3	21	polar aprotic
dimethyl sulfoxide (DMSO)	CH_3SOCH_3	46	polar aprotic
dimethyl formamide (DMF)	$HCON(CH_3)$	37	polar aprotic
acetonitrile (AN)	CH_3CN	36	polar aprotic
nitromethane	CH_3NO_2	36	polar aprotic
hexamethylphosphoramide (HMPA)	$O=P[N(CH_3)_2]_3$	30	polar aprotic
ethyl ethanoate (EtOAc)	$CH_3CO_2CH_2CH_3$	6	nonpolar
ethyl ether (Et$_2$O)	$CH_3CH_2OCH_2CH_3$	4	nonpolar
tetrahydrofuran (THF, cyclic ether)	$(CH_2)_4O$	8	nonpolar
dichloromethane (methylene chloride)	CH_2Cl_2	9	nonpolar
trichloromethane (chloroform)	$CHCl_3$	5	nonpolar
tetrachloromethane (carbon tetrachloride)	CCl_4	2	nonpolar
hexane	$CH_3(CH_2)_4CH_3$	2	nonpolar

*Ethanoic acid violates our arbitray solvent polarity rules but is a strong hydrogen bonding solvent, so is included in the protic solvent group.

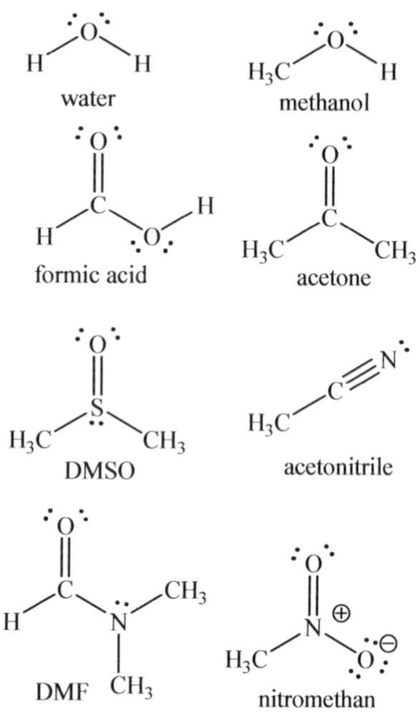

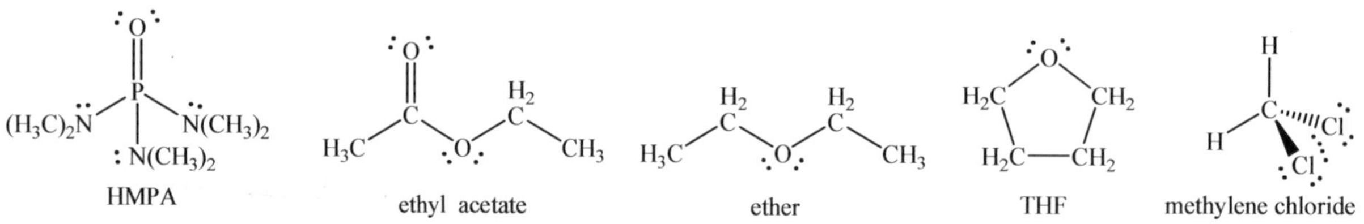

Chapter 2

Problem 8 – Point out the polar hydrogen in methanol. What is it about dimethyl sulfoxide (DMSO) that makes it polar? Draw a simplistic picture showing how methanol interacts with a cation and an anion. Also use DMSO (below) and draw a simplistic picture showing the interaction with cations and anions. Explain the difference from the methanol picture.

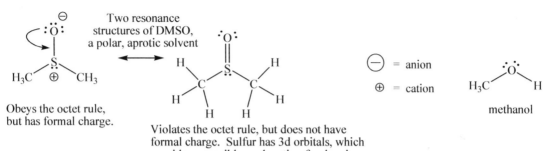

Two resonance structures of DMSO, a polar, aprotic solvent

Obeys the octet rule, but has formal charge.

Violates the octet rule, but does not have formal charge. Sulfur has 3d orbitals, which provides a possible explanation for drawing a double bond. Not all chemists agree.

$\ominus$ = anion
$\oplus$ = cation

methanol

Simplistic Sketch of Mixing Solutes and Solvents

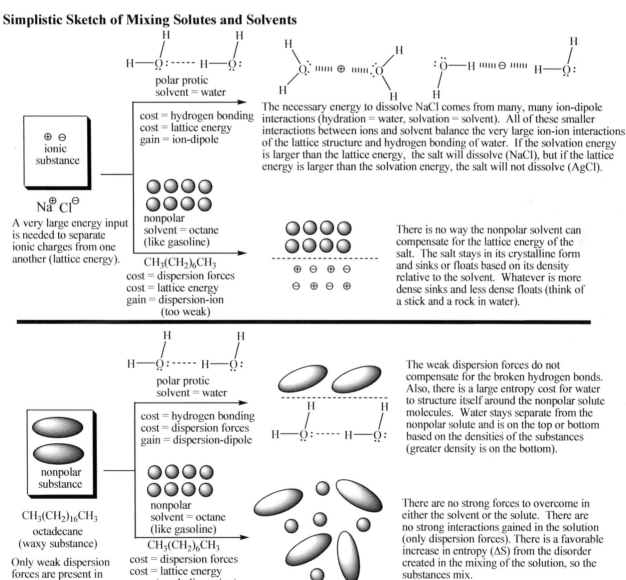

polar protic solvent = water

ionic substance

$Na^{\oplus} Cl^{\ominus}$

A very large energy input is needed to separate ionic charges from one another (lattice energy).

cost = hydrogen bonding
cost = lattice energy
gain = ion-dipole

The necessary energy to dissolve NaCl comes from many, many ion-dipole interactions (hydration = water, solvation = solvent). All of these smaller interactions between ions and solvent balance the very large ion-ion interactions of the lattice structure and hydrogen bonding of water. If the solvation energy is larger than the lattice energy, the salt will dissolve (NaCl), but if the lattice energy is larger than the solvation energy, the salt will not dissolve (AgCl).

nonpolar solvent = octane (like gasoline)

$CH_3(CH_2)_6CH_3$

cost = dispersion forces
cost = lattice energy
gain = dispersion-ion
(too weak)

There is no way the nonpolar solvent can compensate for the lattice energy of the salt. The salt stays in its crystalline form and sinks or floats based on its density relative to the solvent. Whatever is more dense sinks and less dense floats (think of a stick and a rock in water).

polar protic solvent = water

nonpolar substance

$CH_3(CH_2)_{16}CH_3$
octadecane
(waxy substance)

Only weak dispersion forces are present in pentane.

cost = hydrogen bonding
cost = dispersion forces
gain = dispersion-dipole

The weak dispersion forces do not compensate for the broken hydrogen bonds. Also, there is a large entropy cost for water to structure itself around the nonpolar solute molecules. Water stays separate from the nonpolar solute and is on the top or bottom based on the densities of the substances (greater density is on the bottom).

nonpolar solvent = octane (like gasoline)

$CH_3(CH_2)_6CH_3$

cost = dispersion forces
cost = lattice energy
(weak dispersion)
gain = dispersion-dispersion
(weak)

There are no strong forces to overcome in either the solvent or the solute. There are no strong interactions gained in the solution (only dispersion forces). There is a favorable increase in entropy (ΔS) from the disorder created in the mixing of the solution, so the substances mix.

Chapter 2

Problem 9 –
a. Hexane (density = 0.65 g/ml) and water (density = 1.0 g/ml) do not mix. Which layer is on top?
b. Carbon tetrachloride (density = 1.59 g/ml) and water (density = 1.0 g/ml) do not mix. Which layer is on top?

Problem 10 - The melting point of NaCl is very high (≈ 800°C) and the boiling point is even higher (> 1400°C). Does this imply strong, moderate or weak forces of attraction between the ions? Considering your answer, is it surprising that NaCl dissolves so easily in water? Why does this occur? Consider another chloride salt, AgCl. How does your analysis work here? What changed?

Problem 11 – a. Which solvent do you suspect would dissolve NaCl better, DMSO or hexane? Explain your choice?   b. Which solvent do you suspect would dissolve NaCl better, methanol or benzene? Explain your choice?

Problem 12 – The terms "hydrophilic" and "hydrophobic" are frequently used to describe structures that mix well or poorly with water, respectively. Biological molecules are often classified in a similar vein as water soluble (hydrophilic) or fat soluble (hydrophobic). The following list of well known biomolecules are often classified as fat soluble or water soluble. Examine each structure and place it in one of these two categories. Explain you reasoning.

a 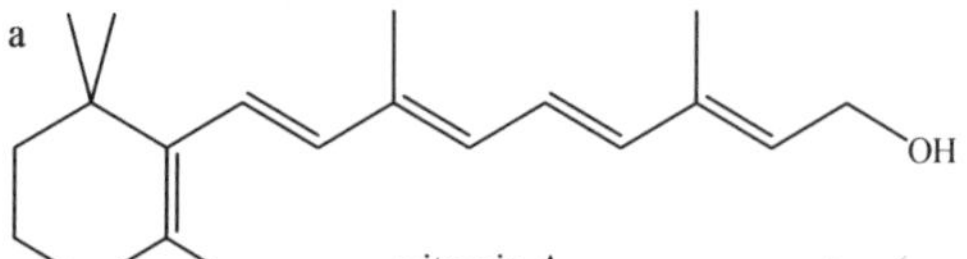
vitamin A

Not going to dissolve in water, only polar where OH is, the rest is non polar

b 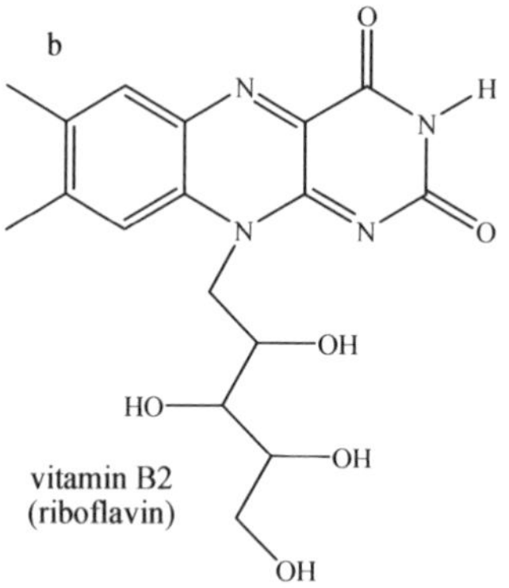
vitamin B2
(riboflavin)

c 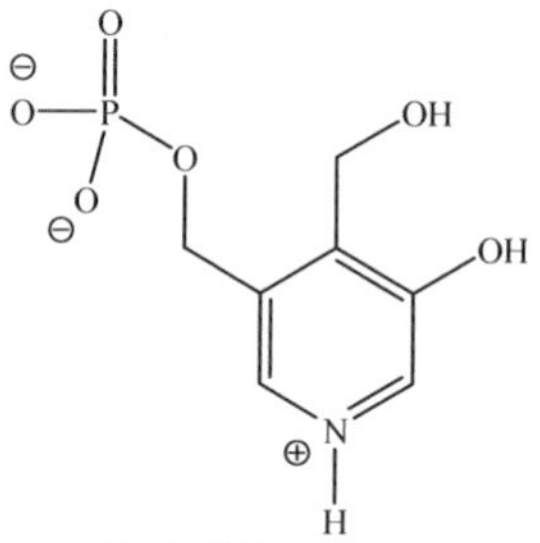
vitamin B6
(pyridoxine)

d 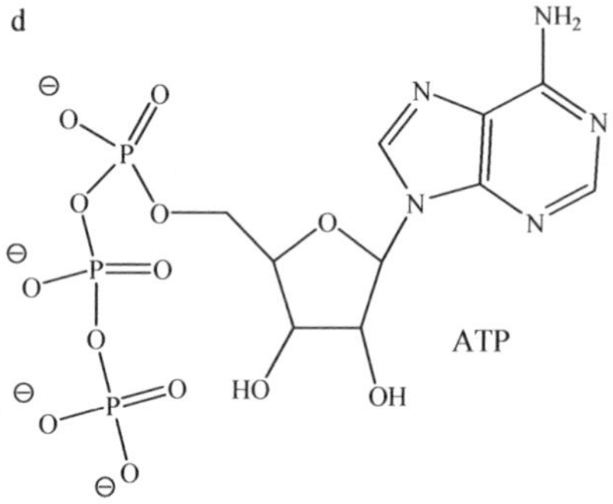
ATP

e 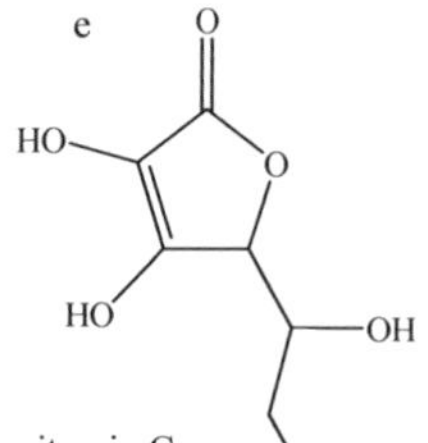
vitamin C
(ascorbic acid)

f 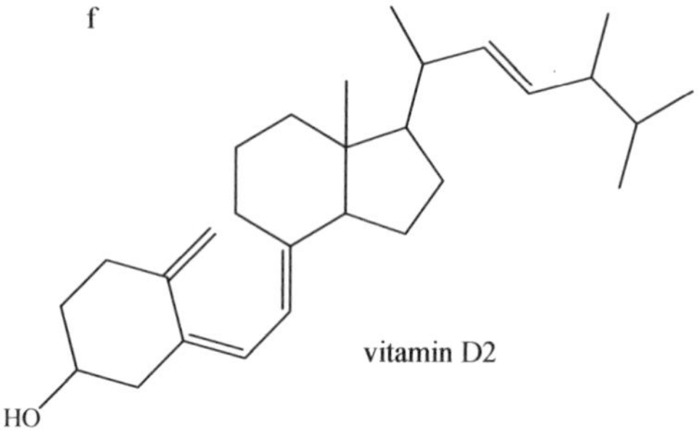
vitamin D2

g 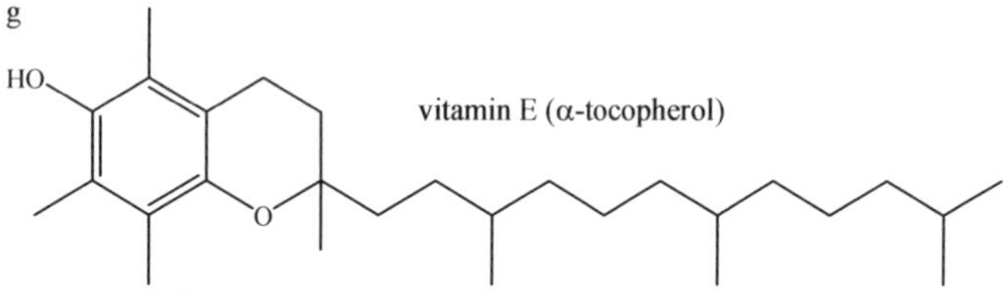
vitamin E (α-tocopherol)

h 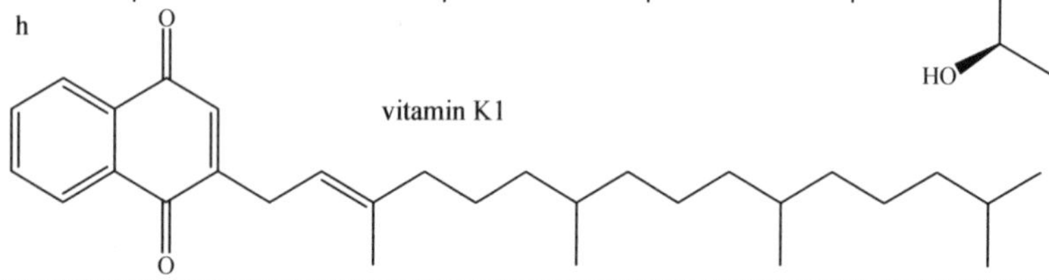
vitamin K1

i 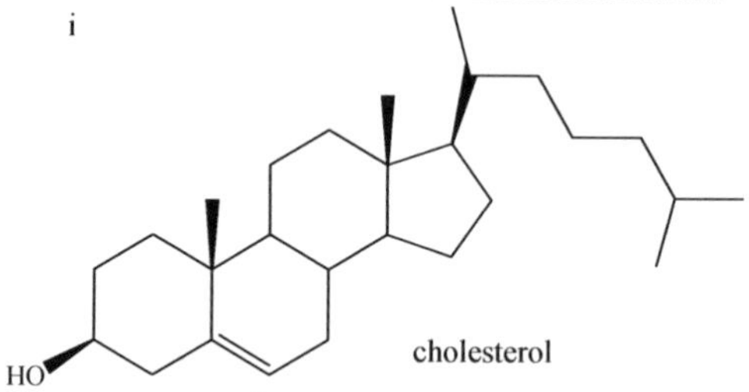
cholesterol

Chapter 2

Problem 13 – a. Carbohydrates are very water soluble and fats do not mix well with water. Below, glucose is shown below as a typical hydrophilic carbohydrate, and a triglyceride is used as a typical hydrophobic fat. Point out why each is classified in the manner indicated.

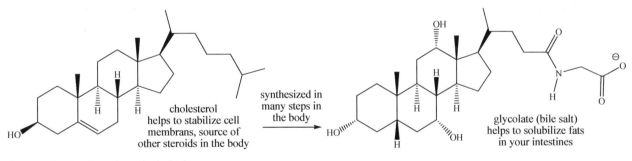

glucose (carbohydrate)   typical saturated triglyceride (fat)

b. All of the "OH" groups in glucose can be methylated. What do you think this will do to the solubility of glucose? Why? One of these structures is soluble in carbon tetrachloride the other one is not. Which one is it and why?

methylation reaction

glucose   methylated glucose

Problem 14 – Bile salts are released from your gall bladder when hydrophobic fats are eaten to allow your body to solubilize the fats, so that they can be absorbed and transported in the aqueous blood. The major bile salt glycolate, shown below, is synthesized from cholesterol. Explain the features of glycolate that makes it a good compromise structure that can mix with both the fat and aqueous blood. Use the 'rough' 3D drawings below to help your reasoning, or better yet, build models to see the structures for yourself (though it's a lot of work).

cholesterol helps to stabilize cell membrans, source of other steroids in the body

synthesized in many steps in the body

glycolate (bile salt) helps to solubilize fats in your intestines

1. source for steroid syntheses in the body
2. important constituent of cell membranes
2. transported in blood to delivery sites via VLDL   LDL   HDL

VLDL = very low density lipoprotein, has high cholesterol concentration
LDL = low density lipoprotein, has medium cholesterol concentration
HDL = high density lipoprotein, has low cholestero contcentration

All polar groups are on the same face. Which side faces water and which side faces fat molecules? (See structures below.)

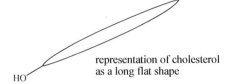

representation of cholesterol as a long flat shape

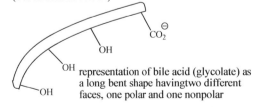

representation of bile acid (glycolate) as a long bent shape havingtwo different faces, one polar and one nonpolar

Chapter 2

CHAPTER 3 – HOW MANY STRUCTURES CAN YOU DRAW FROM A SIMPLE FORMULA?

We have learned how to draw 2D structures and 3D structures from condensed line structures, using a number of functional groups as examples. It is clear that there are a lot of ways to connect even a small number of atoms together. Even in a single structure, there may be considerable variation where electrons are drawn because of parallel, overlapping 2p orbitals in resonance structures. Instead of looking at all of the variety possible in different formulas, let's look at the variety of possible structures in a single formula. The results may surprise you and along the way we will discover a few more facets about organic chemistry.

What possibilities are found within a simple, fully condensed formula like C_7H_{14} (one degree of unsaturation)? There are only two types of elements, carbon and hydrogen, and just a mere seven carbon atoms to arrange in different skeletal patterns. We don't have a clue about what those connectivities are from the formula, as given. It turns out there are 56 ways to connect those atoms! What is clear from this example is that a fully condensed formula can have an awful lot of ways that it can be put together. However, the number of possible structures from this simple formula barely hints at the variety waiting for us in organic chemistry and biochemistry. There is really no easy way to know if you have drawn all of the possible structures from a formula and that will not be our goal. In and of itself there is really no point in being able to draw every possible structure from a formula. However, exercises like this, using simple formulas, will make you aware of the infinite variety found in organic chemistry (and biochemistry). You should also be convinced of the shear impossibility of memorizing the vast expanse of this subject. I will suggest a strategy for drawing possible structures, but there is no reason for you to memorize it and you may even come up with a plan that works better for your way of thinking.

We will begin with formulas that are saturated. In organic chemistry, "saturated" means that there are no pi bonds or rings in a structure. Every possible bonding position is filled with an atom or group that forms single bonds. Initially, we will limit ourselves to just two elements, carbon and hydrogen, though even with this limitation, the possibilities are unlimited. Hydrocarbons that have only single bonds are called alkanes. The carbon skeleton of an alkane is covered with a skin of hydrogen atoms. Chemically, alkanes are fairly inert and we don't study too many reactions involving them. However, the reactions they do undergo are pretty important reactions, including combustion and free radical halogenation. Later, when we include other atoms different than carbon and hydrogen (nitrogen, oxygen and halogens) and degrees of unsaturation, the number of possible structures from a single formula will shoot through the roof.

There is a certain amount of trial and error in our approach to this topic, though it is helpful if you develop a "quasi" systematic approach. A reasonable strategy is to start with a straight chain using all of the carbon atoms. Next, shorten it by one carbon atom, which is then used to make a one carbon branch at nonterminal positions. That branch is then moved down the chain until duplicate structures start being formed. Next, take a second carbon off and make two one carbon branches or a single two carbon branch and connect the branches in every unique way. This continues with three carbons and four carbons, etc. until you are repeating earlier structures. Watch out when adding your branches that you don't repeat a longer chain already generated. As the formulas get bigger, it is also easy to leave out possible isomers. Isomers are molecules that have the same formula, but differ in connectivity of the atoms or their orientation in space. Isomer literally means equal unit. We will discover there are many types of isomers in organic chemistry. Let's start with one carbon and work our way up.

Chapter 3

1. methane: CH₄

A single carbon atom has four bonding positions, completely saturated by the four hydrogen atoms. There is only one possible arrangement of the atoms. CH₄ is completely unambiguous.

condensed formula   2D formula   3D formula

CH₄

methane

2. ethane: C₂H₆

Ethane, as with methane, has only one possible bonding arrangement. Once the two carbons are connected, there are only six additional bonding sites and these are filled by the six hydrogen atoms. Ethane is a saturated molecule. C₂H₆ is completely unambiguous.

condensed formula   2D formula   3D formula

C₂H₆

condensed line formula

CH₃CH₃

ethane

You can twist about this single bond generating different conformations. Build a model and try it.

There is a little more variety in the structure of ethane. The carbon-carbon single bond allows rotation of one group of three C-H sigma bonds past the other group of three C-H sigma bonds. Using a molecular model of ethane, fix one carbon with one hand and spin the other carbon with you other hand. The different arrangements of the atoms as they rotate past one another are called conformations. Conformations are the result of differences in a structure from rotation about single bonds. A single bond is much like the axle on a vehicle and the other groups at the carbon atoms can rotate like the spokes on a wheel. We will study conformations more in a later topic.

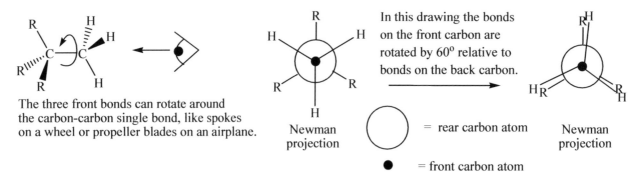

The three front bonds can rotate around the carbon-carbon single bond, like spokes on a wheel or propeller blades on an airplane.

Newman projection

In this drawing the bonds on the front carbon are rotated by 60° relative to bonds on the back carbon.

Newman projection

○ = rear carbon atom

● = front carbon atom

A straight-on view shows the three bonds as lines connecting to the front carbon atom, written as a small dot. The rear carbon is usually drawn as a large circle and the three bonds to that carbon are also drawn as lines, but only down to the circle. Rotation of the front bonds is easily seen in two different structures. These are called Newman projections and we will see more of them later.

Chapter 3

3. propane: C_3H_8

Once again, there is only one possible arrangement of the bonding atoms. The third carbon has to be attached to either of the other two carbons forming a three carbon chain with eight additional bonding sites, each bonded to a hydrogen atom. Propane is a saturated molecule. C_3H_8 is completely unambiguous. This seems pretty easy, doesn't it?

condensed formula

2D formula

3D formula

 C_3H_8

condensed line formula

 $CH_3CH_2CH_3$

propane

You can twist about either single bond generating different conformations. Build a model and try it.

Because the backbone of the carbon skeleton (plus the two hydrogen atoms at the end positions) goes up and down, this shape is sometimes referred to as a zig-zag shape. It can also be called a wedge and dash formula.

bond-line formula

Because single bonds allow rotation, there are a number of ways (conformations) that propane can be drawn using slightly different representations. If you have models, now is a time to use them. Keep them handy. You'll want to use them frequently.

This bond was twisted down.

This bond was twisted down.

This bond was twisted up.

This bond was twisted up.

The point you should get from this example is that three sequentially attached sp³ carbon atoms can be drawn in a variety of ways, but are still the same structure. (Twist your arm around at your elbow, your shoulder and both at the same time. How many shapes can they assume?) With additional carbon atoms (4, 5, 6...) the possibilities increase, but the feature to focus on when deciding if two structures are identical or different is the length of the carbon chain and the length and positions of its branches. Whether you have 3,4,5, or more carbon atoms zig-zagged in any manner possible, you can always redraw it in a simpler straight chain form. The straight chain representation is easier for you to work with, so choose it as your method of drawing a 2D structure (even if it is changing its shape thousands of times per second). Drawing your structures with a straight chain, as much as possible, will give you the best chance of not overlooking a structure. We will use this approach in our subsequent examples.

4. butane and 2-methylpropane (isobutane): C$_4$H$_{10}$

We finally encounter an example where there is more than one possible carbon skeleton. C$_4$H$_{10}$, by itself, is ambiguous. First draw the longest possible carbon chain, which would be...four carbons (you guessed it!). There are ten additional bonding sites that are all saturated with hydrogen atoms.

A four carbon chain can be drawn and rotated in a variety of ways, but there is no change effected in the skeletal connectivity of the atoms. No matter how the bonds are twisted, it's still a four carbon chain.

If you twirl your arm (or arms) at your side, do flips or cartwheels or peddle your legs riding a bike, you are still connected in the same way. A picture of each body position would appear differently, but it's still the same you. The same is true for the connection of carbon atoms.

Let's now shorten our chain by one carbon atom so that the longest chain is three carbon atoms. Can we attach our remaining carbon atom in such a way so as to not make a four carbon chain? If we attach this carbon to either end we have a....four carbon chain, and this repeats the example just above. By attaching the fourth carbon atom to the middle carbon of our three carbon chain, the longest chain is only three carbon atoms in any direction, and we have a one carbon atom branch in the middle. This is a different carbon skeleton.

Again, there are a lot of ways to write this as a 2D carbon skeleton on a piece of paper.

These are all representations of the same structure, but are different from our first four carbon structure. Both molecules have exactly the same formula C$_4$H$_{10}$ but they are not at all the same in their physical properties. The melting points, boiling points, densities and heats of formation are given below to emphasize this. Their spectra and chemistry would not be expected to be just alike either (and they're not). This should not be surprising. Think about how you would interact with the world if one of your arms was attached where your belly button is.

	melting point	boiling point	density	heat of formation
butane, $CH_3CH_2CH_2CH_3$	-138	-1	0.579	-30.4
2-methylpropane, $CH_3CH(CH_3)_2$ (isobutane)	-159	-12	0.549	-32.4

We refer to these two different compounds, having identical molecular formulas, as ***skeletal*** or ***constitutional isomers*** (isomers = equal units). Isomers can be different in the connectivity of the atoms; that is their skeletons are different. Or, isomers can have the same connectivity of atoms, but differ in their orientation in space (***stereoisomers***). To try and visualize stereoisomers, think of your left hand and your right hand. Both hands have the same types of fingers, but pointing in different directions in space, with important consequences for how you interact with the world. Does being left handed or right handed make a difference? Of course it does. We will encounter even other types of isomers as we progress on our journey through organic chemistry (more on this topic later).

5. pentane isomers: C_5H_{12}

So far this hasn't been too difficult. By itself, C_5H_{12} is saturated, but it is ambiguous. As before, we will begin with a five carbon straight chain isomer. There are 12 additional bonding sites, all filled with a hydrogen atom.

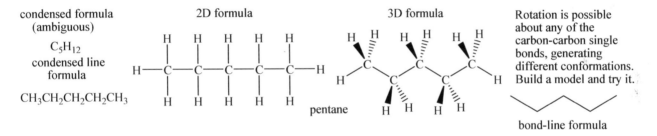

Each C-C single bond can be rotated so, again, there are a variety of ways to represent these five carbons, some of which are shown.

We can shorten this chain by one carbon atom (the longest chain is now four carbon atoms) and attach the one extra carbon atom at a nonterminal position to create a second isomer. If we attach the carbon atom to a terminal position, our chain would be five carbons long and that repeats the first example. Attachment at either of the two possible internal carbons produces the same isomer. There is only one additional isomer resulting from this operation.

$$\begin{array}{cccc} 1 & 2 & 3 & 4 \\ C & \!\!\!-\!\!\! & C & \!\!\!-\!\!\! & C & \!\!\!-\!\!\! & C \end{array}$$

↑ ↑
a b

Add one carbon branch at internal positions down the straight chain until structures are repeated (just past the half-way point).

a →

$$\begin{array}{ccccc} & & H & & \\ & & | & & \\ H_3C & \!\!-\!\! & \underset{2}{\overset{}{C}} & \!\!-\!\!CH_2 & \!\!-\!\!CH_3 \\ 1 & & | & 3 & 4 \\ & & CH_3 & & \end{array}$$

These are identical structures. Make models and turn one of them around and superimpose them on top of one another.

b →

$$\begin{array}{ccccc} & & H & & \\ & & | & & \\ H_3C & \!\!-\!\!H_2C & \!\!-\!\!C & \!\!-\!\!CH_3 \\ & & | & & \\ & & CH_3 & & \end{array}$$

2-methylbutane
(isopentane)

condensed line formula

$(CH_3)_2CHCH_2CH_3$

Shortening the four carbon chain by an additional carbon atom, leaves us with a longest chain of three carbon atoms. Two carbon atoms remain to be added on. We have to avoid the terminal carbons because any additions to these increases the chain length to four carbons. We only have a single internal carbon position and two carbons to add on.

If we add the two carbon atoms together, as a single branch, we have a four carbon chain, which we just did above. However, if we add on each of our two extra carbon atoms as one carbon branches to the internal carbon, then our longest chain is only three carbon atoms and this is yet another isomer.

$$\begin{array}{ccc} 1 & 2 & \\ C & \!\!-\!\!C & \!\!-\!\!C \\ & | & \\ & C & 3 \\ & | & \\ & C & 4 \end{array}$$

The four carbon chain is too long. We have already consider this possibility.

$$\begin{array}{ccc} & CH_3 & \\ & | & \\ H_3C & \!\!-\!\!C & \!\!-\!\!CH_3 \\ & | & \\ & CH_3 & \end{array}$$

2,2-dimethylpropane
(neopentane)

This is a new isomer. The longest carbon chain in any direction is three carons long.

condensed line formula

$(CH_3)_4C$

or

$C(CH_3)_4$

We cannot shorten our carbon chain to two carbon atoms because as soon as we add on our next carbon atom we are back up to a three carbon chain, and we just did that. So there are three different skeletal or structural or constitutional isomers for our saturated formula of C_5H_{12}.

Drawing all possible isomers of the five examples, thus far, has not proven particularly difficult. This looks like a piece of cake.

CH_4	C_2H_6	C_3H_8	C_4H_{10}	C_5H_{12}
1 isomer	1 isomer	1 isomer	2 isomers	3 isomers

6. hexane isomers: C_6H_{12}

From this point on we have to be extra careful to be systematic in our approach. Following steps similar to our above examples we first write out the six carbon atom chain with its 14 additional bonding sites, all filled with hydrogen (saturated with hydrogen atoms).

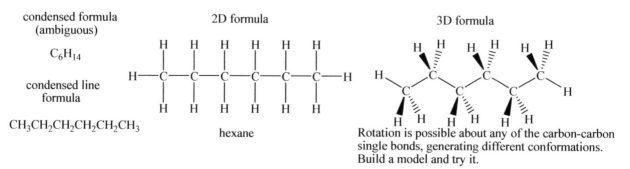

condensed formula (ambiguous)

C_6H_{14}

condensed line formula

$CH_3CH_2CH_2CH_2CH_2CH_3$

2D formula

hexane

3D formula

Rotation is possible about any of the carbon-carbon single bonds, generating different conformations. Build a model and try it.

Chapter 3

Next, we shorten this by one carbon and attach that carbon to nonequivalent, internal positions. We avoid the end carbons because this just regenerates our six carbon atom chain. Adding the one carbon branch to C2 is identical to adding it to C4, however, adding it to C3 is different from either of those possibilities, so there are two additional isomers with a longest chain of five carbons.

Adding the one carbon branch to C2 or C4 produces identical structures (a = c). Adding the one carbon branch to C3 forms a different isomer (b).

```
 1   2   3   4   5
 C — C — C — C — C
     ↑   ↑   ↑
     a   b   c
```

Add one carbon branch at internal positions down the straight chain until structures are repeated, just past the half-way point.

a H_3C—$\overset{H}{\underset{CH_3}{C}}$—$\overset{H_2}{C}$—$\overset{H_2}{C}$—$CH_3$

b H_3C—$\overset{H_2}{C}$—$\overset{H}{\underset{CH_3}{C}}$—$\overset{H_2}{C}$—$CH_3$

c H_3C—$\overset{H_2}{C}$—$\overset{H_2}{C}$—$\overset{H}{\underset{CH_3}{C}}$—$CH_3$

condensed line formulas

$(CH_3)_2CHCH_2CH_2CH_3$
2-methylpentane
(isohexane)

$CH_3CH_2CH(CH_3)CH_2CH_3$
3-methylpentane

same as a

Next, we shorten our six carbon chain by two carbon atoms so that the longest carbon chain is four carbon atoms long. There are two carbon atoms left to add, which can be attached as a single two carbon branch or two one carbon branches. Adding a two carbon branch produces a five carbon chain that we already considered. However, adding two one carbon branches will produce new isomers. As before, we have to avoid the terminal carbon atoms. We can put both one carbon branches at the second carbon (or third carbon, which would be identical). Finally, we can keep one carbon branch in place and move the other carbon branch down the chain (avoiding the other end, of course).

Structures "a" and "b" are identical. Structure "c" is a different isomer.

```
 1   2   3   4
 C — C — C — C
```

Adding two one carbon branches to C2 is the identical to adding two one carbon branches to C3. Adding a one carbon branch to C2 and C3 forms a different isomer.

a H_3C—$\overset{CH_3}{\underset{CH_3}{C}}$—$\overset{H_2}{C}$—$CH_3$

b H_3C—$\overset{H_2}{C}$—$\overset{CH_3}{\underset{CH_3}{C}}$—$CH_3$

c H_3C—$\overset{CH_3}{\underset{H}{C}}$—$\overset{H}{\underset{CH_3}{C}}$—$CH_3$

condensed line formulas
↓
$(CH_3)_3CCH_2CH_3$
2,2-dimethylbutane
(neohexane)

$(CH_3)_2CHCH(CH_3)_2$
2,3-dimethylbutane

We cannot shorten the carbon chain any more (i.e., a three carbon chain). When we try to add on the additional three carbons we are forced to draw at least a four carbon chain, which we just considered. So our six carbon atom formula produced five different skeletal isomers. This still seems like a reasonable result.

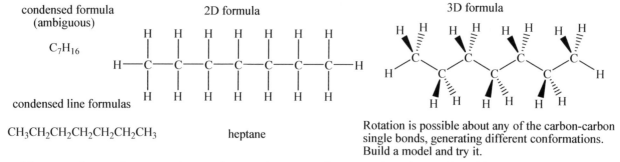

There are five C_6H_{14} skeletal isomers.

7. heptane isomers: C_7H_{16}

As a final example, we will look at the isomers of heptane using a more abbreviated approach. We begin with the straight chain isomer, having 16 bonding positions, all hydrogen here.

condensed formula
(ambiguous)

C_7H_{16}

2D formula

3D formula

condensed line formulas

$CH_3CH_2CH_2CH_2CH_2CH_2CH_3$ heptane

Rotation is possible about any of the carbon-carbon single bonds, generating different conformations. Build a model and try it.

Next, we shorten by one carbon and move that one carbon to unique internal positions on the six carbons chain. Notice substitution on C2 duplicates substitution on C5, while substitution on C3 duplicates substitution on C4.

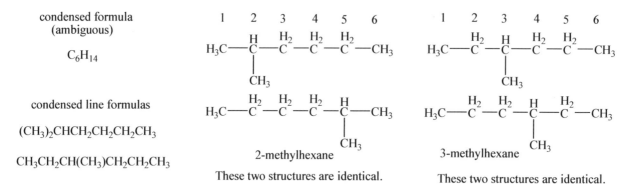

condensed formula
(ambiguous)

C_6H_{14}

condensed line formulas

$(CH_3)_2CHCH_2CH_2CH_2CH_3$

$CH_3CH_2CH(CH_3)CH_2CH_2CH_3$

2-methylhexane

These two structures are identical.

3-methylhexane

These two structures are identical.

Chapter 3

We shorten the carbon chain to five carbon atoms, leaving two carbons to add on as a single two carbon branch or two one carbon branches. The two carbon branch can only be added to C3, or else a six carbon chain is formed (already considered above).

condensed formula
(ambiguous)

C_6H_{14}

condensed line formula

$(CH_3CH_2)_2CHCH_2CH_3$

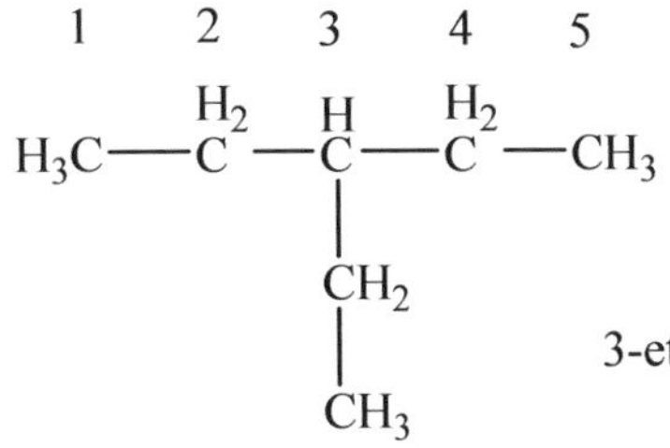

3-ethylpentane

We can also add the extra two carbon atoms as two one-carbon branches. Beginning with both of them on C2 and moving them together to C3 forms a different isomer. Moving them both to C4 duplicates the C2 isomer. Leaving one of the one carbon branches at C2 and moving the other one carbon branch to C3 and then to C4 generates two additional isomers. All together this produces four additional isomers.

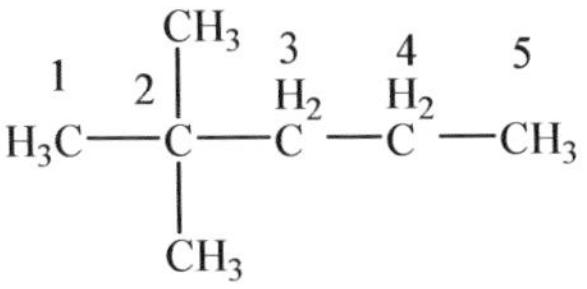

2,2-dimethylpentane

$(CH_3)_3CCH_2CH_2CH_3$

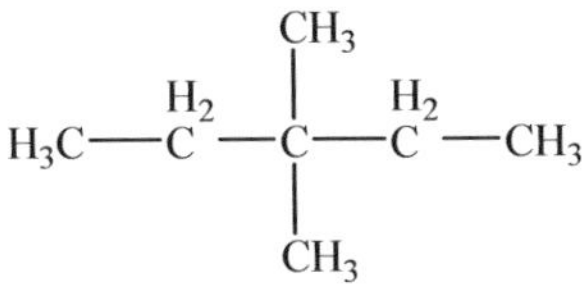

3,3-dimethylpentane

$CH_3CH_2C(CH_3)_2CH_2CH_3$

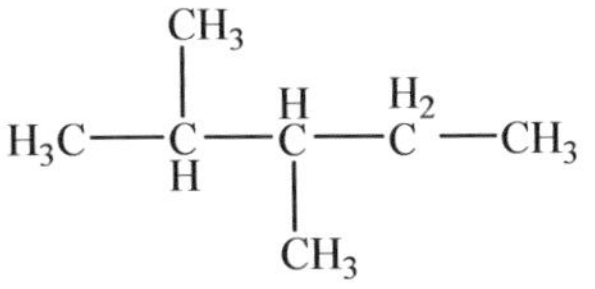

2,3-dimethylpentane

$CH_3CH(CH_3)CH(CH_3)CH_2CH_3$

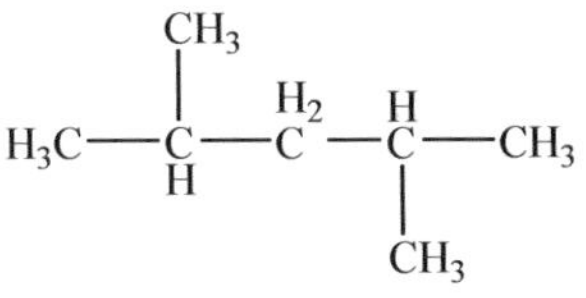

2,4-dimethylpentane

$CH_3CH(CH_3)CH_2CH(CH_3)CH_3$

condensed line formulas

We next shorten the chain to four carbon atoms, leaving three carbon atoms to add on. Addition of a three carbon branch or a two carbon branch is not possible because either chain formed is too long (and already considered above). There is one additional isomer having three one carbon branches at C2 and C3. This is as far as we can go with a formula of C_7H_{16}.

condensed line formula

$(CH_3)_3CCH(CH_3)_2$

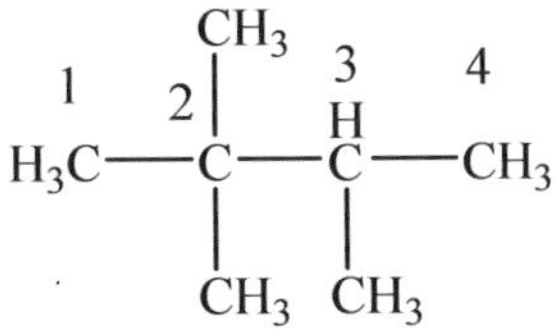

2,2,3-trimethylbutane

There are nine different structural isomers. Their physical properties are clearly different, and they are distinct structures. Notice, too, they all have different names. That's another way we can tell them apart (Topic 4).

Chapter 3

C_7H_{16} alkane isomers	name	melting point (°C)	boiling point (°C)	density (g/cm³)	refractive index	heat of formation (kcal/mole)
1.	heptane	-91	98	0.684	1.3878	-44.88
2.	2-methylhexane	-118	90	0.677	1.3848	-41.66
3.	3-methylhexane	-119	92	0.686	1.3887	-41.02
4.	2,2-dimethylpentane	-124	79	0.674	1.3822	-49.27
5.	3,3-dimethylpentane	-134	86	0.694	1.3909	-48.17
6.	2,3-dimethylpentane	?	90	0.695	1.3919	-47.62
7.	2,4-dimethylpentane	-119	81	0.673	1.3815	-48.28
8.	3-ethylpentane	-119	94	0.698	1.3934	-45.33
9.	2,2,3-trimethylbutane	-24	81	0.690	1.3894	-48.95

As the number of carbon atoms goes up, the number of isomers grows beyond comprehension. Take a look at the numbers below using 'simple' alkane formulas.

formula	number of structural isomers
CH_4	1
C_2H_6	1
C_3H_8	1
C_4H_{10}	2
C_5H_{12}	3
C_6H_{14}	5
C_7H_{16}	9
C_8H_{18}	18
C_9H_{20}	35
$C_{10}H_{22}$	75
$C_{20}H_{42}$	366,319
$C_{30}H_{62}$	4,111,846,763 (that's billion)
$C_{40}H_{82}$	around 62,000,000,000,000 (and that's trillion!)

Chapter 3

Problem 1 - Generate the 18 possible structural isomers of C_8H_{18}. (We'll save $C_{40}H_{82}$ for another lifetime. If you generated one isomer per second, it would take you about 2,000,000 years. We better make that several lifetimes.)

All we've been looking at are molecules that have only carbon and hydrogen and we are talking 2,000,000 years to do a problem. What happens if one little hydrogen atom is replaced with a halogen atom? This seems like a pretty minor change. After all, halogen atoms only form one bond, just like hydrogen atom? No pi bonds or rings can form because our structure is still saturated. We want an example that is complex enough to demonstrate the logic of what is changing, but not so complicated that we have to invest 2,000,000 years. C_5H_{12} should give us the insight we need, yet only has three different structural formulas. First, let's redraw our three carbon skeletons.

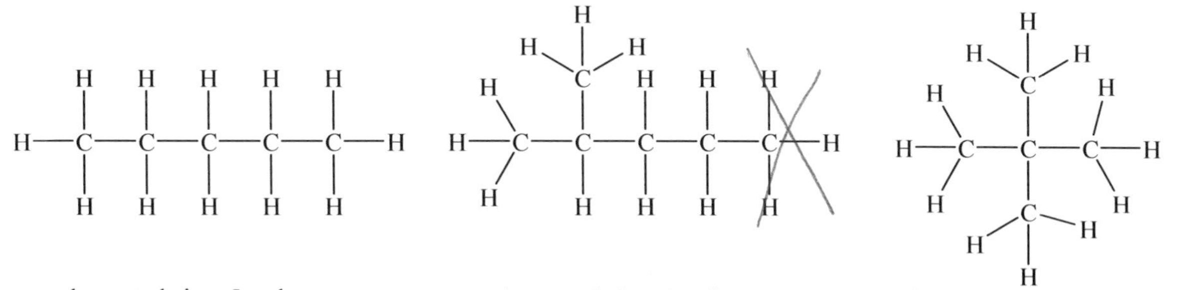

longest chain = 5 carbons   longest chain = 4 carbons   longest chain = 3 carbons

The five carbon atom chain has three different bonding positions where we could switch out a hydrogen atom for a halogen atom (let's use a chlorine atom). The chlorine atom can bond at C1 (identical to C5), or at C2 (identical to C4), or at C3 (unique). There are four different kinds of bonding positions in the four carbon chain (C1 = C5, C2, C3 and C4). Finally, there is only one kind of bonding position in the three carbon chain (…that's right, all 12 C-H bonds are equivalent).

Carbon atoms are sometimes categorized by the number of other carbon atoms attached to them. If only one carbon is attached to a carbon of interest, it is classified as a primary carbon and sometimes symbolized by (1°). If two carbons are attached to a carbon of interest, the classification is secondary (2°). When three carbons are attached to a carbon of interest the classification is tertiary (3°) and finally, if four carbons are attached to a carbon of interest the classification is quaternary (4°). Each of the C_5H_{12} isomers is shown below with designation of primary, secondary or tertiary for the carbon atom bonded to the chlorine atom. The simple little chlorine atom changed three isomers into eight isomers. Notice, all the isomers have different names. If we considered stereoisomers (think of your hands), there would be even more possibilities. We'll wait a bit to introduce those ideas (Chapter 7).

Eight chloroalkane isomers.

Five carbon chain.   C1 = C5 and   C2 = C4

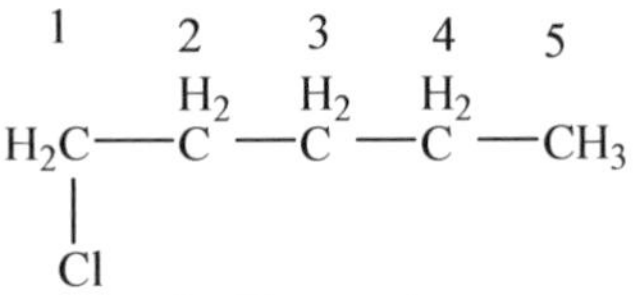

H_2C—$\overset{H_2}{C}$—$\overset{H_2}{C}$—$\overset{H_2}{C}$—CH_3 H_3C—$\overset{H}{C}$—$\overset{H_2}{C}$—$\overset{H_2}{C}$—CH_3 H_3C—$\overset{H_2}{C}$—$\overset{H}{C}$—$\overset{H_2}{C}$—CH_3

|Cl |Cl |Cl

1-chloropentane   2-chloropentane   3-chloropentane

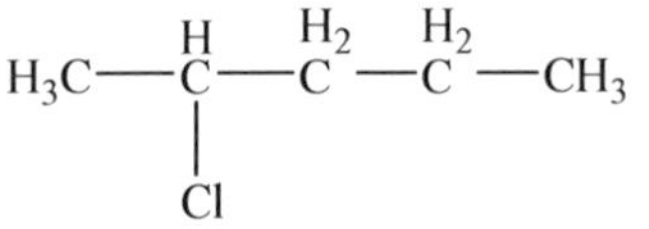

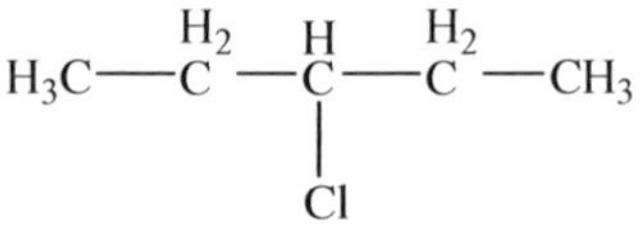

Carbon atom with Cl is a   Carbon atom with Cl is a   Carbon atom with Cl is a
primary carbon, 1° RCl.   secondary carbon, 2° RCl.   secondary carbon, 2° RCl.

Four carbon chain.   C1 = C5

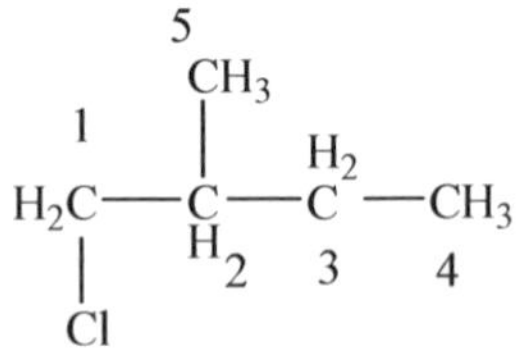

5
CH_3   CH_3   CH_3   CH_3
1 | | | |
H_2C—$\overset{}{C}$—$\overset{H_2}{C}$—CH_3 H_3C—C—$\overset{H_2}{C}$—CH_3 H_3C—$\overset{}{C}$—C—CH_3 H_3C—$\overset{}{C}$—$\overset{H_2}{C}$—CH_2
| $\overset{H_2}{}$ | | H | H |
Cl   3 4   Cl   Cl   Cl

1-chloro-2-methylbutane   2-chloro-2-methylbutane   2-chloro-3-methylbutane   1-chloro-3-methylbutane

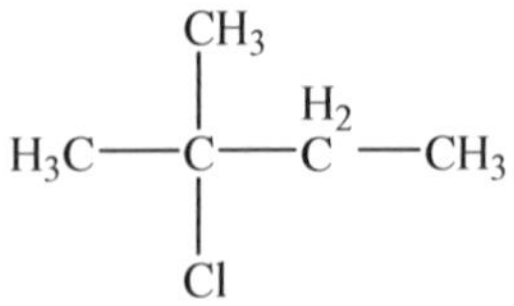

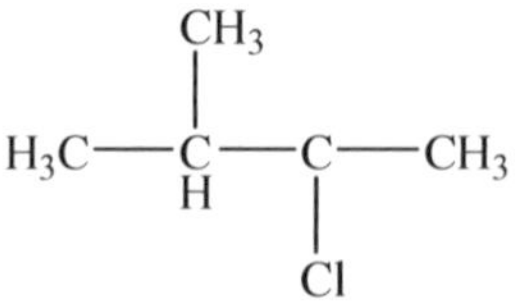

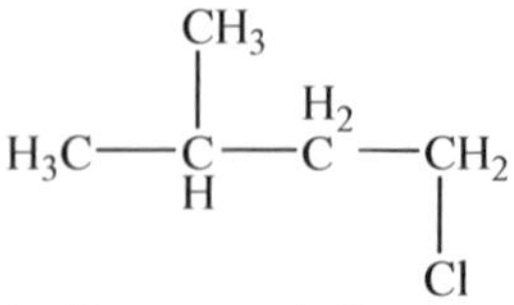

Carbon atom with Cl is a   Carbon atom with Cl is a   Carbon atom with Cl is a   Carbon atom with Cl is a
primary carbon, 1° RCl.   tertiary carbon, 3° RCl.   secondary carbon, 2° RCl. primary carbon, 1° RCl.

Three carbon chain. C1 = C3 = C4 = C5

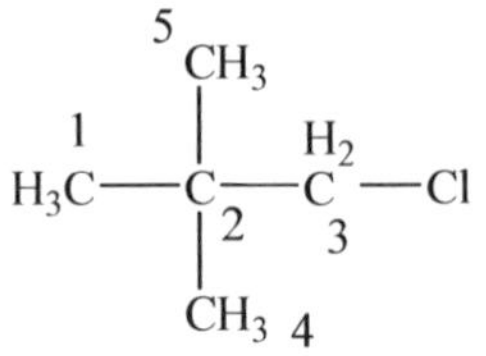

5
CH_3
1 | H_2
H_3C—C—C—Cl Carbon atom with Cl is a   C2 is a quaternary carbon, 4° and cannot
| 2   3   primary carbon, 1° RCl.   form a bond with a chlorine atom.
CH_3 4

1-chloro-2,2-dimethylpropane

Problem 2 – Draw all of the possible isomers of C_4H_9Br. Hint: There should be four. If you feel ambitious, try and draw all of the possible isomers of $C_6H_{13}F$. There should be about 17 of them.

   What happens when an oxygen is added to a formula? Oxygen forms two bonds, so there is a new possibility we haven't had to consider thus far. First, when an oxygen atom bonds to the surface of the carbon skeleton, by inserting itself between a carbon atom and a hydrogen atom, an alcohol functional group is created (ROH). Since the "OH" substituent, as a group, only forms one bond, the number of isomers possible is no different than the chlorine example above. If we used the same five carbon skeletons above, there would be eight possible isomeric alcohols. (Switch out Cl for HO.)
   Oxygen atoms don't have to bond to the surface of a carbon skeleton, they can insert themselves in between two carbon atoms. It gets a little trickier to try and consider all of the possibilities for inserting oxygen between two carbons, since one carbon skeleton is becoming two carbon skeletons. When oxygen is surrounded on both sides by a simple carbon atom, the functional group is an ether (ROR). We can do this as C1 (one way) + C4 (four ways) or C2 (one way) + C3 (two ways), for six total ethers.
   The alcohol and ether functional groups are shown below, along with the classification of primary, secondary or tertiary based on the number of carbon atoms bonded to the carbon atom attached to the oxygen. An isolated CH_3 substituent does not fall under any of the classifications (1°, 2°, 3° or 4°). CH_3 groups go by the name of methyl. Notice, they all have different nomenclature names.

Chapter 3

Eight alcohol isomers.

Five carbon chain.   C1 = C5 and   C2 = C4

$$H_2C \overset{1}{} \overset{2}{\underset{H_2}{C}} \overset{3}{\underset{H_2}{C}} \overset{4}{\underset{H_2}{C}} CH_3 \quad (OH)$$

pentan-1-ol

pentan-2-ol

pentan-3-ol

Carbon atom with OH is a
primary carbon, 1° ROH.

Carbon atom with OH is a
secondary carbon, 2° ROH.

Carbon atom with OH is a
secondary carbon, 2° ROH.

- -

Four carbon chain.   C1 = C5

2-methylbutan-1-ol

2-methylbutan-2-ol

3-methylbutan-2-ol

3-methylbutan-1-ol

Carbon atom with OH is a
primary carbon, 1° ROH.

Carbon atom with OH is a
tertiary carbon, 3° ROH.

Carbon atom with OH is a
secondary carbon, 2° ROH.

Carbon atom with OH is a
primary carbon, 1° ROH.

- -

Three carbon chain. C1 = C3 = C4 = C5

Carbon atom with OH is a
primary carbon, 1° ROH.

2,2-dimethylpropan-1-ol

- -

Six ether isomers. (Combos of 4C + 1C (four ways) and 3C + 2C (two ways)

1-methoxybutane

2-methoxybutane

1-methoxy-2-methylpropane

Insert oxygen between C1 and C2
(same as between C4 and C5).

carbon to left of oxygen is methyl,
carbon to right of oxygen is a
primary carbon

Insert oxygen between C1 and C2
(same as between C5 and C2).

carbon to left of oxygen is methyl,
carbon to right of oxygen is a
secondary carbon

Insert oxygen between C3 and C4.

carbon to left of oxygen is primary,
carbon to right of oxygen is a
methyl

- -

2-methoxy-2-methylpropane

1-ethoxypropane

2-ethoxypropane

Insert oxygen between C1 and C2.
carbon to left of oxygen is a methyl,
carbon to right of oxygen is tertiary

Insert oxygen between C2 and C3
(same as between C4 and C3).

carbon to left of oxygen is primary,
carbon to right of oxygen is also a
primary carbon

Insert oxygen between C2 and C3.

carbon to left of oxygen is secondary,
carbon to right of oxygen is a
primary carbon

- -

A few examples of simple sp^3 oxygen patterns in organic chemistry.

$$H-O-H \quad H_3C-O-H \quad R-\overset{H_2}{\underset{}{C}}-O-H \quad R-\overset{H}{\underset{R}{C}}-O-H \quad R-\overset{R}{\underset{R}{C}}-O-H \quad R-O-R$$

water   methyl   primary   secondary   tertiary   ether
           alcohol   alcohol   alcohol   alcohol

Smaller alcohols are like organic water.

Problem 3 – Draw all of the possible isomers of $C_4H_{10}O$. Hint: There should be seven. If you feel ambitious, try and draw all of the possible isomers of $C_6H_{14}O$. There should be about 32 of them (17 alcohols and 15 ethers, if I counted correctly).

What happens when a nitrogen is added to the formula? Nitrogen forms three bonds, so there is again a new possibility we haven't had to consider before. When simple carbon atoms are bonded to simple nitrogen atoms with only single bonds between them, the functional group is called an amine. Amine nitrogen atoms are also classified by how many carbon atoms are attached to the nitrogen atom. Amines can be primary (1°) when one carbon is attached, secondary (2°) when two carbons are attached or tertiary (3°) when three carbon atoms are attached to the nitrogen atom. If four carbon atoms are attached, a positive formal charge is required and the nitrogen is identified as a quaternary ammonium ion (4°). As with oxygen, amine nitrogen atoms can be attached to the surface of a carbon skeleton as primary amines, RNH_2. If the nitrogen atom is inserted between carbon atoms, it will be a secondary amine, R_2NH, or tertiary amine. R_3N. If we use the same five carbon structures above, there will be 17 possible isomeric amines (quite a change from the simple alkane carbon skeletons we started with)! These are drawn below, along with the classifications of primary, secondary or tertiary amine. If there is an odd number of nitrogen atoms, the number of protons will be odd and the molecular weight will be odd.

Seventeen amine isomers.

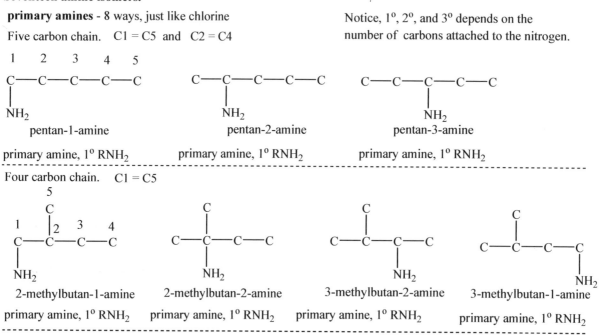

primary amines - 8 ways, just like chlorine

Five carbon chain.   C1 = C5   and   C2 = C4

Notice, 1°, 2°, and 3° depends on the number of carbons attached to the nitrogen.

Chapter 3

Three carbon chain. C1 = C3 = C4 = C5

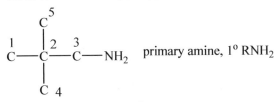

2,2-dimethylpropan-1-amine

- -

secondary amines - 6 ways, just like oxygen

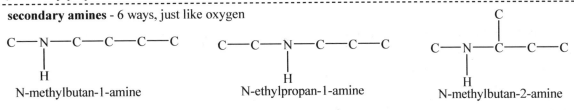

- -

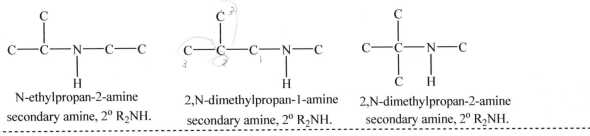

- -

tertiary amines - 3 ways, not previously possible

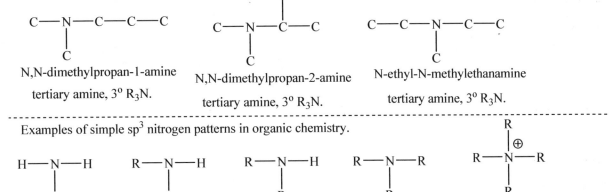

- -

Examples of simple sp^3 nitrogen patterns in organic chemistry.

H—N—H	R—N—H	R—N—H	R—N—R	R—N—R (⊕)
H	H	R	R	R
ammonia	primary amine	secondary amine	tertiary amine	quaternary ammonium ion

Problem 4 – Draw all of the possible isomers of $C_4H_{11}N$. Hint: There should be eight. If you feel ambitious, try and draw all of the possible isomers of $C_6H_{15}N$. There should be about 39 of them. (Yikes!)

All of the bonding patterns above have full saturation, meaning every bonding position possible is filled with an individual atom or group.

Chapter 3

Degrees of Unsaturation (or Hydrogen Deficiency)

Recall that alkanes are an example of saturated molecules. Every single bonding position on the carbon skeleton is filled with a hydrogen atom. If a halogen atom or an OH of an alcohol of NH_2 of an amine are substituted in for one of the hydrogen atoms, the molecule is still saturated. The number of bonding positions on an alkane carbon skeleton is easy to calculate. There are two positions on every carbon atom plus two additional positions at the end carbons.

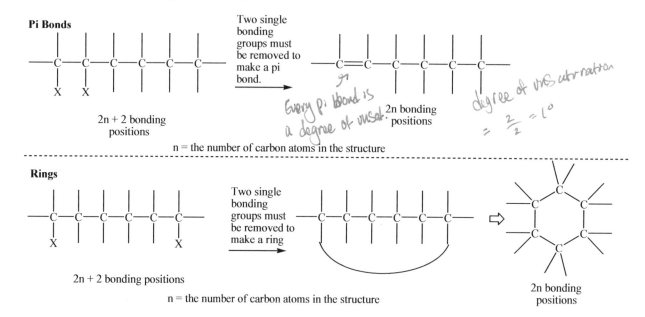

*Saturated means you can't put any more bonds

Maximum number of bonding positions on a saturated carbon skeleton $= 2n + 2$

n = the number of carbon atoms in the structure

Two bonding positions on every carbon, plus two more at the end positions.

Maximum number of bonding positions on a six carbon skeleton $= 2(6) + 2 = 14$

When a pi bond or a ring is present, the number of single bonding groups will be less than this maximum number ($2n + 2$). To make a pi bond, two hydrogen atoms must be removed from adjacent atoms, so that a second bond can form between those two atoms. Making a ring also requires that two hydrogen atoms to be removed in order for the chain of atoms to connect back on itself. The maximum number of bonding positions decreases by two for every degree of unsaturation. A pi bond or a ring is referred to as a degree of unsaturation (some books call this hydrogen deficiency).

Pi Bonds

Two single bonding groups must be removed to make a pi bond.

$2n + 2$ bonding positions

n = the number of carbon atoms in the structure

Every pi bond is a degree of unsat.

2n bonding positions

degree of unsaturation $= \frac{2}{2} = 1°$

Rings

Two single bonding groups must be removed to make a ring

$2n + 2$ bonding positions

n = the number of carbon atoms in the structure

2n bonding positions

We restricted ourselves in our initial examples, above, to saturated structures in order to limit the number of possible isomers. Adding just a single degree of unsaturation will greatly increase this number.

Let's see what happens to the two C_4H_{10} formulas when a single degree of unsaturation is introduced, either as a pi bond or as a ring? The new formula will have two fewer hydrogen atoms, C_4H_8. There are two different alkane carbon skeletons and we can use those as starting points to decide where pi bonds and rings can be placed.

Chapter 3

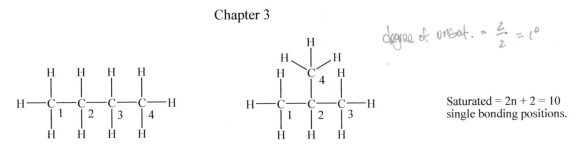

degree of unsat. = $\frac{2}{2}$ = 1°

Saturated = 2n + 2 = 10
single bonding positions.

Either the straight chain or the branched chain will allow a pi bond between any two adjacent atoms. In the straight chain a pi bond between C1 and C2 is identical to a pi bond between C4 and C3. A pi bond between C2 and C3 actually generates two different isomers, because of the restricted rotation about the pi bond. If the two CH$_3$ groups (methyl groups) are on the same side of the pi bond, the isomer is a *cis* alkene and if the two methyl groups are on opposite sides of the pi bond, the isomer is a *trans* alkene. Three alkenes are possible from the straight chain skeleton. In the branched carbon skeleton, a pi bond of C2 with any of the other carbon atoms forms an identical alkene, having two methyl groups on one carbon and two hydrogen atoms on the other carbon. No cis/trans isomers are possible when two identical groups are on a single carbon.

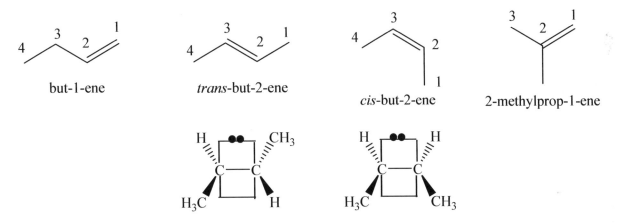

but-1-ene   *trans*-but-2-ene   *cis*-but-2-ene   2-methylprop-1-ene

No rotation is possible without breaking the pi bond. This is too energetically expensive at room temperature. Restricted rotation about the pi bond makes these two alkenes different.

There are two possible ways to make one ring using the four carbon atoms. All four carbon atoms will form a four carbon ring (cyclobutane) and three of the carbon atoms can form a ring having a one carbon branch (methylcyclopropane). The two carbon skeletons of C$_4$H$_{10}$ becomes six carbon skeletons of C$_4$H$_8$ when there is one degree of unsaturation.

cyclobutane   methylcyclopropane

The five alkane isomers of C$_6$H$_{14}$ becomes 17 alkene isomers and 16 cycloalkane isomers (...if I found them all). These are presented below without comment. A single degree of unsaturation really changes things a lot.

sp - 180°
sp² - 120°
sp³ = 109.5°

Chapter 3

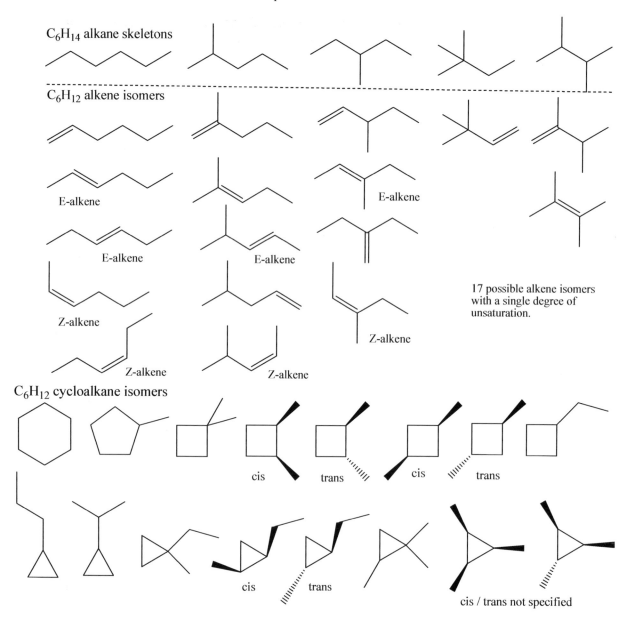

C_6H_{14} alkane skeletons

C_6H_{12} alkene isomers

E-alkene

E-alkene

E-alkene

E-alkene

Z-alkene

Z-alkene

Z-alkene

Z-alkene

Z-alkene

17 possible alkene isomers with a single degree of unsaturation.

C_6H_{12} cycloalkane isomers

cis   trans   cis   trans

cis   trans

cis / trans not specified

16 possible cycloalkane isomers with a single degree of unsaturation. There are actually more, but only cis/trans stereoisomers are included.

Problem 5 – Calculate the degree of unsaturation in the hydrocarbon formulas below. Draw one possible structure.

a.
C_7H_8

b.
$C_{10}H_8$

c.
C_8H_{10}

d.
$C_{12}H_{26}$

e.
C_6H_{12}

f.
C_6H_{10}

g.
C_6H_8

h.
C_6H_6

i.
C_6H_4

j.
C_6H_2

Problem 6 – Draw all of the alkene and cycloalkane isomers of C_5H_{10}. I calculate that there should be six alkenes and six cycloalkanes. Start with the three alkane skeletal isomers.

Chapter 3

What would two degrees of unsaturation do to the four carbon alkane? We have to consider more possible combinations, including two pi bonds, a pi bond and a ring or two rings. Each is possible as is shown below.

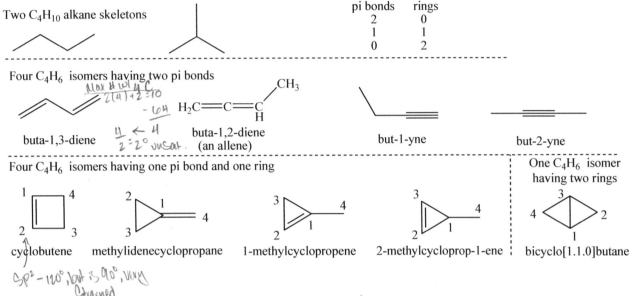

pi bonds	rings
2	0
1	1
0	2

Two C$_4$H$_{10}$ alkane skeletons

Four C$_4$H$_6$ isomers having two pi bonds

buta-1,3-diene   buta-1,2-diene (an allene)   but-1-yne   but-2-yne

Four C$_4$H$_6$ isomers having one pi bond and one ring   One C$_4$H$_6$ isomer having two rings

cyclobutene   methylidenecyclopropane   1-methylcyclopropene   2-methylcycloprop-1-ene   bicyclo[1.1.0]butane

Problem 7 – If you feel daring, try and draw some of the isomers of C$_5$H$_8$. I found four dienes (2 pi), three alkynes (2 pi), three allenes (2 pi), ten ring and pi bond isomers and four isomers with two rings. It's quite possible that I missed some. Why not shoot for two of each possibility.

How do heteroatoms affect our degree of unsaturation calculation? Any pi bond or ring uses up two bonding positions and introduces a degree of unsaturation, no matter what atoms are bonded together.

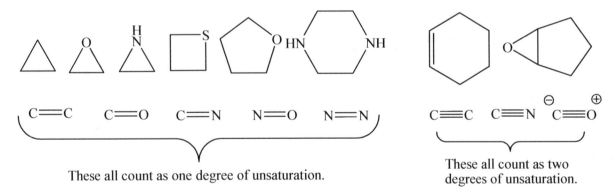

These all count as one degree of unsaturation.

These all count as two degrees of unsaturation.

Other Atoms and Degrees of Unsaturation (halogens, oxygen and nitrogen)

Halogen atoms are similar to hydrogen atoms in that they only form single bonds. They are added to the hydrogen atom count to obtain a total number of single bonding groups and this number is compared to the maximum number of single bonding positions (2n + 2). Every two groups short of the maximum number of bonding positions is a degree of unsaturation (could be a ring or a pi bond).

Chapter 3

Problem 8 – Calculate the degree of unsaturation in the formulas below. Draw one possible structure for each formula.

a.   b.   c.   d.   e.

C_7H_7FBrCl $C_{10}H_6F_2$ C_5H_4FCl C_6Br_6 $C_6H_{12}I_2$

are similar, in the sense that both types of atoms only occupy a single bonding position. X = F, Cl, Br, I

How do oxygen atoms change the calculation of degree of unsaturation? The answer is oxygen atoms do not change the calculation at all. Because oxygen atoms form two bonds, the bond they take away by connecting to the carbon skeleton they give back through their second bond. Oxygen atoms act like spacers between carbon and hydrogen atoms or between two carbon atoms. Just act like the oxygen atoms aren't there and perform the calculation just as you did above.

Oxygen atoms do not increase of decrease the number of bonding position.

6 bonding positions   still 6 bonding positions

Problem 9 – Calculate the degree of unsaturation in the formulas below. Draw one possible structure for each formula.

a.   b.   c.   d.   e.

C_2H_6O C_3H_6O C_4H_5ClO $C_6H_4F_2O_2$ $C_6H_{12}O_6$

How do nitrogen atoms change the calculation of degree of unsaturation? Nitrogen atoms add an extra bonding position for every single nitrogen atom present. Nitrogen atoms make three bonds. One bond is used in connection to the carbon skeleton. They give back two additional bonds in the place of that single bond, so there is one extra bond that wasn't there before. For every nitrogen atom present, you have to add an extra bonding position.

⁕ Each nitrogen atom increases the number of bonding positions by one.

one position becomes...   ...two positions

The complete formula we use to calculate degrees of unsaturation has to be modified as follows. You then compare the actual number of single bonding groups (H and halogens for us) to this maximum number. Subtract the actual number of single bonding groups from the maximum number and divide by 2 to determine the degree of unsaturation.

Maximum number of single bonding groups on a carbon skeleton = 2(#C) + 2 + #N	#C = the number of carbon atoms #N = the number of nitrogen atoms

$$\text{degrees of unsaturation} = \frac{(2(\#C) + 2 + \#N) - (\text{total single bonding groups})}{2}$$

R = "C" , "H"

Chapter 3

Problem 10 – Calculate the degree of unsaturation in the formulas below. Draw one possible structure for each formula.

a. C_2H_3N b. C_5H_5N c. $C_4H_4N_2$ d. $C_6H_6F_2N_2O$ e. $C_6H_{13}NO_2$

Problem 11– The following functional groups are unsaturated. Draw as many structures as you can for each of the examples below. Consider the alkane skeletons with the same number of carbon atoms as your starting points.

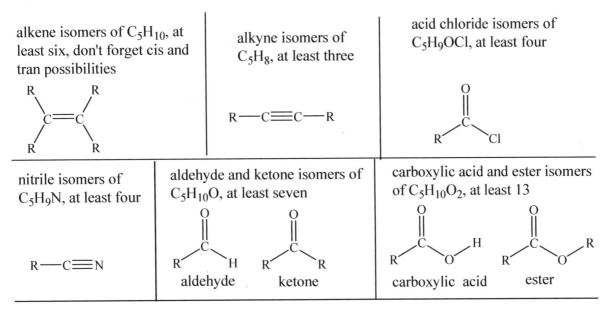

primary, secondary and tertiary amide isomers of $C_5H_{11}NO$, at least 17

Problem 12– Draw several isomers of molecular formula = CH_3NO_2? (There are many. How many? I don't know.)

    There are many ways to represent structures in organic chemistry and we will introduce several of them. Some common ways include condensed, 2D structural, 3D structural, sawhorse, bond-line, Newman, Fischer, Haworth, etc. Most molecular formulas can generate many different isomeric forms. Isomers are different compounds with the same molecular formula. These can include skeletal, positional, functional group, conformational isomers, enantiomers and diastereomers. We will study all of these.

Chapter 3

CHAPTER 4 – NOMENCLATURE

Organic chemistry requires a language that is precise, concise and systematic. Not only do we have to communicate in the written and oral manner, but we must catalogue and store our information in a manner that allows us to retrieve it when searching through the literature. With the explosion of information that is occurring in our time, this becomes more important with each passing day. As a foreign language, organic is relatively simple. But as a scientific language, organic is probably more complicated than in any other science (I'm considering organic and biochemistry as a part of the same language). To our benefit, organic is also probably more organized than any other scientific language.

Our nomenclature topic covers about 20 functional groups of organic chemistry. A few more are listed for reference purposes and we will not cover many additional, but less frequently used functional groups. While common names are not stressed in these topics, there are a few that any student of organic chemistry should be familiar with. These will be given in the relevant sections. Don't be intimidated by these trivial names. They are included so you have a list of the more common examples (for consultation, not for memorization). A few aromatic derivatives (substituted benzene derivatives, etc.) are also included in the relevant functional group sections and at the end. If you have questions as you work through this chapter, visit an office hour for clarification. My hope is that you will be able to speak "basic organic" after finishing these pages. This nomenclature topic has been written with the goal that you can teach yourself too.

The following sequences of rules should usually produce an acceptable name for commonly encountered organic compounds. Accessible guidelines to a more complete strategy of nomenclature are provided in Section C of the CRC "Handbook of Physics and Chemistry". An even more complete compilation of rules of organic nomenclature is provided in "A Guide to IUPAC Nomenclature of Organic Compounds, Recommendations, 2013", (IUPAC = International Union of Pure and Applied Chemistry). (link: http://www.acdlabs.com/iupac/nomenclature/)

As we proceed through organic chemistry we will often focus our interest on a limited portion of a structure when the remainder of the structure is not important to the topic of discussion. Symbolic representations for generic portions of a structure are given below and are commonly used. We will often use these representations in this and the coming chapters.

R = any general carbon group (it sometimes includes hydrogen too)
Ar = any general aromatic group, (when more specificity than 'just' R is desired)

The foundation of organic nomenclature requires an ability to name alkanes, alkenes and alkynes. Learning the rules for these groups will be your biggest nomenclature challenge. Many functional group patterns have a prefix (if lower priority) and a suffix (if higher priority) to specify their presence. Numbers on the longest specified chain will identify their locations. It's all very organized, but it does require practice. As usual in school, you are the only one who can do that. Pencil, paper and your hard work provides the most reliable path to success.

I. Nomenclature Rules For Alkanes and Cycloalkanes

The following list provides the names for carbon straight chains of various lengths. They must be memorized (through C_{19} for problems in this book, memorizing $C_{11} - C_{19}$ is easier than you think).

CH_4	methane (C_1)	$C_{11}H_{24}$	undecane (C_{11})	$C_{21}H_{44}$	henicosane (C_{21})
C_2H_6	ethane (C_2)	$C_{12}H_{26}$	dodecane (C_{12})	$C_{22}H_{46}$	doicosane (C_{22})
C_3H_8	propane (C_3)	$C_{13}H_{28}$	tridecane (C_{13})	$C_{23}H_{48}$	tricosane (C_{23})
C_4H_{10}	butane (C_4)	$C_{14}H_{30}$	tetradecane (C_{14})	$C_{24}H_{50}$	tetraicosane (C_{24})
C_5H_{12}	pentane (C_5)	$C_{15}H_{32}$	pentadecane (C_{15})	$C_{25}H_{52}$	pentaicosane (C_{25})
C_6H_{14}	hexane (C_6)	$C_{16}H_{34}$	hexadecane (C_{16})	$C_{26}H_{54}$	hexaicosane (C_{26})
C_7H_{16}	heptane (C_7)	$C_{17}H_{36}$	heptadecane (C_{17})	$C_{27}H_{56}$	heptaicosane (C_{27})
C_8H_{18}	octane (C_8)	$C_{18}H_{38}$	octadecane (C_{18})	$C_{28}H_{58}$	octaicosane (C_{28})
C_9H_{20}	nonane (C_9)	$C_{19}H_{40}$	nonadecane (C_{19})	$C_{29}H_{60}$	nonaicosane (C_{29})
$C_{10}H_{22}$	decane (C_{10})	$C_{20}H_{42}$	icosane (C_{20})	$C_{30}H_{62}$	triacontane (C_{30})

Chapter 4

Basic Steps to Name an Alkane

1. Locate the longest carbon chain present. This becomes the parent name. Make sure to check at each branch point for the longest chain path. (Unless it is obvious, count at each branch point through all possible paths.)

Where is the longest chain? (R = substituent branch)

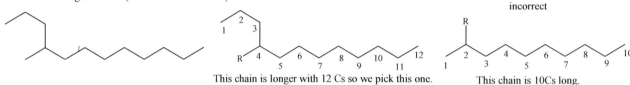

This chain is longer with 12 Cs so we pick this one.   This chain is 10Cs long.

If there are several branches radiating out from a central carbon, you can count how long those branches are and use the longest two plus the central carbon and add them all together.

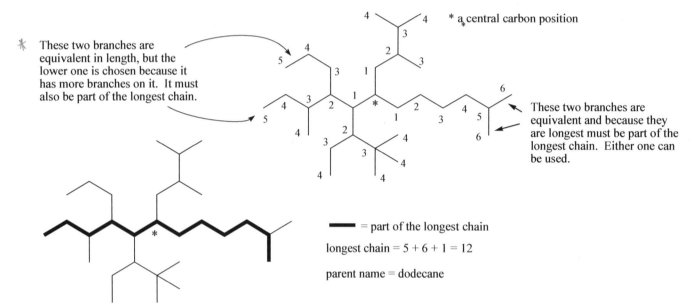

✳ These two branches are equivalent in length, but the lower one is chosen because it has more branches on it. It must also be part of the longest chain.

* a central carbon position

These two branches are equivalent and because they are longest must be part of the longest chain. Either one can be used.

— = part of the longest chain

longest chain = 5 + 6 + 1 = 12

parent name = dodecane

Number the longest chain from the end nearest a branch point or first point of difference. (The lowest first branch number decides which end of the chain you number from in an alkane.)

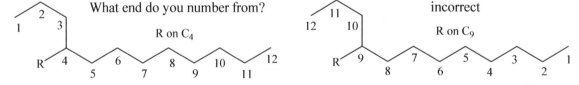

What end do you number from?

R on C_4

incorrect

R on C_9

A lower number for the substituent branch, R, is preferred. C_4 is better than C_9 so number from the left end.
The number will be used to specify the position of the alkyl branch

If additional branches are present, the lowest number of the first branch determines the numbering direction on the longest chain.

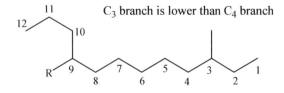

C_3 branch is lower than C_4 branch

The direction of the numbering is reversed because the new branch would get a lower number, C_3, at the first point of difference, than C_4 if the numbering occurred from the opposite direction.

Chapter 4

2. When an alkane portion is present as a substituent or branch (i.e. it is not part of the longest carbon chain) one drops the -ane suffix of a similar length alkane and adds the suffix -yl. Alkane becomes alkyl when it is a substituent; (ethane → eth + -yl → ethyl.) These substituent names are placed in front of the parent name, as prefixes, with their designating numbers immediately in front of them. Use the numbers obtained from rule 2 to show the location(s) of any substituent(s) or branch(es). Each substituent gets a number, even if it is identical to another substituent and on the same carbon. Hyphens are used to separate the numbers from the letters. Separate substituent position numbers from one another with commas (if the numbers are adjacent). The substituents are listed in alphabetical order. The numerical prefixes (see rule 3) do not count in deciding the alphabetical order (unless they are inside parentheses).

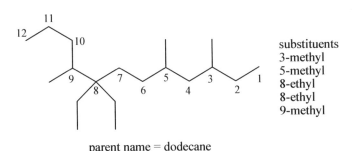

substituents
3-methyl
5-methyl
8-ethyl
8-ethyl
9-methyl

parent name = dodecane

There are three one carbon branches on C3, C5 and C9 carbons and two two carbon branches, both on the C8 carbon. The first branch at C3 determines the direction of numbering because it generates the lowest possible number at the fist point of difference.

1C branch = methyl	7C branch = hyptyl
2C branch = ethyl	8C branch = octyl
3C branch = propyl	9C branch = nonyl
4C branch = butyl	10C branch = decyl
5C branch = pentyl	11C branch = undecyl
6C branch = hexyl	12C branch = dodecyl

3. For identical substituents, use the prefixes di-, tri-, tetra-, penta-, hexa-, etc. to indicate 2, 3, 4, 5, 6, etc. of these substituents. These prefixes are <u>not</u> considered in deciding the alphabetical order of each substituent (unless inside parentheses).

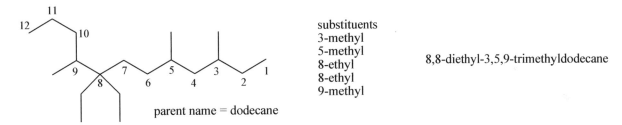

substituents
3-methyl
5-methyl
8-ethyl
8-ethyl
9-methyl

parent name = dodecane

8,8-diethyl-3,5,9-trimethyldodecane

4. With two or more possible longest chains of identical length, choose as the parent name the one with the greater number of substituents. This will produce simpler substituent names.

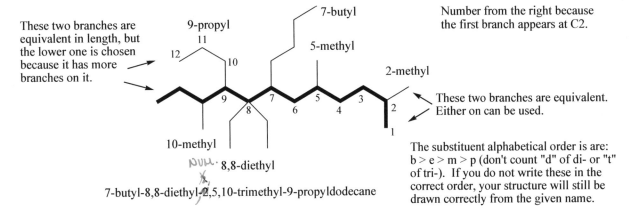

These two branches are equivalent in length, but the lower one is chosen because it has more branches on it.

9-propyl

7-butyl

5-methyl

2-methyl

10-methyl

8,8-diethyl

7-butyl-8,8-diethyl-2,5,10-trimethyl-9-propyldodecane

Number from the right because the first branch appears at C2.

These two branches are equivalent. Either on can be used.

The substituent alphabetical order is are: b > e > m > p (don't count "d" of di- or "t" of tri-). If you do not write these in the correct order, your structure will still be drawn correctly from the given name.

5. For complex substituents (substituents that have substituents on themselves), follow the above rules for alkanes except:

   i. The -ane suffix of the subparent name is changed to -yl (see rule 2 above)

   ii. The longest chain of the complex branch always uses the carbon directly attached to the parent chain as C_1. Starting at this position one would count the longest substituent chain possible, as shown below.

   iii. Parentheses are used to separate the entire complex substituent name, its numbers, its branches, and its subparent name, from the principle parent name. A number and a hyphen precede the entire complex substituent name in parentheses to indicate its location on the parent chain. If the complex branch has a common name, this can be used and no parentheses are necessary, but you have to memorize these.

   iv. Prefixes <u>do</u> count in alphabetizing the branch names when part of a complex substituent name and inside parentheses. This is not true for simple substituents on the parent chain.

2,10-dimethyl

5-(2-methylpropyl)

7-(1-ethylbutyl)

8-(1,2-dimethylpropyl)

9-propyl

parent chain = dodecane

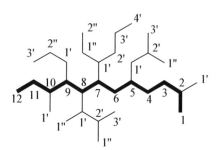

— = part of the longest chain

1' = point of attachment to longest chain

1" = point of attachment to longest branch chain

Outside the parentheses the numerical prefix does not count for alphabetical order, but inside the parentheses the numerical prefix does count for alphabetical order. If you do not write these prefix names in the correct order, your structure will still be drawn correctly from the given name.

8-(1,2-dimethylpropyl)-7-(1-ethylbutyl)-2,10-dimethyl-5-(2-methylpropyl)-9-propyldodecane

alphabetical = dimethylpropyl > ethylbutyl > dimethyl > methylpropyl > propyl
"d" (parens)   "e"   "m" (no parens)   "m-p"   "p"

Examples

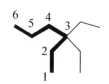

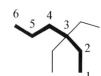

Rule 1 - The longest chain is six carbons.
Rule 2 - Number from the end closer to the two two carbon branches, as C_3 positions instead of C_4 positions.
Rule 3 - Name two carbon branches as 3-ethyl and 3-ethyl.
Rule 4 - Use the di prefix since there are two ethyl substituents.

All of the above are: 3,3-diethylhexane

There are many ways to get this name in this particular compound.

3,3-diethyl-2-methylhexane

In this example, the parent name would still be hexane (6 carbons long). This would be the only acceptable parent "hexane". The chain would have to be numbered this way because the methyl substituent gets a lower number than any other possibility, 2-methyl. The 3,3-diethyl would come alphbetically before 2-methyl, and would appear first in the name.

Choose longest Chain w/ most Branches

Chapter 4

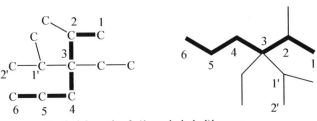

A slight additional modification produces two identical parent chains. There is now a complicated branch (a branch that has a branch) at the C3 position and parentheses are needed.

3-ethyl-2-methy-3-(1-methylethyl)hexane

Additional Examples

1.

parent = octane
C_2,C_3,C_5 substituents = 2,3,5-trimethyl
C_4,C_5 substituents = 4,5-dipropyl
alphabetical = methyl > propyl

2,3,5-trimethyl-4,5-dipropyloctane

2.

parent = undecane
C_2 substituent = 2-methyl
C_4 substituent = 4-(2-methylpropyl)
C_5 substituent = 5-(1,1-dimethylethyl)
C_8 substituent = 8-(1-methylethyl)

alphabetical = C_5 (dimethylethyl) > C_2 methyl > C_8 (methylethyl) > C_4 (methylpropyl)

5-(1,1-dimethylethyl)-2-methyl- 8-(1-methylethyl)-4-(2-methylpropyl)undecane

3.

parent = dodecane
C_6 substituent = 6-pentyl
C_6 substituent = 6-(1,2-dimethylbutyl)
C_5 substituent = 5-methyl
C_4 substituent = 4-(1-methylethyl)

alphabetical = C_6 (dimethylbutyl) > C_5 (methyl) > C_4 (methylethyl) > C_6 (pentyl)

6-(1,2-dimethylbutyl)-5-methyl-4-(1-methylethyl)-6-pentyldodecane

Problem 1 – Provide an acceptable name for the following structures.

a.

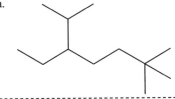

b.

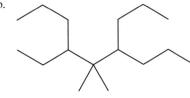

c.

d.

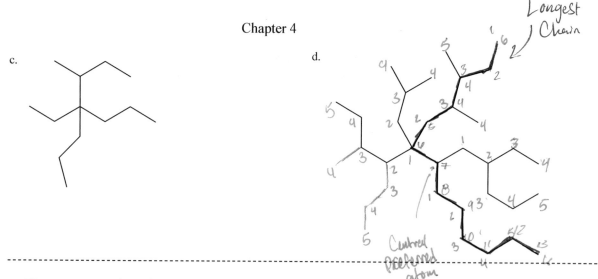

There are a number of specialized terms used to refer to the number of carbon structures and branches bonded to a specific atom (carbon or nitrogen below). For nomenclature purposes we will introduce them here, but you will also have to know some of them to predict patterns of reactivity when you start studying organic reactions later in the course. Learning these terms is memorization, just like learning a new friend's name. You have to use them to become familiar with them.

1. **Primary patterns (1°)** - If a carbon has only one other carbon attached to it, it is called a primary carbon (1°), and if a nitrogen has only one carbon attached to it, it is called a primary amine (1°). Similar terms are used with amides.

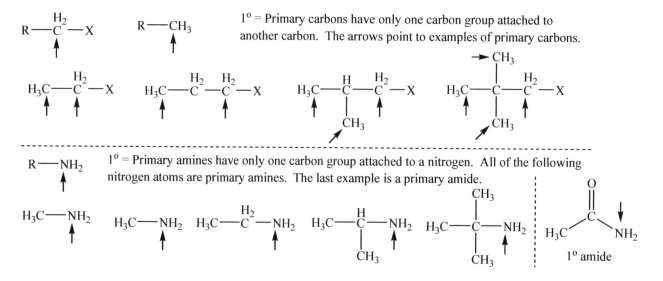

2. **Secondary patterns (2°)** - If a carbon has only two carbons attached to it, it is called a secondary carbon (2°), and if a nitrogen has only two carbons attached to it, it is called a secondary amine (2°). Similar terms are used with amides.

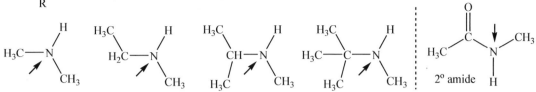

R—N—H with R below   2° = Secondary amines have two carbon groups attached to a nitrogen. All of the following
nitrogen atoms are secondary amines. The last example is a secondary amide.

3. Tertiary patterns (3°) - If a carbon has three carbons attached to it, it is called a tertiary carbon (3°), and if a nitrogen has three carbons attached to it, it is called a tertiary amine (3°). Similar terms are used with amides.

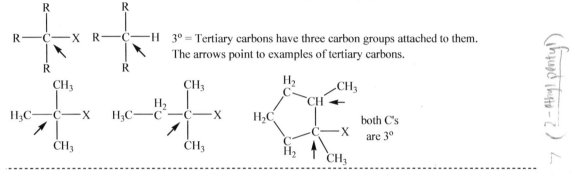

R—C—X   R—C—H (with R top and bottom)   3° = Tertiary carbons have three carbon groups attached to them.
The arrows point to examples of tertiary carbons.

(2 - ethyl pentyl)

R—N—R with R below   3° = Tertiary amines have three carbon groups attached to a nitrogen. All of the following
nitrogen atoms are secondary amines. The last example is a tertiary amide.

4. Quaternary patterns (4°) - If a carbon has four carbons attached to it, it is called a quaternary carbon (4°), and if a nitrogen has four carbons attached to it, it is called a quaternary ammonium ion and has a positive formal charge.

R—C—R (with R top and bottom)   4° = Quaternary carbons have four carbon groups attached to them.
The arrows point to examples of quaternary carbons.

R—N⊕—R (with R top and bottom)   4° = Quaternary ammonium ions have four carbon groups attached to a nitrogen. Such a
nitrogen has a positive formal charge. Both of the following nitrogen atoms are quaternary
ammonium ions.

There are also three special names for one carbon groups having different numbers of hydrogen atoms attached to them.

$$—CH_3$$

methyl,
has three hydrogen
atoms attached

$$\diagdown CH_2 \diagup$$

methylene,
has two hydrogen
atoms attached

$$—C—H$$

methine, (long i, common nomenclature
one hydrogen attached, methine is used
more frequently than the official
"methylidene" nomenclature

Frequently encountered common names for alkyl branches

1. The letter *n* (= *normal)* signifies a straight chain branch, with no subbranches on itself. If used, the "*n*" is not part of the alphabetical ordering, i.e. *n*-propyl is alphabetized under "p". Also the *n* is italicized in typing or underlined in writing. (We won't follow this rule.) It is generally assumed that if the *n* is not present, a straight chain branch is present. Its use is disappearing and we will rarely use it (common commercial product that we use: n-butyl lithium).

$$CH_3CH_2CH_2—X \qquad CH_3CH_2CH_2CH_2—X \qquad CH_3CH_2CH_2CH_2CH_2—X \qquad CH_3CH_2CH_2CH_2CH_2CH_2—X$$

n-propyl "X"   *n*-butyl "X"   *n*-pentyl "X"   *n*-hexyl "X"

2a. The "iso" pattern is a simple three carbon chain with only one branch or substituent on the middle carbon. The "iso" prefix is part of the name and is used in alphabetizing the substituent name. It is not italicized. If the "iso" pattern is present with additional carbons in a straight chain the root name includes the total number of carbons.

$$H-\underset{\underset{CH_3}{|}}{\overset{\overset{CH_3}{|}}{C}}—$$

"iso" pattern

$$H-\underset{\underset{CH_3}{|}}{\overset{\overset{CH_3}{|}}{C}}—X$$

isopropyl "X"

$$H-\underset{\underset{CH_3}{|}}{\overset{\overset{CH_3}{|}}{C}}—CH_2—X$$

isobutyl "X"

$$H-\underset{\underset{CH_3}{|}}{\overset{\overset{CH_3}{|}}{C}}—CH_2—CH_2-X$$

isopentyl "X"

$$H-\underset{\underset{CH_3}{|}}{\overset{\overset{CH_3}{|}}{C}}—CH_2—CH_2—CH_2—X$$

isohexyl "X"

"iso" is part of the name and used to alphabetize

Note - This pattern can be written and twisted in a variety of ways, but all are still called "iso".

$$H-\underset{\underset{CH_3}{|}}{\overset{\overset{CH_3}{|}}{C}}—X$$

$$H\underset{\underset{CH_3}{|}}{\overset{\overset{CH_3}{|}}{C}}—X$$

$$CH_3-\underset{\underset{CH_3}{|}}{CH}—X$$

$$CH_3-\overset{\overset{CH_3}{|}}{CH}—X$$

$$(CH_3)_2CH—X$$

b. A substituent at a secondary carbon of the four carbon straight chain is sometimes designated with special notation. Such a substituent is prefaced with "*sec*", which stands for secondary. This designation is only unambiguous for the four carbon chain. With five carbons or more it becomes ambiguous and is not used. The "*sec*" portion is italicized when typed or underlined when written. (Again, we won't follow this rule.) Also "*sec*" is not used in alphabetizing the substituent.

$$H_3C-\underset{\underset{X}{|}}{CH}—CH_2—CH_3$$

only one type of "*sec*"
position in *sec*-butyl "X"

$$H_3C-\underset{\underset{X}{|}}{CH}—CH_2—CH_2—CH_3$$

2-"X"

$$H_3C—H_2C-\underset{\underset{X}{|}}{CH}—CH_2—CH_3$$

3-"X"

sec-pentyl "X" is ambiguous

(handwritten notes:) Only used in 4 Carbon Chains

(handwritten notes:) Sec is not Part of the word, always denote w/ "sec-"

Chapter 4

3. The *tertiary* or "*tert-*" or "*t-*" pattern has three methyl groups attached to a central carbon with a fourth bond to a carbon or substituent group. The "*tert-*" or "*t-*" prefix is italicized (or underlined) and is not alphabetized in a manner similar to "*sec*". (Again, we won't follow this rule.) If the "*t*" pattern is present with additional carbons in a straight chain the <u>root</u> name includes the total number of carbons (like iso).

$$
\begin{array}{cccc}
\text{CH}_3 & \text{CH}_3 & \text{CH}_3 & \text{CH}_3 \\
| & | & | & | \\
\text{CH}_3\!-\!\text{C}\!-\! & \text{CH}_3\!-\!\text{C}\!-\!\text{X} & \text{CH}_3\text{CH}_2\!-\!\text{C}\!-\!\text{X} & \text{CH}_3\text{CH}_2\text{CH}_2\!-\!\text{C}\!-\!\text{X} \\
| & | & | & | \\
\text{CH}_3 & \text{CH}_3 & \text{CH}_3 & \text{CH}_3 \\
\end{array}
$$

"*t-*" pattern   *t*-butyl "X"   *tert*-pentyl "X"   *t*-hexyl "X"

4. "Neo" groups have three methyls and an additional fourth methylene carbon group around a central 4° carbon atom (a minimum of five carbons). The "neo-" prefix is part of the substituent name (just like iso) and is used to alphabetize the substituent. It is not italicized or underlined. The central carbon with four other carbons attached is a quaternary carbon. If the "neo" pattern is present with additional carbons in a straight chain the <u>root</u> name includes the total number of carbons (like iso).

$$
\begin{array}{cccc}
\text{CH}_3 & \text{CH}_3 & \text{CH}_3 & \text{CH}_3 \\
| & | & | & | \\
\text{CH}_3\!-\!\text{C}\!-\!\text{CH}_2\!-\! & \text{CH}_3\!-\!\text{C}\!-\!\text{CH}_2\!-\!\text{X} & \text{CH}_3\!-\!\text{C}\!-\!\text{CH}_2\!-\!\text{CH}_2\!-\!\text{X} & \text{CH}_3\!-\!\text{C}\!-\!\text{CH}_2\!-\!\text{CH}_2\!-\!\text{CH}_2\!-\!\text{X} \\
| & | & | & | \\
\text{CH}_3 & \text{CH}_3 & \text{CH}_3 & \text{CH}_3 \\
\end{array}
$$

"neo" pattern   neopentyl "X"   neohexyl "X"   neoheptyl "X"

"neo" is part of the name and used to alphabetize

    If there is a choice (e.g. you are naming the compound), it is preferable to use systematic nomenclature and to not use the trivial names above. However, when familiar complex substituents are present as branches on larger parent chains, the temptation to use the simple common names is often overwhelming to avoid the complex names with their multiple numbers, branches and parentheses. Also, other people use the trivial names and you may need to interpret such nomenclature (*n*-butyl lithium, *sec*-butyl lithium, *t*-butyl lithium). Learning the above terms is memorization, just like learning a new friend's name. You have to use them to become familiar with them. Work your way around the following oval several times and you will know them.

Iso - All but one Carbon form cont. chain, iso is part of the name.

Neo - All but two " , neo is part of the name

Sec - Only on a carbon chain & a group is bonded to Secondary Carbon, "Sec -"

Tert - only when Func. group is bonded to tertiary Carbon, "t -"

$$
\begin{array}{ccccc}
\text{CH}_3 & \text{CH}_3 & & & \text{CH}_3 \\
| & | & & & | \\
\text{H}-\text{C}- & \text{CH}_3-\text{C}-\text{CH}_2- & \text{H}_3\text{C}-\text{CH}-\text{CH}_2-\text{CH}_3 & & \text{CH}_3-\text{C}- \\
| & | & | & & | \\
\text{CH}_3 & \text{CH}_3 & \text{X} & & \text{CH}_3 \\
\end{array}
$$

"iso"   "neo"   "Sec -"   "t -"
or
"tert-"

Chapter 4

Problem 2 - Identify each of the substituent patterns below by its common name. Point out an example of a 1°, 2°, 3° and 4° carbon and nitrogen. Also, point out an example of a methyl, methylene and methine (methylidene) position.

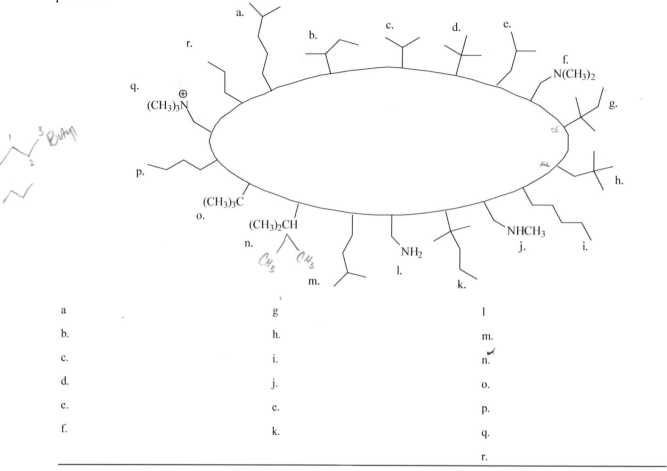

a	g	l
b.	h.	m.
c.	i.	n.
d.	j.	o.
e.	e.	p.
f.	k.	q.
		r.

Naming Cycloalkanes (Rings)

A ring is formed when a chain of atoms connects back on itself. This is possible with as few as three atoms and can be of unlimited size. There are ring structures in the literature numbering in the hundreds. There are only minor differences in the nomenclature of cycloalkanes and acylic alkanes. Often a ring can be drawn in more than one way. You just have to count the carbons to know what the parent name will be.

1. Attach a prefix of cyclo- to the name of a cyclic alkane possessing the same number of carbons as the straight chain name. A cycloalkane has the carbons connected in a ring. The smallest possible ring has 3 carbons. The following are unsubstituted cycloalkanes up to C_{12}.

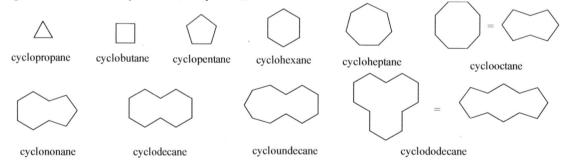

Chapter 4

2. With only one alkyl branch, name the structure as an "alkylcycloalkane". No number is necessary since there is only one substituent present.

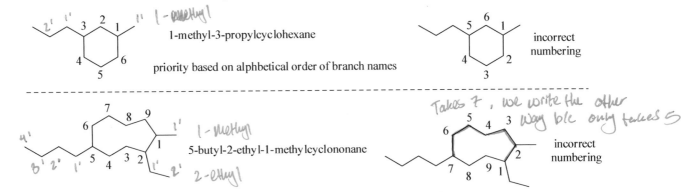

t-butylcyclopentane

1,1-dimethylethylcyclopentane

pentylcyclohexane

3. With two or more alkyl substituents, number about the ring so as to obtain the lowest number possible at the first point of difference. In cycloalkanes, alphabetic order is used to determine the number one position if the numbers are similar. In the second example below, numbering is based on lowest number, not alphabetically.

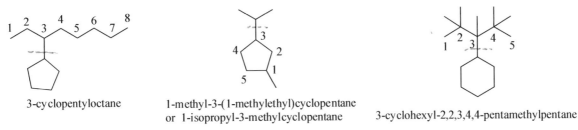

1-methyl-3-propylcyclohexane

priority based on alphbetical order of branch names

incorrect numbering

5-butyl-2-ethyl-1-methylcyclononane

incorrect numbering

Priority based on lowest possible number at first point of difference (5 versus 7 for butyl).
Alphbetical order is used for the branch names

4. If a chain component has more carbons than the ring, then the chain may become the parent name and the cyclic portion a branch (ends in "-yl"). However, if it is easier to name as a cycloalkane, that is acceptable.

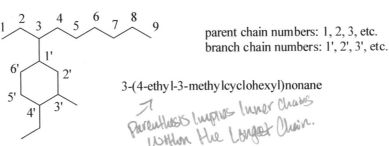

3-cyclopentyloctane

1-methyl-3-(1-methylethyl)cyclopentane
or 1-isopropyl-3-methylcyclopentane

priority based on alphbetical
order of branch names

3-cyclohexyl-2,2,3,4,4-pentamethylpentane

5. If the ring is classified as the substituent, the C_1 carbon of the ring (for numbering purposes) is the carbon directly attached to the parent chain. Number around the ring in the direction to give the lowest number at the first point of difference (or the lowest number to the highest priority feature in the ring, once functional groups are included).

parent chain numbers: 1, 2, 3, etc.
branch chain numbers: 1', 2', 3', etc.

3-(4-ethyl-3-methylcyclohexyl)nonane

6. When two substituent groups are on the same side of a cyclic structure (both on top or both on bottom), the "*cis*" prefix is used. When two groups are on opposite sides (one on top, one on bottom) and on different carbons, the "*trans*" prefix is used. These terms go just before the designating number and are separated from the number by a hyphen. The terms "*cis*" and "*trans*" are italicized (or underlined, if written) and not used in alphabetizing. (We won't follow this rule.)

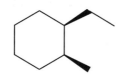

cis-1-ethyl-2-methylcyclohexane   *trans*-1,2-dimethylcyclohexane   ✳ cis = same side   ✳ trans = opposite sides

7. We won't use the cis/trans rule when more than two substituents are present in a ring at different carbons.

<u>Problem 3</u> – Provide an acceptable name for each of the following structures.

a.   b.   c.   d.

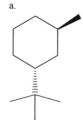

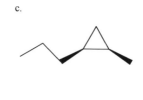

II. Nomenclature Rules For Alkenes

1. The parent name will be the longest carbon chain that contains both carbons of the double bond. Drop the -ane suffix of the alkane name and add the –ene suffix. Never name the double bond as a prefix. If a double bond is present, you have an alkene, not an alkane.

$$\text{alk\sout{ane}} + \text{-ene} = \text{alkene}$$

2. Begin numbering the chain at the end nearest the double bond. Always number through the double bond and identify its position in the longest chain with the <u>lower</u> number. In the older IUPAC rules (American system) the number for the double bond was placed in front of the stem name with a hyphen. Under the newer rules (European or British system), the number for the double bond is placed right in front of "ene", with hyphens. We will use the newer rules for specifying the location of pi bonds.

$$
\begin{array}{cccccc}
1 & 2 & 3 & 4 & 5 & 6
\end{array}
$$
$$H_3C\!-\!CH\!=\!CH\!-\!CH_2\!-\!CH_2\!-\!CH_3$$

hex-2-ene (newer rules)

2-hexene (older rules)

Chapter 4

3. Indicate the position of any substituent group by the number of the carbon atom in the parent (longest) chain to which it is attached.

$$\underset{1}{H_3C}-\underset{2}{CH_2}-\underset{3}{CH}=\underset{4}{CH}-\underset{5}{CH}-\underset{6}{\overset{CH_3}{CH}}-\underset{7}{CH_3}$$
$$\underset{6}{\overset{|}{CH_3}}$$

5,6-dimethylhept-3-ene (newer rules)

5,6-dimethyl-3-heptene (older rules)

Numbering is determined by the double bond, not the branches, because the double bond has higher priority than any alkyl branch.

4. Number cycloalkenes so that the double bond is 1,2 (number through the double bond). Number in the direction about the ring so that the lowest number is used at the first point of difference.

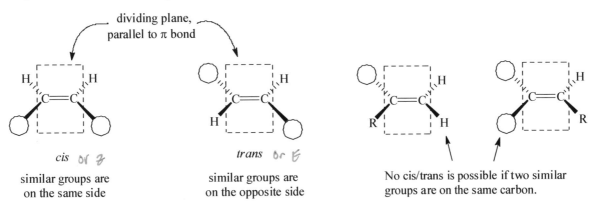

always number through the double bond.
B/c of Priority Group.

Started here, b/c closer to the Branch.

cyclohexene

(No number is needed because it has to be "1".)

1-methylcyclohex-1-ene

(Numbers are optional here.)

3-methylcyclohex-1-ene

(Numbered to give the methyl the lowest number.)

1,6-dimethylcyclohex-1-ene

(The first methyl decides the direction of the numbering.)

5. One method to designate the geometry about a carbon-carbon double bond having two different groups on each carbon is by using "*cis*" (if two similar groups are on the same side) or "*trans*" (if two similar groups are on opposite sides). The two sides referred to are determined by a plane running through the two, connected carbons of the double bond, parallel to the p orbitals of the pi bond. If *cis/trans* identifiers are used, the substituent groups are usually obvious. Note that if there are two identical groups attached to the same carbon of the double bond, then *cis/trans* structures do not exist (they are not possible). Cis/trans isomers can also be called "geometric isomers". They represent a special type of stereoisomer and can also be called diastereomers. We will study distereomers in the chapter on stereochemistry.

dividing plane, parallel to π bond

cis or Z

similar groups are on the same side

trans or E

similar groups are on the opposite side

No cis/trans is possible if two similar groups are on the same carbon.

If four different groups are attached to the carbons of the double bond, you need to use the E/Z nomenclature convention (you can also use this system even if only two groups are present). In the E/Z system of classification the atoms attached to each carbon of a double bond are assigned a **priority** based on atomic number. Higher atomic number atoms are assigned a higher priority. In organic chemistry we are mostly concerned with the halogens, oxygen, nitrogen, hydrogen and occasionally sulfur and phosphorous. It would seem like nothing could be lower than a hydrogen, but a lone pair of electrons is the lowest of all (...so I > Br > Cl > S > P > F > O > N > C > H > lone pair of electrons). If the attached groups at carbon are different, but the directly attached atoms are identical, the highest priority path in each group must be followed until a distinction can be made.

* The Big Groups Are On the Same Side.

The relative positions of the high priority groups at each carbon of the double bond are compared. If the two high priority groups (on different carbons) are on the same side of the double bond, they are classified as Z from the German word **zussammen**, meaning together (…notice "same" in the middle of zussamen?…or think "Z-BIG groups are on Z-SAME side", said with a French accent). If the high priority groups (on different carbons) are on opposite sides of the double bond, then the double bond is classified as E from the German word **"entegegan"**, for opposite. To interconvert one geometric isomer to the other requires enough energy to break the pi bond (typically, over 60 kcal/mole). Background thermal energy is about 20 kcal/mole, so this interchange is very, very slow. Each carbon of the double bond with the two different groups is called a **stereogenic center (or atom)**, because if the two groups are switched, a new stereoisomer is formed. The symbols E and Z specify the **configuration** of the double bond (configuration can be thought of as a relatively stable stereochemical arrangement of atoms in our world, not to be confused with conformation = single bond rotations).

Priority is the atomic #. Highest atomic # wins

Z stereochemistry
high priority groups
on the same side

E stereochemistry
high priority groups
on opposite sides

① = higher priority group

② = lower priority group

1 & 2 are equal, so No E/Z

The priority rules are quite extensive to cover all of the possibilities (to classify every E/Z scenario…and later, R/S configurations). We will not look at all such possibilities, but will focus on some that are more common. These rules arose from a state of confusion that existed up to about the middle 1950's, when three chemists cleared things up with a set of defining rules. Sidney Cahn, Christopher Ingold and Vladimir Prelog were the chemists and that has become the name of the defining rules ("CIP" for short).

Some basics rules of the CIP (Cahn-Ingold-Prelog) sequence rules for defining priorities are listed below (E/Z here and R/S in the stereochemistry topic). We will elaborate these rules in our chapter on stereochemistry.

1. Higher atomic number = higher priority (typical possibilities in organic chemistry: $I > Br > Cl > S > P > F > O > N > C > H >$ lone pair)

2. If two (or more) directly attached atoms are the same, then compare the atoms attached to them in order of decreasing priority, (based on atomic number). Make a priority designation at the first point of difference. Always follow the highest priority path along each atom to make the designation. A single higher priority atom takes precedence over any number of lower priority atoms.

3. Double and triple bonds are treated as if they are bonded to the same atoms two or three times, respectively. However, those hypothetical atoms do not continue beyond that connection, as real atoms would.

Example

$C_1(F,C,H)$ ①
$C_2(H,H,H)$

decided on first carbon

$C_1(C,H,H)$ ②
$C_2(Br,Br,Br)$

$C_1(C,C,C)$
$C_2(C,H,H)$ ②

decided on second carbon

$C_1(C,C,C)$ ①
$C_2(C,C,H)$

Chapter 4

<u>Problem 4</u> - Classify the order of priorities among each group below (1 = highest). The "*" specifies attachment to a stereogenic center.

a..   * —NH$_2$ * —CH$_2$F   * —CH$_2$CH$_2$Cl   * —CH$_2$CH$_2$CH$_2$I

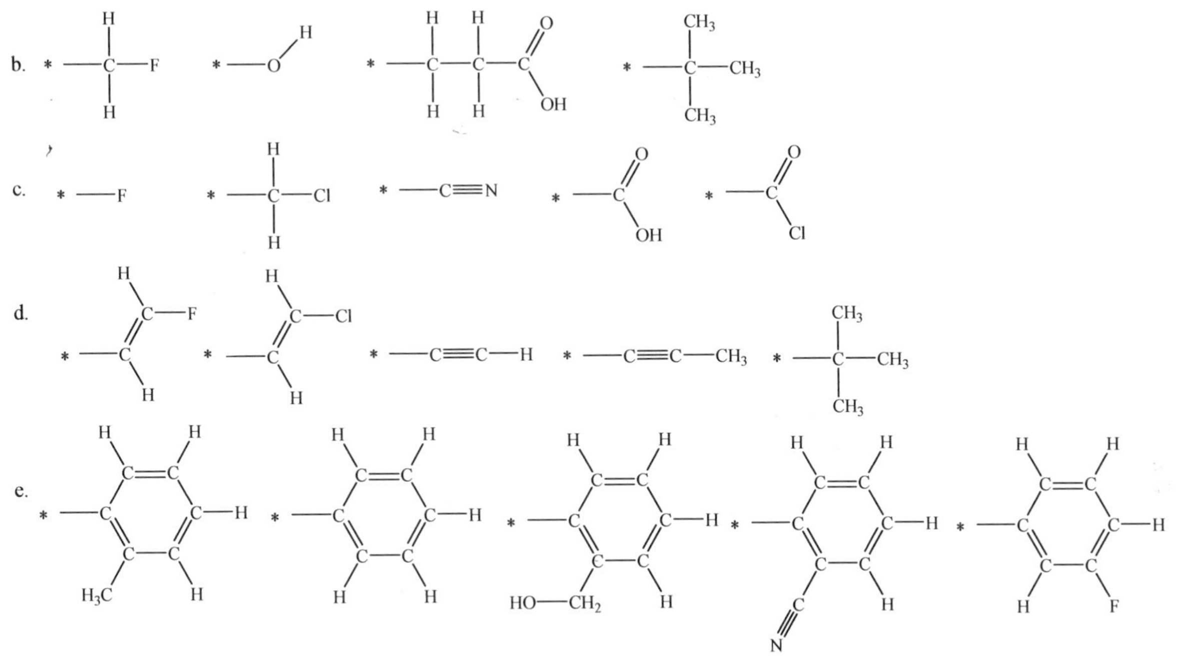

b. * —C(H)(H)—F   * —O—H   * —C(H)(H)—C(H)(H)—C(=O)OH   * —C(CH$_3$)(CH$_3$)—CH$_3$

c. * —F   * —C(H)(H)—Cl   * —C≡N   * —C(=O)—OH   * —C(=O)—Cl

d. * —C(H)=C(H)—F   * —C(H)=C(H)—Cl   * —C≡C—H   * —C≡C—CH$_3$ * —C(CH$_3$)(CH$_3$)—CH$_3$

e. (five substituted benzene ring structures)

<u>Problem 5</u> - Classify each alkene, below, as E, Z or no stereochemistry present.

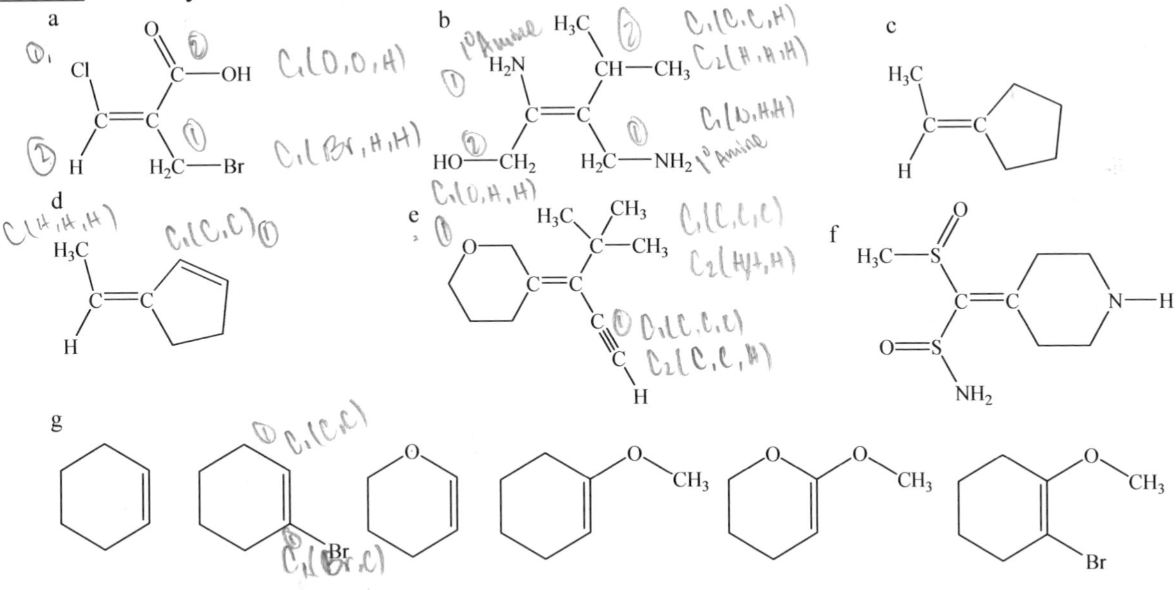

a   ② / ①   C,(O,O,H)   C,(Br,H,H)   C,(O,H,H)   C(H,H,H)   C,(C,C,C) ②

b   1° Amine   ②   C,(C,C,H)   C$_2$(H,H,H)   ①   C,(D,H,H)   ① 1° Amine   C,(O,H,H)

c

d   ① C(H,H,H)   C,(C,C) ①

e   ②   C,(C,C,C)   C$_2$(H,H,H)   ① C,(C,C,C)   C$_2$(C,C,H)

f

g   ① C,(C,C)   C,(Br,C)

All small rings are 'cis' inside the ring, but not all are Z. The E/Z terms are part of a classification system.
If you change any atom in a structure you may alter the priorities and change the absolute configuration (E or Z).

6. Rings having a double bond and seven carbons (atoms) or less are only stable with a *cis* geometry (in the ring). The *cis* designation is almost never explicitly written in such rings, but understood to be required.

These are all known to be cis (in the ring), so no identifier is required.

cyclopropene   cyclobutene   cyclopentene   cyclohexene   cycloheptene

Even though "cis" in the ring, the stereochemistry can still be E.

It is too difficult for trans groups to connect to one another until there are at least 8 atoms. Even in a cyclooctene, the trans stereoisomer is about 10 kcal/mole less stable than the cis isomer. However, if the ring gets big enough, the trans isomer becomes the more stable stereoisomer, as is our usual expectation.

1Z-alkene
alk-1Z-ene

cyclooct-1Z-ene

cyclooct-1E-ene

= rear orbital

= front orbital

Small ring trans-cycloalkenes are less stable because of loss of overlap of p orbitals in the pi bond due to distortion of the connecting bridge, which raises their potential energy and makes them less stable.

7. If more than one double bond is present, the prefixes di-, tri-, tetra-, etc. are used for 2, 3, 4, etc. of the double bonds. An "a" is added between the "alk" stem and the "ene" suffix part for better phonetics (i.e., alkadiene, alkatriene, etc.) A number is required for each double bond present. Place any numbers right in front of the "ene" prefix. (Remember, one always uses the lower of the two numbers possible for each occurrence of a double bond.) The *cis/trans* nomenclature should be indicated just previous to each number, when relevant. A hyphen separates the letters from the numbers. *Cis* and *trans* are italicized when typed or underlined when written, and they are not alphabetized when applicable. (We won't follow this rule.) If E/Z terms are used to specify the pi bond configuration, they can be written just after the number or all of the configurations can be placed in parentheses in the front of the name.

$$\text{alk} \; + \; \text{a} \; + \; \text{diene} \; = \; \text{alka-\#,\#-diene}$$

$$\text{alk} \; + \; \text{a} \; + \; \text{triene} \; = \; \text{alka-\#,\#,\#-triene}$$

$$\text{alk} \; + \; \text{a} \; + \; \text{tetraene} \; = \; \text{alka-\#,\#,\#,\#-tetraene}$$

3-ethyl-3-methylocta-1,4Z,6E-triene

3-ethyl-3-methylocta-1-*cis*-4-*trans*-6-triene

(4Z,6E)-3-ethyl-3-methylocta-1,4,6-triene

hepta-1,3Z,5E-triene

hepta-1-*cis*-3-*trans*-5-triene

(3Z,5E)-hepta-1,3,5-triene

cycloocta-1,3,5,7-tetraene

(1Z,3Z,5Z,7Z)-cycloocta-1,3,5,7-tetraene

Chapter 4

8. When double bonds are present in lower priority substituent branches, they are named as indicated above with the C_1 carbon being the point of attachment to the parent chain. As a substituent, the name will end in "yl" as in alkenyl, not alkene. Two common alkene branch names you should know are "vinyl" and "allyl".

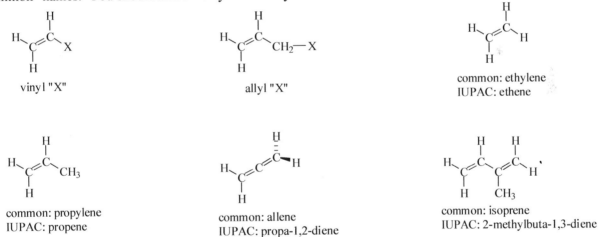

Very important

high priority
chain = ⌁⌁⌁⌁⌁

#

#-ethenyl
or #-vinyl

#

#-(prop-2-enyl)
or   #-allyl

#

#-(prop-*trans*-1-enyl)
#-(prop-1E-enyl)

#

#-(prop-*cis*-1-enyl)
#-(prop-1Z-enyl)

#

#-(5,5-dimethylcyclopent-2-enyl)

No E/Z is necessary because
of the small size of the ring.

#

#-(4-methylpent-*trans*-2-enyl)
#-(4-methylpent-2E-enyl)

The following are some examples of frequently encountered alkene substituent and compound "trivial" or "common" names. You should know "vinyl" and "allyl" substituents.

vinyl "X"

allyl "X"

common: ethylene
IUPAC: ethene

common: propylene
IUPAC: propene

common: allene
IUPAC: propa-1,2-diene

common: isoprene
IUPAC: 2-methylbuta-1,3-diene

Examples You should know how to name all of the following.

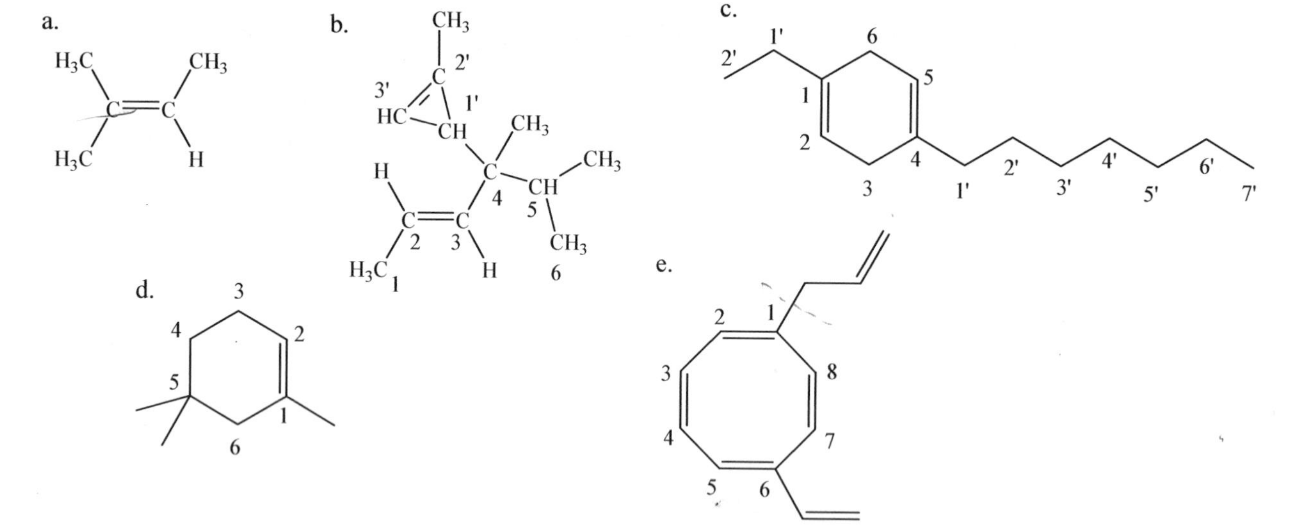

a.

b.

c.

d.

e.

a. 2-methylbut-2-ene   (no *E/Z* or *cis/trans* possible with two identical groups on one carbon)

b. 4-(2-methylcycloprop-2-enyl)-4,5-dimethylhex-2*E*-ene

c. 1-ethyl-4-heptylcyclohexa-1,4-diene   (numbers based on alphabetical order)

d. 1,5,5-trimethylcyclohex-1-ene   (numbers based on the first methyl group and number through C=C)

e. 1-allyl-6-vinylcycloocta-1,3,5,7-tetraene or 1-ethenyl-6-(prop-2-enyl)cycloocta-1,3,5,7-tetraene

Problem 6 – Provide an acceptable name for each of the following structures.

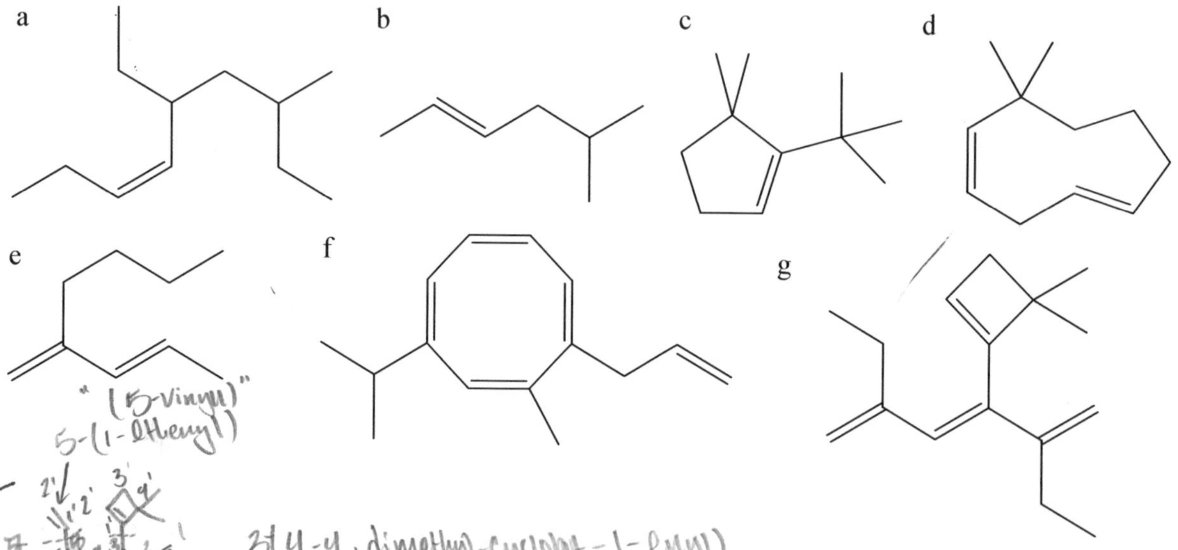

a

b

c

d

e

f

g

"(5-vinyl)"
5-(1-ethenyl)

10/17

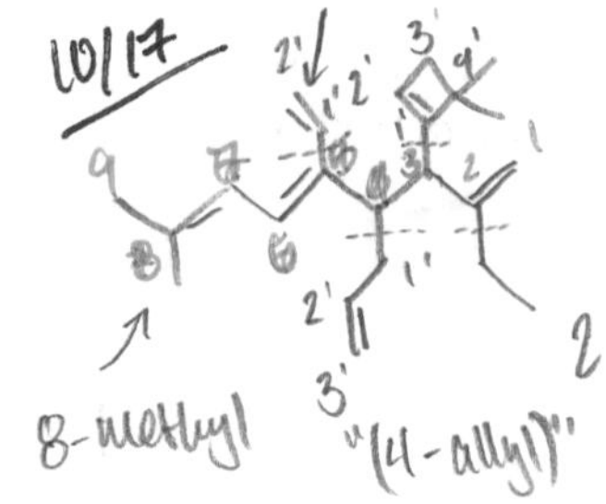

3(4-4,dimethyl-cyclobut-1-enyl)

2 ethyl

nona-1,3Z,5E,7-tetraene

8-methyl

"(4-allyl)"

4-(prop-2-enyl)

Chapter 4

III. Nomenclature Rules For Alkynes

1. Pick the longest carbon chain which contains the triple bond as the parent name. Drop the -ane suffix of the alkane name and add -yne.

> alk~~ane~~ + yne = alkyne

2. Begin numbering the chain at the end nearest the triple bond to assign it the lowest possible number. Write the number immediately preceding "yne" with hyphens on both sides. Triple bonds do not have E/Z stereoisomer possibilities, because of their linear shape, so these terms are not needed.

> $R—C≡C—R$ No E/Z stereoisomers possible

3. Indicate any substituent groups by the number of the carbon atom in the longest chain to which they are attached.

7-methyl
6(1-methylethyl)

6-ethyl-7-methyloct-3-yne

4. Number cycloalkynes so that the triple bond is 1,2 (number through the triple bond). Number in the direction about the ring so that the lowest number is used at the first point of difference. Like the cycloalkenes, there are no stable cycloalkynes until there are at least 8 atoms in the ring (cyclooctyne). It takes that many atoms to reach around and attach the ends to one another. Even with eight atoms in the ring the linear shape of the triple bond is distorted, producing a higher potential energy state and more reactive triple bond. (Build a model and look at it.)

At least 8 atoms are needed in a ring containing a triple bond to connect the opposite sides of the triple bond and even then there is a fair amount of strain energy present in the ring.

front-on view

top-down view

5. If more than one triple bond is present, a numerical prefix is used to indicate the appropriate number of triple bonds. Just as with alkenes, an "a" is added between the "alk" stem and the "yne" suffix part for better phonetics (i.e., alkadiyne, alkatriyne, etc.) A position number is provided for each occurrence. Hence, alka-#,#-diyne, alka-#,#,#-triyne, alka-#,#,#,#-tetrayne, etc. are used for 2, 3, 4, etc. of the triple bonds.

> alk + a + diyne   =   alka-#,#-diyne
>
> alk + a + triyne   =   alka-#,#,#-triyne
>
> alk + a + tetrayne =   alka-#,#,#,#-tetrayne

6. When triple bonds are present in lower priority substituent branches, they are named as indicated above with the C_1 carbon being the point of attachment to the parent chain. As a substituent, the name will end in "yl" as in alkynyl, (not alkyne).

1-methyl-3-(4-methylpent-2-ynyl)cyclohexa-1,3-diene

The following are some examples of frequently encountered compounds and one alkynyl substituent "trivial" or "common" names. You should know "propargyl".

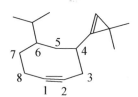

H—C≡C—CH$_2$

propargyl "X"

H—C≡C—H

common: acetylene
IUPAC: ethyne

R—C≡C—R

common: dialkylacetylene

CH$_3$—C≡C—CH$_2$CH$_3$

common: ethylmethylacetylene
IUPAC: pent-2-yne

Examples You should know all of these.

a.

b.

c.

d.

e.

a. 4-ethyl-5-methylhex-2-yne

b. 4-(3,3-dimethylcycloprop-1-enyl)-6-isopropylcyclooct-1-yne

c. 2,2,7-trimethylocta-3,5-diyne

d. 1-cyclopentyl-3-ethyl-4,4-dimethylpent-1-yne

e. 5-(1-methylpent-2-ynyl)-1-(2-methylprop-2-enyl)cyclohex-1-ene

Problem 7 - Provide an acceptable name for each of the following structures. (Know how to do all of these.)

a

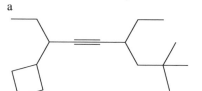

b

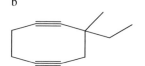

c

d

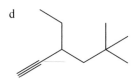

e

Chapter 4

IV. Nomenclature Rules If Both Alkene(s) and Alkyne(s) Are Present

If both double and triple bonds are present, the parent name becomes alk-#-en-#-yne. The alkene part always comes before the alkyne part. Numbers are required for every pi bond. The numbers for any alkene(s) are placed directly in front of "ene" and separated from the name with hyphens and the number for any alkyne(s) are placed prior to "yne", also separated by hyphens. Because the trailing "yne" begins with a vowel (y), the final "e" of "ene" is dropped. If more than one double or triple bond is present, the appropriate numerical prefixes are also used (di-, tri-, etc.). If an alkene has stereogenic centers use the appropriate E/Z designation with its location number.

* when talking about Ene vs yne
Ene > yne unless yne can
be lower than Ene for Numbering.

alk-#-en-#-yne

alka-#,#-dien-#,#,#-triyne

All other things being equal, <u>the alkene is higher priority than the alkyne for numbering purposes</u>. The structures below would all start with the #1 position at the first carbon of the alkene pi bond.

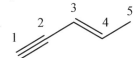

non-1-en-8-yne

hexa-1-*trans*-3-dien-5-yne

(3E)-hexa-1,3-dien-5-yne

hexa-1,3E-dien-5-yne

cycloundec-*cis*-1-en-4-yne

(1Z)-cycloundec-1-en-4-yne

cycloundec-1Z-en-4-yne

However, if a lower initial number can be used by assigning the triple bond a higher priority, then it takes precedence. Alternatively, if an alkene and an alkyne are equally positioned, a substituent can determine the direction of the numbering in order to assign it the lowest possible number. The following examples should clarify this point.

pent-3E-en-1-yne
pent-*trans*-3-en-1-yne

The priority changes because
a lower first number can be
assigned to the triple bond.

4,4-dimethylnon-8-en-1-yne

The priority changes due to
lower numbers for the methyls.

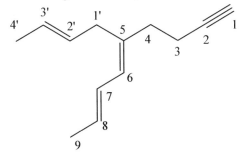

5-(but-2E-enyl)nona-5Z7E-dien-1-yne

The priority changes because a lower first number can be assigned
to the triple bond. The nine carbon chain with two double bonds is
chosen over the nine carbon chain with one double bond.

5-ethenylocta-1,3Z-dien-6-yne

The 6,7 triple bond is chosen over the
6.7 double bond because it is a longer chain.

Chapter 4

Examples of " ene " plus " yne " hydrocarbons (You should know all of these.)

a.

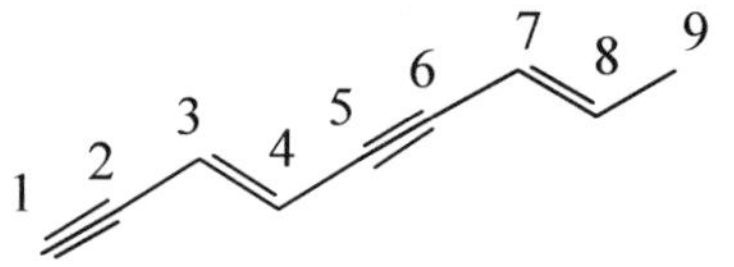

b.

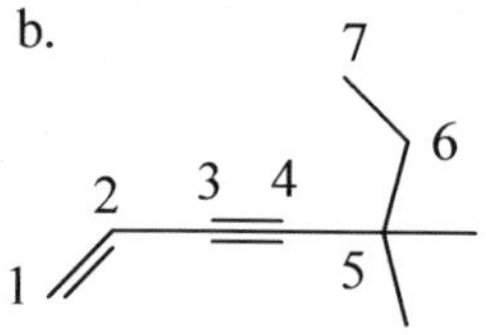

c.

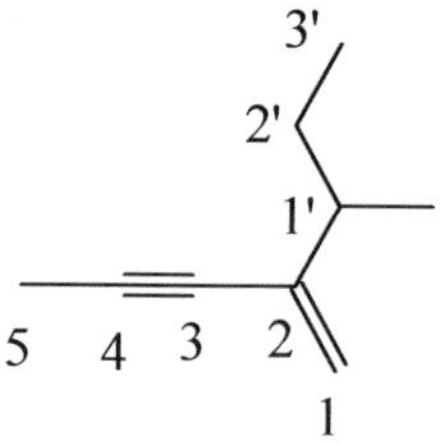

d.

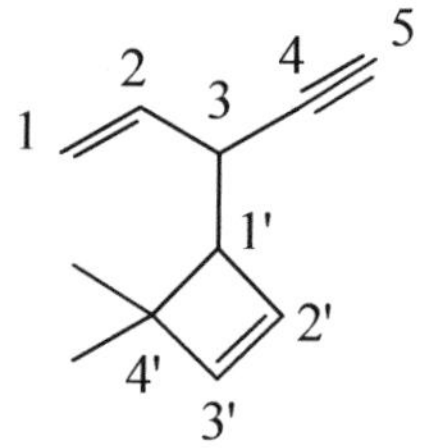

a.   nona-3E,7E-dien-1,5-diyne   or   (3E,7E)- nona-3,7-dien-1,5-diyne

b.   5,5-dimethylhept-1-en-3-yne

c.   2-(1-methylpropyl)pent-1-en-3-yne   or   2-*sec*-butylpent-1-en-3-yne

d.   3-(4,4-dimethylcyclobut-2-enyl)pent-1-en-4-yne   (a tie goes to the alkene)

Problem 8 - Provide an acceptable name for each of the following structures.

a.

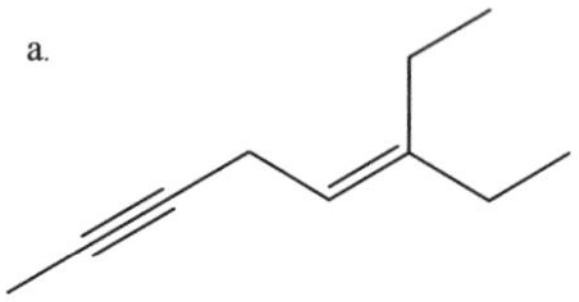

b.

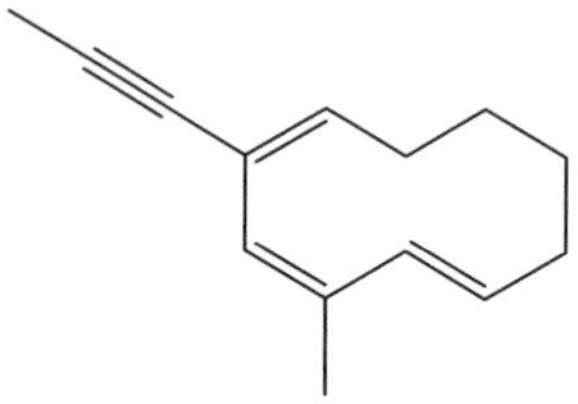

c.

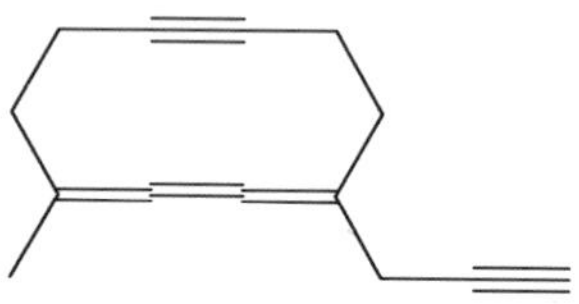

d.

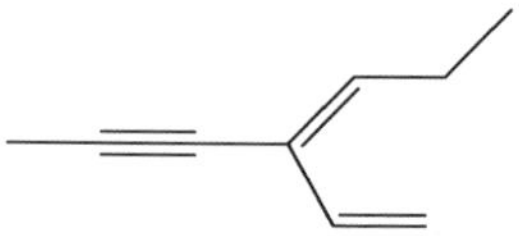

e.

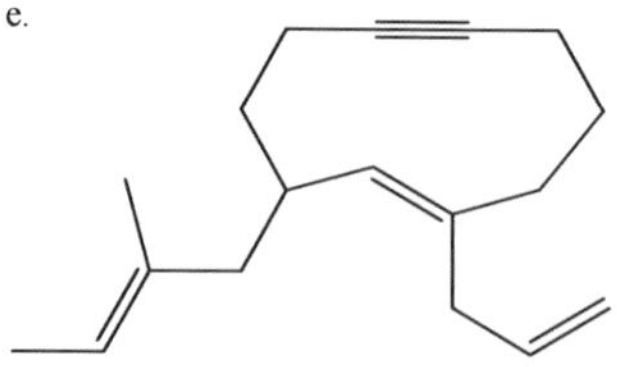

f.

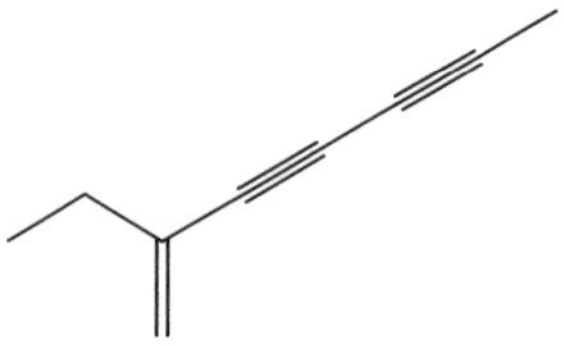

g.

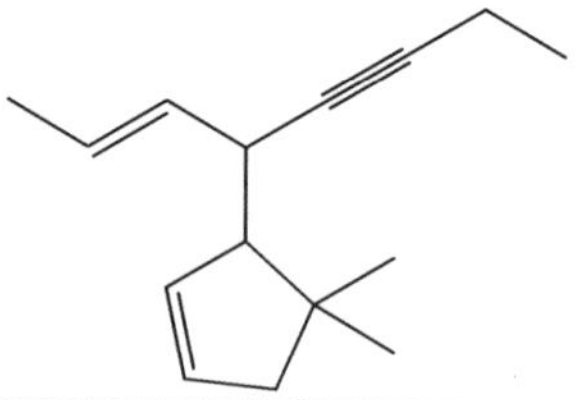

V. Functional Group Nomenclature

Many commonly encountered functional groups in organic chemistry are discussed below. Only a "t bones" list is provided showing the functional groups in their relative priorities from top to bottom (highest priority = first on the list to lowest priority = last on the list). A very brief overview of their nomenclature is provided showing the prefix and suffix terms associated with each functional group. The highest priority group in a structure is indicated by a suffix and a designating number. (If the number can only be #1 it is usually left out.) In compounds where more than one functional group is present, lower priority functional groups are indicated with prefixes preceded by numbers identifying each substituent's position on the parent chain. Alkene and alkyne functionality is always indicated at the end of the name (as a suffix), even if both are present. This means there could be as many as three suffix components if an alkene, alkyne and a high priority functional group are all present.

To form the parent name, drop the -e ending if the suffix of the highest priority group begins with a vowel (most do) from the alkane, alk-#-ene, alk-#-yne or alk-#-en-#-yne name. If the suffix begins with a consonant, retain the final e of the parent name and simply append the suffix. In this book the only two functional group examples like this are nitrile and thiol.

Try to adopt a systematic (repetitive) strategy to approach each nomenclature problem. If you do this every time you work on a nomenclature problem, it won't take many efforts until nomenclature problems become a routine effort, devoid of any mystery.

Basic Strategy for Naming Structures – Approach every nomenclature problem the same way.

1. Identify the highest priority group and name the longest continuous chain of carbons containing the highest priority group so that it is assigned the lowest number possible. Lesser priority groups are still considered when secondary choices can be made at branches to decide what is the longest chain. Include these lesser priority groups in the longest chain when possible.

2. Use the numbers in the longest chain to indicate the position of all branches and substituents. Lower priority groups are indicated with prefixes at the beginning of the name, along with any necessary numbers to indicate their position(s). The prefixes should be placed in the name in alphabetical order for referencing purposes. If the prefixes are not in alphabetical order, a correct structure can still be drawn, however, you might have a problem looking up a structure in alphabetized lists in the literature.

3. Indicate any double bonds and triple bonds with suffixes, in addition to the high priority group suffix. Each occurrence of these will get a number preceding it to indicate its position on the parent chain. If more than one multiple bond or group is present, indicate this with the numerical prefix of the suffix name (i.e., two double bonds = -#,#-diene, three triple bonds = -#,#,#-triyne, four alcohol groups = -#,#,#,#-tetraol, etc.). If both a double and triple bond are present, the parent name is alk-#-en-#-yne. If two doubles, three triples and an alcohol group are present, the parent is alka-#,#-dien-#,#,#-triyn-#-ol). Stereochemical identifiers (E/Z and R/S) can be grouped together in parentheses in front of the name with their appropriate numbers.

In addition to several alkane, alkene and alkyne branch names listed earlier there are two commonly encountered aromatic branch names we use in our course.

Chapter 4

Common Functional Groups

Most of the following functional groups have more details than is specified in the examples provided in this topic. More groups are provided in the priority list below than are emphasized in our course. Also, there are even more functional groups not discussed by us. Only the more commonly encountered groups are included here. The functional groups you are responsible for are listed in **bold**. There are 12 functional groups bolded, plus 5 lower priority substituents at the end. The first functional group you are responsible for in this list is the carboxylic acid functional group, number 4.

Order of Priorities of Organic Functional Groups (You are only responsible for those in **bold**.)

1. Free Radicals– not covered by us.

General Structure	Prefix Name (if lower priority)	Suffix Name (if higher priority)	Example
	none given	alkyl radical	2-methyl-2-butyl radical

2. Cationic Compounds– not covered by us.

a. carbon– not covered by us.

General Structure	Prefix Name (if lower priority)	Suffix Name (if higher priority)	Example
	none given	alkyl cation	2-methyl-2-butyl cation

b. Nitrogen– not covered by us.

General Structure	Prefix Name (if lower priority)	Suffix Name (if higher priority)	Example
	ammonio-	-ammonium	ethyltrimethylammonium

c. Oxygen– not covered by us.

General Structure	Prefix Name (if lower priority)	Suffix Name (if higher priority)	Example
	oxonio-	-oxonium	triethyloxonium

3. Anionic Compounds– not covered by us.

a. Carbon– not covered by us.

General Structure	Prefix Name (if lower priority)	Suffix Name (if higher priority)	Example
	none given	alkyl anion / -ide	2-methyl-2-butyl anion / 2-methyl-2-butanide

b. Nitrogen– not covered by us.

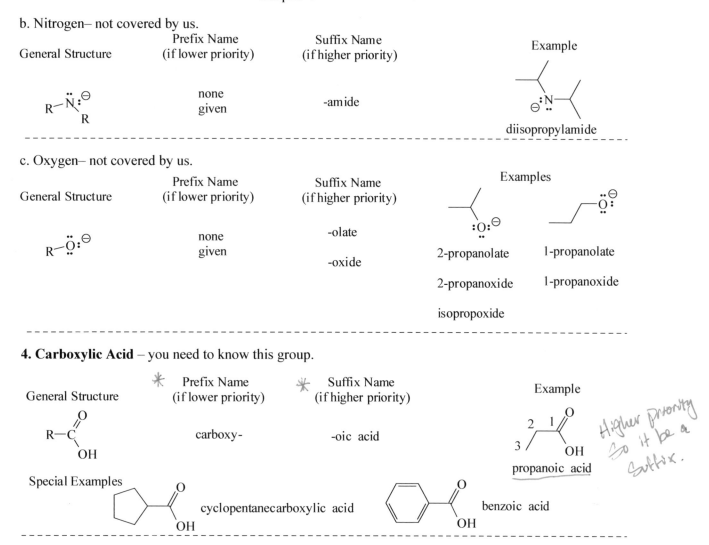

General Structure | Prefix Name (if lower priority) | Suffix Name (if higher priority) | Example

none given | -amide | diisopropylamide

c. Oxygen– not covered by us.

General Structure | Prefix Name (if lower priority) | Suffix Name (if higher priority) | Examples

none given | -olate / -oxide | 2-propanolate 1-propanolate / 2-propanoxide 1-propanoxide / isopropoxide

4. Carboxylic Acid – you need to know this group.

General Structure | * Prefix Name (if lower priority) | * Suffix Name (if higher priority) | Example

carboxy- | -oic acid | propanoic acid

Higher priority so it be a suffix.

Special Examples: cyclopentanecarboxylic acid   benzoic acid

Drop the -e ending of the longest chain having the carboxylic acid functional group and add the **-oic acid** suffix. The acid functional group is usually at the #1 position of the longest chain and does not require a position number. Since it is our highest priority group we will never use the prefix name. Cover the right side, below, and try to name each structure on your own.

methanoic acid
(common: formic acid)

2-methylbutanoic acid

2-ethyl-3-methylpent-2E-enoic acid

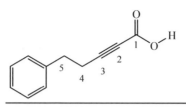

5-phenylpent-2-ynoic acid

---

5. Peroxycarboxylic Acids – not covered by us.

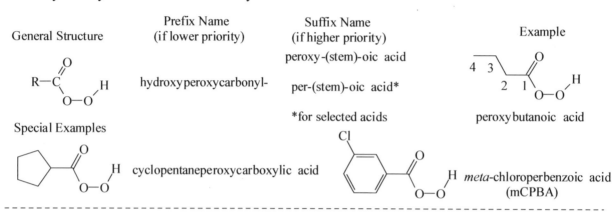

General Structure	Prefix Name (if lower priority)	Suffix Name (if higher priority)	Example

peroxy-(stem)-oic acid

hydroxyperoxycarbonyl- per-(stem)-oic acid*

*for selected acids

peroxybutanoic acid

Special Examples

cyclopentaneperoxycarboxylic acid *meta*-chloroperbenzoic acid (mCPBA)

- -

6. Sulfonic Acid – not covered by us.

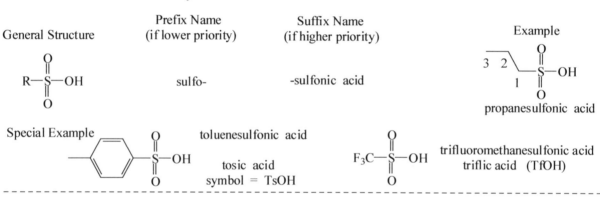

General Structure	Prefix Name (if lower priority)	Suffix Name (if higher priority)	Example

sulfo- -sulfonic acid

propanesulfonic acid

Special Example

toluenesulfonic acid

tosic acid

symbol = TsOH

trifluoromethanesulfonic acid
triflic acid (TfOH)

- -

7. Anhydride– you need to know this group.

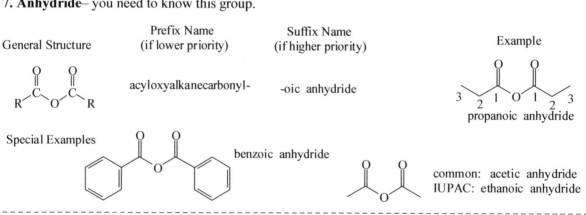

General Structure	Prefix Name (if lower priority)	Suffix Name (if higher priority)	Example

acyloxyalkanecarbonyl- -oic anhydride

propanoic anhydride

Special Examples

benzoic anhydride

common: acetic anhydride
IUPAC: ethanoic anhydride

- -

Drop the -e ending of the longest chain having the anhydride functional group and add the **-oic anhydride** suffix. Use just one name if the anhydride is symmetrical. If the anhydride is not symmetrical, name each part as –oic and then a single "anhydride" to finish the name. If the anhydride is the high priority group, each of the carbonyl carbons will be the #1 carbons of their respective chains and do not require a location number. The anhydride prefix (when it is a lower priority group) is too complicated for a first time course.

ethanoic anhydride
(common: acetic anhydride)

2-(1-methylethyl)pentanoic but-3-ynoic anhydride

2-isopropylpentanoic but-3-ynoic anhydride

2-ethyl-3-methoxypent-3Z-enoic propanoic anhydride

8. Ester– you need to know this group.

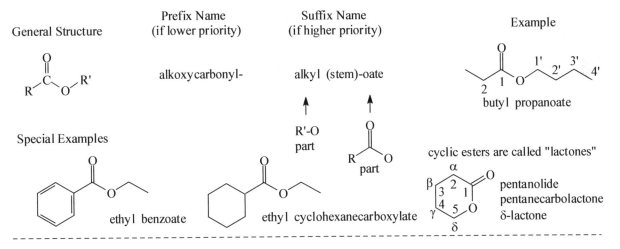

General Structure	Prefix Name (if lower priority)	Suffix Name (if higher priority)	Example

alkoxycarbonyl-

alkyl (stem)-oate

R'-O part

R part

butyl propanoate

Special Examples

cyclic esters are called "lactones"

ethyl benzoate

ethyl cyclohexanecarboxylate

pentanolide
pentanecarbolactone
δ-lactone

Drop the -e ending of the longest chain having the carbonyl portion (C=O) and add the **-oate** suffix. The other carbon chain, attached to oxygen with only a single bond, is <u>named at the very beginning of the name as an **alkyl** substituent and placed as a separate word</u>. If the ester is the high priority group, the carbonyl carbon will be the #1 carbon and does not require a location number. If the ester is a lower priority group it is named as "alkoxycarbonyl" with a locator number.

ethyl ethanoate
(common: ethyl acetate)

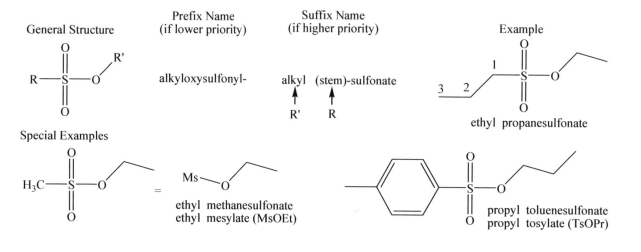

ethyl 3-methylbut-2-enoate

but-3-ynyl 2-(2-methylpropyl)pentanoate

but-3-ynyl 2-isobutylpentanoate

cyclopent-2-enyl 2-benzyl-3-hydroxy-4-phenylhex-4E-enoate

5-methoxycarbonylheptanoic acid

9. Sulfonate Ester - not covered by us.

General Structure	Prefix Name (if lower priority)	Suffix Name (if higher priority)	Example

alkyloxysulfonyl-

alkyl (stem)-sulfonate

R'   R

ethyl propanesulfonate

Special Examples

ethyl methanesulfonate
ethyl mesylate (MsOEt)

propyl toluenesulfonate
propyl tosylate (TsOPr)

10. Acid Halide– you need to know this group.

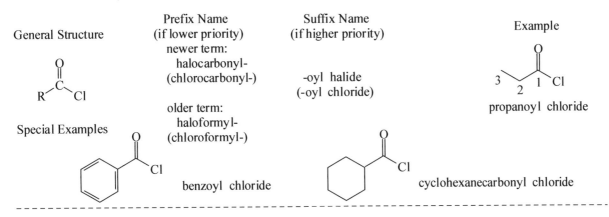

General Structure	Prefix Name (if lower priority)	Suffix Name (if higher priority)	Example

newer term:
halocarbonyl-
(chlorocarbonyl-)

-oyl halide
(-oyl chloride)

propanoyl chloride

older term:
haloformyl-
(chloroformyl-)

Special Examples

benzoyl chloride

cyclohexanecarbonyl chloride

- -

Drop the -e ending of the longest chain having the functional group and add **-oyl chloride**. If the acid chloride is the high priority group, the carbonyl carbon will be the #1 carbon and does not require a location number. If the acid chloride is a lower priority group it is named as "chlorocarbonyl" with a locator number.

ethanoyl chloride
common: acetyl chloride

2-(2-methylcyclopropyl)pent-3-ynoyl chloride

3-benzyl-5-phenylpent-2Z-enoyl bromide

3-chlorocarbonylheptanoic acid

Chapter 4

11. Sulfonyl Chloride - not covered by us.

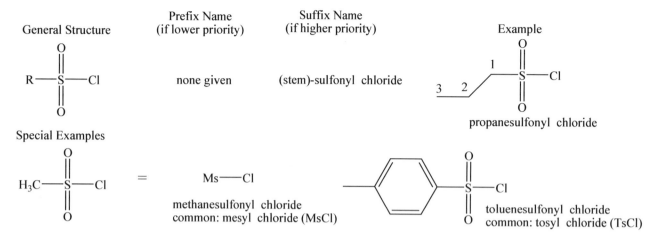

12. Amide – you need to know this group. (1°, 2°, 3° are possible, refers to the number of carbons attached to the nitrogen atom)

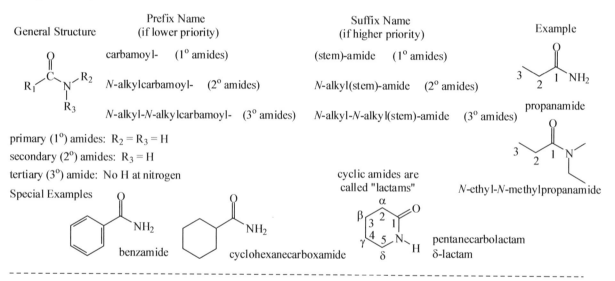

Drop the -e ending of the longest chain having the C=O functional group and add **amide**. If nitrogen is substituted with an alkyl group indicate so with **N-** preceding the substituent name. If the nitrogen is substituted with two alkyl groups, use an **N-** before each of the substituents. An amide nitrogen with only the carbonyl carbon (C=O) attached is a primary amide (1°). If two carbons are attached to the nitrogen, it is a secondary amide (2°) and if three carbons are attached, it is a tertiary amide (3°). If the amide is the high priority group, the carbonyl carbon will be the #1 carbon and does not require a location number. If the amide is lower in priority it can be named using either the "carbamoyl" or "amido" prefix (with a locator number).

3-aminoheptanamide

1° amide

N-ethyl-2-methoxyhex-2Z-en-4-ynamide

2° amide

N-ethyl-N-methyl-2-hydoxy-3-(6-methylcyclohex-3-enyl)propanamide

3° amide

N,N-dimethylethanamide

common: N,N-dimethylacetamide

3° amide

3-carbamoylheptanoic acid

3-amidoheptanoic acid

---

13. Sulfonamide - not covered by us.

General Structure	Prefix Name (if lower priority)	Suffix Name (if higher priority)	Example

R—S—NH₂ : sulfonylamoyl : -sulfonamide

propanesulfonamide

Special Examples

cyclopentanesulfonamide

toluenesulfonamide

14. Nitrile – you need to know this group.

General Structure	Prefix Name (if lower priority)	Suffix Name (if higher priority)	Example

-nitrile

R—C≡N : cyano-

(Don't drop the final "e" of the stem name because the suffix begins with a consonant.)

propanenitrile

Special Examples

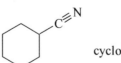

benzonitrile

cyclohexanecarbonitrile

Chapter 4

Do not drop the -e ending. Since nitrile begins with a consonant the -e is retained in the root name and **nitrile** is added to this name. If the nitrile is the high priority group, the carbon with the nitrogen will be the #1 carbon and does not require a location number. If a higher priority group is present use the **cyano** prefix with a location number. Examples are provided.

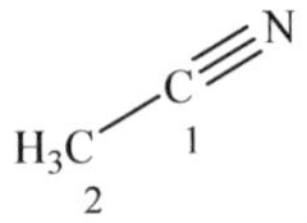

ethanenitrile

common: acetonitrile

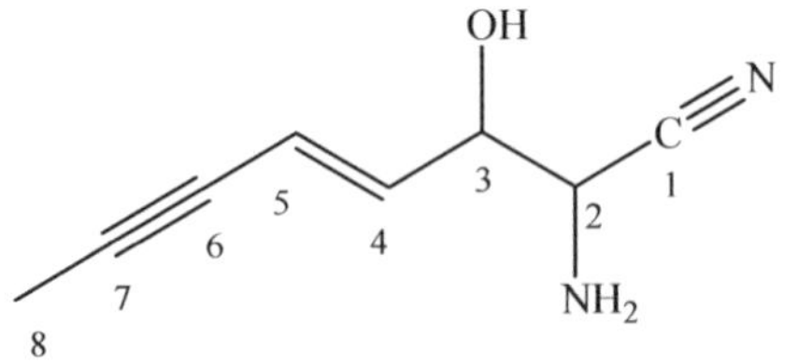

2-amino-3-hydroxyoct-4E-en-6-ynenitrile

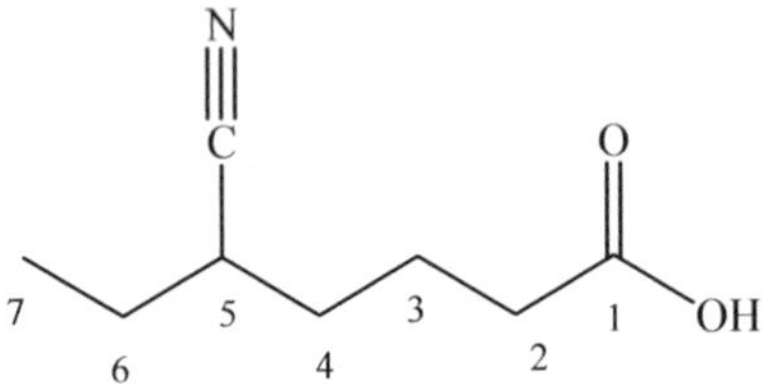

5-cyanoheptanoic acid

15. Aldehyde– you need to know this group.

General Structure	Prefix Name (if lower priority)	Suffix Name (if higher priority)	Example
	oxo- (if at the end of the longest chain)	-al	
	formyl- (if a branch off the longest chain)		

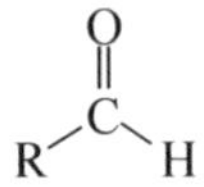

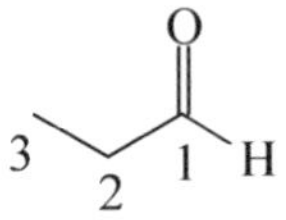

propanal

Special Examples

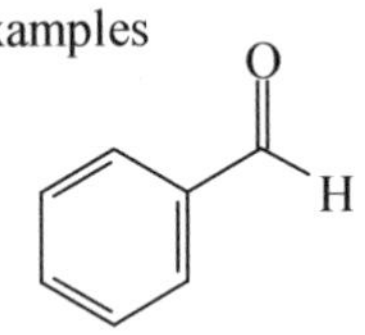

benzaldehyde

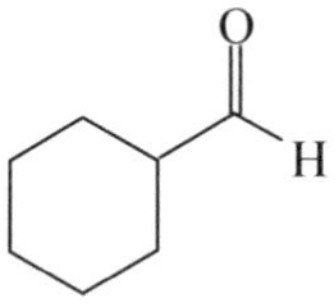

cyclohexanecarboxaldehyde

Drop the -e ending of the longest chain having the functional group and add -**al**. If the aldehyde is the high priority group, the carbonyl carbon will be the #1 carbon and does not require a location number. If a higher priority group is present, use the **oxo**- prefix with a location number when the aldehyde is part of the longest chain. If the aldehyde is a branch off of the longest chain it gets a location number and the prefix "formyl" Examples are provided.

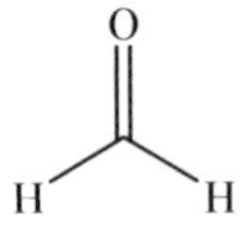

methanal

common: formaldehyde

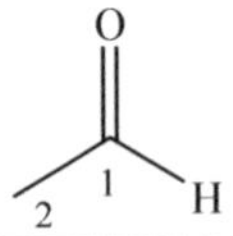

ethanal

common: acetaldehyde

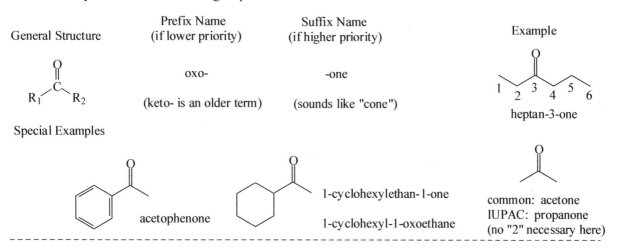

2-ethenyl-3-(prop-2-enyl)oct-6-ynal

2-vinyl-3-allyloct-6-ynal

5-formyl-7-oxoheptanoic acid

16. Ketone– you need to know this group.

General Structure	Prefix Name (if lower priority)	Suffix Name (if higher priority)	Example
	oxo-	-one	
	(keto- is an older term)	(sounds like "cone")	heptan-3-one

Special Examples

acetophenone

1-cyclohexylethan-1-one

1-cyclohexyl-1-oxoethane

common: acetone
IUPAC: propanone
(no "2" necessary here)

Drop the -e ending of the longest chain having the functional group and add **-one** (sounds like cone, tone, bone). If the ketone is the highest priority group, it will usually require a location number. If a higher priority group is present, use the **oxo-** prefix (same as the aldehydes) with a location number. Examples are provided. (Older ketone nomenclature used the "keto" prefix. We will not use this prefix.)

propanone

common: acetone

3,4-dimethylpent-2-one

4-bromo-1-phenylhex-4Z-en-2-one

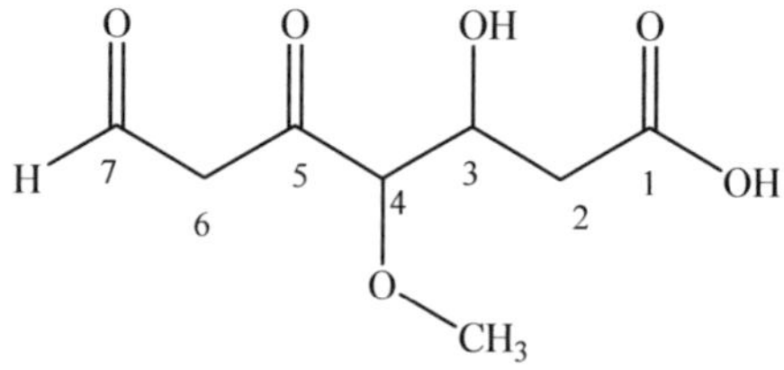

3-hydroxy-4-methoxy-5,7-dioxoheptanoic acid

---

17. Alcohol – you need to know this group. (1°, 2°, 3° are possible, refers to the number of carbons attached to the alcohol carbon atom)

General Structure	Prefix Name (if lower priority)	Suffix Name (if higher priority)	Example
R—CH$_2$—OH primary alcohol (1°)			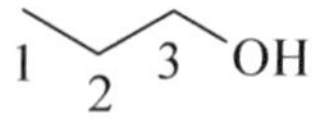 propan-1-ol
R—CH—OH | R secondary alcohol (2°)	hydroxy-	-ol	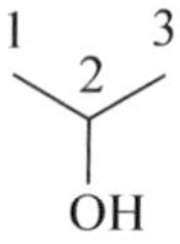 common: isopropyl alcohol IUPAC: propan-2-ol
R—C—OH | R tertiary alcohol (3°)			

Special Examples

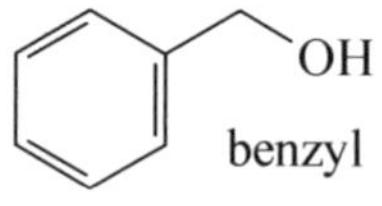

benzyl alcohol 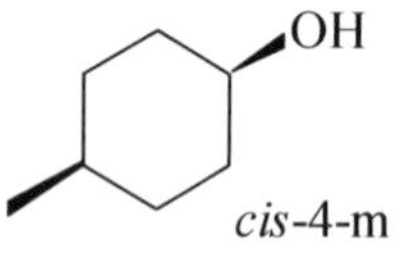   cis-4-methylcyclohexan-1-ol

CH$_3$OH   common: methyl alcohol   IUPAC: methanol

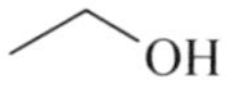

common: ethyl alcohol   IUPAC: ethanol

Drop the -e ending of the longest chain having the functional group and add **-ol**. If a higher priority group is present, use the **hydroxy-** prefix with a location number. If one carbon is attached to the alcohol carbon, it is a primary alcohol (1°), if two carbons are attached to the alcohol carbon, it is a secondary alcohol (2°) and if three carbons are attached to the alcohol carbon, it is a tertiary alcohol (3°). Examples are provided.

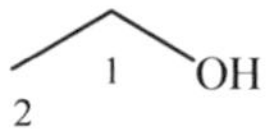

ethanol

common: ethyl alcohol

1° alcohol

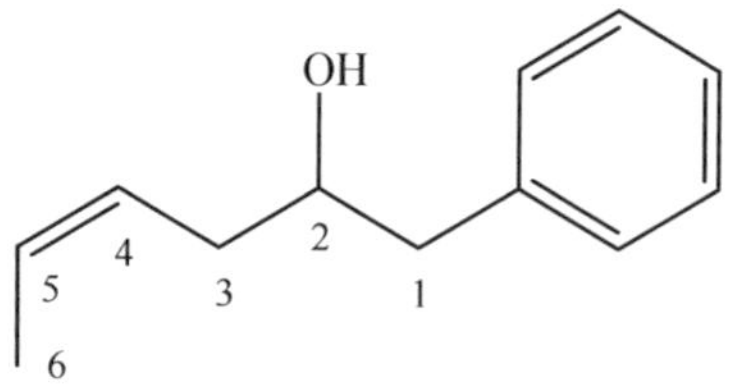

1-phenylhex-4Z-en-2-ol

2° alcohol

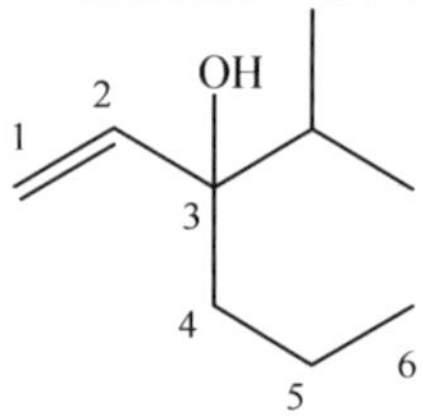

3-(1-methylethyl)hex-1-en-3-ol
3-isopropylhex-1-en-3-ol

3° alcohol

Chapter 4

3,5,7-trihydroxy-4-(1,1-dimethylethyl)heptanoic acid

3,5,7-trihydroxy-4-t-butylheptanoic acid

one 1° alcohol and two 2° alcohols

18. Thiol – you need to know this group. (1°, 2°, 3° are possible, refers to the number of carbons attached to the thiol carbon atom)

General Structure	Prefix Name (if lower priority)	Suffix Name (if higher priority)	Example
R—CH₂—SH primary thiol (1°)	sulfanyl- or mercapto-	-thiol	propane-1-thiol
R—CH—SH | R secondary thiol (2°)		(Don't drop the final "e" in the stem name.)	
R—C—SH (R, R) tertiary thiol (3°)		CH₃SH common: methyl mercaptan IUPAC: methanethiol	SH common: isopropyl mercaptan IUPAC: propane-2-thiol

$R-CH_2-SH$ primary thiol $(1°)$

$R-CH-SH$ secondary thiol $(2°)$ (with R below)

$R-C-SH$ tertiary thiol $(3°)$ (with R above and below)

Do not drop the -e ending of the longest chain having the functional group and add -**thiol** because it starts with a consonant If a higher priority group is present, use the **mercapto-** or **sulfanyl** prefix with a location number. If one carbon is attached to the thiol carbon, it is a primary thiol (1°), if two carbons are attached to the thiol carbon, it is a secondary thiol (2°) and if three carbons are attached to the thiol carbon, it is a tertiary thiol (3°). Examples are provided.

2,3,8-trimethylnon-3Z-en-6-yne-5-thiol

3-(cycloocta-2Z,4E-dienyl)nonane-4-thiol

3-(cycloocta-2Z,4E-dienyl)-6-(1-ethyl-2-methylbutyl)non-6Z-ene-4-thiol

3-mercapto-4-(1-methylprop-2-enyl)-5-methoxy-7-hydroxyheptanoic acid

3-sulfanyl-4-(1-methylprop-2-enyl)-5-methoxy-7-hydroxyheptanoic acid

3-mercapto-4-(1-methylprop-2-ynyl)-5-formyl-7-cyanoheptanoic acid

3-sulfanyl-4-(1-methylprop-2-ynyl)-5-formyl-7-cyanoheptanoic acid

19. Amine – you need to know this group. (1°, 2°, 3° are possible, refers to the number of carbons attached to the nitrogen atom). Two nomenclature systems are shown below (CA = Chemical Abstracts and IUPAC = International Union of Pure and Applied Chemistry). Both are commonly used.

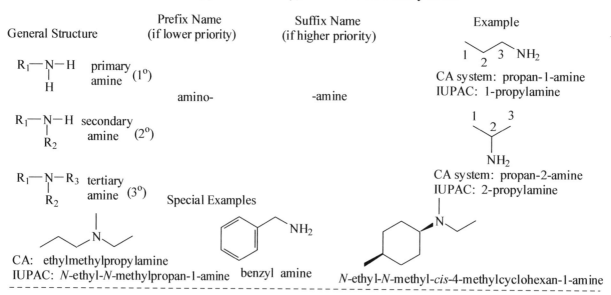

General Structure	Prefix Name (if lower priority)	Suffix Name (if higher priority)	Example

R_1—N—H primary amine (1°)

R_1—N—H secondary amine (2°)
R_2

R_1—N—R_3 tertiary amine (3°)
R_2

amino-

-amine

CA system: propan-1-amine
IUPAC: 1-propylamine

CA system: propan-2-amine
IUPAC: 2-propylamine

Special Examples

CA: ethylmethylpropylamine
IUPAC: *N*-ethyl-*N*-methylpropan-1-amine

benzyl amine

N-ethyl-*N*-methyl-*cis*-4-methylcyclohexan-1-amine

Drop the -e ending of the longest chain having the functional group and add **amine**. If nitrogen is substituted with another alkyl group, indicate so with **N**- preceding the substituent name. If the nitrogen is substituted with two additional alkyl groups, use an **N**- before each of the substituents. An amine nitrogen with only one carbon attached is a primary amine (1°). If two carbons are attached to the nitrogen, it is a secondary amine (2°) and if three carbons are attached, it is a tertiary amine (3°). Amines can also be named as though the largest carbon portion is alkylamine and any other alkyl chains are named as **N-alkyl**. If a higher priority group is present, use the **amino-** prefix with a location number. Examples are provided.

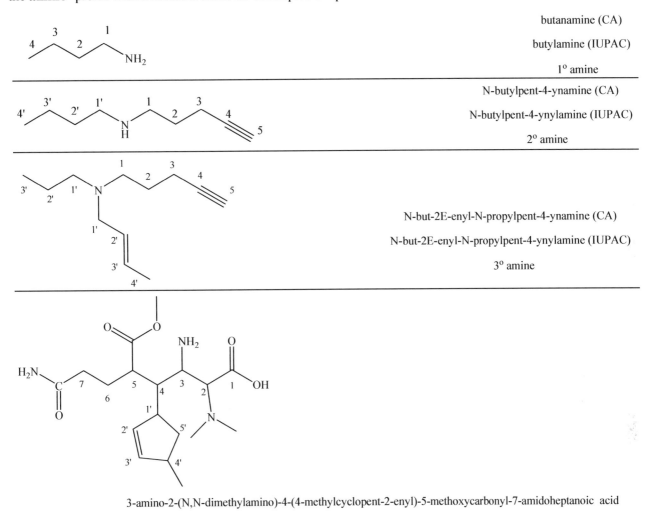

butanamine (CA)

butylamine (IUPAC)

1° amine

N-butylpent-4-ynamine (CA)

N-butylpent-4-ynylamine (IUPAC)

2° amine

N-but-2E-enyl-N-propylpent-4-ynamine (CA)

N-but-2E-enyl-N-propylpent-4-ynylamine (IUPAC)

3° amine

3-amino-2-(N,N-dimethylamino)-4-(4-methylcyclopent-2-enyl)-5-methoxycarbonyl-7-amidoheptanoic acid

3-amino-2-(N,N-dimethylamino)-4-(4-methylcyclopent-2-enyl)-5-methoxycarbonyl-7-carbamoylheptanoic acid

20. Ether– you need to know this group.

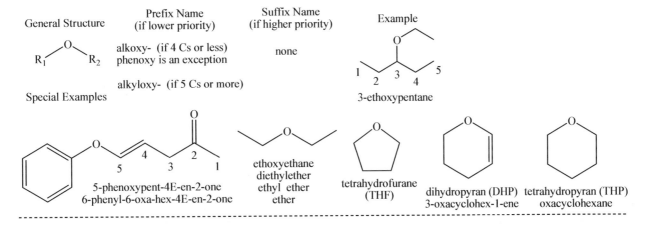

General Structure	Prefix Name (if lower priority)	Suffix Name (if higher priority)	Example
R₁–O–R₂	alkoxy- (if 4 Cs or less) phenoxy is an exception	none	3-ethoxypentane
	alkyloxy- (if 5 Cs or more)		

Special Examples

5-phenoxypent-4E-en-2-one
6-phenyl-6-oxa-hex-4E-en-2-one

ethoxyethane
diethylether
ethyl ether
ether

tetrahydrofurane
(THF)

dihydropyran (DHP)
3-oxacyclohex-1-ene

tetrahydropyran (THP)
oxacyclohexane

Noncyclic ethers have two carbon chains to name (like the anhydrides and the esters). Usually, the smaller alkyl portion is named as a substituent on the higher priority portion. The smaller (lower priority) group is named as

Chapter 4

alkoxy (if 4 or fewer carbons are present) or alkyloxy- (if more than 5 carbons are present, we won't use this one). Ether substituents are always named as prefixes. When two simple alkyl groups are present ethers are sometimes named as alkyl alkyl ether. Symmetrical ethers are often named with a single alkyl name followed by ether. The word "ether" alone usually refers to diethyl ether (ethoxyethane). It is also possible to name a compound as though all the atoms were carbon and use a location number and the prefix "oxa" for the location of an oxygen. Try not confuse "oxa" with the "oxo" prefix used for ketones and aldehydes.

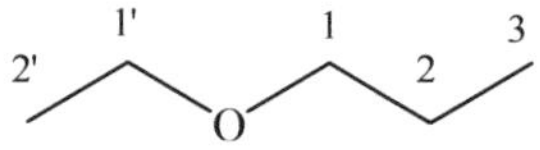

1-ethoxypropane

ethyl propyl ether

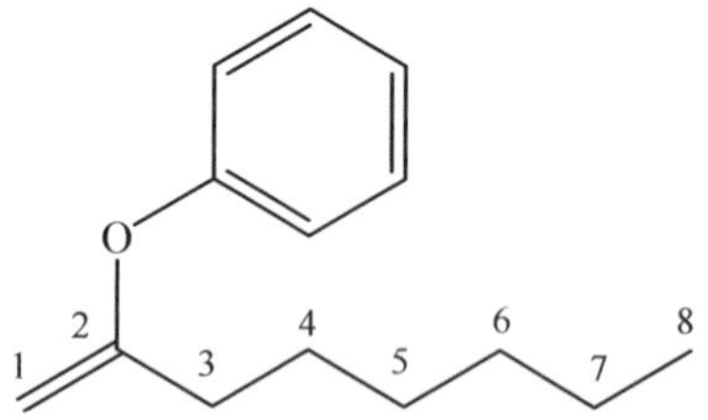

2-phenoxyoct-1-ene

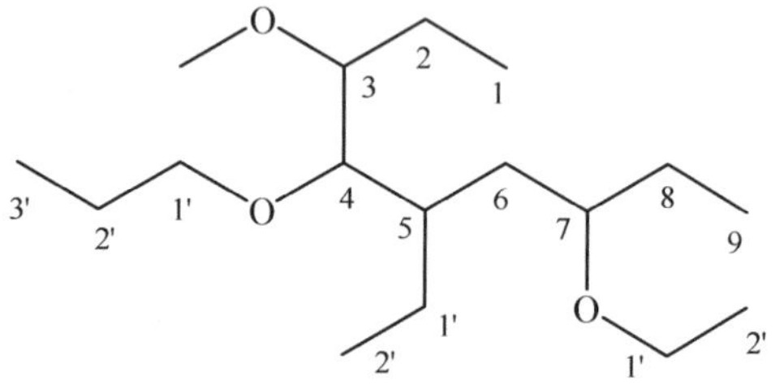

5-ethyl-7-ethoxy-3-methoxy-4-propoxynonane

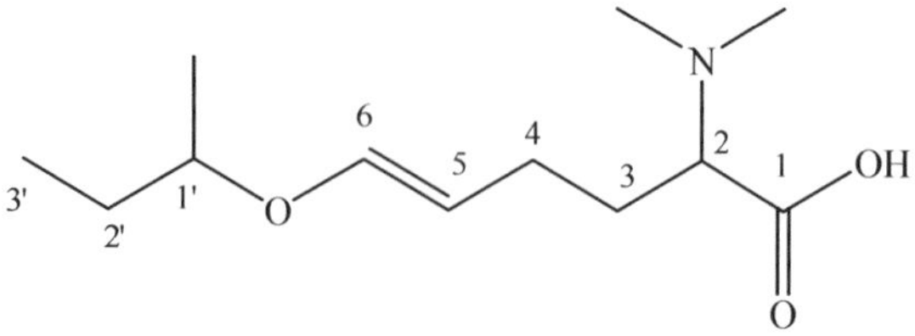

2-(N,N-dimethylamino)-6-(1-methylpropoxy)hex-5E-enoic acid

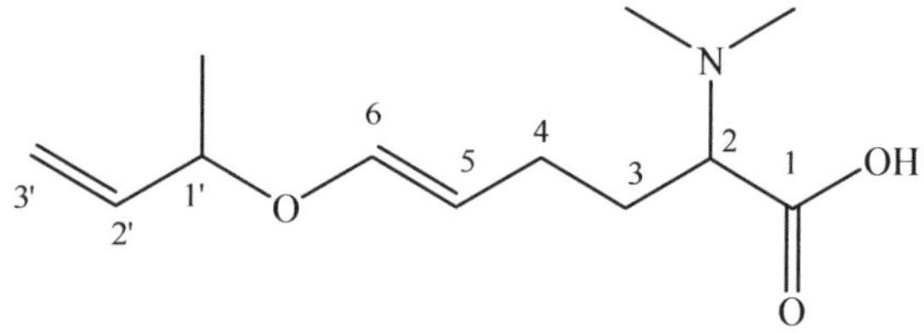

2-(N,N-dimethylamino)-6-(1-methylprop-2-enoxy)hex-5E-enoic acid

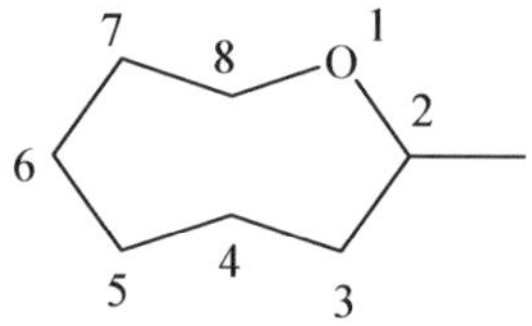

2-methyl-1-oxacyclooctane

21. Sulfide - not covered by us.

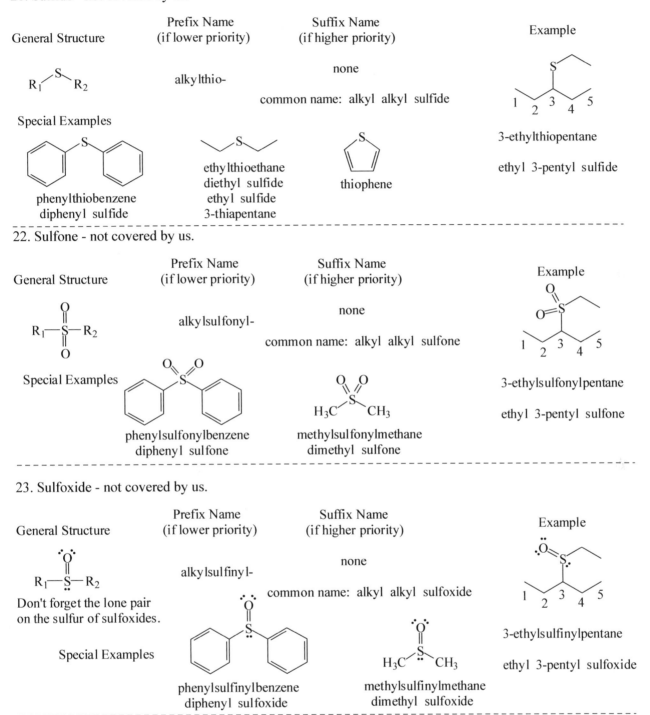

General Structure	Prefix Name (if lower priority)	Suffix Name (if higher priority)	Example

Special Examples

phenylthiobenzene
diphenyl sulfide

ethylthioethane
diethyl sulfide
ethyl sulfide
3-thiapentane

thiophene

3-ethylthiopentane

ethyl 3-pentyl sulfide

- -

22. Sulfone - not covered by us.

General Structure	Prefix Name (if lower priority)	Suffix Name (if higher priority)	Example

alkylsulfonyl-

common name: alkyl alkyl sulfone

Special Examples

phenylsulfonylbenzene
diphenyl sulfone

methylsulfonylmethane
dimethyl sulfone

3-ethylsulfonylpentane

ethyl 3-pentyl sulfone

- -

23. Sulfoxide - not covered by us.

General Structure	Prefix Name (if lower priority)	Suffix Name (if higher priority)	Example

alkylsulfinyl-

common name: alkyl alkyl sulfoxide

Don't forget the lone pair
on the sulfur of sulfoxides.

Special Examples

phenylsulfinylbenzene
diphenyl sulfoxide

methylsulfinylmethane
dimethyl sulfoxide

3-ethylsulfinylpentane

ethyl 3-pentyl sulfoxide

- -

24. Hydrocarbon branches and miscellaneous substituents named as prefixes.

The hydrocarbon branches were discussed in prior sections.

Special Substituents (Know these five.)

There are several groups that occupy one bonding position on the carbon skeleton and are named only with a prefix and a number to indicate their position. Some of the more common ones are listed below. They should be listed in alphabetical order with their designating number. Priority among these groups, if no high priority functional group is present, is based on the lowest possible number at the first point of difference. Some of these groups must be drawn with formal charge and have resonance structures, if full Lewis structures are drawn.

1. Halogen Compounds– you need to know this group.

General Structure	Prefix Name	Examples	
R—F	fluoro-	H_3C—F	fluoromethane
R—Cl	chloro-	Cl	chloroethane
R—Br	bromo-	Br	1-bromopropane
R—I	iodo-	I	1-iodobutane

Halogens are officially named with the **halo-** prefix name specific to the halogen present (**fluoro-, chloro-, bromo-, iodo-**) and a location number for each halogen present. If one carbon is attached to the halogen carbon, it is a primary halide (1°), if two carbons are attached to the halogen carbon it is a secondary halide (2°) and if three carbons are attached to the halogen carbon it is a tertiary halide (3°). Common nomenclature uses alkyl halide (like a salt) for simple names (e.g. methyl chloride). Examples are provided.

chloromethane
common: methyl chloride

dichloromethane
common: methylene chloride

trichloromethane
common: chloroform

tetrachloromethane
common: carbon tetrachloride

2-fluoro-3-(2,2-dibromocyclobutyl)-4-iodohept-2Z-en-5-yne

5-chloro-8-chlorocarbonyl-8-fluoro-2-(1-iodoethyl)-7-oxooct-5Z-enoic acid

2. Azido Compounds – you need to know this group. (requires formal charge and resonance structures)

General Structure   Prefix Name   Examples

$R-N_3$ $\left(\begin{array}{c} R \\ N=N=N: \\ \oplus \quad \ominus \end{array} \longleftrightarrow \begin{array}{c} R \\ N-N\equiv N: \\ \ominus \quad \oplus \end{array} \right)$ azido-

4-azidopent-1-ene

3. Diazo Compounds – you do not need to know this group. (requires formal charge and resonance structures)

General Structure   Prefix Name   Examples

$R_2C=N_2$ $\left(\begin{array}{c} R \\ C=N=N: \\ R \quad \oplus \quad \ominus \end{array} \longleftrightarrow \begin{array}{c} R \\ \ominus C-N\equiv N: \\ R \quad \oplus \end{array} \right)$ diazo-

4-diazopent-1-ene

4. Nitro Compounds – you need to know this group. (requires formal charge and resonance structures)

General Structure   Prefix Name   Examples

$R-NO_2$ $\left(\begin{array}{c} :O: \\ R-N\oplus \\ :O: \ominus \end{array} \longleftrightarrow \begin{array}{c} :O: \ominus \\ R-N\oplus \\ :O: \end{array} \right)$ nitro-

4-nitropent-1-ene

5. Nitroso Compounds – you need to know this group.

General Structure   Prefix Name   Examples

$R-NO$ $R-N\overset{:O:}{=}$ nitroso-

4-nitrosohex-2E-ene

Examples (You should be able to name all of these.)

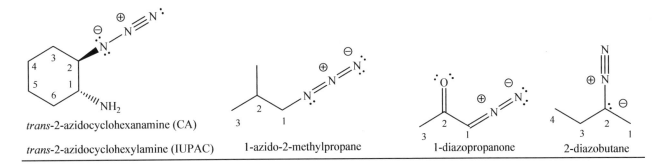

trans-2-azidocyclohexanamine (CA)

trans-2-azidocyclohexylamine (IUPAC)

1-azido-2-methylpropane

1-diazopropanone

2-diazobutane

1-methyl-1-nitrocyclopentane   4-nitro-5-oxohexanal   3-nitro-7-nitrosocyclooct-1Z-ene   3-nitrosoheptan-2,6-diol

Aromatic versions of various functional groups. Aromatic compounds have many special names. To learn them is mostly brute force memorization. If you use them, you will know them, if you don't, you won't. The names below are provided as examples. You are not required to know these names in this course. They are included for reference only.

benzoic acid   benzoic anhydride   methyl benzoate   benzoyl chloride

benzamide   benzonitrile   benzaldehyde   acetophenone   benzyl alcohol

benzyl amine   benzyl bromide   toluene   aniline   phenol   styrene

anisole   acetanilide   salicylic acid   acetylsalicylic acid aspirin   acetaminaphen Tylenol

o-xylene   m-xylene   p-xylene   cumene   cymene

Chapter 4

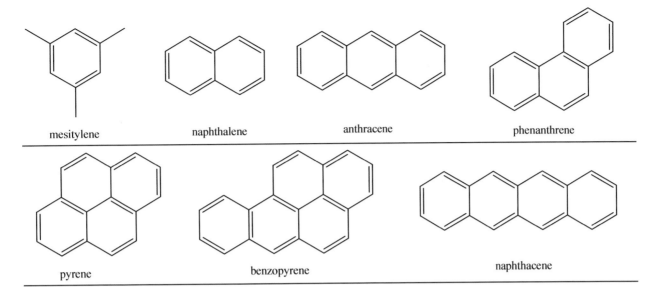

mesitylene   naphthalene   anthracene   phenanthrene

pyrene   benzopyrene   naphthacene

Chapter 4

General Strategy For Naming Simple Organic Compounds (Bare bones summary sheets)

1. Find the highest priority group. These are listed in order of priority in the table of functional groups (next page).

2. Find the longest chain containing the highest priority group. You should know carbon chains of length C_1-C_{19} (listed in the table).

3. Number the longest chain containing the highest priority group to give the highest priority group the lowest number possible in numbering the longest chain. For the first seven functional groups, the functional group carbon will be number 1 (if it is the highest priority group) and the "1" can be omitted, since it is understood that it has to be this way.

4. Usually the highest priority group is named as a suffix at the end of alkane, alk-#-ene or alk-#-yne. The final e is dropped if the suffix begins with a vowel and it is retained if the suffix begins with a consonant (only two for us, "nitrile" and "thiol"). A number will be present in front of the suffix name unless its position is unambiguously clear (e.g. carboxyl groups, aldehydes, nitriles, etc. always = 1, if highest in priority). If there is a C/C pi functional group to identify (alkene or alkyne), the number in front of its part of the name describes its position (see rule 6 below). If both a pi bond and a high priority substituent are present, then two (or more) numbers may be necessary, one for each functionality.

5. Lower priority groups are named with their prefix names and their location numbers based on the numbering of the parent chain (always true for substituents numbered 12 on the next page). The lower priority substituents should be listed in alphabetical order. Some parts of prefix names count in this regard and some don't. We will not emphasize this aspect in this course.

6. Double bonds and triple bonds are named as alk-#-ene or alk-#-yne, respectively. If both are present, name as alk-#-en-#-yne. Multiple pi bonds (or other substituents) use the prefixes di, tri, tetra, penta etc. with a number for each occurrence. In such cases, an "a" is added in front of the numerical prefix for better phonetics.

    (alka-#,#-diene or alka-#,#,#-triyne, alka-#,#-dien-#,#-diyne, etc.)

The essential functional groups to know (for our course) and their prefixes and suffixes are given in the table on the next page. In this table the term "alkan-#-suffix" is a generic term for any alkane with a functional group suffix, and it must be replaced with the correct parent stem name based on the number of carbons in the longest chain (C_1-C_{19} for us). If there is a double bond, the name will change to "alk-#-en-#-suffix and if there is a triple bond, the name will change to "alk-#-yn-#-suffix.

(2Z,4S)-3-ethoxy-4-methyl-8-oxooct-2-en-6-ynoic acid

R and S are possible

E and Z are possible

Notice that no "1" is used for the carboxylic acid group, because it has to be "1" in this structure.

FG - Priorities!

Should know for Exam! :)

The part of each name specific to the functional group is in **bold** and underlined to help you see those features. They are not part of the names.

Functional Group		prefix	suffix
1. Carboxylic acid	$R-\overset{O}{\overset{\|}{C}}-OH$	not considered	alkan**oic acid**
2. Anhydride	$R-\overset{O}{\overset{\|}{C}}-O-\overset{O}{\overset{\|}{C}}-R$	not considered	alkan**oic anhydride** (if symmetrical)
3. Ester	$R-\overset{O}{\overset{\|}{C}}-O-R'$	**#-alkoxycarbonyl**	**alkyl** alkan**oate** (R') (RCO$_2$)
4. Acid halide	$R-\overset{O}{\overset{\|}{C}}-Cl$	**#-chlorocarbonyl**	alkan**oyl chloride**
5. Amide	$R-\overset{O}{\overset{\|}{C}}-NH_2$	**#-carbamoyl** (or **#-amido**)	alkan**amide**
6. Nitrile	$R-C\equiv N$	**#-cyano**	alkan**enitrile**
7. Aldehyde	$R-\overset{O}{\overset{\|}{C}}-H$	**#-oxo**	alkan**al**
8. Ketone	$R-\overset{O}{\overset{\|}{C}}-R'$	**#-oxo** (older = **#-keto**)	**#-alkanone**
9. Alcohol	$R-OH$	**#-hydroxy**	**#-alkanol**
10. Thiol	$R-SH$	**#-mercapto** (or **#-sulfanyl**)	**#-alkanethiol**
11. Amine	$R-NH_2$	**#-amino**	**#-alkylamine** **#-alkanamine**
12. Ether	$R-O-R'$	**#-alkoxy** (if more than 5C's, then #-alkyloxy) (can also use "#-oxa" prefix and count as carbon in longest chain)	
12. Halogen	$R-X$	**#-fluoro, #-chloro, #-bromo, #-iodo**	
12. Azide*	$R-N_3$	**#-azido**	
12. Diazo*	$R-N_2$	**#-diazo**	
12. Nitro*	$R-NO_2$	**#-nitro**	
12. Nitroso	$R-NO$	**#-nitroso**	
12. Carbon branches	$R-$	#-alkyl, #-(alk-#-enyl), #-(alk-#-ynyl)	

"R" = carbon chains		
# carbons	alkane chain name	alkyl branch name
1	methane	methyl
2	ethane	ethyl
3	propane	propyl
4	butane	butyl
5	pentane	pentyl
6	hexane	hexyl
7	heptane	heptyl
8	octane	octyl
9	nonane	nonyl
10	decane	decyl
11	undecane	undecyl
12	dodecane	dodecyl
13	tridecane	tridecyl
14	tetradecane	tetradecyl
15	pentadecane	pentadecyl
16	hexadecane	hexadecyl
17	heptadecane	heptadecyl
18	octadecane	octadecyl
19	nonadecane	nonadecyl

Low in Priority so No Suffixes.

always prefixes (no suffix names)

* = formal charge is necessary in these Lewis structures and there are two reasonable resonance structures

stereoisomerism	prefixes	parent stem	C/C pi bonds	high priority suffix
↑ R/S and E/Z	↑ branches and low priority functional groups	↑ see box above	↑ -ene -yne	↑ see list above

Chapter 4

Problem 9 – Provide an acceptable name for each of the following structures.

a

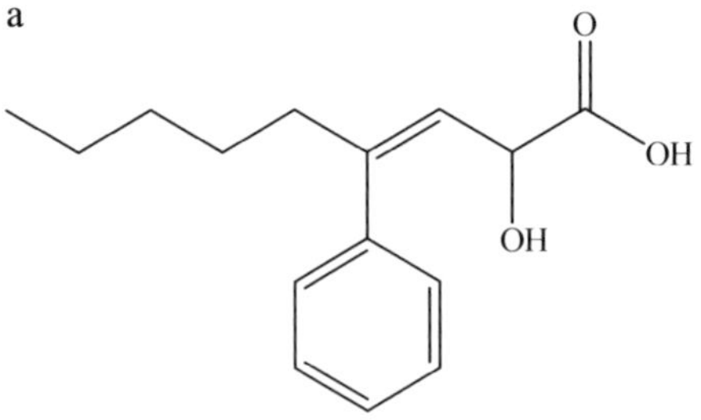

b

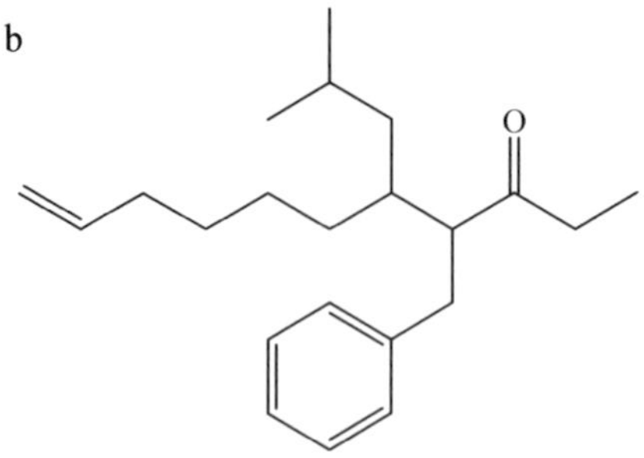

c

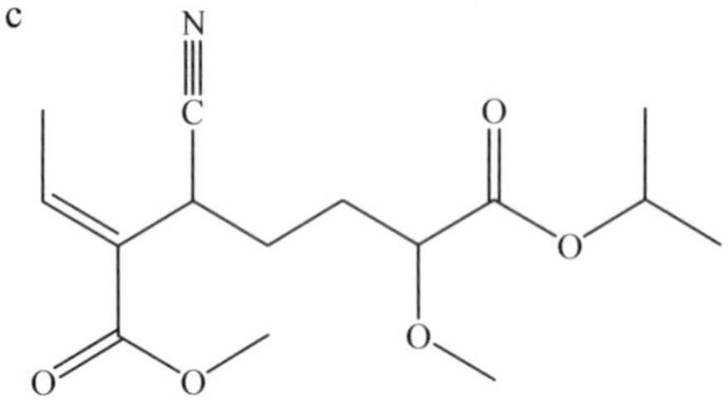

d

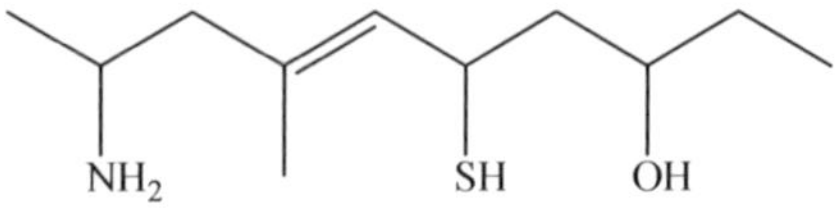

e

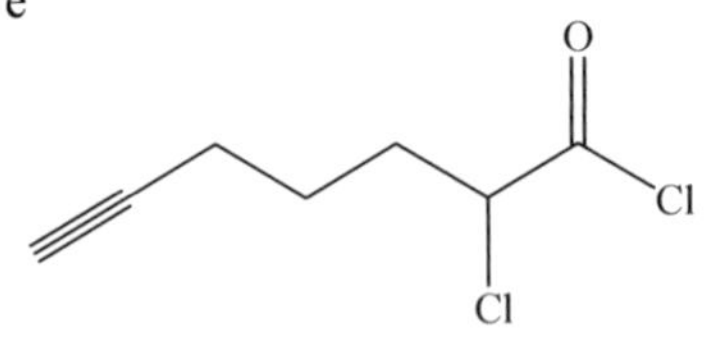

f

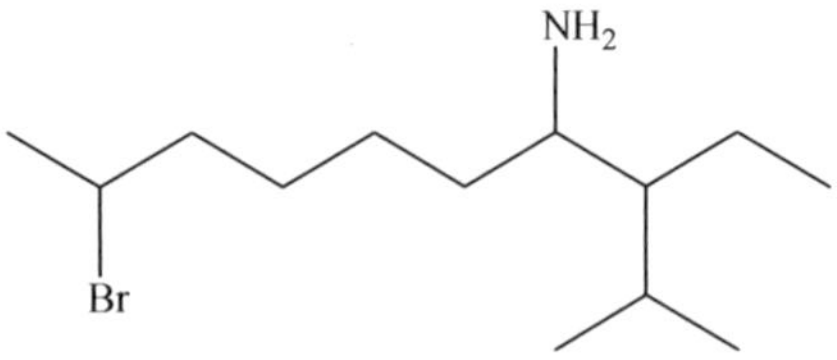

g

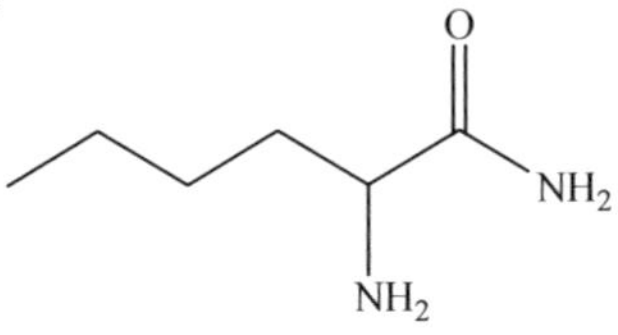

h

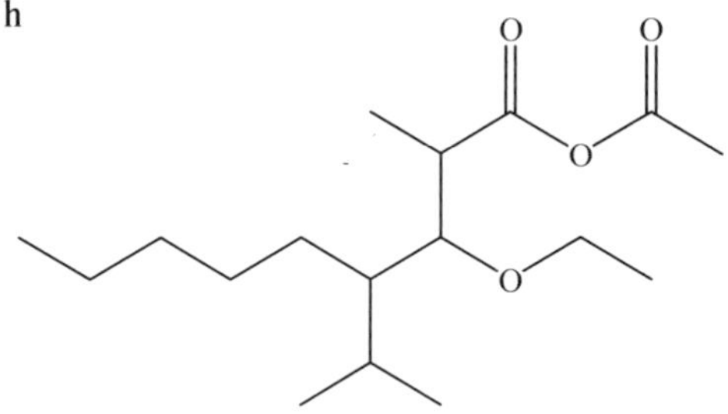

i

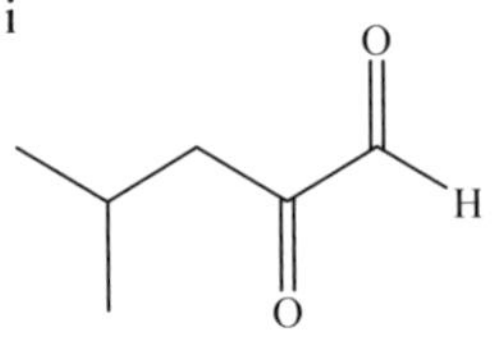

j

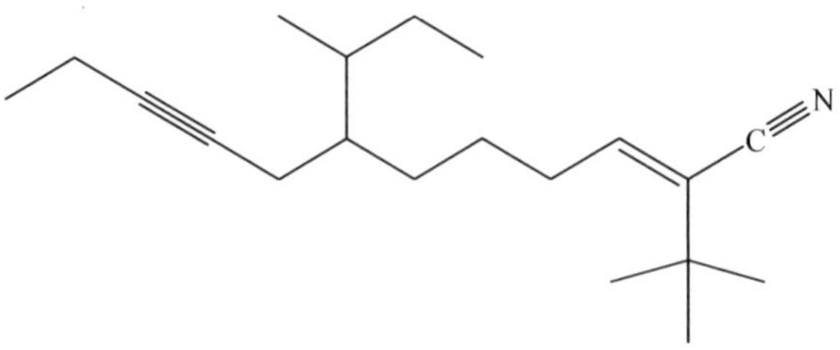

Chapter 4

Problem 10 – Provide an acceptable name for each of the following structures.

a

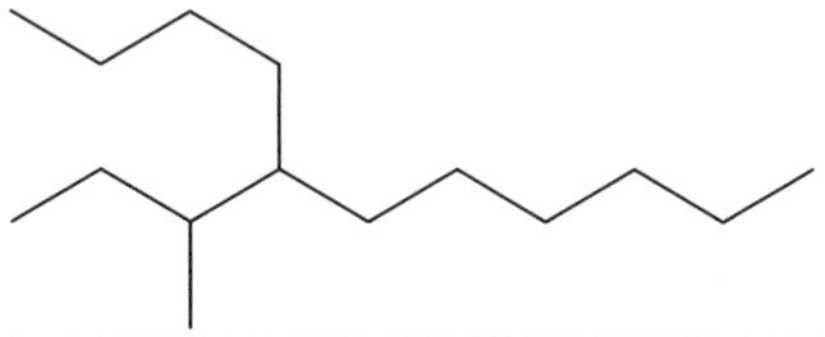

b

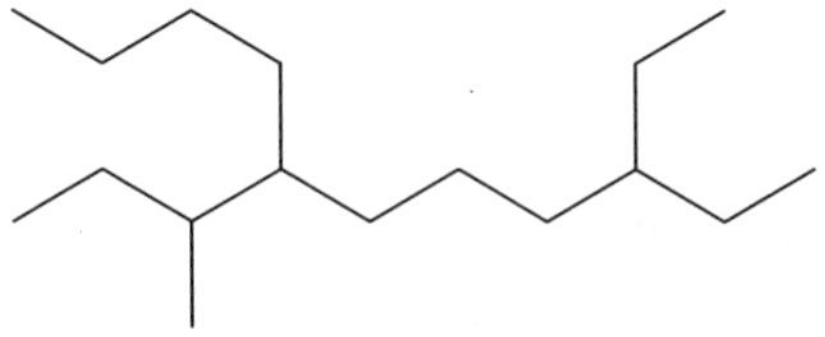

c

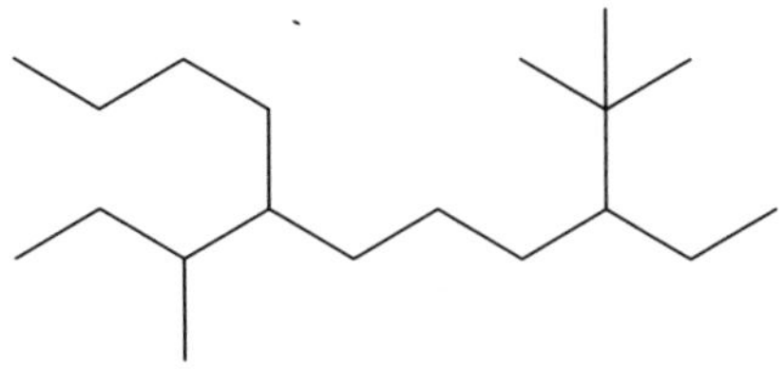

d

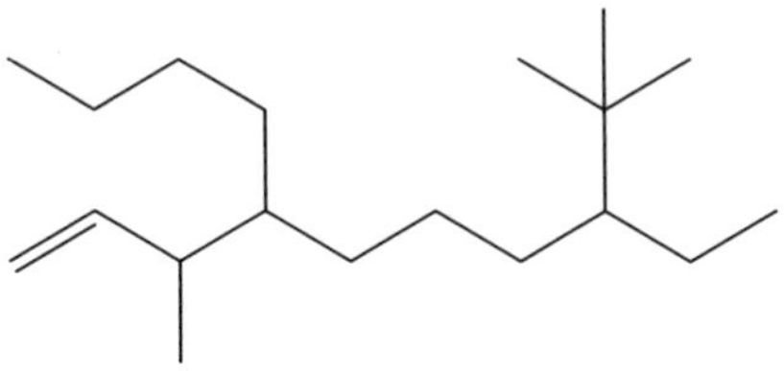

e

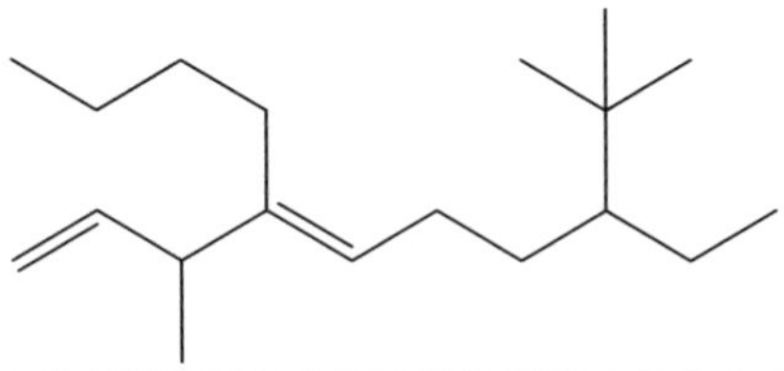

f

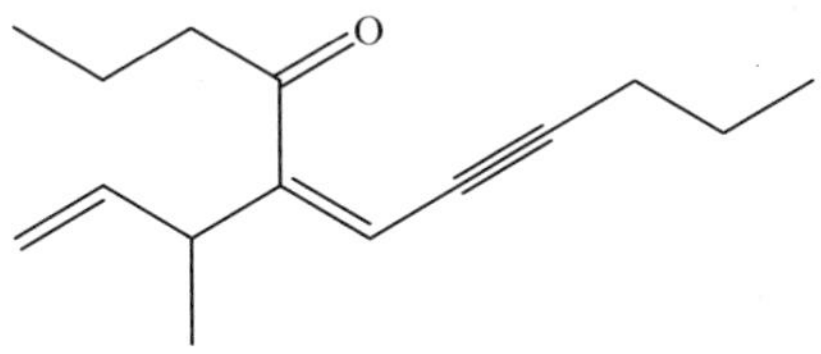

g

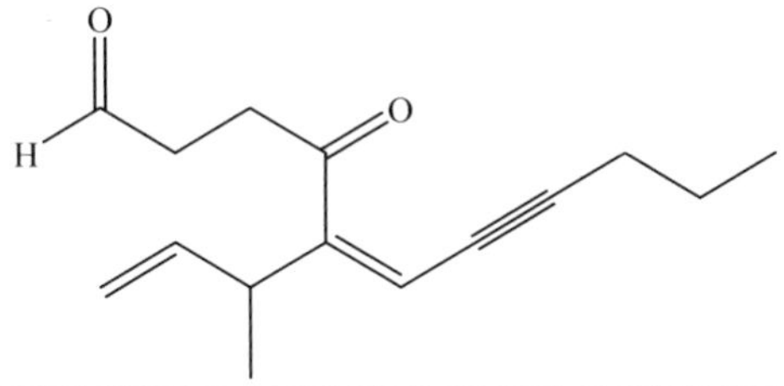

h

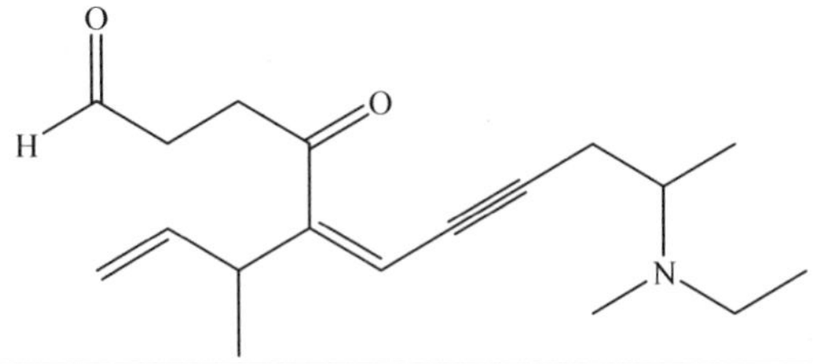

i

j

k

l

m

n

Ether

Alcohol
OH

Aldehyde

O

10

9

8

Ketone

2'

3'

1'

7

6 5

Nitrile

4 3 2 1

Acid Chloride

O

Cl

N

3° Amine

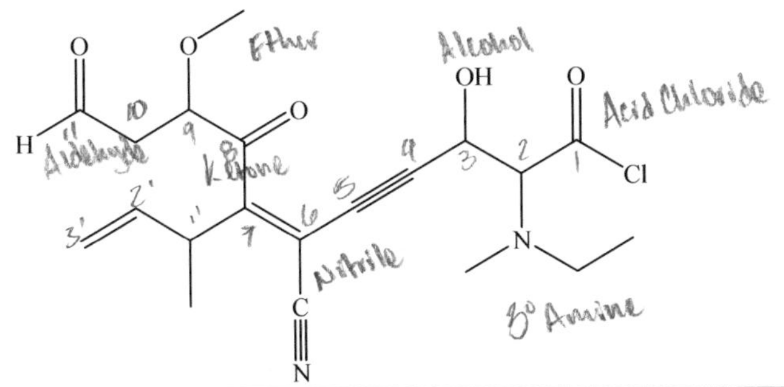

2 - (N-methyl - N - Ethyl(amino))

3 - hydroxy

6 - Cyano

7 - (1-methylprop-2-enyl)

8 - Oxo

9 - methoxy

11 - Oxo

o

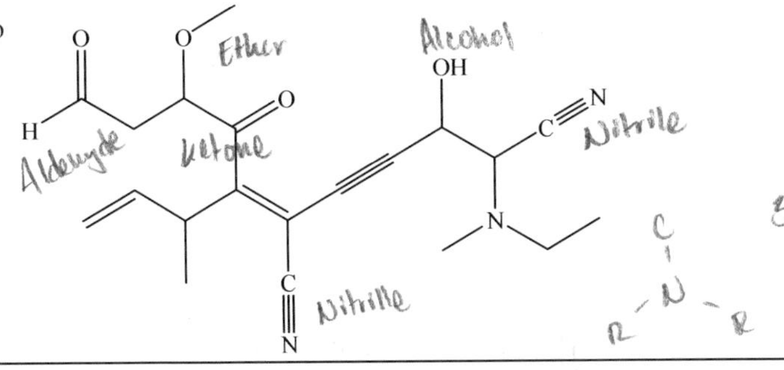

Ether

Alcohol
OH

Aldehyde

O

Ketone

C≡N

Nitrile

N

3° Amine

C
|
R — N — R

Nitrile
N

Chapter 4

Final Examples from alkane to poly functional group.

1

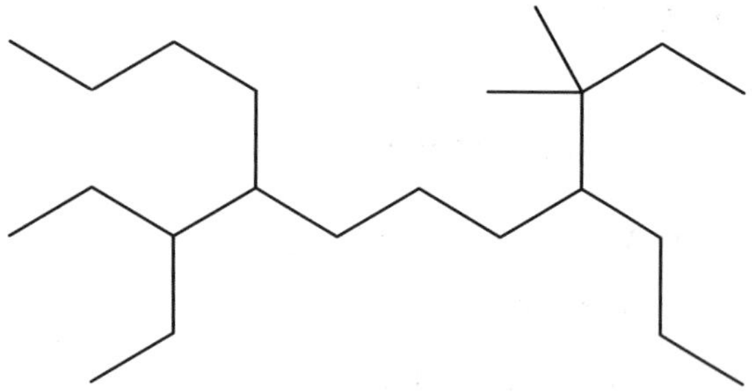

2

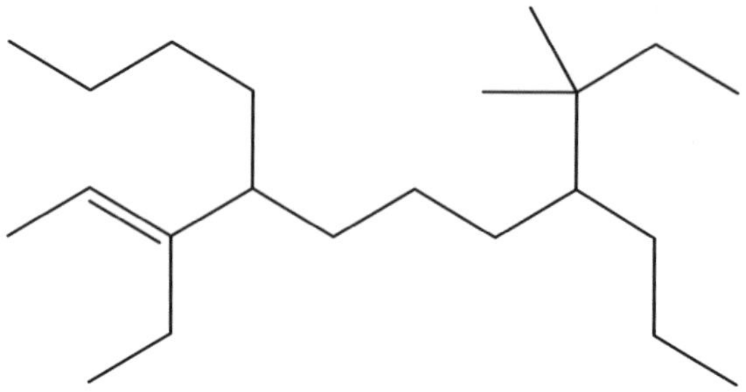

3

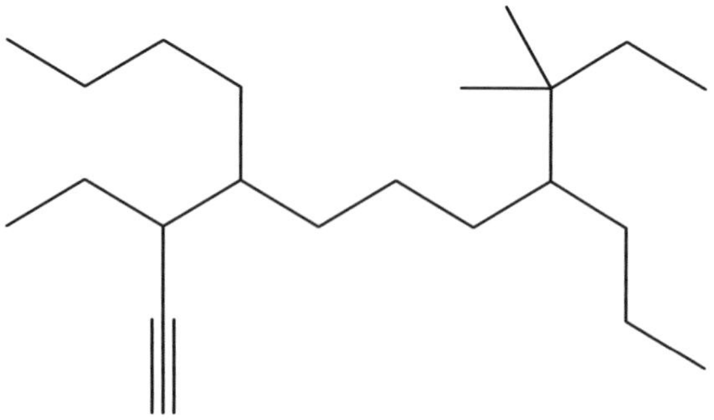

4

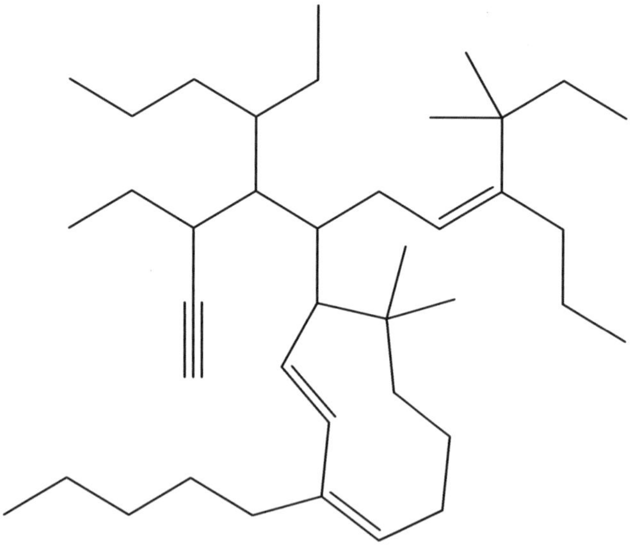

5

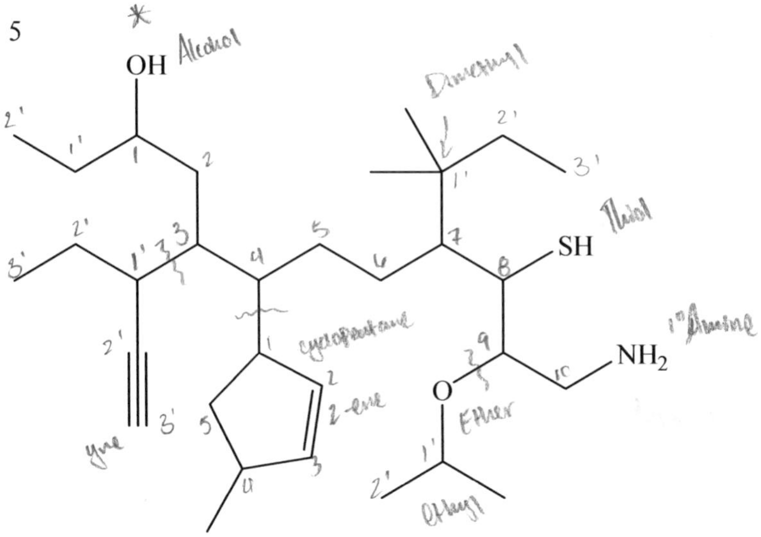

Alcohol
OH
2' 1'
Dimethyl
2'
3'
5
7 8 SH Thiol
4 6
2' 3
3' 1' 3 1
1
cyclopentane
2
2-ene
yne 3' 5
3
4

1 - ethyl
3 - (1 - Ethylprop-2-ynyl)
4 - (4 - methyl cyclopent-2-enyl)
7 - (1,1 - dimethyl propyl)
8 - mercapto-
9 - (1methyl ethoxy)
10 - Amino

Amine
NH₂
9
O
Ether
2' 1'
Ethyl

6

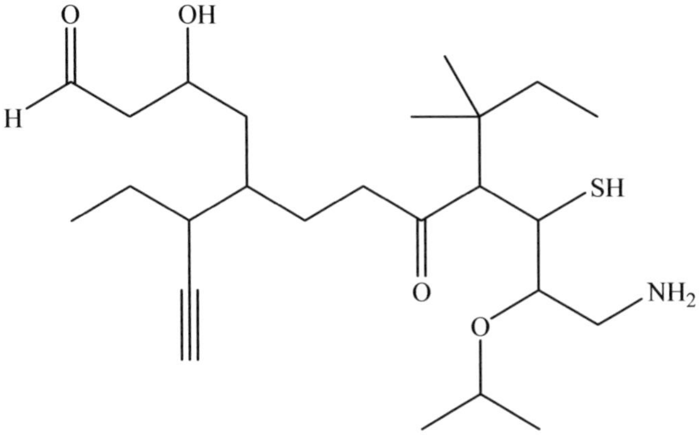

O
H
OH
SH
O
NH₂
O

7

Br
Cl
I
F
O
O
O
H
OH
O
NH₂
SH
O
NH₂
O

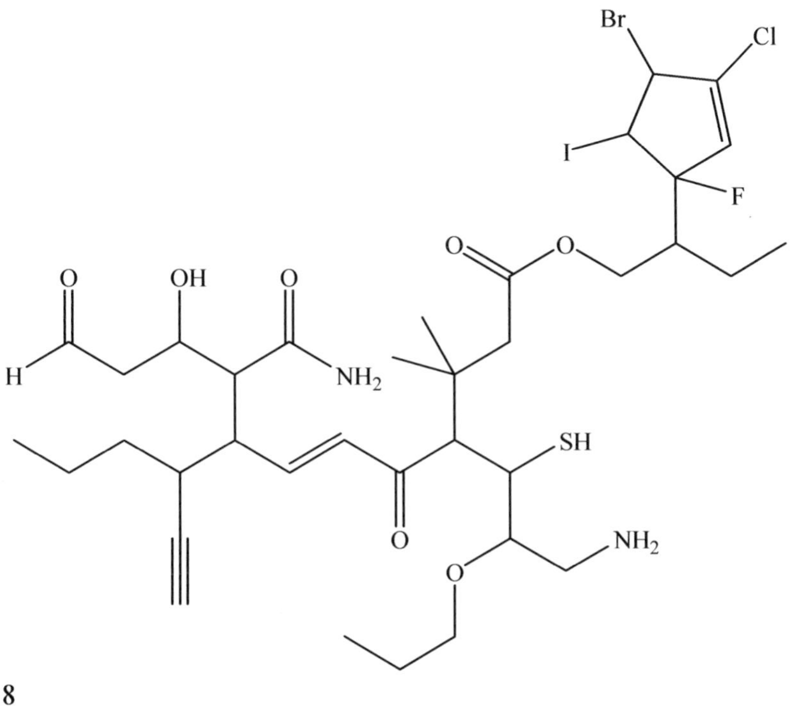

8

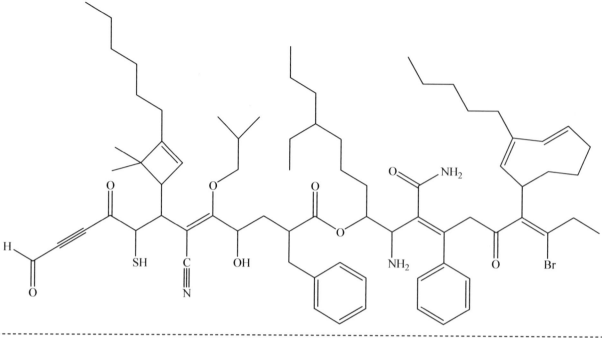

Chapter 4

CHAPTER 5 – CONFORMATIONS OF CHAINS

Types of Hydrocarbon Isomers

Alkanes represent one of four hydrocarbon families (having only carbon and hydrogen). These families include alkanes, alkenes, alkynes, and aromatics. Alkanes have only single bonds (C-C and C-H), while the pi bond is a key feature in the other three groups. On paper the alkenes and aromatics look very similar, however, their chemistry is so different that they are considered as separate functional groups. For purposes of classification, we will also include cumulenes in the alkene family even though they have a mix of sp^2 and sp hybridization.

The alkane family will provide our main examples for a discussion of the variations of shape in chains and rings resulting from rotation around single bonds. Without a pi bond, the alkanes are less reactive than the other hydrocarbon families and they undergo a more limited number of reactions. The generic term *aliphatic* is often used to indicate hydrocarbons that are not aromatic. If rings are present, this is often modified to *alicyclic*.

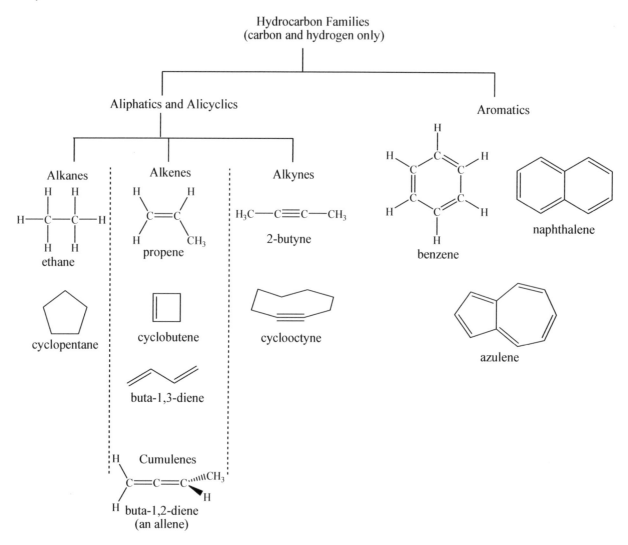

Chapter 5

Single bonds allow a wide range of motions depending on such factors as bond angles, bond lengths, polarities and whether an open chain or a ring is present. The different possible shapes that result from rotations about single bonds are called conformational changes and the different structures are referred to as *conformational isomers*. The energy necessary to cause these changes in conformation is usually in the range of 1-15 kcal/mole. This is lower than ambient thermal background energy (approximately 20 kcal/mole at room temperature), and conformational changes tend to occur very rapidly (typically, many thousands of times per second). Organic molecules therefore exist in a dynamic world of constant change.

In general, chains have a wider range of rotational motions possible than do rings. Chains have the potential to rotate 360° about their single bonds, while rings are limited in rotation by attachments to other ring atoms. However, as chains get very long they are less able to rotate freely, due to entanglements with nearby molecules. As rings get very large, they are more able to rotate, even as far as 360°.

Can't go 360°, only wiggle back & forth.

chain   ring   both chain and ring

As we have seen earlier, more than one structural (constitutional) isomer may exist for a particular formula. Within each specific constitutional isomer there exists many variations of conformation due to rotations about the single bonds. Usually the variations of conformation occur so fast, there is no chance of separating one conformational isomer from another.

Ethane					rotate C-C bond (infinite possible rotations)	
C_2H_6	CH_3CH_3					
condensed formula	condensed line formula	2D Lewis structure	3D structure conformation "a"			3D structure conformation "b"

When drawn on the page, our organic structures have at least two problems. They appear to be flat, but they are really three dimensional. Also they appear fixed or static, but they really are constantly undergoing a wide range of motions. While the connectivity of the atoms remains the same when rotations occur, the three dimensional representations of conformers are different. Since conformational isomers interconvert so easily (and quickly), an equilibrium distribution is generally present with the more stable arrangements predominating. The relative free energies ($\Delta G° = \Delta H° - T\Delta S°$) of each possible structure will determine how much of each conformation is present. ($\Delta G°$ = free energy, $\Delta H°$ = enthalpy ≈ relative bond energies, $\Delta S°$ = entropy ≈ relative disorder or probability)

Possible reaction pathways may be different or similar, depending on how a reaction occurs and the structural and conformational features present. One conformation may be reactive while another is nonreactive, or both conformations may be reactive but lead to different products. The shapes of a molecule can be very important in directing the course of a chemical reaction. This is true not only of conformational effects considered in this chapter, but stereochemical relationships considered in a later chapter (Chapter 7). Conformations and stereochemistry are not completely separate subjects from one another, and there is some overlap between them. They are presented separately here, for learning purposes.

Chapter 5

Newman Projections

Newman projections provide a particularly useful way of viewing structures to illustrate the three dimensional differences we are emphasizing above. A Newman projection focuses on two, attached atoms. The viewing perspective is straight on, looking down the linear bond connecting the two atoms. Sighting down the C-C bond in ethane is a simple, first example. The front carbon (or atom) is represented as a dot while the rear carbon (or atom) is represented as a large circle. Artistic perspective is actually backwards. What's close is drawn small and what's far away is drawn larger, because we want to be able to see both of atoms and the bonds that are connecting any other groups.

ethane

The front carbon in ethane has three other attached groups; in this case they are all hydrogen atoms. Additionally, with the front-on view we lose our three dimensional perspective and instead of seeing the 109° bond angles from the hydrogen atoms leaning forward, we see a flat 360°, divided three ways. The angles between the hydrogen atoms are drawn as though they are 120° apart. The angles are not really 120°, but we can't see this from the flat perspective of the page. Since the dot is the front carbon and nothing is blocking its view, we draw in the bond connections all the way down to the dot.

ethane

The rear carbon of ethane also has three additional attached groups, all hydrogen atoms. All three hydrogen atoms are leaning away. We don't see this in a flat Newman projection, and once again the angles between the rear hydrogen atoms appear to be 120°. Since these hydrogen atoms are behind the rear carbon, the lines showing these bonds only go down as far as the circle and then stop, disappearing behind the rear carbon.

ethane

Finally, we fill in whatever the three groups are on the front carbon atom and whatever the three groups are on the rear carbon atom. In ethane, there are six hydrogen atoms at these positions.

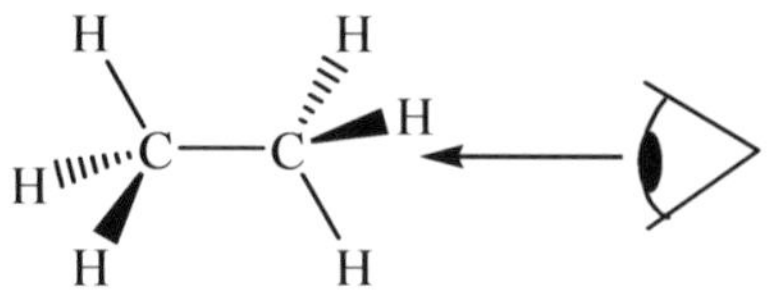

Chapter 5

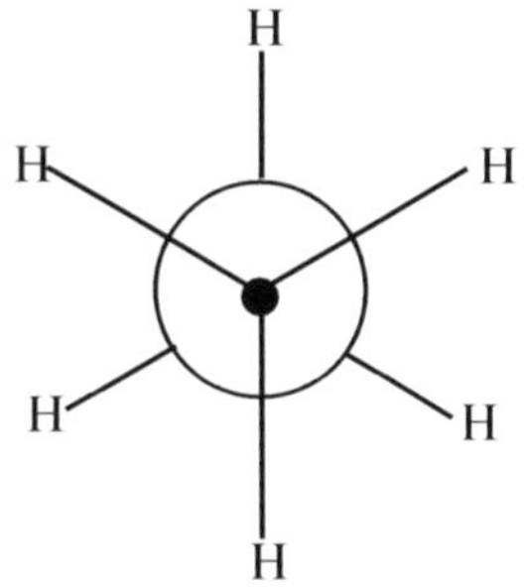

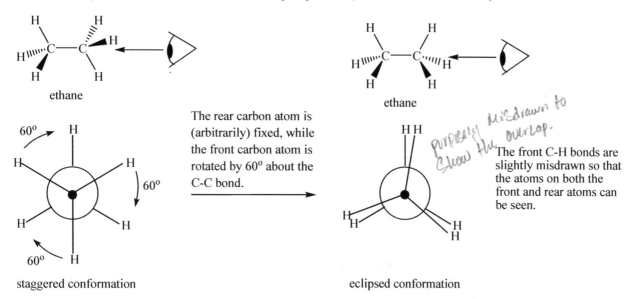

In addition to the bond connecting the two atoms of a Newman projection (behind the dot), there are three additional bonds on the front carbon atom and three additional bonds on the rear carbon atom.

ethane

staggered conformation

Models are almost essential for really understanding this subject. It would be of great benefit here for you to quickly assemble a molecular model of ethane and view it in this manner. (...I'm waiting.)

A special perspective was chosen in this example. The position of the groups on the front carbon bisect the positions of the groups on the rear carbon in our straight-on view. In ethane, this is the most stable conformation. It goes by the name of *staggered conformation*. However, it is by no means the only conformation possible. There are an infinite number of possible representations depending on how much you choose to rotate one of the carbons relative to the other (front and back can be rotated). Newman projections are used because they provide a particularly useful way of viewing these relationships.

In drawing Newman projections, what is usually done is to select only certain conformations of extreme potential energy (both high energy and low energy) and emphasize these. We have already viewed the low energy staggered conformation in ethane. To view the high energy conformation we must rotate one of the carbon atoms by 60° in our flat two dimensional perspective (I'll choose the front one).

ethane

ethane

The rear carbon atom is (arbitrarily) fixed, while the front carbon atom is rotated by 60° about the C-C bond.

purposely misdrawn to show the overlap.

The front C-H bonds are slightly misdrawn so that the atoms on both the front and rear atoms can be seen.

staggered conformation

eclipsed conformation

This is clearly a different arrangement of the molecule. Since the change resulted from rotations about single bonds, it is a conformational isomer. The front C-H bonds, when viewed down the connecting C-C bond are directly on top of the rear C-H bonds. This arrangement of groups about the Newman projection is called an *eclipsed conformation* (like an eclipse of the sun or the moon). It is also the highest potential energy conformation (least stable).

If we mark one of the rear hydrogen atoms and one of the front hydrogen atoms (with a square below) to distinguish them from the others, we can see that energetically equivalent, but technically different structures are possible by rotating one hydrogen through 360° (front atom in this case). In doing this we have arbitrarily fixed the rear carbon atom and its attached groups (all hydrogen atoms here).

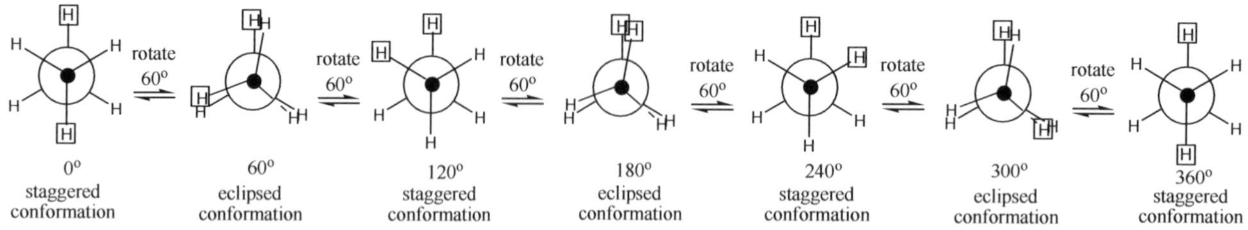

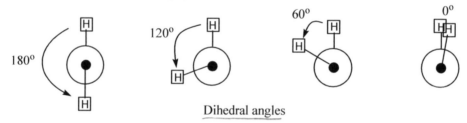

0°	60°	120°	180°	240°	300°	360°
staggered	eclipsed	staggered	eclipsed	staggered	eclipsed	staggered
conformation	conformation	conformation	conformation	conformation	conformation	conformation

Fixing the position of one of the atoms and its substituents is a good practice for you to adopt. It will reduce the number of variables you have to keep track of when drawing Newman projections to just those on the front atom. Remember, the rear atom is not fixed in an absolute sense, but only in a relative sense (this makes our life easier).

The front-on angle between any substituent on the front atom relative to any substituent on the rear atom is called the *dihedral angle*. The hydrogen atoms in the first structure below have a dihedral angle of 180°. In the second, third and fourth structures they have dihedral angles of 120°, 60° and 0°. The dihedral angle is frequently used to specify a relationship between two groups in a particular conformation. (We won't consider if the dihedral angle is to the left or to the right).

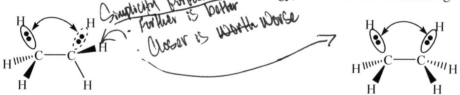

Dihedral angles

As stated above, the potential energy of the molecule changes with rotation through these conformational isomers. Repulsion of electrons in the bonds on the front carbon with electrons in the bonds of the rear carbon raises the potential energy of the molecule. The staggered conformation allows the electrons to be farther apart than the eclipsed conformation and this reduces the potential energy of the staggered conformation. There are a number of other factors that affect the conformational energy, but we are not considering them.

In the stagered conformation electron-electron repulsion is minimized. Electron pairs are farther apart and the potential energy is lower and the molecule is more stable.

In the eclipsed conformation electron-electron repulsion is maximized. Electron pairs are closer together and the potential energy is higher and the conformation is less stable.

Rotating the three hydrogen atoms (in front) past the three hydrogen atoms (in back) raises the potential energy by 2.9 kcal/mole over the staggered arrangements. Room temperature thermal background energy is approximately 20 kcal/mole, which allows the 2.9 kcal/mole energy fluctuations to occur many tens of thousands of times per second. The two CH_3 groups can be viewed as spinning propeller blades (only the bonds are spinning faster). The increase in energy due to eclipsing interactions is referred to as *torsional strain*. Since there are three such H-H interactions in ethane, the increase in energy for each must be (2.9 kcal/mole) / (3) ≈ 1.0 kcal/mole.

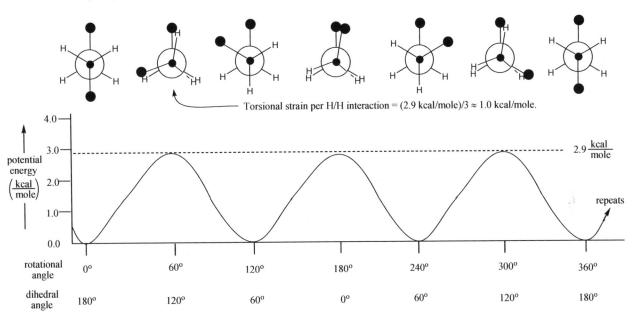

● = symbol for a hydrogen atom to keep track of rotations

Torsional strain per H/H interaction = (2.9 kcal/mole)/3 ≈ 1.0 kcal/mole.

Torsional strain specifically refers to an increase in potential energy due to eclipsing interactions of groups on adjacent atoms. In a more general sense, electron-electron repulsions through space are sometimes referred to as *steric effects*. Steric effects are more severe when groups are larger, sometimes having many atoms and pairs of electrons. Such groups can occupy a larger volume element and crowd other groups. The interfering groups need not be on adjacent atoms, but do have to be spatially close to cause steric crowding. Steric effects are one of the major arguments invoked in organic chemistry to rationalize structure and reactivity. Steric effect arguments were one of the earliest theories used by organic chemists to explain how and why things happened. Before the turn of the 19th century almost every rationalization involved some aspect of steric effects. Steric effects preceded the other major theories of inductive effects, resonance effects, solvent effects and molecular orbital arguments.

The equilibrium percentages can be roughly calculated if the relative energy differences of each conformational isomer are known (we are assuming $\Delta G° \approx \Delta H°$). In the case of ethane, a 2.9 kcal/mole potential energy difference predicts a preponderance of staggered conformations. Over 99% of the molecules will be staggered in ethane.

3 Main Ideas in CHEM

1. Inductive effects
2. Resonance "
3. Steric "

$$K = 10^{\frac{-\Delta G}{2.3RT}} \approx 10^{\frac{-\Delta H}{2.3RT}} = 10^{\frac{-(2900\ cal/mol)}{2.3(2\ cal/mol\text{-}K)(298K)}} = 10^{-2.12} = \frac{1}{130} = \frac{eclipsed}{staggered}$$

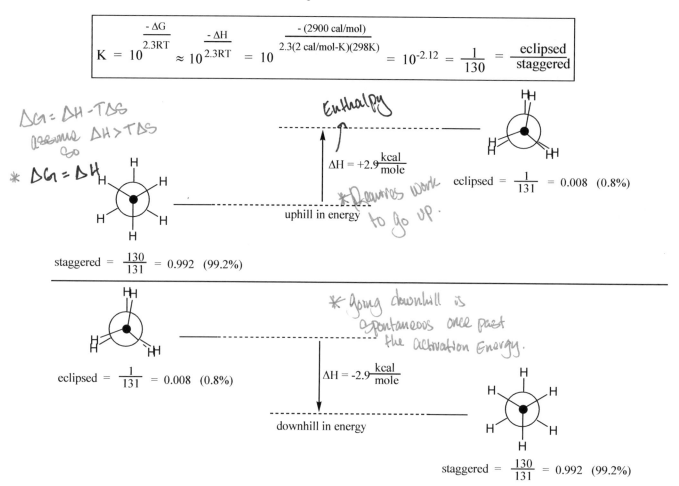

ΔG = ΔH - TΔS
assume ΔH > TΔS
so
* ΔG = ΔH

Enthalpy

ΔH = +2.9 kcal/mole

uphill in energy

*Requires work to go up.

eclipsed = $\frac{1}{131}$ = 0.008 (0.8%)

staggered = $\frac{130}{131}$ = 0.992 (99.2%)

eclipsed = $\frac{1}{131}$ = 0.008 (0.8%)

* going downhill is spontaneous once past the activation energy.

ΔH = -2.9 kcal/mole

downhill in energy

staggered = $\frac{130}{131}$ = 0.992 (99.2%)

The molecular collection of conformations is a dynamic equilibrium. Interchange is constantly going on tens of thousands of times per second. An instantaneous snap shot, however, would show approximately 130 molecules in the staggered conformation and only one in the eclipsed conformation (quickly on its way to being a staggered conformation).

Problem 1 - If chlorine atoms (Cl) are substituted for all hydrogen atoms in ethane (perchloroethane), the energy barrier to rotation is 10.8 kcal/mole. The increase in the energy barrier is thought to be, in part, due to the polar nature of the C-Cl bond (repulsion from the similar bond dipoles). Draw Newman projections and show the potential energy changes associated with them for a complete 360° rotation cycle. Label each conformation as staggered or eclipsed. Calculate the approximate percent of staggered versus eclipsed conformations.

Substitution of Hydrogen with other groups

Switching a hydrogen atom with a methyl group barely changes our analysis of the conformations possible. We do have one new decision to make and that's whether to sight down the C1→C2 direction or the C2→C1 direction. From the C1→C2 perspective, the extra methyl group will appear on the rear carbon atom, while it will appear on the front carbon atom from the C2→C1 perspective.

Chapter 5

Strategy to draw Newman projections.

   1. Find the two connected atoms to be used. Decide what direction of viewing will be used?

   2. The dot represents the front carbon and the circle represents the rear carbon.

   3. Decide what are the three additional bonds at the front carbon (draw lines connecting to the dot) and three additional bonds at the rear carbon (draw lines only down to the circle). Add in the bonded groups to the end of each line.

   4. Keep the rear carbon fixed and only rotate the front carbon.

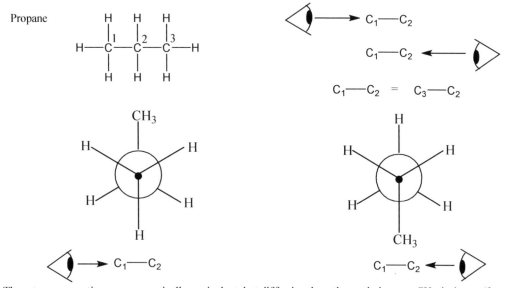

These two perspectives are energetically equivalent, but differ in where the methyl group, CH_3, is drawn (front or back).

   Using either perspective for propane, all of the staggered conformations will be equal in potential energy and all of the eclipsed conformations will be equal (and higher) in potential energy. You might expect the difference in energy between the staggered and eclipsed conformations would be greater because of the larger size of a methyl group over a hydrogen atom, and you would be correct. The eclipsed conformation is 3.4 kcal/mole higher in potential energy than the staggered conformation. Using 1.0 kcal/mol as the energy cost of H/H eclipsing interactions from the ethane calculation (there are two of them), we can calculate a value of about 1.4 kcal/mol for the energy cost of a CH_3/H eclipsing interaction.

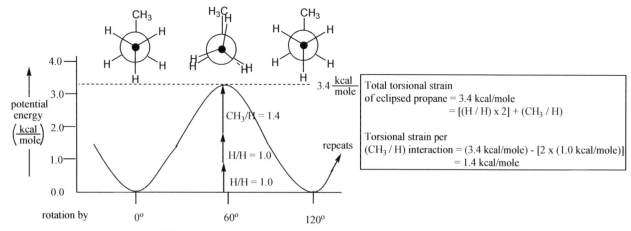

   As with ethane, the equilibrium percentages can be roughly calculated from the value of ΔH° (assuming $\Delta G^\circ \approx \Delta H^\circ$). For propane, a 3.4 kcal/mole potential energy difference predicts an even greater preponderance of staggered conformations (99.7%) than was found in ethane (99.2%).

Chapter 5

$$K \approx 10^{\frac{-\Delta H}{2.3RT}} = 10^{\frac{-(3400 \text{ cal/mol})}{2.3(2 \text{ cal/mol-K})(298K)}} = 10^{-2.48} = \frac{1}{302} = \frac{\text{eclipsed}}{\text{staggered}}$$

$$\text{staggered} = \frac{302}{303} = 0.997 \quad (99.7\%)$$

$$\text{eclipsed} = \frac{1}{303} = 0.003 \quad (0.3\%)$$

Butane is our next example. In butane we have four attached carbon atoms and we also have even more choices in how to draw the Newman projections.

Remember, we have to sight down two directly attached atoms. We could use the C_1-C_2 bond or the C_2-C_3 bond. (Why not include the C_3-C_4 bond?) In addition we could view our projection from either perspective $C_1 \rightarrow C_2$ or $C_2 \rightarrow C_1$ and $C_2 \rightarrow C_3$ or $C_3 \rightarrow C_2$. In butane, the $C_1 \rightarrow C_2$ and $C_2 \rightarrow C_1$ perspectives produce slightly differing views, while the views of $C_2 \rightarrow C_3$ and $C_3 \rightarrow C_2$ are equivalent because they are at the center of a symmetrical molecule.

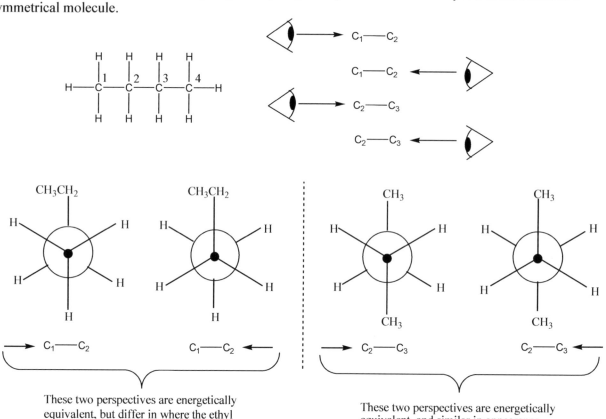

These two perspectives are energetically equivalent, but differ in where the ethyl group, CH_3CH_2, is drawn (front or back).

These two perspectives are energetically equivalent, and similar in appearance.

Our approach is to limit the perspectives to only staggered and eclipsed conformations, focusing on the extreme potential energy positions. In this book, staggered conformations are always lower in potential energy and eclipsed conformations are always higher in potential energy. Each of these can be set up in advance with the indicated degrees of rotation. The following set up works for all sp^3-sp^3 Newman projections. Once you decide which attached atoms you are using and which direction you are sighting from, you merely fill in the six attached groups from the name or a 2D drawing of the structure.

Chapter 5

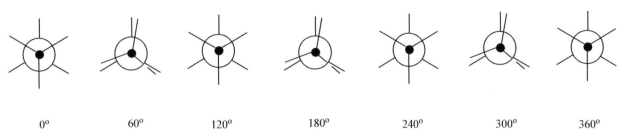

0°   60°   120°   180°   240°   300°   360°

Using the C2→C3 bond of butane and 60° rotations, we need to add the three additional groups at each of the carbon atoms. On the C_2 carbon atom of butane, two hydrogen atoms and a methyl group are attached. On C_3 there are also two hydrogen atoms and a methyl group attached. If we arrange these groups in a staggered conformation with one methyl pointing up (the rear carbon) and the other methyl pointing down (the front carbon), we obtain the lowest energy staggered conformation. If you build molecular models, this will likely be intuitively obvious to you. The larger methyl groups are as far apart in space as they can possibly get. To simplify our analysis we will fix the rear carbon atom, while we rotate the front carbon atom in 60° increments to obtain several additional conformational isomers. We can generate six different Newman projections before we begin to repeat ourselves.

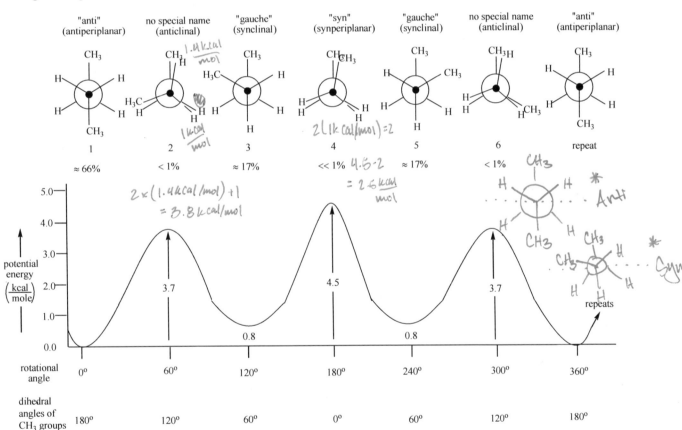

Are any of these conformations equivalent? The answer depends on what we mean by equivalent. If we mean they are exactly the same and can be superimposed on top of one another with no differences, then the answer is no. None of the conformational isomers are identical (until we begin to repeat our structures at 360°). Some of the structures appear to be very similar; for example 2 and 6 or 3 and 5. However, they are nonsuperimposable mirror images. If you have models, you can try and superimpose them on one another, but it just won't work.

Use your hands to help you understand this difference. Your hands look very similar, but are they identical? From your lifetime experience using a left hand and a right hand, you know they are different. The comparisons we are making between conformations 2 and 6 or 3 and 5 are similar to the comparison between your hands. The Greek word for hand is *chiral* and these conformations are referred to as being chiral. The pairs above (2,6) and (3,5) each represent a pair of nonsuperimposable mirror images called *enantiomers* (Greek = opposite unit). This property of stereoisomerism is developed more fully in the stereochemstry chapter.

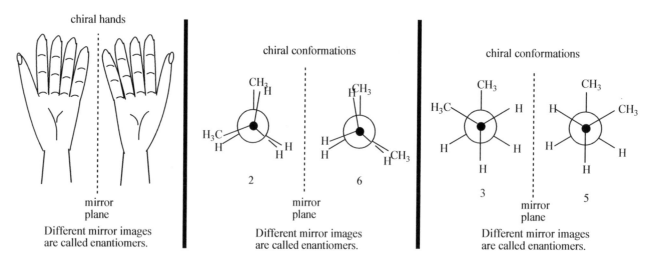

Even though structures (2 and 6) or (3 and 5) are not identical, they are equal in potential energy. In both 2 and 6 there are two eclipsing interactions of CH_3/H and one eclipsing interaction of H/H. Intuitively, most of us would view the eclipsing interaction of a methyl group, with its larger number of atoms and electrons, as more sterically demanding than a hydrogen atom. But we don't have to be intuitive here. We calculated this value in propane as 1.4 kcal/mole. The value estimated for eclipsing H/H interactions was 1.0 kcal/mole. Adding these three values together totals 3.8 kcal/mole [2x(CH_3/H) + 1x(H/H)], which is close to the experimental value of 3.7 kcal/mole for conformational isomers 2 or 6.

We can also use this new energy value to double check our previous calculation of the CH_3/H eclipsing energy. The two CH_3/H eclipsing interactions add an additional 3.7 - 2.9 = 0.8 kcal/mole of torsional strain energy over and above the H/H eclipsing interactions of ethane (0.4 kcal/mol per CH_3/H). Roughly, this totals 1.0 kcal/mole plus an extra 0.4 kcal/mole for each CH_3/H interaction = 1.4 kcal/mole, essentially the same value of 1.4 kcal/mole from propane.

There is an additional new and unique eclipsing conformation observed in butane that has the two methyl groups eclipsing one another. The potential energy of this conformation is raised by 4.5 kcal/mole over the most stable staggered conformation. The CH_3/CH_3 eclipsing interaction adds an additional 2.5 kcal/mole of torsional strain energy, as shown in the calculation below.

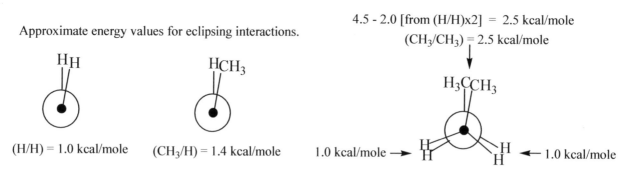

Each of the above butane conformations has a special identifying term. The lowest potential energy staggered conformation with a dihedral angle of 180° for the two methyl groups is call *antiperiplanar* or more commonly the "anti" conformation. The staggered arrangement where the CH_3 / CH_3 dihedral angle is 60° are called *synclinal*, or more commonly the "gauche" conformation. The two eclipsed conformations with 120° dihedral angles for the CH_3's are called *anticlinal*, but usually no specifying term is used. Finally, the highest potential energy conformation which has a dihedral angle of 0° for the two CH_3 groups is called *synperiplanar*

Chapter 5

or, more commonly, "syn". All other intermediate conformations are referred to as being skewed conformations and we usually ignore these.

anti
(antiperiplanar) → CH$_3$

staggered conformation (anti)
(usually the most stable conformation)

gauche
(synclinal) → CH$_3$

H$_3$C

staggered conformation as gauche (usually more stable than eclipsed conformations, but not as stable as "anti")

no common name
(anticlinal) → H$_3$C

H$_3$C

eclipsed conformation

syn
(synperiplanar)   H$_3$CCH$_3$

eclipsed conformation
(usually the least stable conformation)

H$_3$C

H$_3$C

H$_3$C

CH$_3$

skewed conformations are all other conformatons
than staggered or eclipsed (infinite possibilities)

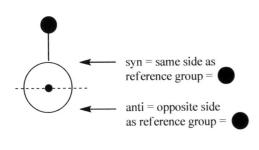

syn = same side as
reference group = ●

anti = opposite side
as reference group = ●

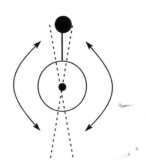

periplanar = approximately in the
same plane as the reference group

clinal = approximately sloping
away from the reference group

anti- opp. as Ref. group
Syn- Same as Ref.
group

Periplanar

Clinal

Chapter 5

There is an additional kind of strain energy present in butane, which sometimes goes by the name of *van der Waals strain*. Torsional strain occurs when groups repel each other while on adjacent atoms in eclipsed conformations. In contrast, van der Waals strain occurs when the groups crowd one another on more distantly separated atoms than adjacent atoms. In van der Waals strain, electron clouds from an atom or group are forced into the volume element of another atom or group and the repulsion energy rapidly goes up when two groups try to occupy one another's space. (What would you do if the person next to you tried to sit in your chair...while you were still in it?) A short list of van der Waals radii is provided below.

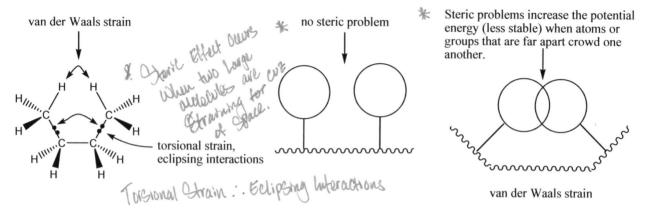

van der Waals radii in A (10^{-8} cm), atom or group

H (1.2 A)	C (1.85A)	N (1.55 A)	O (1.40 A)	F (1.35 A)	smaller
		P (1.9 A)	S (1.85 A)	Cl (1.80 A)	More electron shells increase the size of an atom.
		As (2.0 A)	Se (2.0 A)	Br (1.95 A)	
CH$_3$ (2.00A)			Te (2.2 A)	I (2.15 A)	larger

larger ← smaller

Higher Z_{eff} tends to contract the electron shells.

Problem 2 – Draw Newman projections for all staggered and eclipsed conformations of 2-methylpropane. Sight down the $C_1 \rightarrow C_2$ bond. The energy barrier to rotation is 3.9 kcal/mole. What value does this suggest for a CH$_3$/H eclipsing interaction? Is this reasonable from the values calculated earlier using butane or propane?

```
  1      2      3
         H
         |
H₃C——C——CH₃
         |
        CH₃
```

Problem 3 – Use the $C_1 \rightarrow C_2$ bond to draw a Newman projection for 2-methylbutane. What are the three additional groups on C_1? What are the three additional groups on C_2? Do the same using the $C_2 \rightarrow C_3$ bond. For each part, start with the most stable conformation and rotate through 360°. Point out all gauche, anti and syn methyl interactions and provide an estimate of what the relative energy for each conformation is (use the butane example as your guide). Start your energy scale at the lowest energy calculated. Assume an ethyl/H eclipsing interaction is 1.5 kcal/mole and a methyl/H eclipsing interaction is 1.4 kcal/mole and an ethyl/H gauche is 0.1 kcal/mole. Estimate the equilibrium ratio between the high and low energy conformations in the first example. In the second example estimate the equilibrium ratio between the lowest two staggered conformations.

Problem 4 – Methanamine (CH_3NH_2) has rotational barriers of 1.98 kcal/mole and methanol (CH_3OH) has rotational barriers of 1.07 kcal/mole. Write out the staggered and eclipsed conformational structures along with a potential energy versus angle of rotation diagram using Newman projections. What are the 3 additional groups on "N" and "O" in the Newman projections? Do your diagrams suggest that a lone pair of electrons has more or less torsional strain than a sigma bond with a hydrogen atom? Fill in the energy data in the table provided. Estimate the equilibrium ratios between the high and low energy conformations.

Eclipsing Groups	Increase in Potential Energy
H / H	
CH_3 / H	
CH_3 / CH_3	
lone pair / H	
gauche CH_3's	

Problem 5 – a. Draw the Newman projections and dash/wedge 3D structures to show the possible rotations about the C_2-C_3 bond of 2,3-dimethylbutane. (Does it make any difference which perspective you use?) Rotate through 60° increments and start with the most stable conformation. Estimate the relative potential energies of the conformations using the energy values from the previous problem.

Ellipsing group	Inc. in PE
H / H	1.0
CH_3 / N	1.4
CH_3 / CH_3	2.5
Lone pair H	0.0
gauche CH_3's	0.8

b. The highest energy barrier for rotation of 2,3-dimethylbutane is experimentally about 4.3 kcal/mole above the lowest energy conformation. Reevaluate your Newman projections for complete 360° rotation and state whether this seems reasonable with energy values above. (What is the energy difference between the most stable and least stable conformations?)

For more complex structures, having more variety, we will use the following table of eclipsing and gauche energies to calculate energy differences and predict relative amounts in conformational isomer problems.

Approximate Eclipsing Energy Values (kcal/mole)						
	H	Me	Et	i-Pr	t-Bu	Ph
H	1.0	1.4	1.5	1.6	3.0	1.7
Me	1.4	2.5	2.7	3.0	8.5	3.3
Et	1.5	2.7	3.3	4.5	10.0	3.8
i-Pr	1.6	3.0	4.5	7.8	13.0	8.1
t-Bu	3.0	8.5	10.0	13.0	23.0	13.5
Ph	1.7	3.3	3.8	8.1	13.5	8.3

$$\Delta G \approx \Delta H$$
$$K_{eq} = 10^{\frac{-\Delta H}{2.3RT}}$$

Approximate Gauche Energy Values (kcal/mole)						
	H	Me	Et	i-Pr	t-Bu	Ph
H	0	0	0.1	0.2	0.5	0.2
Me	0	0.8	0.9	1.1	2.7	1.4
Et	0.1	0.9	1.1	1.6	3.0	1.5
i-Pr	0.2	1.1	1.6	2.0	4.1	2.1
t-Bu	0.5	2.7	3.0	4.1	8.2	3.9
Ph	0.2	1.4	1.5	2.1	3.9	2.3

Problem 6 – Calculate the relative conformation energies of the conformations of 2,4-dimethylhexane shown below. Draw an energy pictures of the rotations. Use the $C_4 \rightarrow C_3$ perspective.

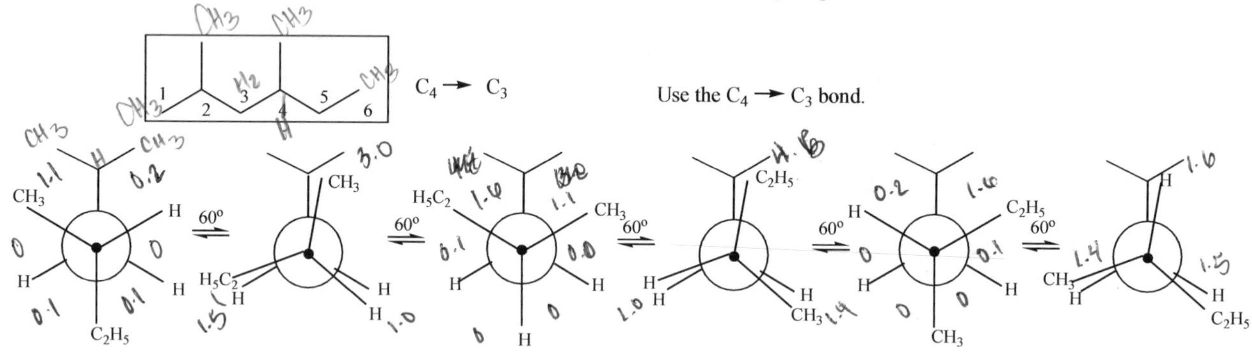

$C_4 \rightarrow C_3$ Use the $C_4 \rightarrow C_3$ bond.

1 most stable   2   3   4   5   6

Problem 7 – Calculate the relative conformation energies of the conformations of 2,2-dimethyl-4-phenylhexane shown below. Draw an energy pictures of the rotations. Use the $C_3 \rightarrow C_4$ perspective.

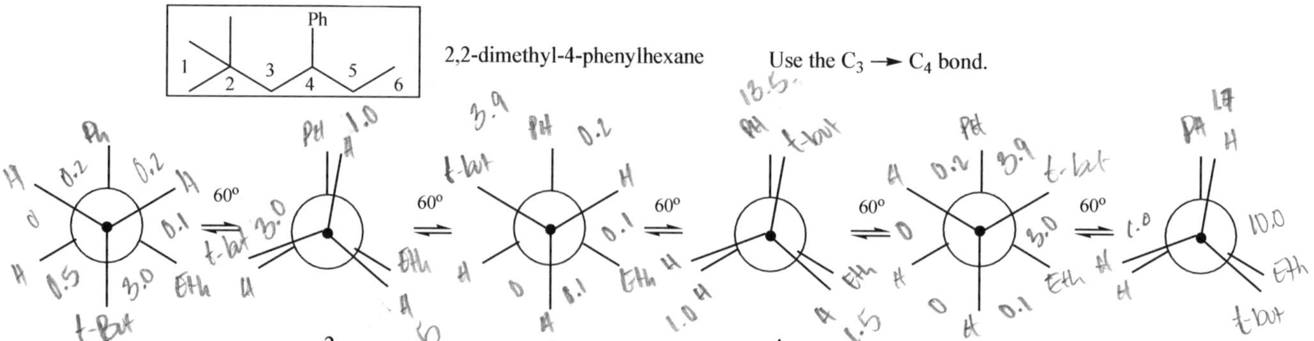

2,2-dimethyl-4-phenylhexane   Use the $C_3 \rightarrow C_4$ bond.

1 most stable   2   3   4   5   6

IN Class Example! :')

Ex1: Calc the Relative Conformation Energies of Conf. of
2-methyl-3-(1-methylethyl)4-phenylhexane. Show PE
Changes for Rotations around the $C_3 \rightarrow C_4$ bond.

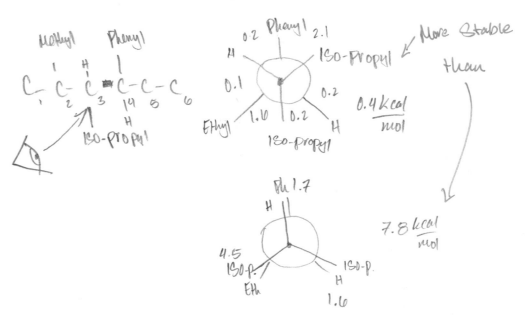

0.4 $\frac{kcal}{mol}$

7.8 $\frac{kcal}{mol}$

Chapter 6

CHAPTER 6 – CONFORMATIONS OF CYCLOHEXANE

Cyclohexane rings (six atom rings in general) are the most well studied of all ring systems. They have a limited number of, almost strain free, conformations. Because of their well defined conformational shapes, they are frequently used to study effects of orientation or steric effects when studying chemical reactions. Additionally, six atom rings are the most commonly encountered rings in nature. For these reasons, and our time is limited, we will only study cyclohexane rings.

If cyclohexane rings were flat (as Baeyer had originally assumed back in 1885), there would definitely be angle strain. The geometrical internal angles of a hexagon are 120°, much wider than the 109° bond angles expected for sp³ carbon. The bonds in a cyclohexane ring would have to point inward (if this were the case), resulting in a loss of overlap of electron density. Of course a flat structure would also produce a large torsional strain energy from eclipsing interactions at all six carbons and we would not be emphasizing cyclohexane structures.

In a hypothetical flat cyclohexane the sp³ hybrid orbitals would protrude into the center of the ring resulting in ring strain.

The real C-C-C bond angle in cyclohexane is 111.4, very close to the straight chain angle.

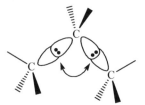

Typical straight chain C-C-C bond angle is 112°.

However, cyclohexane does not choose to be flat. Slight twists at each carbon atom allow cyclohexane rings to assume much more comfortable conformations, which we call *chair conformations*. (Chairs even sound comfortable.) The chair has an up and down shape all around the ring, sort of like the zig-zag shape seen in straight chains (...time for models!).

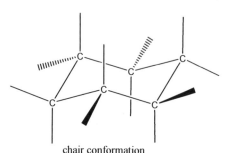

chair conformation

lounge chair - used to kick back and relax while you study your organic chemistry

Cyclohexane rings are flexible and easily allow partial rotations (twists) about the C-C single bonds. There is minimal angle strain since each carbon can approximately accommodate the 109° of the tetrahedral shape. Torsional strain energy is minimized in chair conformations since all groups are staggered relative to one another. This is easily seen in a Newman projection perspective. An added new twist to our Newman projections is a side-by-side view of parallel single bonds. If you look carefully at the structure above or use your model, you should be able to see that a parallel relationship exists for all C-C bonds across the ring from one another.

 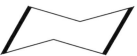 

There are three sets of parallel C-C bonds in cyclohexane rings. Any 'set' could be used to draw two parallel Newman projections.

Two simultaneous Newman projection views are now possible as shown below. Remember, that any two bonds on opposite sides of the ring can be used and they can viewed from front or rear directions.

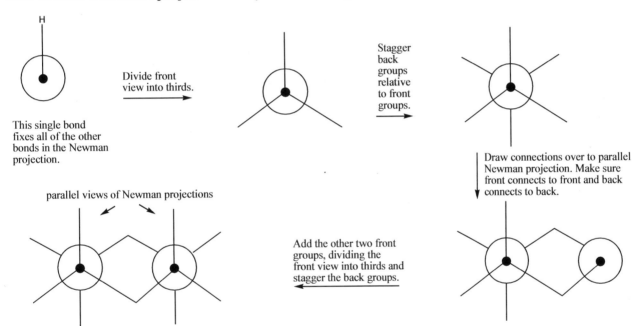

Chair conformations can be viewed down any two parallel bonds (from either side) as two parallel Newman projections. In this example, the view is from the left side.

All groups are staggered in chair conformatios of cyclohexane rings.

Once the first bond is drawn in a Newman projection of a chair conformation of cyclohexane ring, all of the other bonds (axial and equatorial) are fixed by the staggered arrangements and the cyclic connections to one another. Once you have drawn all of the bonds in a Newman projection, you merely "fill in the blanks" at the end of each line, based on the substituents that are present (determined from the name or a 2D drawing).

How to draw a Newman projection of a cyclohexane ring.

This single bond fixes all of the other bonds in the Newman projection.

Divide front view into thirds.

Stagger back groups relative to front groups.

Draw connections over to parallel Newman projection. Make sure front connects to front and back connects to back.

parallel views of Newman projections

Add the other two front groups, dividing the front view into thirds and stagger the back groups.

Ideal dihedral angles of cyclohexane (the real ones are slightly different)

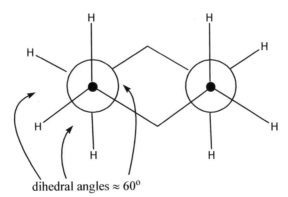

dihedral angles ≈ 60°

Chapter 6

Problem 1 - How many possible Newman perspectives of chairs are possible on the numbered cyclohexane below? Consider all numbers and front/back views. Write them as, C1→C2, C5→C4, etc.)

A Newman projection of a chair conformation of cyclohexane clearly shows that torsional strain is minimized, since all groups are staggered. Additionally, either bond-line or Newman formulas reveal that the two hydrogen atoms at each CH_2 are not the same. Three hydrogen atoms point straight up and three H's point straight down (at alternate positions) relative to the approximate plane of the carbon atoms. Those six parallel C-H bonds are also parallel to the highest axis of symmetry, (which we are ignoring in this book). Because these six C-H bonds are parallel to an axis of symmetry, they are called the axial positions (axial = along an axis). There is one axial position at every carbon atom in the ring and, again if you look carefully, you will see that axial positions alternate pointing up and then pointing down. You should also observe that when they point up they are on the top face of the ring, and when they point down they are on the bottom face of the ring.

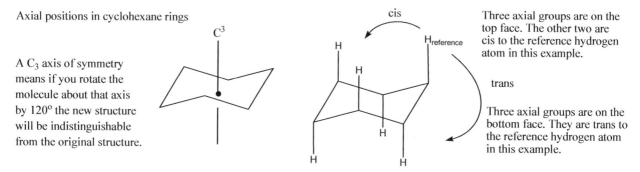

Axial positions in cyclohexane rings

A C_3 axis of symmetry means if you rotate the molecule about that axis by 120° the new structure will be indistinguishable from the original structure.

cis

Three axial groups are on the top face. The other two are cis to the reference hydrogen atom in this example.

trans

Three axial groups are on the bottom face. They are trans to the reference hydrogen atom in this example.

There are six other hydrogen atoms present, one at each of the carbon atoms in the ring. However, these hydrogen atoms are approximately in the plane of the ring of carbon atoms. These hydrogen atoms all point out, around the perimeter of the ring (…or around the "equator"). These positions are called *equatorial positions* (equator ≈ equatorial). The equatorial positions also alternate around the ring, being on top when axial points down and being on the bottom when axial points up.

These terms (axial and equatorial) create a common point of confusion for students. When students hear that two groups are axial, they often assume they must be the same in all respects, i.e. on the same side (cis). This is true at alternate positions about the ring, but it is not true at adjacent positions or directly across the ring, where they are trans.

Equatorial positions

O = top side, trans to reference hydrogen atom

* = bottom side, cis to reference hydrogen atom

Drawing a Cyclohexane

You will need to learn how to draw a cyclohexane ring, with all of its axial and equatorial positions, if you are to ever understand cyclohexane and related ring structures. Once you can draw cyclohexane rings, it is simply "fill in the blank" to add the substituent atoms (using a name as your guide). You can generate an endless number of substituted possibilities. If you switch out a carbon atom for an oxygen atom then you are set to study the pyranoses of biochemistry (e.g. cyclic glucose). As you work through this section, keep your models by your side, and use them.

First you have to generate the outline of the ring. (You should actually practice these steps several times.) This can be done by drawing two parallel lines, slightly off center, tracing slightly up or slightly down. Next a V shape is added on both sides so that it points in an opposite sense to the direction of the parallel lines. These two chair conformations are the most common shapes of cyclohexane rings, and interconvert with one another (thousands of times per second).

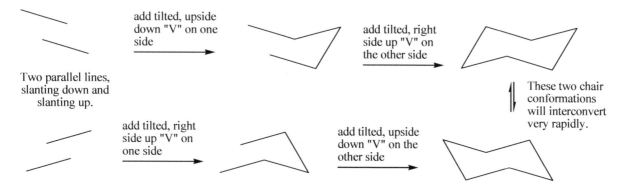

An alternate strategy is to draw a slanted V shape first, then add the parallel lines in an opposite direction. Then the other V shape is added in the opposite direction to the parallel lines. Use whatever works best for you. Make sure you practice until you can do this with ease.

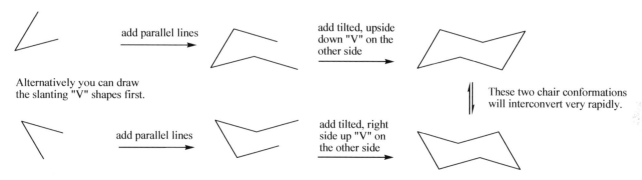

Our next task is to draw in the axial and equatorial positions. The ring is very helpful to us in this regard. The ring always points to the axial positions. If a ring carbon atom is in an up position, the axial position is pointing straight up (on the top of the ring). If a ring carbon atom is in a down position, the axial position is pointing straight down (on the bottom of the ring). The axial positions can now be added to our chair below.

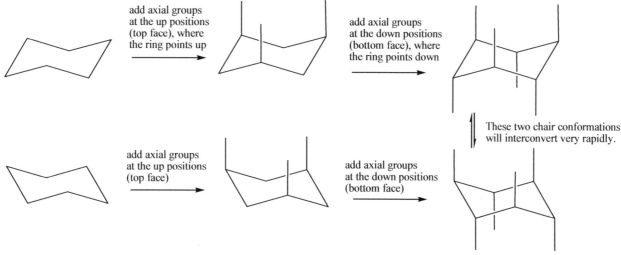

Chapter 6

The equatorial positions are added last, about the perimeter (or equator) of the ring. The two opposite positions, in the plane of the paper are the easiest to add since our perspective is a side-on view and they follow the zig-zag shape we expect in straight chains. Notice that one of these equatorial positions is on the top of the ring and one of them is on the bottom. They should both be drawn parallel to the ring bonds that are parallel to the plane of the paper.

Add equatorial positions in the plane of the page.

equatorial bond on the top

equatorial bond on the bottom

○ = equatorial position

□ = axial position

Axial and equatorial positions switch places in the two conformations.

Add equatorial positions in the plane of the page.

equatorial bond on the bottom

equatorial bond on the top

○ = axial position

□ = equatorial position

The other equatorial positions require slightly better artistic skills. To represent the equatorial positions coming out of the plane of the page, toward the viewer, wedges are drawn; if projecting behind the page, away from the viewer, dashed lines are drawn. These bonds will be parallel to bonds in the ring already drawn and in fact they have an anti relationship with those parallel bonds (use models). Use the ring bonds to guide you in drawing the equatorial bonds in the ring.

bold lines indicate parallel bonds

bold lines indicate parallel bonds

Heavy lines indicate parallel equatorial and ring bonds (above). Use the ring bonds to help you draw the equatorial bonds and then add the proper 3D perspective as simple lines, wedged lines or dashed lines (below).

Equatorial positions in the two chair conformations of a cyclohexane ring.

This is what you need to be able to draw to analyze six atom rings.

conformations rapidly interconvert

Equatorial and axial positions in the two chair conformations of a cyclohexane ring.

If you practice this several times, you should be able to generate a chair cyclohexane structure in seconds. Substituents can be added as indicated from a chemical name. These two chair conformations are the most common and comfortable of all the conformational possibilities available to a cyclohexane ring. Almost all of your work with cyclohexanes will involve chair conformations.

Chapter 6

If you don't acquire this skill, you may draw amusing, misshapen cyclohexanes that are good for a chuckle, but will usually confuse you and your audience as you try to fill in appropriate substituents. You won't be able to analyze expected conformational effects in chemical reactions. Get out your pencil and paper and take the time to develop your chemical art skills.

Boat and Twist Boat Conformations

There are additional conformations of cyclohexane rings: boat, twist boat and half-chair conformations. These are high potential energy conformations that are encountered when one chair rotates to the other chair conformation. Torsional strain and van der Waals strain cause an increase in the potential energy of these conformations. We will take a limited look at the boat conformation (but not the twist boat or half-chair). Occasionally these conformations are found in rigid, fixed structures having cyclohexane subunits, either by design or in nature.

Boat Conformation

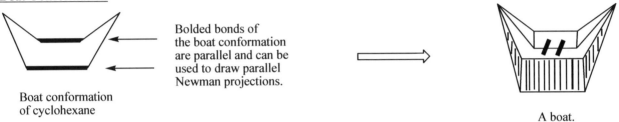

Bolded bonds of the boat conformation are parallel and can be used to draw parallel Newman projections.

Boat conformation of cyclohexane

A boat.

A Newman projection viewing down the two parallel bonds of the boat conformation clearly shows the increased torsional strain from eclipsing interactions.

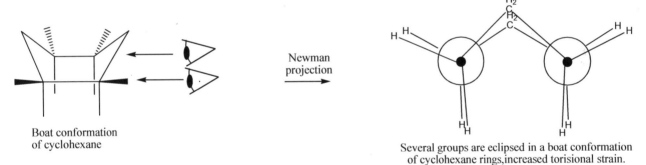

Newman projection

Boat conformation of cyclohexane

Several groups are eclipsed in a boat conformation of cyclohexane rings, increased torisional strain.

An additional destabilizing feature present in the boat is van der Waals strain between the two hydrogen atoms pointing inward, towards each. These interfering groups are called "flagpole" interactions, and also contribute to the overall ring strain of the boat conformation.

The van der Waals radius for a hydrogen atom is 1.2 A and represents the radius of a sphere of space occupied by the electron cloud of a hydrogen atom. The nuclei of two hydrogen atoms would be expected to approach no closer than 2.4 A (= 2 x 1.2 A) without considerable repulsion. The flagpole hydrogen atoms are 1.8 A apart in the boat conformation. We classify this as a steric effect.

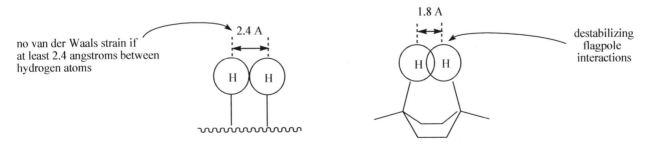

no van der Waals strain if at least 2.4 angstroms between hydrogen atoms

2.4 A

1.8 A

destabilizing flagpole interactions

Chapter 6

A chair does not immediately become a boat, and then the other chair. There is an even higher transitory conformation, which is called a half chair. Due to the torsional strain and bond angle strain this conformation is even higher in potential energy than the boat. We can consider the half chair as a transition state that exists only on the way to a chair or boat conformation. The potential energy of the half chair is estimated to be about 11 kcal/mole. A potential energy diagram for interconverting the two possible chair conformations would look something like the following.

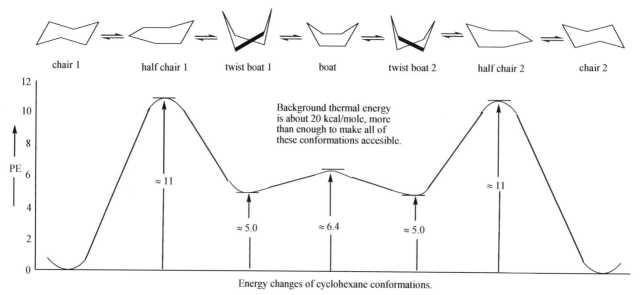

Energy changes of cyclohexane conformations.

However, we will view the interconversion of the two cyclohexane conformations, simplistically, as chair 1 in equilibrium with chair 2 via the boat conformation.

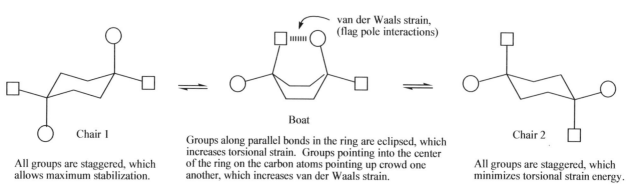

Equilibration, back and forth, between the two chair conformations is rapid at room temperature and occurs on the order of 80,000 times per second. At lower temperatures interconversion is much slower (there is less thermal background energy). At -40°C the interconversion occurs about 40 times per second, at -120°C interconversion occurs about once every 23 minutes and at -160°C it is estimated to occur once every 23 years. The equilibration can occur in either of two directions as shown below.

Chapter 6

Notice that all axial positions (top and bottom) become equatorial positions with the flip-flop of two chairs. Of course, all equatorial positions become axial positions at the same time.

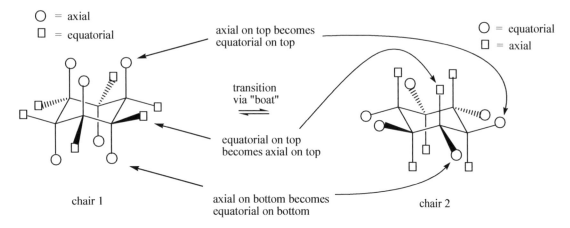

Since the two chair conformations are so much more stable than the boat or half chair over 99.99% of the molecules are in one of the two chair conformations. If all of the ring substituents are hydrogen atoms, there should nearly be a 50/50 mixture of two indistinguishable chair structures.

$$K = \frac{boat}{chair} = 10^{\frac{-\Delta G}{2.3RT}} = 10^{\frac{-(6400 \text{ cal/mole})}{(2.3)(2 \text{ cal/mol-K})(298 \text{ K})}} = 10^{-4.60} = 2.1 \times 10^{-5} = \frac{0.002\%}{99.998\%}$$

$$K = \frac{chair\ 1}{chair\ 2} = 10^{\frac{-\Delta G}{2.3RT}} = 10^{\frac{-(0 \text{ cal/mole})}{(2.3)(2 \text{ cal/mol-K})(298 \text{ K})}} = 10^0 = \frac{1}{1} = \frac{50\%}{50\%}$$

Real molecules actually exist that illustrate most of the cyclohexane conformations mentioned, above, as substructures (chair, boat and twist boat). Adamantane has four chair cyclohexane rings in its complicated tricyclic arrangement. Twistane has a good example of a twist boat conformation and is isomeric with adamantane. Norbornane has a boat conformation locked into its rigid bicyclic framework with a bridging CH_2 holding the boat shape in place. Each of these is highlighted below. Your models can help you see this a lot more clearly than these pictures.

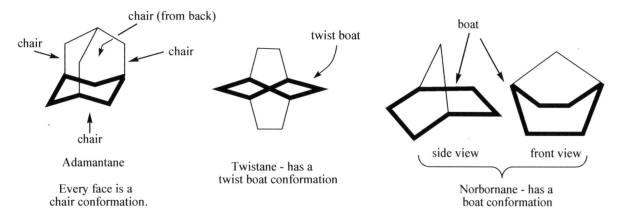

Adamantane

Every face is a
chair conformation.

Twistane - has a
twist boat conformation

Norbornane - has a
boat conformation

Chapter 6

Mono Substituted Cyclohexanes

A single substitution of a hydrogen atom with another group complicates the analysis of cyclohexane conformations. There are still two rapidly equalibrating chair conformations, but they are no longer equal in potential energy.

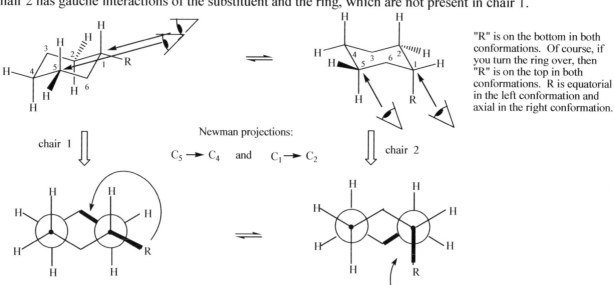

"R" is on the bottom in both conformations. Of course, if you turn the ring over, then "R" is on the top in both conformations. R is equatorial in the left conformation and axial in the right conformation.

A Newman perspective using the $C_1 \rightarrow C_2$ and $C_5 \rightarrow C_4$ bonds can help to evaluate which of these two conformations is more stable. Both chair conformations have an all staggered orientation about the ring, but chair 2 has gauche interactions of the substituent and the ring, which are not present in chair 1.

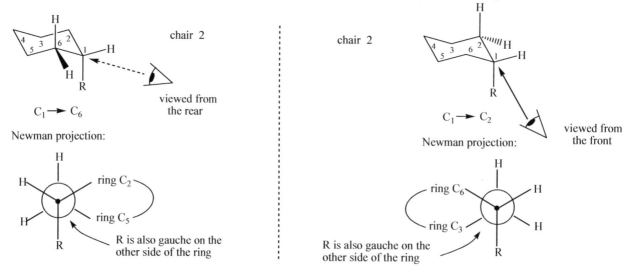

"R" is on the bottom in both conformations. Of course, if you turn the ring over, then "R" is on the top in both conformations. R is equatorial in the left conformation and axial in the right conformation.

chair 1

Newman projections:

$C_5 \rightarrow C_4$ and $C_1 \rightarrow C_2$

chair 2

Equatorial "R" is anti to the ring on two sides. Only one shows in this Newman projection.

Axial "R" is gauche to the ring on two sides. Only one shows in this Newman projection.

If we view down the $C_1 \rightarrow C_6$ bond and the $C_1 \rightarrow C_2$ bond, it is clear that an axial R substituent is gauche with both sides of the ring. Any axial substituent in a cyclohexane ring really has two gauche interactions.

chair 2

chair 2

$C_1 \rightarrow C_6$

viewed from the rear

$C_1 \rightarrow C_2$

viewed from the front

Newman projection:

Newman projection:

ring C_2

ring C_5

R is also gauche on the other side of the ring

ring C_6

ring C_3

R is also gauche on the other side of the ring

Chapter 6

We have previously seen that a CH_3/CH_3 gauche interaction raises the potential energy about 0.8 kcal/mole.

anti CH_3/CH_3 = 0 kcal/mole   gauche CH_3/CH_3 = 0.8 kcal/mole

Two such relationships should raise the energy by approximately (2)x(0.8 kcal/mole) = 1.6 kcal/mole. The actual value observed for an axial CH_3 in cyclohexane relative to equatorial is 1.7 kcal/mole, very close to what we expect.

$$K = \frac{\text{chair 2}}{\text{chair 1}} = 10^{\frac{-\Delta G}{2.3RT}} = 10^{\frac{-(1700 \text{ cal/mole})}{(2.3)(2 \text{ cal/mol-K})(298 \text{ K})}} = 10^{-1.2} = 0.058 = \frac{1}{17} = \frac{6\%}{94\%} = \frac{\text{axial } CH_3}{\text{equatorial } CH_3}$$

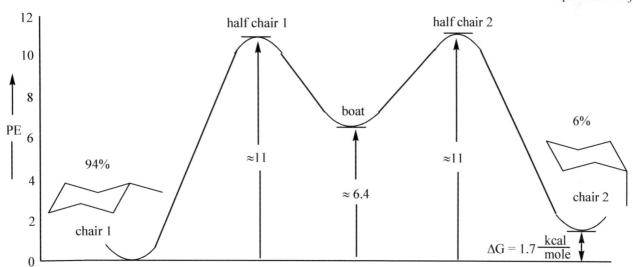

Energy changes of methylcyclohexane conformations.

The two gauche interactions of axial R substituents and the ring carbon atoms in cyclohexane structures, are called 1,3-diaxial interactions. In the axial position, the substituent R, is forced close to the other two axial groups on the same side of the ring. Since these are both three atoms away from the ring carbon atom with the R substituent, the 1,3-diaxial descriptor is appropriate.

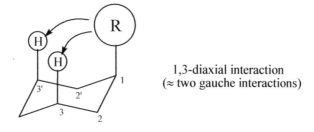

1,3-diaxial interaction
($\approx$ two gauche interactions)

Chapter 6

Problem 2 - Which boat conformation would be a more likely transition state in interconverting the two chair conformations of methylcyclohexane...or are they equivalent? Explain your answer.

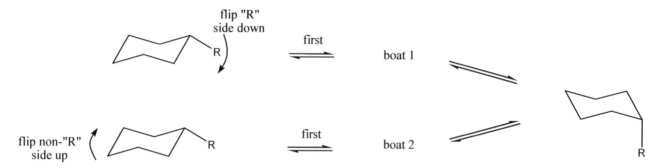

A wide range of substituents has been studied and almost all show a preference for the equatorial position. Several examples are listed in the table below. The actual energy difference between equatorial substituents and axial substituents is often called the A value (axial strain). Larger A values indicate a greater equatorial preference for the substituent due to larger destabilizing 1,3-diaxial interactions (or double gauche interactions).

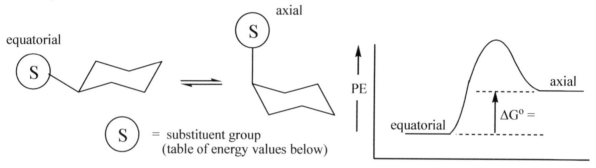

Substituent	ΔG^o (A value)		Substituent	ΔG^o (A value)
-H	0.0		-CH$_2$OH	1.8
-CH$_3$	1.7		-CH$_2$Br	1.8
-CH$_2$CH$_3$	1.8		-CF$_3$	2.4
-CH(CH$_3$)$_2$	2.1		-O$_2$CCH$_2$CH$_3$	1.1
-C(CH$_3$)$_3$	> 5.0		-OH	0.9
-F	0.3		-OCH$_3$	0.6
-Cl	0.5		-SH	1.2
-Br	0.5		-SCH$_3$	1.0
-I	0.5		-SC$_5$H$_6$	1.1
-CH=CH$_2$	1.7		-SOCH$_3$	1.2
-CH=C=CH$_2$	1.5		-SO$_2$CH$_3$	2.5
-CCH	0.5		-SeC$_6$H$_5$	1.0
-CN	0.2		-TeC$_6$H$_5$	0.9
-C$_6$H$_5$ (phenyl)	2.9		-NH$_2$	1.2(C$_6$H$_5$CH$_3$), 1.7(H$_2$O)
-CH$_2$C$_6$H$_5$ (benzyl)	1.7		-N(CH$_3$)$_2$	1.5 (C$_6$H$_5$CH$_3$), 2.1(H$_2$O)
-CO$_2$H	0.6		-NO$_2$	1.1
-CO$_2\ominus$	2.0		-HgBr	0.0
-CHO	0.7		-HgCl	-0.2
			-MgBr	0.8

Chapter 6

The energy differences in the table between equatorial and axial positions for substituent groups from methyl through ethyl, isopropyl and *t*-butyl appear puzzling. Ethyl definitely is larger than methyl and isopropyl is larger than either of those two, but they all have similar A values. The A value for the t-butyl group, on the other hand, suddenly jumps to a much higher energy. Examination of the three dimensional axial conformation provides insight into this observation.

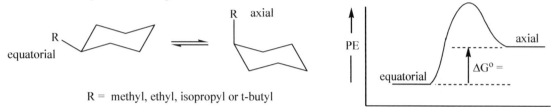

R = methyl, ethyl, isopropyl or t-butyl

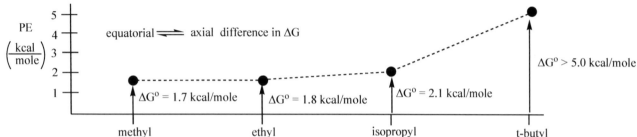

Compare axial interactions on the top face of the two conformations.

Rotation is possible around this bond, which allows different "R" groups to face in towards the center of the ring.

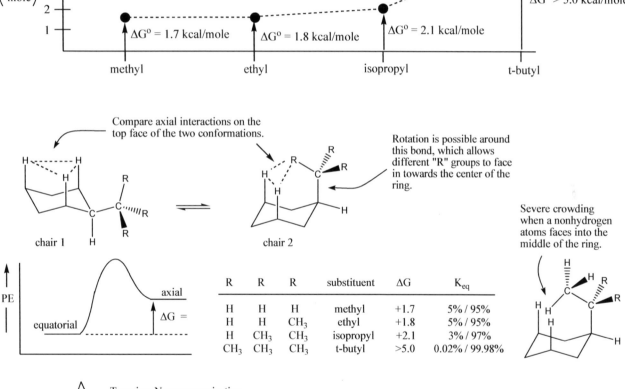

chair 1

chair 2

Severe crowding when a nonhydrogen atoms faces into the middle of the ring.

R	R	R	substituent	ΔG	K_{eq}
H	H	H	methyl	+1.7	5% / 95%
H	H	CH_3	ethyl	+1.8	5% / 95%
H	CH_3	CH_3	isopropyl	+2.1	3% / 97%
CH_3	CH_3	CH_3	t-butyl	>5.0	0.02% / 99.98%

Top view Newman projection.

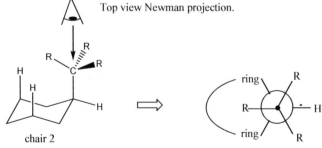

chair 2

R can be a hydrogen atom pointing into the center of the ring, when the axial substituent is methyl, ethyl and isopropyl. It is not possible when the ring substituent is t-butyl. If the t-butyl substituent is axial, a methyl has to point into the center of the ring. This is very destabilizing and for the most part, t-butyl substituents are in the equatorial position.

In the case of methyl, ethyl or isopropyl there is always a hydrogen atom that can (and most often will) point inward toward the center of the ring. Free rotation is possible about the R_3C-C(ring) bond. Since van der Waals strain is lower for a hydrogen atom pointing toward the middle of the ring than methyl, any substituent from methyl (R = H, R = H, R = H) to ethyl (R = H, R = H, R = CH_3) to isopropyl (R = H, R = CH_3, R = CH_3) will choose an inward pointing hydrogen as its preferred conformation. When the substituent is *t*-butyl (R = CH_3, R = CH_3, R = CH_3), there is no such option. All possible orientations of an axial *t*-butyl point a CH_3 toward the center of the ring, which is extremely destabilizing (easy to see with a model) and chair 1 is overwhelmingly preferred.

Even though an equatorial substituent is preferred for all of the examples above, both chair conformations are present in rapid equilibrium. What varies is the percent of each chair conformation that is present. Even with *t*-butyl as the substituent, a tiny fraction of cyclohexane rings will have a transient, axial *t*-butyl (see the table above).

Problem 3 – a. Propose an explanation for why ethenyl (-CH=CH_2) has a larger preference for the equatorial position than ethynyl (-CCH).

substituent =

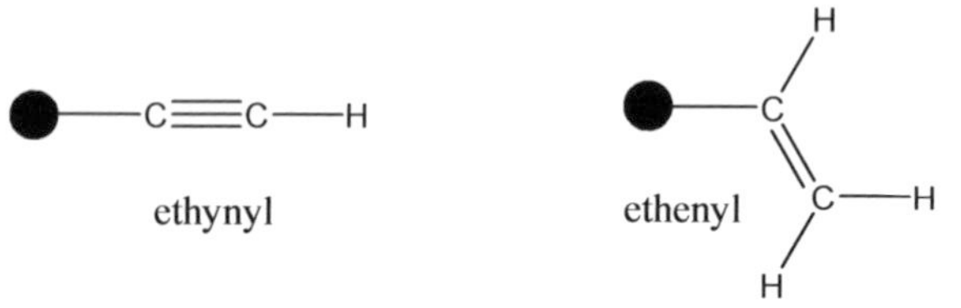

● = attachment to the ring

b. Propose an explanation for why ethenyl (-CH=CH_2) has a smaller preference for the equatorial position than phenyl (-C_6H_5).

substituent =

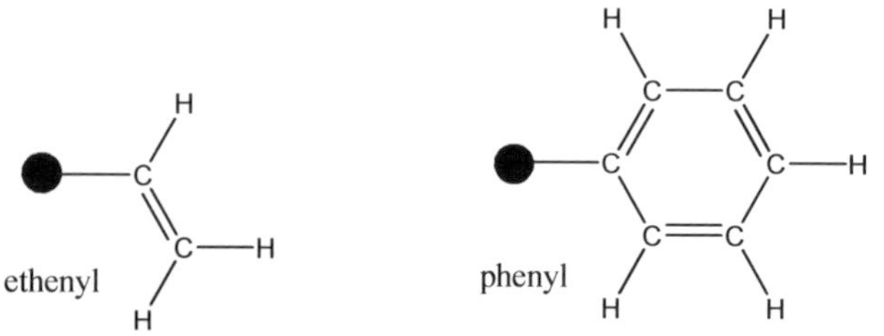

● = attachment to the ring

c. Benzyl would seem to be a larger group than phenyl, but has a smaller A value. Propose a possible explanation.

substituent =

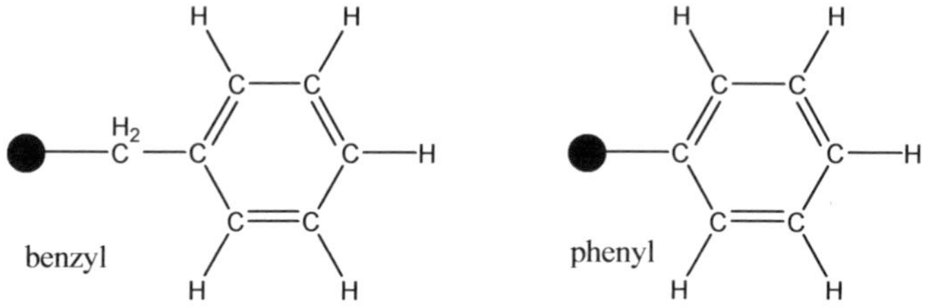

● = attachment to the ring

Disubstituted Cyclohexanes

Here is an excellent example of why memorization does not work in organic chemistry. Not only is it easier to learn a limited number of basic principles and logically use them in newly encountered situations, it is down right impossible to memorize every possible situation you might encounter.

When we add a second substituent to monosubstituted cyclohexane rings, the possibilities increase tremendously. Seven flat ring structures are drawn below, which just emphasize the top and bottom positions on the ring. Flat structures can be drawn as time average approximations between two interconverting chairs, although in actuality none of the structures are flat. These structures are only used as a short hand to show that substituents are on the same side or are on opposite sides. In biochemistry they are sometimes called Haworth projections and commonly used with cyclic sugar molecules.

Chapter 6

Haworth projections: $R_1 = R_2$ $R_1 \neq R_2$ gauche interactions are possible when substituents are 1,2-substituted

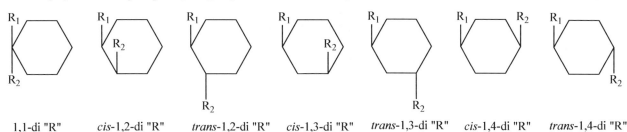

| 1,1-di "R" | cis-1,2-di "R" | trans-1,2-di "R" | cis-1,3-di "R" | trans-1,3-di "R" | cis-1,4-di "R" | trans-1,4-di "R" |

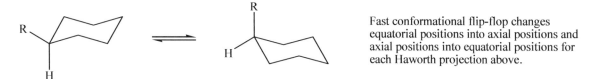

Fast conformational flip-flop changes equatorial positions into axial positions and axial positions into equatorial positions for each Haworth projection above.

   Even though we have gone from a single monosubstituted cyclohexane to seven disubstituted cyclohexane rings, the situation is even more complicated yet! Just as our monosubstituted cyclohexane had two chair conformations, there are two possible chair conformations for each of the flat structures shown above. In some cases the two conformations are equivalent in energy, but in other cases one conformation is preferred.

   We need a systematic method of analysis or we will quickly become hopelessly lost in the wilderness of flip-flopping cyclohexane rings. I recommend the following strategy for every cyclohexane analysis.

Possible systematic approach to Analyze Cyclohexane Conformations

1. Draw a chair cyclohexane ring framework as a bond-line formula.
2. Add both axial and equatorial positions. Axials point straight up or down in alternating fashion (the ring points to the axial positions). Equatorials are off to the side (use the axial positions to guide you as to top and bottom positions). Both alternate on top or bottom of the face of the ring as you move around the ring. Use parallel bonds in the ring to guide you where to draw the equatorial bonds.
3. Add in the necessary substituents according to the name of the structure (fill in the blank). It is generally easier to visualize substituents drawn on the extreme left or extreme right carbon atoms of the ring because those bonds will be in the plane of the paper, so these are good places to draw your first substituent.
4. Draw the other conformation by flipping one side up and flip the other side down. All of the axial and equatorial positions will interchange, but the top will still be the top and the bottom will still be the bottom.
5. In addition to 1,3-diaxial interactions look for an extra gauche interactions when substituents are substituted 1,2 (vicinal substitution). We will use gauche values from the table in Chapter 5 (page 160) or 0.8 if not available.
6. Evaluate which is the preferred conformation using the available energy values. Use ΔH to calculate a ratio.

$$K = 10^{\frac{-\Delta G}{2.3RT}} \approx 10^{\frac{-\Delta H}{2.3RT}}$$

Chapter 6

Example 1   *cis* 1,4-dimethylcyclohexane

1.Draw the bond line formula of a chair.

   *cis* 1,4-dimethylcyclohexane

2. Add axial and equatorial positions (ring points axial).

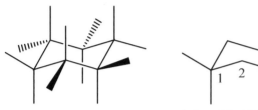

...or just this, if 1,4-disubstituted.

You can simplify this step somewhat by just adding the axial and equatorial positions of the substituted carbon atoms. Include both axial and equatorial positions at substituted positions, even if a hydrogen atom occupies one of them. If you can draw this template structure, you should be able to draw any possible structure.

3. Add the indicated substituents. Generally the easiest positions to fill are the positions on the far left and far right carbon atoms (1 and 4, just above), since their representation is drawn in the plane of the page. At least the first substituent should be drawn at one of these positions and then fill in any other substituent(s) as required from the name.

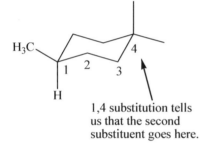

1,4 substitution tells us that the second substituent goes here.

cis = same side, so both CH$_3$ groups should be on the same side (top side as written in this example). If you turned the structure over, the two methyl groups would still be on the same side, but that side would be the bottom.

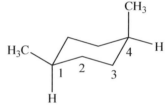

One CH$_3$ is axial and one CH$_3$ is equatorial, but both are on the same side.

4. Draw the other conformation (flip up, flip down). Axial/equatorial interchange occurs but the structure is still *cis*.

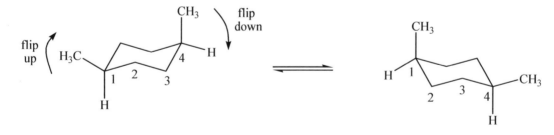

5. There is one axial CH$_3$ (+1.7 kcal/mole) and one equatorial CH$_3$ in each conformation.

6. 1,2-gauche interactions are not applicable in this example, since there are no vicinal substituents.

7. Since the two conformations are equivalent in energy we would expect a 50/50 mixture.

$$K = \frac{\text{chair 2}}{\text{chair 1}} = 10^{\frac{-\Delta G}{2.3 RT}} = 10^{\frac{-(0\ \text{cal/mole})}{(2.3)(2\ \text{cal/mol-K})(298\ K)}} = 10^0 = \frac{1}{1} = \frac{50\%}{50\%}$$

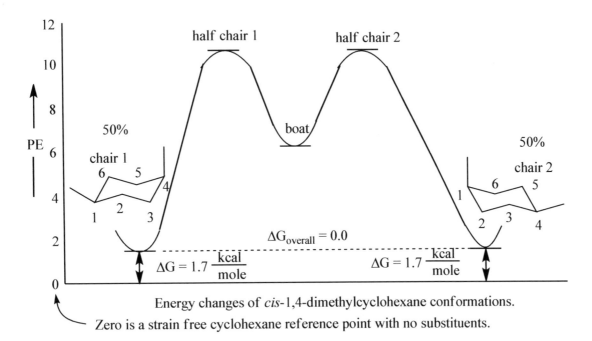

Energy changes of *cis*-1,4-dimethylcyclohexane conformations.
Zero is a strain free cyclohexane reference point with no substituents.

Newman Projections

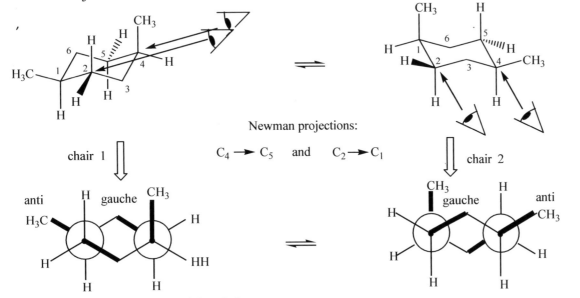

Newman projections: $C_4 \rightarrow C_5$ and $C_2 \rightarrow C_1$

chair 1

One axial methyl and one equatorial methyl

chair 2

One axial methyl and one equatorial methyl

Chapter 6

Example 2   *trans*-1,2-dimethylcyclohexane

1. See above.

2. Add axial and equatorial positions wherever substituents are present (the ring points to axial).

trans-1,2-dimethylcyclohexane

3. Add the indicated substituents.

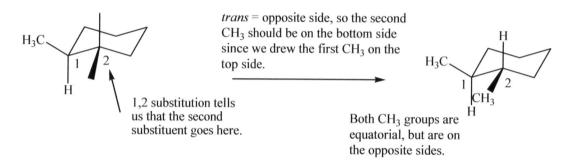

trans = opposite side, so the second CH₃ should be on the bottom side since we drew the first CH₃ on the top side.

1,2 substitution tells us that the second substituent goes here.

Both CH₃ groups are equatorial, but are on the opposite sides.

4 & 5. Draw the other conformation and estimate the energy expense of each conformation.

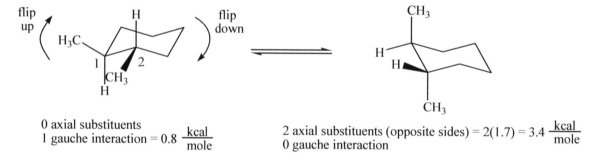

0 axial substituents
1 gauche interaction = 0.8 $\frac{kcal}{mole}$

2 axial substituents (opposite sides) = 2(1.7) = 3.4 $\frac{kcal}{mole}$
0 gauche interaction

6. Chair 1 has a gauche relationship between the two CH₃ substituents, which increases energy of that conformation by +0.8 kcal/mole.

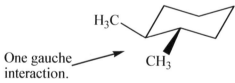

One gauche interaction.

7. The potential energy difference between the two conformations is $(3.4 - 0.8) = 2.6$ kcal/mole. Use this value to calculate $K_{equilibrium}$.

$$K = \frac{\text{chair 2}}{\text{chair 1}} = 10^{\frac{-\Delta G}{2.3RT}} = 10^{\frac{-(2,600 \text{ cal/mole})}{(2.3)(2 \text{ cal/mol-K})(298 \text{ K})}} = 10^{-1.9} = \frac{1}{79} = \frac{1.3\%}{98.7\%}$$

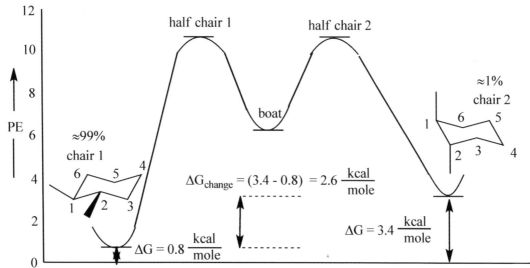

Energy changes of *trans*-1,2-dimethylcyclohexane conformations.

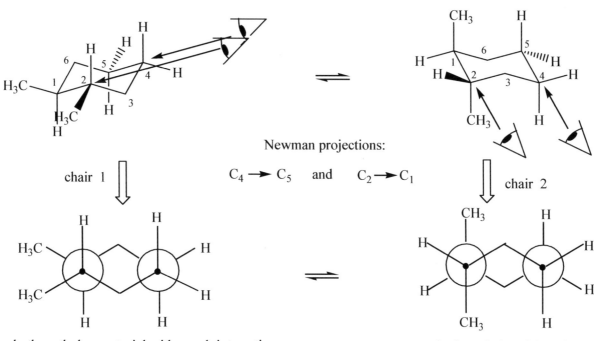

both methyls equatorial with gauch interaction   both methyls axial, anti

Example 3   *trans*- 1-t-butyl-3-methylcyclohexane

1. See above.

2. Add axial and equatorial positions wherever substituents are present (the ring points to axial).

trans- 1-t-butyl-3-methylcyclohexane

3. Add the indicated substituents.

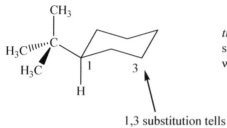

trans = opposite side, so the CH₃ should be on the bottom side since we drew the t-butyl on the top side

1,3 substitution tells us that the second substituent goes here.

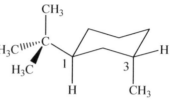

The t-butyl group is equatorial and the CH₃ group is axial, and they are on opposite sides.

4 & 5. Draw the other conformation and estimate the energy expense of each conformation.

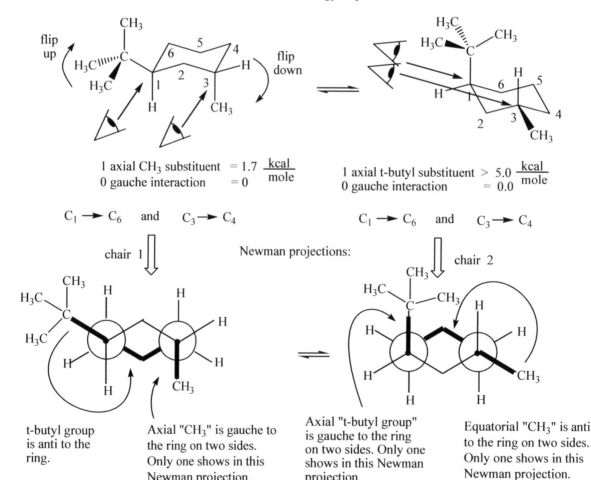

1 axial CH₃ substituent = 1.7 $\frac{kcal}{mole}$
0 gauche interaction = 0

1 axial t-butyl substituent > 5.0 $\frac{kcal}{mole}$
0 gauche interaction = 0.0

$C_1 \rightarrow C_6$ and $C_3 \rightarrow C_4$

$C_1 \rightarrow C_6$ and $C_3 \rightarrow C_4$

chair 1   Newman projections: chair 2

t-butyl group is anti to the ring.

Axial "CH₃" is gauche to the ring on two sides. Only one shows in this Newman projection.

Axial "t-butyl group" is gauche to the ring on two sides. Only one shows in this Newman projection.

Equatorial "CH₃" is anti to the ring on two sides. Only one shows in this Newman projection.

6. Gauche relationships are not applicable because substituents are not 1,2 (vicinal).

7. The potential energy difference between the two conformations is (5.0 – 1.7) = 3.3 kcal/mole. Use this value to calculate $K_{equilibrium}$.

$$K = \frac{chair\ 2}{chair\ 1} = 10^{\frac{-\Delta G}{2.3RT}} = 10^{\frac{-(3,300\ cal/mole)}{(2.3)(2\ cal/mol\text{-}K)(298\ K)}} = 10^{-0.004} = \frac{1}{255} = \frac{0.3\%}{99.7\%}$$

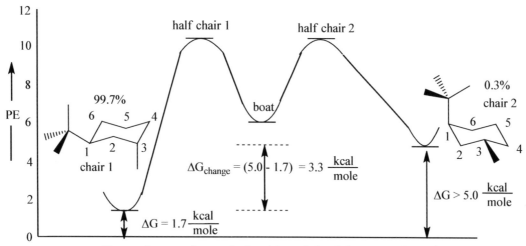

Energy changes of *trans*-1-t-butyl-3-methylcyclohexane conformations.

Chapter 6

Problem 4 – Draw all isomers of dimethylcyclohexane and evaluate their relative energies and estimate an equilibrium distribution for the two chair conformations. Use the given energy values for substituted cyclohexane rings. If the two substituents on the ring were different, all 14 conformations below would have different energies!

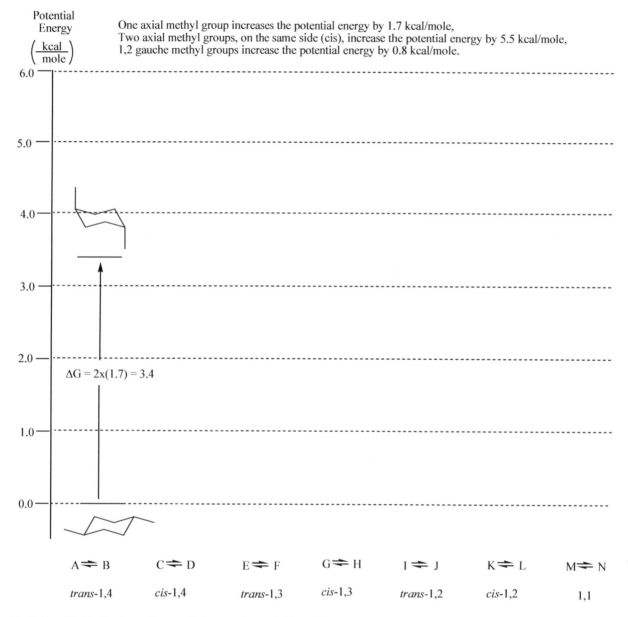

Potential Energy $\left(\dfrac{kcal}{mole}\right)$

One axial methyl group increases the potential energy by 1.7 kcal/mole,
Two axial methyl groups, on the same side (cis), increase the potential energy by 5.5 kcal/mole,
1,2 gauche methyl groups increase the potential energy by 0.8 kcal/mole.

$\Delta G = 2 \times (1.7) = 3.4$

A ⇌ B	C ⇌ D	E ⇌ F	G ⇌ H	I ⇌ J	K ⇌ L	M ⇌ N
trans-1,4	*cis*-1,4	*trans*-1,3	*cis*-1,3	*trans*-1,2	*cis*-1,2	1,1

Problem 5 - Both *cis* and *trans* 1-bromo-3-methylcyclohexane can exist in two chair conformations. Evaluate the relative energies of the two conformations in each isomer (use the energy values from the table presented earlier). Estimate the relative percents of each conformation at equilibrium using the difference in energy of the two conformations from you calculations. Draw each chair conformation in 3D bond line notation and as a Newman projection using the $C_1{\rightarrow}C_6$ and $C_3{\rightarrow}C_4$ bonds to sight down.

1-bromo-3-methylcyclohexane

Chapter 6

Problem 6 – Estimate a value for the strain energy of the trimethylcyclohexane structures provided. Use the energy values below.

One axial methyl group increases the potential energy by 1.7 kcal/mole,
Two axial methyl groups, on the same side (cis), increase the potential energy by 5.5 kcal/mole,
Three axial methyl groups, on the same side, increase the potential energy by 12.9 kcal/mole and
1,2 gauche methyl groups increase the potential energy by 0.8 kcal/mole.

1. Reference compound (no features with strain)	2	3	4
5	6	7	8
9	10	11	12
13	14	15	16
17	18	19	20
21	22	23	24

Chapter 6

CHAPTER 7 - STEREOCHEMISTRY

Stereochemistry is a fascinating and important topic. All of life is built on its subtlety. In the past few decades we have come to understand that the three dimensional arrangement of proteins, carbohydrates, fats, and nucleic acids are essential to their function. The slightest change in three dimensional structure can change the smell of mint to caraway, change a plant growth accelerator to just another compound, change a sweet taste into a bitter one, and alter a medicinally effective pharmaceutical drug into a teratogen (a chemical that causes birth defects). Each of the pairs of molecules below has a very subtle difference. They are different mirror images or are cis/trans isomers.

What once was a quaint curiosity has become an essential concern in the pharmaceutical and agrochemical businesses totaling in the hundreds of billions of dollars per year. The stereochemical guidelines used by the Food and Drug Administration (FDA) are currently in a state of flux, as ever higher standards and expectations are required. The symbols, R and S allow us to draw exact 3D shapes of tetrahedral atoms. E and Z allow us to draw exact substitution patterns around double bonds. The symbols + and – (or d and l), indicate pairs of isomers that are mirror images of one another, but different, much like your two hands. All of these terms, and more are explained in this chapter.

carvone enantiomers

caraway   mirror images   mint

thalidomide enantiomers

(-)-(S)-thalidomide
(powerful sedative)

mirror images

(+)-(R)-thalidomide
(teratogen)

asparagine enantiomers

(S)-asparagine
(bitter)

mirror images

(R)-asparagine
(sweet)

cinnamic acid diastereomers

Z-cinnamic acid
(plant growth accelerator)

E-cinnamic acid
(inactive)

Stereochemistry is too big a subject to learn every facet of this challenging area from a short chapter. Instead, our goal will be to build a solid foundation that will allow you to further explore this topic, should you need to do so. There exists a remarkable book of 1267 pages (published in 1994) called Stereochemistry of Organic Compounds, written by Ernest Eliel and Samuel Wilen. Almost any conceivable stereochemical topic you might encounter will probably be discussed in detail in that book.

You will also find that molecular models are extremely useful in stereochemistry. Your understanding here will be greatly enhanced if you take the time to build, look at, and study with models. If you just can't bring yourself to build molecular models, then you will have to resort to your ever present model set - your hands. Stereochemistry involves all three dimensions of our world, so the two dimensional page will work only if your mind can fill in the extra dimension to the flat page. With models, hands, pencil and paper, and practice you will be able to train your mind in these important ideas.

Chapter 7

Isomer Overview

The following table summarizes the various types of isomers we will encounter as we proceed through organic chemistry.

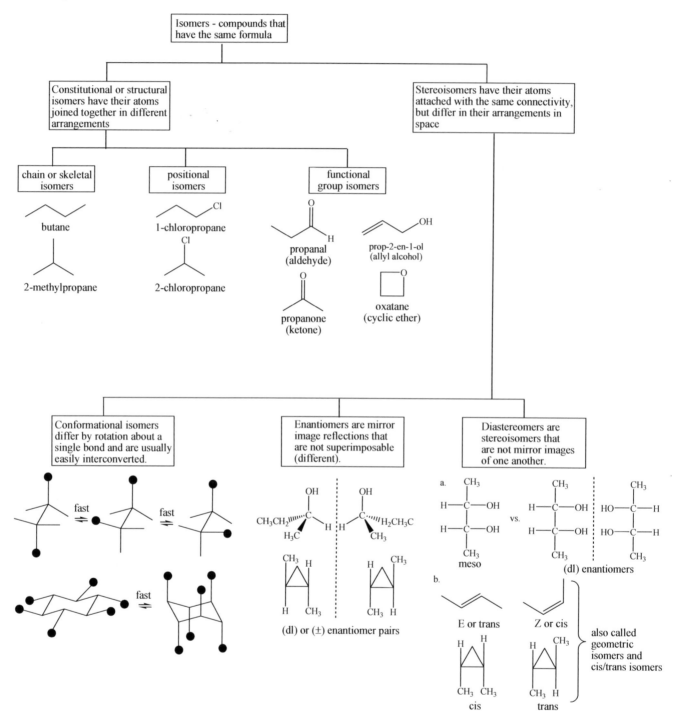

Chapter 7

Essential Vocabulary of Stereochemistry – terms to know for this topic

Chiral center – a tetrahedral atom with four different groups attached, a single chiral center is not superimposable on its mirror image, also called a stereogenic center, and asymmetric center

Stereogenic center – any center where interchange of two groups produces stereoisomers (different structures), could be a chiral center producing R/S differences, or an alkene producing E/Z differences or cis/trans in a ring

Absolute configuration (R and S) – the specific 3 dimensional configuration about a tetrahedral shape, determined by assigning priorities (1-4) based on the highest atomic number of an atom at the first point of difference. When the lowest priority group is away, a circle is traced through the highest 3 priorities (1 → 2 → 3) in a clockwise direction (= R) or a counterclockwise direction (=S) R - Rectus
S - Sinister

E/Z Nomenclature – a system to unambiguously identify the stereochemistry at alkenes based on the priorities of the attached groups at each carbon. E has the highest two priority groups on the opposite side of the double bond and Z has the highest two priority groups on the same side of the double bond.

Chiral Molecules – a molecule that is not superimposable upon its mirror image

Achiral Molecules – a molecule that is superimposable upon its mirror image

Enantiomers – stereoisomers that are different mirror images of one another, there can only be one enantiomer

Diastereomers – stereoisomers that are not mirror images of each other, can be a result of either (R/S) differences, (E/Z) differences in alkenes, cis/trans differences in rings or any combination of all those kinds of differences; there may be many, many diastereomers.

Meso Compounds – compounds that contain two or more chiral centers, yet are identical with their mirror images (they are achiral and optically inactive). A mirror plane cuts through the middle of the molecule.

Optical Activity – angle of rotation of plane polarized light using a polarimeter as it passes through a solution of chiral compound, achiral compounds do not rotate plane polarized light and are optically inactive (no rotation).

d (+) = dextrorotatory – clockwise rotation of plane polarized light in a polarimeter (only with chiral compounds)
l (-) = levorotatory – counterclockwise rotation of plane polarized light in a polarimeter (only with chiral compounds)

Racemic Mixture – a equal mixture of two enantiomers. Racemic mixtures are optically inactive even though there are chiral molecules present. Each enantiomer cancels out the optical rotation of the other enantiomer.

Fischer Projections – a method for representing stereogenic centers in chains, with the stereogenic carbon at the intersection of vertical and horizontal lines, the horizontal lines are consider to be coming forward in front of the plane (wedges) and the vertical lines are considered to be going backward behind the plane (dashes).

Haworth Projections – a method for representing cyclic structures, drawn flat with vertical lines to show top and bottom faces of the ring, analogous to Fischer projections for straight chains. Commonly used for cyclic sugars in biochemistry.

Prochiral – molecules are those that can be converted from achiral to chiral in a single step (*re* and *si* faces in pi bonds).

Enantiomeric excess (ee) is a measure for how much more of one enantiomer is present compared to the other (50% ee in R is 75% R and 25% S)

Other types of chirality include *axial (helical) chirality* (allenes, DNA) and *planar chirality* (E-cyclooctene).

How many butan-2-ols are there? At first, it seems there must be only one. But let's build models and compare.

OH
|
H_3C—C—$\underset{H}{\overset{H_2}{C}}$—$CH_3$
|
H

2D representation
of butan-2-ol

mirror plane

S absolute configuration   R absolute configuration

3D representation of butan-2-ol
reveals two different mirror images
(called enantiomers)

Three dimensional structures of 2-butan-2-ol reveal that there are two *different* butan-2-ols which are mirror images of one another. Different mirror images are called **enantiomers**. This property can be observed in enantiomeric molecules using plane polarized light. Rotation of plane polarized light can rotate either to the right (**dextrorotatory** = d = +) or to the left (**levorotatory** = l = -) depending on the substance. Such samples are called 'optically active' and have a similar relationship to one another as your left and right hands.

Plane Polarized Light

The first observation of handedness in chiral molecules was made using plane polarized light in the early 1800s, before chemists knew what atoms and molecules were. For plane polarized light to see a difference between enantiomers, it has to be chiral itself.

The plane polarized light can be considered as two circularly polarized helices which have an equal component turning in opposite directions. This seems like a needless complication but it is useful for understanding the chiral nature of plane polarized light. Picturing two screws, one with right-handed threads and one with left-handed threads might help. Imagine how your right or left hand would have to move to screw in these different screws. The achiral screwdriver would make your chiral hands perform different motions to twist the chiral screw in. The combination of the screw and either hand is diastereomeric (left hand plus right threaded screw versus right hand plus right threaded screw).

A similar relationship exists between two enantiomers interacting with one type of plane polarized light. The plane of polarized light rotates when the electrical vector component interacts with the electrons in the bonds of the chiral molecule.

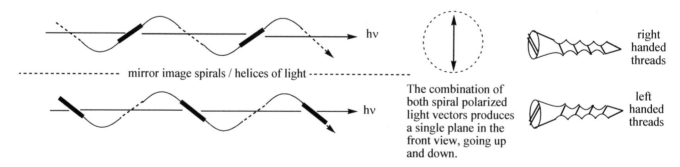

mirror image spirals / helices of light

hv

hv

The combination of both spiral polarized light vectors produces a single plane in the front view, going up and down.

right handed threads

left handed threads

The light from the light source is initially random in its propagation vectors (see figure below). Completely random light has electrical vector components at all angles through 0-360°. When it passes through the Nicol polarizing lens, only one plane of the light (two opposite turning helices) continues through the sample compartment that is parallel to the gratings in the lens. All other electrical vectors are blocked.

Chapter 7

Random electrical vectors of light has planes in all 360° of a circle, viewed straight on.

Straight on view of many randomly oriented electrical vectors of light (all 360° are possible).

A sketch of the basic components of a polarimeter instrument are shown below. The observed angle of rotation out the backside of the sample cell is called α (alpha). We will not emphasize the quantitative nature of α or polarimeters in this book. However, you might try your own qualitative experiment the next time you are at the supermarket. View the store lights through the lenses of two different polarized sunglasses. Hold them parallel and then twist one pair perpendicular to the other and see what happens to the amount of light passing through to your eyes.

A polarimeter selects out just one plane of random light and calls that 0° rotation.

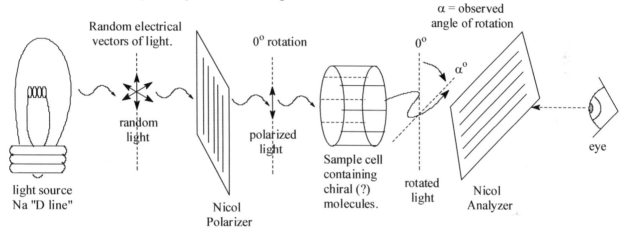

Hold your fingers straight up in front of you and try passing your pen or pencil between them at various angles. Only one orientation will work, which has the pencil parallel to your fingers. The plane of light passing through the polarizing lens works something like that.

A pen or pencil at just the right angle can pass between the fingers of a hand. With any other rotation, this cannot happen.

When plane polarized light exits the sample cell and is rotated to the right, its rotation is called dextrorotatory (Latin = right in the sense of direction, R = rectus = right in the sense of correct). This is symbolically represented with a small "d" or "+" sign. When the light is rotated to the left, its rotation is called levorotatory (Latin = left in the sense of direction) and represented with a small "l" or "-" sign. Do not confuse these terms with R and S, which are used to define how the drawing of a single chiral center will look or will be built with a model. The symbols "d/l" or "+/-" are experimental observations of how light rotates after interacting with an entire molecule. A chiral molecule can have zero chiral atoms (we don't discuss these) to

Chapter 7

hundreds of chiral atoms (with each atom having an R or S configuration). However, there will only be a single "d/+" or "l/-" for an entire molecule.

dextrorotatory = d = + = plane polarized light rotates to the right after
passing through mixture having chiral moleclues

levorotatory = l = - = plane polarized light rotates to the left after passing
through mixture having the opposite chirality

When enantiomers are present in equal amounts, there is no net rotation of light. For every chiral molecule, there is its exact mirror image also present in a 50/50 mixture. The right rotation by the "d/+" enantiomer is exactly cancelled by the left rotation of the "l/-" enantiomer. Such a mixture has a special name of **racemic mixture** (a 50/50 mixture of the two enantiomers). Yet another mixture that does not rotate plane polarized light is any type of achiral molecule.

Chiral Centers and Molecules Have "Handedness"

The property of handedness will be observed at any tetrahedral center with four different groups present.

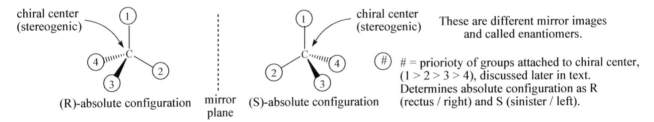

chiral center (stereogenic)
(R)-absolute configuration   mirror plane   (S)-absolute configuration

chiral center (stereogenic)

These are different mirror images and called enantiomers.

= priorioty of groups attached to chiral center, (1 > 2 > 3 > 4), discussed later in text. Determines absolute configuration as R (rectus / right) and S (sinister / left).

Any tetrahedral carbon with four different groups about it is called a chiral center and a stereogenic center. In an actual molecule, there may be only one chiral center or 10s, 100s, 1000s or more chiral centers.

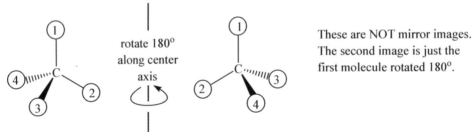

rotate 180° along center axis

These are NOT mirror images. The second image is just the first molecule rotated 180°.

If a 50/50 racemic mixture of different mirror image structures (enantiomers) is present, this will lead to no observed rotation of polarized light because the right and left rotations cancel.

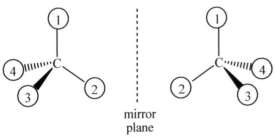

mirror plane

Racemic mixture = 50/50 mixture of enantiomers

A racemic mixture occurs when both mirror images are present in equal amounts. Each rotation cancels the other rotation for no net observed rotation

If there are no chiral centers then the mirror image structures are really identical, and this leads to no net observed rotation of polarized light (any right rotation is cancelled by an equal amount of left rotation).

The mirror images are identical (achiral), so there is no net rotation of polarized light.

These two molecules are identical.

Molecular models you always carry with you – They're called hands!

This subtle difference in structure can be very difficult to visualize on paper or even with models. However, it turns out that nature gave us the perfect tools to consider this difficult 3D problem. These tools are called hands (and of course, we have our minds)! Our hands can reveal some of the problems we face in working with enantiomers, and some of the solutions to those problems. Many of us will have a problem picturing abstract molecules in space, especially different mirror image molecules. Almost none of us will have a problem picturing mirror image hands, because of our lifetime experience using them.

Let's draw the outline of a hand. If we don't use more specific nomenclature, we wouldn't know if the hand is a right hand or a left hand? Either of your hands would fit in the first trace of a hand below. What further information do we need? We could say "right" or "left", or we could add some additional details to our drawings.

A plain silhouette does not provide enough information to tell if the hand is face up or face down (left or right).

hand = ?   right hand   left hand

Added features make it easy to determine if the top or the bottom of the hand is towards you (...or you could add a descriptor of left or right).

Our mental image of enantiomers (different mirror images) is rock solid when we're talking about hands. We don't really even need a picture to distinguish right from left. From a lifetime experience of using them, we know that our two hands, while looking similar, are very different from one another. Just try switching hands the next time you are taking notes in organic chemistry. Almost every detail we need to understand about molecules can be modeled with your hands. You can even use your hands as approximations of tetrahedral atoms, using your arm, thumb and first two fingers. Try it using the pictures below as an example. Use your hands to model tetrahedral centers and this subject will be a whole lot easier.

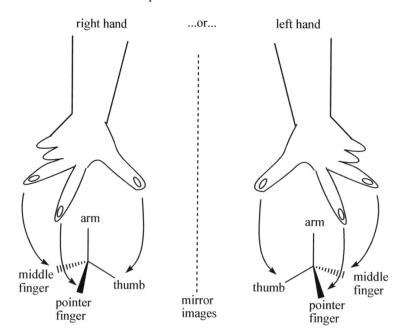

R and S Nomenclature (Absolute configuration as R or S)

In addition to recognizing when enantiomers are present (stereoisomers in general), we also have to correct our nomenclature so that it is specific enough to identify a unique structure. The rules to solve this dilemma were created by three chemists in the mid-1950s, R.S. Cahn, C.K. Ingold and V. Prelog (CIP rules). The absolute configuration of a chiral center was designated as R (rectus = right, Latin) or S (sinister = left, Latin). These two letters specify whether a stereogenic center is the "right" one or the "left" one.

Priorities of the four different groups attached to a chiral atom are assigned, based on atomic numbers. The higher an atomic number is, the higher the priority will be. This is easy in organic chemistry (and biochemistry), because we only have a few atoms to consider (plus lone pairs): I > Br > Cl > S > P > F > O > N > C > H > lone pair of electrons. It is possible for a single element to have different isotopes (e.g. hydrogen = H, deuterium = D and tritium = T). In these cases, the most massive isotope is the highest in priority (e.g. T > D > H). The highest priority group is specified as #1 and the lowest priority group as #4. A specific viewing perspective requires that the low priority group (#4) be pointed away from the viewer (you), while the remaining three groups trace a circle from #1 → #2 → #3, either clockwise (CW) or counterclockwise (CCW). When the circle traces clockwise, the absolute configuration is R and when the circle traces counterclockwise the absolute configuration is S.

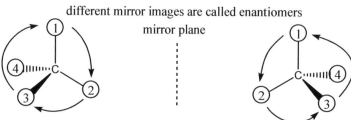

"R" absolute configuration requires that the low priority group, 4, is away from the viewer while a circle traced from 1 to 2 to 3 goes in a clockwise direction (CW).

"S" absolute configuration requires that the low priority group, 4, is away from the viewer while a circle traced from 1 to 2 to 3 goes in a counterclockwise direction (CCW).

Priority numbers are based on atomic numbers in the periodic table. The lowest priority group is a lone pair of electrons.

Now this was set up to be easy, the priority numbers were given and the low priority group was away. The lowest priority group was already away from you, which made the analysis simple (and you didn't have to assign any priorities). In a random example, you have no guarantee that this will be the case. Also, a molecule could have several chiral centers with a variety of viewing perspectives. You may have to redraw a structure or build a model to be able to evaluate R or S (at every chiral center), especially if your 3D visualization skills are weak.

There is a simple alternative, however, and this will make you a pro at making stereochemical assignments. All you need are your hands and the ability to evaluate the priorities. Let's assign the absolute configuration (R or S) in the structure below. First, we need to assign the priorities (double check my assignments). I like to redraw the chiral center and insert the numbers of the priorities. It's a little extra work, but it makes me more accurate in my answers. If we blindly traced our circle through 1 → 2 → 3, we might be tricked into thinking that the absolute configuration shown is R, which is not the case. Our answer is wrong because we haven't oriented the low priority group away from our viewing perspective.

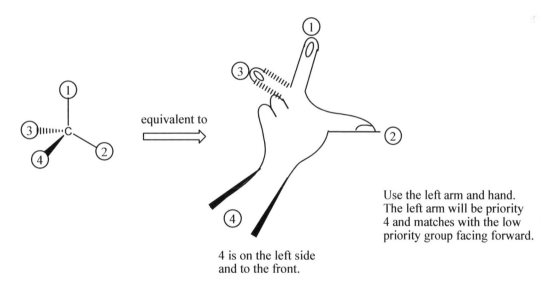

We have to turn the molecule around (in our head, on paper, build a model...), and then evaluate the trace of the circle. However, there is an easier, more reliable way to do this. We can make our arm and fingers into a tetrahedron. The four different groups will be matched with our arm, thumb and first two fingers of either hand. **The arm will always be the low priority group when that group is toward us.** If the low priority group faces towards the front and is on the left side, you will use your left arm. If the low priority group faces front and is on your right side, you will use your right arm. Your thumb, pointer finger and middle finger will complete the tetrahedron and you will assign the appropriate number (1, 2 or 3) to the appropriate finger.

A convenient movement is now possible. We can turn our arm around at the elbow and look at our three fingers with the low priority group (our arm) away from us. Of course you must remember at least which fingers are 1 → 2 to get the correct direction of the circle you are tracing. I don't try to remember all three because then I get mixed up by the time I rotate my arm around. Remembering 1 → 2 is easier. Do not flatten out your hand when you turn it around. It makes it difficult to trace your circle. Keep it in a tetrahedral shape.

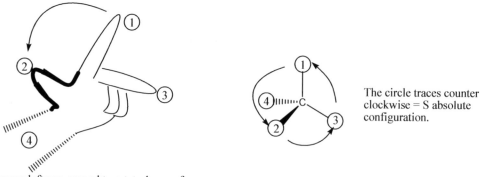

The circle traces counter clockwise = S absolute configuration.

Turn your left arm around to rotate 4 away from you. Remember which fingers are assigned 1 and 2 to start the trace of your circle, counter clockwise in this example.

An alternative way to make the assignment of absolute configuration is to hold one prong of a tetrahedral atom from your molecular model kit with your finger tips and let it represent the chiral tetrahedral atom. Hold it in place to model a chiral center, assign two of the prongs as 1 and 2 priorities, and then turn the tetrahedral model atom around like it was your hand. Trace the circle using priorities 1 and 2 as your guide. The only disadvantage to this approach is you might not have your model atom with you, but you will always have your hands and arms with you.

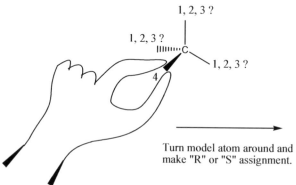

Turn model atom around and make "R" or "S" assignment.

What about the mirror image structure? Since the low priority group faces forward and is on the right side, you will need to use your right arm. Spread the appropriate fingers to model the tetrahedral center and assign the other priorities (1, 2 and 3). As before twist your elbow so that the low priority group (your arm) is away from you and trace the circle from 1 to 2 to 3. In this example the circle traces clockwise and the absolute configuration is R (as it must be because it is the mirror image of an S absolute configuration).

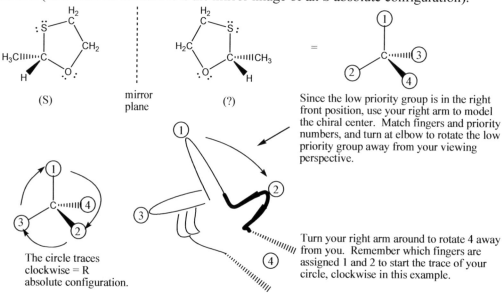

(S)   mirror plane   (?)

Since the low priority group is in the right front position, use your right arm to model the chiral center. Match fingers and priority numbers, and turn at elbow to rotate the low priority group away from your viewing perspective.

The circle traces clockwise = R absolute configuration.

Turn your right arm around to rotate 4 away from you. Remember which fingers are assigned 1 and 2 to start the trace of your circle, clockwise in this example.

Chapter 7

We now have an absolutely certain way to accurately draw or build a three dimensional representation of an sp^3 center having this property of chirality (handedness). The classification of the absolute configuration at a chiral center must be either R or S.

What if a chiral atom has four different attached groups, but two of the directly attached atoms are the same? We encountered just such a problem in our example of butan-2-ol. It is obvious that methyl (CH_3) is not ethyl (CH_2CH_3), however when we examine the atoms attached to the chiral center, we find the two carbon atoms are identical in priority (at the intermediate priority levels of 2 and 3).

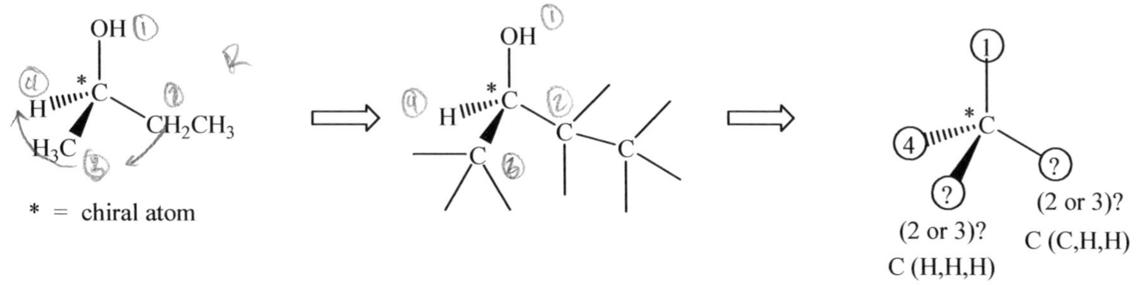

* = chiral atom

Our next step in such a situation is to examine the additional bonded groups to these equivalent atoms. In the case of carbon, that will mean three additional bonds to consider (the fourth bond is the attachment to the chiral center). Our decision of priorities is always based on atomic number. The highest priority atom (highest atomic number) will decide the issue, regardless of what other atoms are present. If the atoms attached in the first sphere of atoms are equivalent then we continue to the second sphere of attached atoms for evaluation. When the pathways begin to branch out in such a case, always trace the highest priority path possible until you come to a difference that will distinguish one path from another. Some examples will prove helpful here.

Example 1

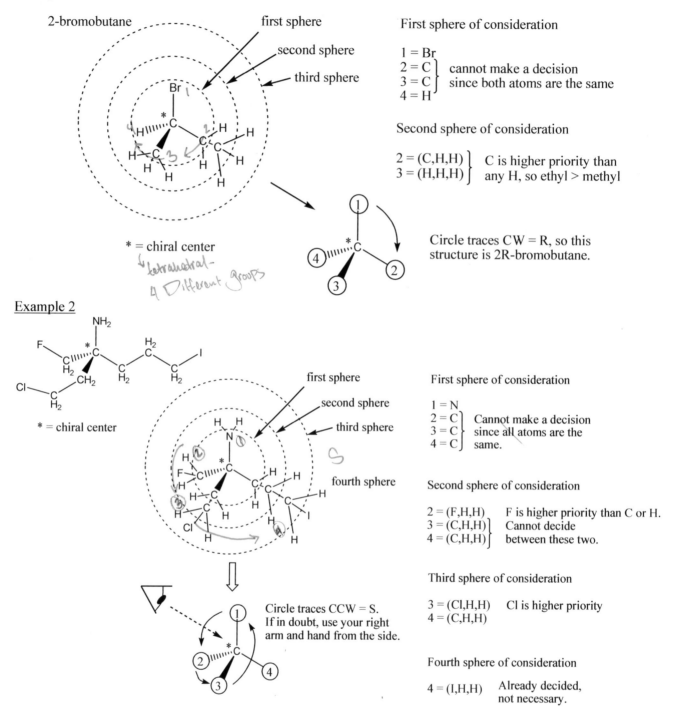

First sphere of consideration

1 = Br
2 = C } cannot make a decision
3 = C } since both atoms are the same
4 = H

Second sphere of consideration

2 = (C,H,H) } C is higher priority than
3 = (H,H,H) } any H, so ethyl > methyl

* = chiral center

Circle traces CW = R, so this structure is 2R-bromobutane.

Example 2

* = chiral center

First sphere of consideration

1 = N
2 = C } Cannot make a decision
3 = C } since all atoms are the
4 = C } same.

Second sphere of consideration

2 = (F,H,H)   F is higher priority than C or H.
3 = (C,H,H) } Cannot decide
4 = (C,H,H) } between these two.

Third sphere of consideration

3 = (Cl,H,H)   Cl is higher priority
4 = (C,H,H)

Circle traces CCW = S.
If in doubt, use your right arm and hand from the side.

Fourth sphere of consideration

4 = (I,H,H)   Already decided, not necessary.

The temptation in this example is to see the iodine and immediately classify it as the highest priority group. However, we have to pass through several spheres of priority before we come to the iodine atom and the priority decision is already decided when we reach it.

Chapter 7

Example 3   One additional (and tricky) example will also prove helpful.

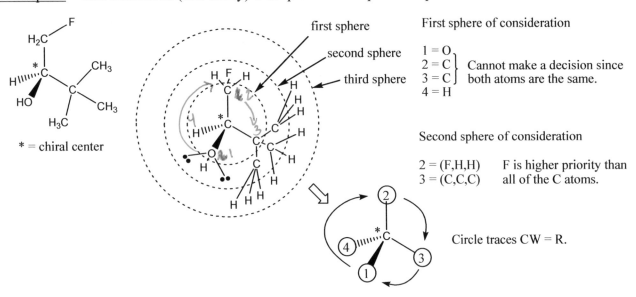

First sphere of consideration

1 = O
2 = C Cannot make a decision since
3 = C both atoms are the same.
4 = H

Second sphere of consideration

2 = (F,H,H)   F is higher priority than
3 = (C,C,C)   all of the C atoms.

Circle traces CW = R.

* = chiral center

first sphere
second sphere
third sphere

        In this example the temptation is to consider three carbon atoms higher in priority than one fluorine atom and two hydrogen atoms. Remember, it is the atom with the highest atomic number that establishes the priority. If all attached groups on one atom are equivalent to all attached groups on another atom, you move to the next sphere of consideration. What we need now is some practice.

Problem 1 - Classify the absolute configuration of all chiral centers as R or S in the molecules below. Use hands (or model atoms) to help you see these configurations whenever the low priority group is facing towards you (the wrong way). Find the chiral centers, assign the priorities and make your assignments.

Chapter 7

Pi Bond Priority

Pi bonds of all kinds are common in organic chemistry. If a double bond is present, it is assumed that each atom of the double bond is duplicated, and if a triple bond is present it is assumed that each atom of the triple bond is triplicated. Since real atoms are obvious, this involves drawing in (or thinking) one additional imaginary atom in double bonds or two additional imaginary atoms in triple bonds. The imaginary atoms are sometimes placed inside parentheses, as illustrated below.

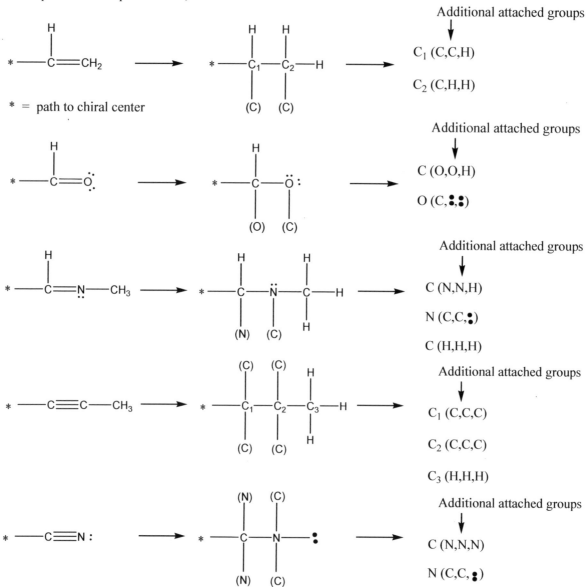

Problem 2 - Evaluate the order of priority in each part from highest (= 1) to lowest (= 4).

a.

* = path to chiral center

| ethynyl | phenyl | 2-propenyl | t-butyl |

b.

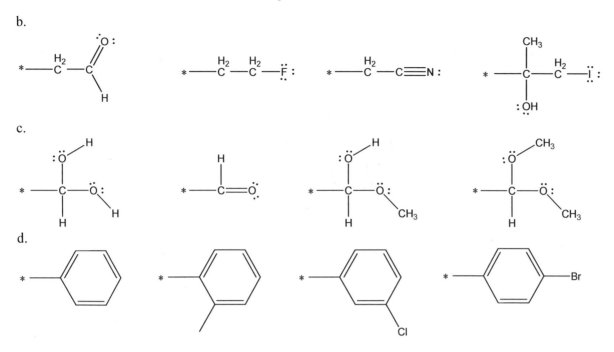

c.

d.

Any stereogenic center has two possible outcomes, R and S (or E and Z at double bonds). It might occur to you that the chance of encountering a specific absolute configuration or its mirror image is the same as getting heads (H) or tails (T) when flipping a coin (a 50/50 proposition). With only one flip you can get only heads (H) or tails (T), but with two flips you have four possible outcomes (H,H), (H,T), (T,H), (T,T) and so forth. This would correspond to absolute configurations of (R,R) (R,S) (S,R) or (S,S). A third flip of a coin doubles this number again to eight. Each time you flip a coin an additional flip or each time there is an additional stereogenic center, the total number of possibilities doubles from the prior number. The result for the possible number of stereoisomers is a maximum of 2^n stereoisomers, where n equals the number of stereogenic atoms. This is a maximum number possible (sometimes it could be less). There may be fewer than this number of stereoisomers, if special symmetry features are present (such as meso structures, discussed later).

Number of Stereogenic Centers	Maximum number of stereoisomers. (There may be fewer than this number if meso compounds are present.)
1	$2^1 = 2$
2	$2^2 = 4$
3	$2^3 = 8$
4	$2^4 = 16$
etc.	

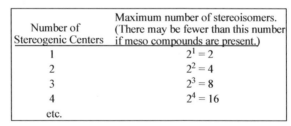

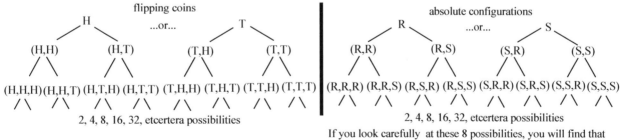

If you look carefully at these 8 possibilities, you will find that there are 4 pairs of enantiomers.

Chapter 7

Fischer Projections – A useful method for representing stereoisomers with more than one chiral center

A reasonably simple molecule with multiple chiral centers to consider is 2-bromo-3-chlorobutane. If we first draw a 2D structure we can identify any chiral centers. Stereogenic atoms are easy to identify because they have four different groups at a tetrahedral center. 2-bromo-3-chlorobutane has two stereogenic atoms, C_2 and C_3, and are marked with asterisks, below.

* = chiral center = stereogenic center

2 chiral centers present

$2^2 = 4$ possible stereoisomers

Any stereogenic center has a 50/50 chance of two possible outcomes, R and S (or E and Z at double bonds). In this example, the possible number of stereoisomers is a maximum of $2^2 = 4$ possible stereoisomers,. There may be fewer than this number of stereoisomers, if special symmetry features are present (such as meso structures, discussed soon).

Fischer projections are a convenient new way to draw stereoisomers having more than one chiral center. These drawings will allow us to quickly and easily evaluate if stereoisomers are enantiomers, diastereomers or meso compounds in comparisons with one another.

Fischer projections require that we place the longest carbon chain in the vertical direction. The highest priority *nomenclature* group is placed in the top half of the drawing. Priority in this instance refers to nomenclature priorities and not the R,S priorities we've just been discussing. This is, unfortunately, a confusing use of the word "priority", but you need to keep the difference straight.

All along that vertical chain are drawn the other two groups at each sp³ atom in horizontal positions, one to the left and one to the right. The horizontal groups along the longest carbon chain backbone are considered to be coming out in front of the page, towards you (the viewer). Vertical groups are considered to be projecting back, behind the page (away from you). With these assumptions, wedges and dashed lines may be dropped and simple lines used in their place. The center atom is omitted as well, and merely understood to exist at the crossing point of horizontal and vertical lines.

To show how they are created and used, it may prove helpful to first represent a molecule using a familiar Newman projection in an eclipsed conformation (higher potential energy), and then with wedges and dashed lines in a tilted up position. Alternatively, you can represent a molecule in an eclipsed sawhorse or 3D display. We will use (2S,3S)-2-bromo-3-chlorobutane as an example (there are actually four possibilities for doing this).

staggered sawhorse projection
* = chiral center

Newman projection (staggered)

(twist rear carbon atom 180°)

Newman projection (eclipsed)

eclipsed sawhorse projection
* = chiral center

Tilt up towards viewer with horizontal groups facing forward, in front of the page.

The vertical groups project backwards, behind the page, and the horizontal groups project forwards, in front of the page. Carbon atoms lie at the crossing points of the lines.

3D details eliminated as in a normal Fischer projection.

Regular Fischer projection

3D Fischer projection, with wedges and dashes

Basic Rules For Drawing a Complete Set of Fischer Projections

1. Place the longest carbon chain in vertical direction, with the highest nomenclature priority group in the top half of your representation.

2. Horizontal groups will project toward the front (in front of the page/surface).

3. Vertical groups will project away from the viewer (in back of the page/surface).

4. A carbon atom is indicated at each intersection of vertical and horizontal lines.

5. As much as possible, place non-hydrogen substituents on the same side in your first structure (I use the right).

6. Immediately draw the mirror image of that structure (groups on the right will move to the left and vice versa).

7. Move the top-most group across to the other side and immediately draw the mirror image.

8. If you can, continue to move single groups across, one at a time, until you reach the maximum number of stereoisomers based on the number of chiral centers.

9. With four or more chiral centers, you can start again at the first structure and move two substituents at a time across, and draw the mirror image until you reach the maximum number of stereoisomers. (Four chiral centers would potentially have 16 stereoisomers.)

As a consequence of placing the longest carbon chain in the vertical position, and having the horizontal groups facing forward, assigning absolute configurations (R/S) will usually require that chiral centers be turned around. If there is a hydrogen atom bonded to a chiral center (often the case), it will always be in a horizontal position. You will find the hand/arm approach to classify chiral centers as R or S works very well in this situation. It is fast, accurate and easy. If you have a whole series of stereoisomers to assign, you can make the initial assignments on the first stereoisomer and use those assignments in all of the rest of the stereoisomers. They will either be the same, or they will be opposite. Just be sure your first assignments are correct!

The example molecule, above, has two chiral centers and therefore has a maximum of $2^2 = 4$ possible stereoisomers. Without drawing a single structure, we know that the absolute configurations will be (R,R) and its mirror image (S,S), and (R,S) and its mirror image (S,R). My approach when drawing a complete set of stereoisomers is to place all of the "different" groups on one side (Br and Cl here). Then, beginning at the top, individually move one group across at a time all the way down the line. For each possible structure, I immediately draw its mirror image and check to see if it is different (probably enantiomers, though if meso, the mirror images will be identical). If there are enough substituents, you may have to begin at the top again, and move two substituents to the other side, and so forth. You should also check that your later drawn structures do not duplicate some earlier drawn structures. (This occasionally occurs in symmetrical molecules). You need some sort of systematic approach to draw all the possibilities. I have used my approach with Fischer projections of 2-bromo-3-chlorobutane, below.

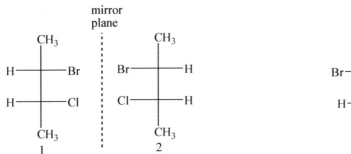

Place both Br and Cl on the same side and then draw the mirror image.

Move Br across and leave Cl in place and then draw the mirror image.

The maximum number of stereoisomers is $2^2 = 4$. The horizontal groups, including hydrogen atoms are toward you. To assign the absolute configuration of any chiral centers will require that the low priority hydrogen atom be turned away to trace the direction of the other priorities, 1 to 2 to 3, as R or S.

The assignment of absolute configurations in all of the stereoisomers can be made by specifying the priorities at each chiral center in the first stereoisomer, (1). In this example we would use the left arm and hand to assign both, since the lowest group, hydrogen, is on the left side in each case. I usually double check myself one time to see if I get the opposite result on the enantiomer (2). If that classification seems correct, I label all of the other absolute configurations by comparison to the first structure (either they are the same or opposite).

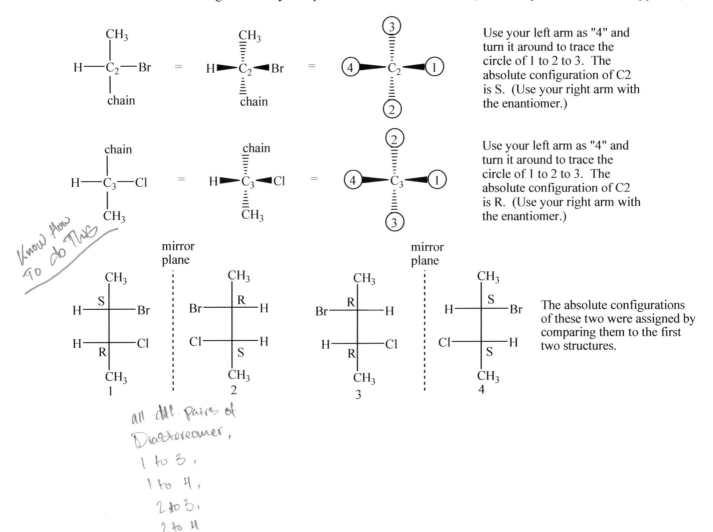

Use your left arm as "4" and turn it around to trace the circle of 1 to 2 to 3. The absolute configuration of C2 is S. (Use your right arm with the enantiomer.)

Use your left arm as "4" and turn it around to trace the circle of 1 to 2 to 3. The absolute configuration of C2 is R. (Use your right arm with the enantiomer.)

The absolute configurations of these two were assigned by comparing them to the first two structures.

Know how to do this

all diff. pairs of Diastereomer,
1 to 3,
1 to 4,
2 to 3,
2 to 4

Stereoisomer 1 is part mirror image and partially identical to both 3 and 4. It is neither identical, nor an enantiomer of these molecules.

These relationships are grouped in the class of diastereomers. Stereoisomer 2 is likewise a diastereomer of both structures 3 and 4. There are two pairs of enantiomers (1,2) and (3,4) and four pairs of diastereomers (1,3), (1,4), (2,3) and (2,4).

It is easy to include chiral centers as part of a name. You can use the same chemical name for all of the stereoisomers, without consideration of stereochemistry, and just put the number and its absolute configuration (R or S) of each stereocenter in parentheses in front of the name.

(2S,3R)-2-bromo-3-chlorobutane	(2R,3S)-2-bromo-3-chlorobutane	(2R,3R)-2-bromo-3-chlorobutane	(2S,3S)-2-bromo-3-chlorobutane

It is important that you remember that the horizontal groups are toward you. You can move any drawing 180° around **in the plane of the page** and the horizontal groups will still be toward you as long as you do not lift the structure off of the plane of the page. This is a maneuver you will occasionally have to do when comparing stereoisomers. This is analogous to you being stood on your head, but always facing the same direction. In either position, we could always recognize your face and know that it was you.

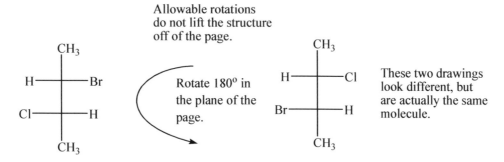

You may <u>NOT</u> turn the molecule over (like flipping pancakes), since that would make the horizontal groups project away from you, (behind the page). This effectively inverts all stereogenic centers and generates the mirror image structure, as represented by a two dimensional Fischer projection. The mirror image structure may or may not be different (an enantiomer or a meso compound) depending on what symmetry features are present.

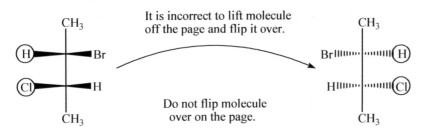

It is incorrect to lift molecule off the page and flip it over.

Do not flip molecule over on the page.

Backwards horizontal groups are away. If drawn as a Fischer projection, this molecule would appear as its mirror image. Everyone looking at it would be fooled.

Our hands can again provide us with a useful analogy. Specifying that horizontal groups are toward us is similar to always specifying the back of the hand as always facing up in a drawing (finger nails up). With this convention we could view a trace of the hand and always know for sure if it was a left or right hand. However, if you drew your hand, palm up, anyone looking at your drawing would be completely confused about what you were trying to represent. Fischer projections are similar.

We can't tell top or bottom without some convention?

This must be a right hand, if the convention tells us we are always looking at the top of a hand.

Our convention tells us this is a left hand, but the added features reveal it is a right hand drawn in violation of convention.

Meso Structures

Let's now look at all possible 2,3-dibromobutanes.

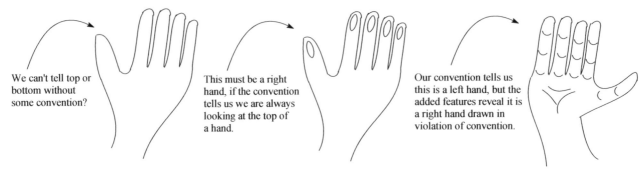

* = chiral center = stereogenic center

2 chiral centers present

2^2 = maximum of 4 possible stereoisomers

mirror plane		mirror plane	
R	S	R	S
R	S	S	R
enantiomers		enantiomers	

We will follow the same systematic approach used with the 2-bromo-3-chlorobutane isomers (above),

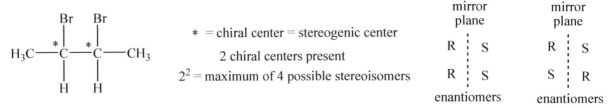

Place both Br's on the same side and then draw the mirror image.

Move Br across and leave the bottom Br in place and then draw the mirror image.

The maximum number of stereoisomers is $2^2 = 4$. The horizontal groups, including hydrogen atoms are toward you. To assign the absolute configuration of any chiral centers will require that the low priority hydrogen atom be turned away to trace the direction of the other priorities, 1 to 2 to 3, as R or S.

As before, we want to be systematic in our approach, so we draw the Fischer projection with both substituents on one side and then move the top bromine over to the other side. In each case we immediately generate the mirror image structure and see if it is different. You should also check that later drawn structures do not duplicate some earlier drawn structures, which occasionally occurs in symmetrical molecules that are larger than this example.

We don't expect more than $2^2 = 4$ stereoisomers and can stop after drawing the four stereoisomers shown above. This is a maximum number possible, not a required number.

We can rotate structure 5 180° in the plane of the page (do not flip it over) to compare it to its mirror image, 6. When we do this, we discover something that we have not seen before in a molecule that has chiral centers. We find that the mirror images (5/6) are identical (superimposable), which means that the two structures are NOT enantiomers. Instead, they are identical. In a similar way, we can compare 7 and 8. This mirror image pair (7/8) is not superimposable (i.e. – they are enantiomers). Even though four stereoisomers can be drawn, only three are actually different structures (5 = 6 and 7 ≠ 8).

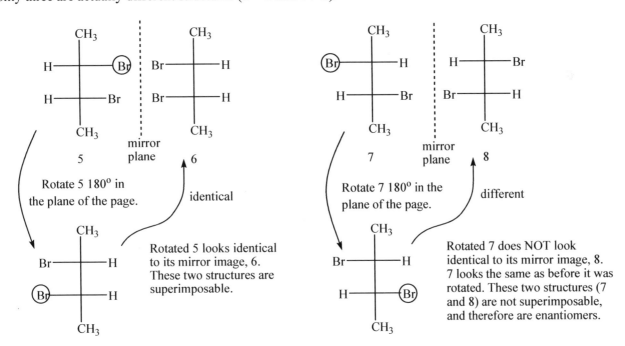

Structures such as 5/6, which have stereogenic centers but are achiral as molecules, have a special name. They are called "*meso*" structures (Greek = middle). The name meso (middle) provides us with an explanation for the lack of chirality. If we cut the molecule 6a in two pieces, horizontally, right in the middle (meso), we discover that the top half is a reflection of the bottom half. The top half also reflects across to its mirror image, structure, 6. Since these are both mirror image representations of the same stereogenic atom, they must be identical. In a similar way, the same is true for the bottom carbon atom. **This is the principle advantage of Fischer projections.** You can do this bisection in an instant, comparing the top half of the molecule to the bottom half. If they are mirror images of one another, then the molecule is meso, and identical to its mirror image.

The bottom chiral center in 5 and the top chiral center in 6 are both mirror reflections of the same chiral center (top in 5). They must, therefore, be identical. The same is true in the reverse direction. The top chiral center in 5 and the bottom chiral center in 6 are both mirror reflections of the same chiral center (bottom in 5) and must, therefore, be identical. Because of the mirror plane bisecting 5, it is a meso structure and identical to its mirror image, even though chiral centers are present.

In the other pair of stereoisomers (7 and 8), there is no mirror plane dividing the two halves. The three groups on the top half do not eclipse the same three groups on the bottom half. These two structures are different and enantiomers.

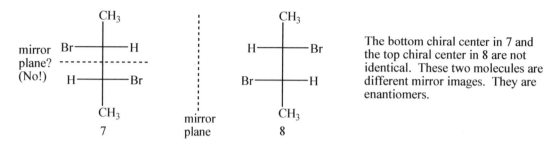

The bottom chiral center in 7 and the top chiral center in 8 are not identical. These two molecules are different mirror images. They are enantiomers.

Molecules 5 (= 6, meso) and 7 or 8 are also stereoisomers (same connectivity but different orientations in space). However, 5 is not the mirror image of 7 or 8. If we compare 5 to 7 we find that the top stereogenic atom is a mirror image and the bottom stereogenic atom is identical. These stereoisomers are diastereomers. This type of stereoisomerism is observed when two molecules have the same overall connectivity of atoms, yet are neither enantiomers nor identical.

CH₃
H———Br
H———Br
CH₃
5 = 6 = meso

mirror image
configuration
←→

identical
configuration
←→

diastereomers

CH₃
Br———H
H———Br
CH₃
7

A similar result is observed with 5 and 8.

CH₃
H———Br
H———Br
CH₃
5 = 6 = meso

identical
configuration
←→

mirror image
configuration
←→

diastereomers

CH₃
H———Br
Br———H
CH₃
8

Instead of the maximum four stereoisomers, we only find three, because of the meso structure. Structures 5 and 6 are identical (meso) and diastereomeric with 7 and 8, which are enantiomers. Naming the structures is easy, because we only change the R and S descriptors in the front of the name.

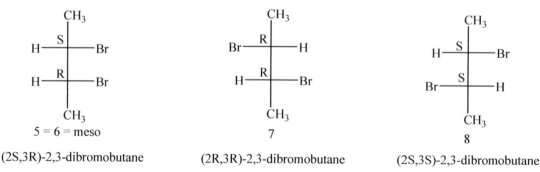

(2S,3R)-2,3-dibromobutane   (2R,3R)-2,3-dibromobutane   (2S,3S)-2,3-dibromobutane

Chapter 7

Diastereomers usually have different physical properties. Unlike enantiomers, which are completely identical in the absence of a chiral environment, diastereomers will have different melting points, boiling points, solubilities, spectra and the like (except by coincidence). Recall that a 50/50 racemic mixture of enantiomers is yet another common arrangement or stereoisomers. Racemic mixtures usually have different physical properties from the other stereoisomer possibilities, as well.

Problem 3 – For the following set of Fischer projections answer each of the questions below by circling the appropriate letter(s) or letter combination(s). Hint: Redraw the Fischer projections with the longest carbon chain in the vertical direction and having similar atoms in the top and bottom portion (highest nomenclature priority in the top half). Classify all chiral centers in the first structure as R or S absolute configuration.

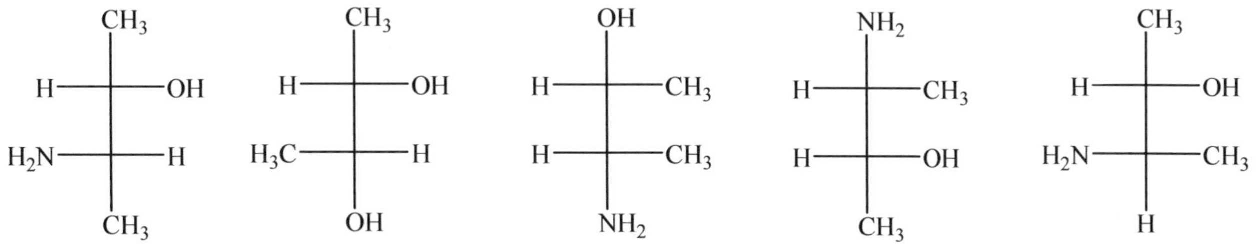

	A	B	C	D	E

a. Which are optically active?   A   B   C   D   E

b. Which are meso?   A   B   C   D   E

c. Which is not an isomer with the others?   A   B   C   D   E

d. Which pairs are enantiomers?   AB   AC   AD   AE   BC   BD   BE   CD   CE   DE

e. Which pairs are identical?   AB   AC   AD   AE   BC   BD   BE   CD   CE   DE

f. Which pairs are diastereomers?   AB   AC   AD   AE   BC   BD   BE   CD   CE   DE

g. Which pairs, when mixed in equal amounts   AB   AC   AD   AE   BC   BD   BE   CD   CE   DE
   will not rotate plane polarized light?

h. Draw any stereoisomers of 3-amino-2-butanol as Fischer projections, which are not shown above.
   If there are none, indicate this.

i. Would anything change if, in compound C, the NH_2 was replaced with a OH group?

j. Circle all chiral centers in a recently discovered Costa Rican fungal compound showing antibacterial properties against vancomycin resistant bacteria. How many stereoisomers are possible with that many chiral centers?

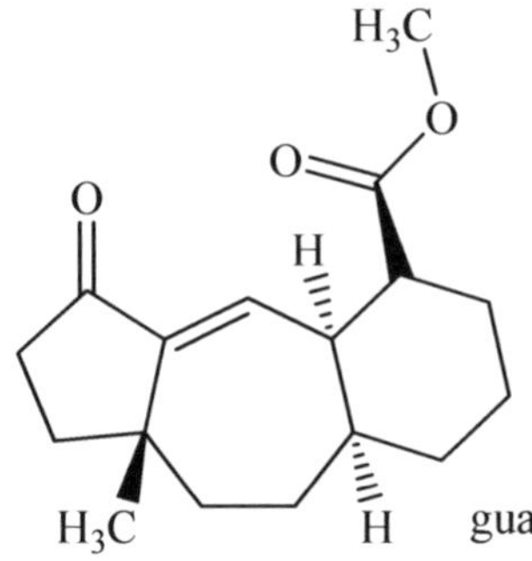

guanacastepene A

Problem 4 - For the following set of Fischer projections answer each of the questions below by circling the appropriate letter(s) or letter combination(s). Hint: Redraw the Fischer projections having the longest carbon chain in the vertical direction and having similar atoms in the top and bottom portion. Classify all chiral centers in the first structure as R or S absolute configuration.

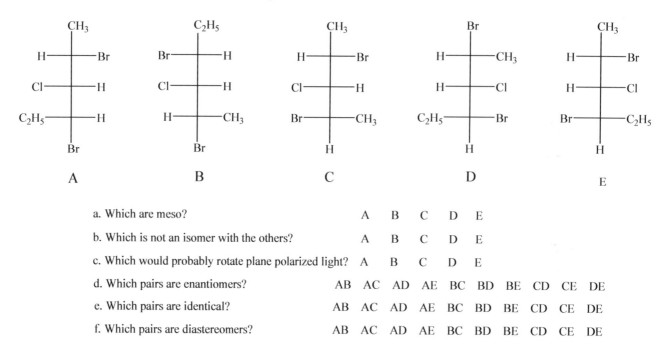

a. Which are meso?   A   B   C   D   E

b. Which is not an isomer with the others?   A   B   C   D   E

c. Which would probably rotate plane polarized light? A   B   C   D   E

d. Which pairs are enantiomers?   AB   AC   AD   AE   BC   BD   BE   CD   CE   DE

e. Which pairs are identical?   AB   AC   AD   AE   BC   BD   BE   CD   CE   DE

f. Which pairs are diastereomers?   AB   AC   AD   AE   BC   BD   BE   CD   CE   DE

e. Draw Fischer projections of any stereoisomers of "A" which are not shown above. If there are none, indicate this.

f. Derivatives of the antitumor steroidal saponin were recently prepared. They are highly potent and selective anticancer compounds. They inhibit Na^+/Ca^{+2} exchange leading to higher Ca^{+2} in the cytosol and mitochrondria causing cell death (apotosis) (Org. Lett. ASAP, 2014). Circle all chiral centers and any other stereogenic features in the partial structure below, and calculate the maximum number of stereoisomers possible.

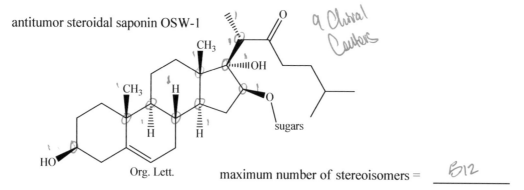

antitumor steroidal saponin OSW-1

9 Chiral Centers

Org. Lett.   maximum number of stereoisomers = ___512___

Problem 5 – For the following set of Fischer projections answer each of the questions below by circling the appropriate letter(s) or letter combination(s). Hint: Redraw the Fischer projections having the longest carbon chain in the vertical direction and having similar atoms in the top and bottom portion. Classify all chiral centers in the first structure as R or S absolute configuration.

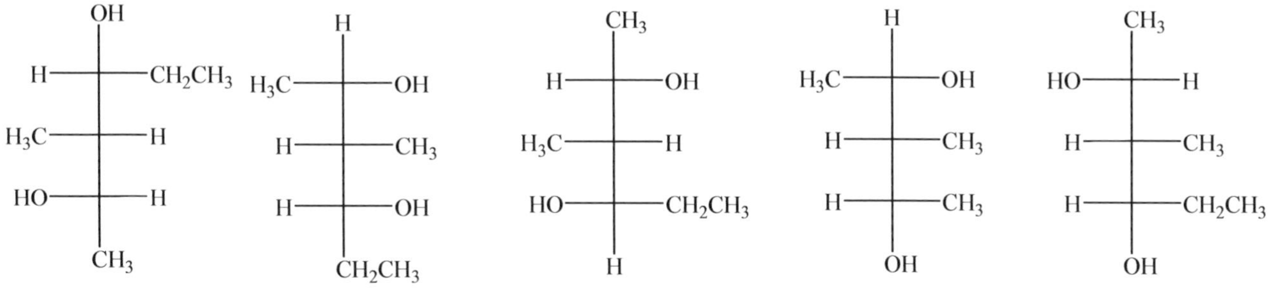

	A	B	C	D	E
a. Which are meso?	A	B	C´	D	E
b. Which is not an isomer with the others?	A	B	C´	D	E
c. Which would probably rotate plane polarized light?	A	B	C	D	E

d. Which pairs are enantiomers?   AB AC AD AE BC BD BE CD CE DE

e. Which pairs are identical?   AB AC AD AE BC BD BE CD CE DE

f. Which pairs are diastereomers?   AB AC AD AE BC BD BE CD CE DE

g. Draw Fischer projections of any stereoisomers of "A" which are not shown above. If there are none, indicate so.

h. Would anything change if the ethyl branch was changed to a methyl branch?

i. The structure of lobophytone A was recently determined (and the absolute configuration of all chiral centers!). It was isolated from the soft coral, *Lobophytum pauciflorum*, found in the South China Sea (Org. Lett. p.2482, 2010). Circle all chiral centers and any other stereochemical features, and calculate the maximum number of stereoisomers possible.

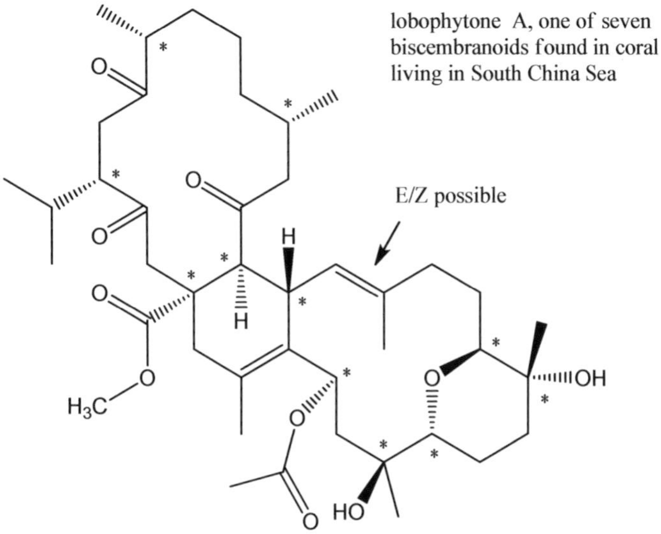

lobophytone A, one of seven biscembranoids found in coral living in South China Sea

E/Z possible

2^{11} possible from chiral centers = 2048

2^{12} including E/Z center = 4096

Chapter 7

Problem 6 - Place glyceraldehyde (2,3-dihydroxypropanal) in the proper orientation to generate a Fischer projection. Can you find any stereogenic centers? Is the molecule as a whole chiral? If so, draw each enantiomer as a 3D representation and classify all stereogenic centers as R or S.

Problem 7 - Draw a 3D Newman projection and a sawhorse representation for each the following Fischer projections. Redraw each structure in a sawhorse projection of a stable conformation. Identify stereogenic atoms as R or S.

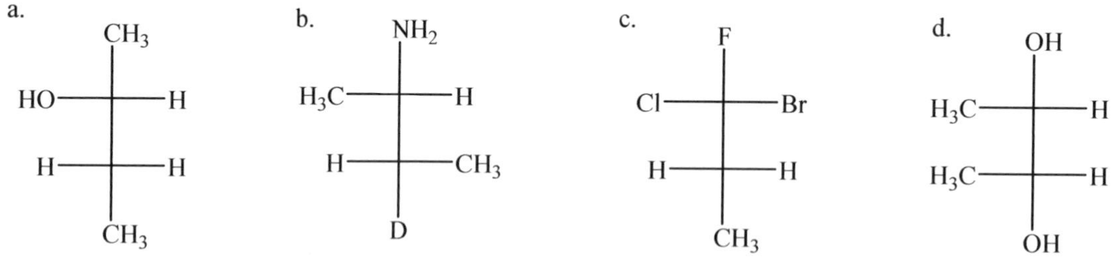

a.

CH_3
HO——H
H——H
CH_3

b.

NH_2
H_3C——H
H——CH_3
D

c.

F
Cl——Br
H——H
CH_3

d.

OH
H_3C——H
H_3C——H
OH

Problem 8 - Rearrange the Fischer projections below to their most acceptable form. You may have to rotate the top and/or bottom atom(s). You also may have to twist a molecule around 180° in the plane of the paper. Assign the absolute configurations of all stereogenic centers. Using your arm and fingers is helpful here. Write the name of the first structure. How many total stereoisomers are possible for each example?

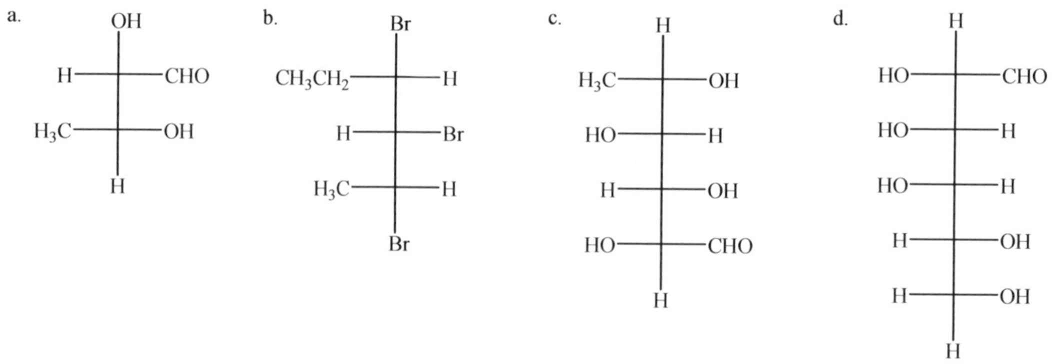

a.

OH
H——CHO
H_3C——OH
H

b.

Br
CH_3CH_2——H
H——Br
H_3C——H
Br

c.

H
H_3C——OH
HO——H
H——OH
HO——CHO
H

d.

H
HO——CHO
HO——H
HO——H
H——OH
H——OH
H

Remember: The absolute configurations, R and S, are defined by an arbitrary system of priorities to allow everyone to draw or think about actual 3D structures for comparisons. They apply to a single chiral center. Optical rotations, (d,l) = (+/-), on the other hand are experimental values determined by observing the rotation of plane polarized light of the entire molecule (right = d = dextrorotatory and left = l = levorotatory). Molecules having one or multiple chiral centers (1, 10s, 100s...) as R and/or S have only a single value for d or l.

Biological examples of chiral molecules – the aldose sugars (aldehydes)

Some of the following carbohydrates show a small sampling of Nature's choices for energy use, structural support and also serve as recognition targets in cells. Nature almost exclusively picks one enantiomer of a pair. When drawn as a Fischer projection it is the stereoisomer having the second to the last OH on the right side. In Biochemistry, this is referred to as the "D" isomer. If the second to last OH were on the left side in a Fischer projection it would be referred to as the "L" isomer. These are different from d and l of optical rotations. There is no logical reason for these designations. They must be memorized, as do the positions of all of the other "OH" groups up the chain. Organic chemistry uses "R" and "S" to identify each chiral center (no memorization required). "Reducing carbohydrates" have an aldehyde carbon at C1 and can be 2Cs, 3Cs, 4Cs, 5Cs, 6Cs or more in length. Glucose is probably the most famous of the carbohydrates. Some of these are shown below. How many chiral centers does each example have? How many stereoisomers are possible for each length of carbon? What are the absolute configurations of any chiral centers? What stereochemical relationship does the first stereoisomer as A (then B, C, etc.) have to the others? Are there any meso structures. (See problem 9 for comparison.)

H - R - OH : OH - R - H
(D - m) : (L - m)

Chapter 7

Three carbon aldose carbohydrates – two possible, all are 2,3-dihydroxypropanal

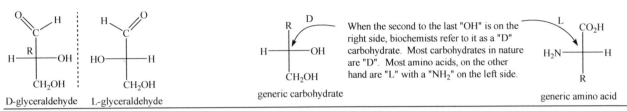

D-glyceraldehyde   L-glyceraldehyde

When the second to the last "OH" is on the right side, biochemists refer to it as a "D" carbohydrate. Most carbohydrates in nature are "D". Most amino acids, on the other hand are "L" with a "NH₂" on the left side.

generic carbohydrate

generic amino acid

Four carbon aldose carbohydrates– four possible, all are 2,3,4-trihydroxybutanal

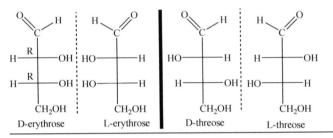

D-erythrose   L-erythrose   D-threose   L-threose

Biochem names are all different. Organic names are easier. All of these are 2,3,4-trihydroxybutanal.

1 = (2R,3R)-2,3,4-trihydroxybutanal
2 = (2S,3S)-2,3,4-trihydroxybutanal
3 = (2S,3R)-2,3,4-trihydroxybutanal
4 = (2R,3S)-2,3,4-trihydroxybutanal

Five carbon aldose carbohydrates – eight possible, all are 2,3,4,5-tetrahydroxypentanal

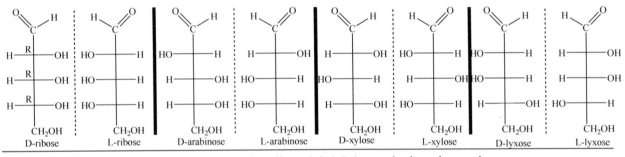

D-ribose   L-ribose   D-arabinose   L-arabinose   D-xylose   L-xylose   D-lyxose   L-lyxose

Six carbon aldose carbohydrates – 16 possible, all are 2,3,4,5,6-pentahydroxyhexanal

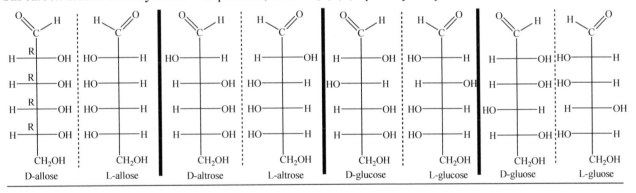

D-allose   L-allose   D-altrose   L-altrose   D-glucose   L-glucose   D-gluose   L-gluose

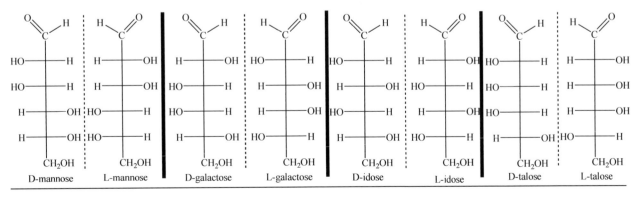

D-mannose   L-mannose   D-galactose   L-galactose   D-idose   L-idose   D-talose   L-talose

Chapter 7

Problem 9 – What would happen to the number of stereoisomers in each case above if the top aldehyde functionality were reduced to an alcohol functionality (a whole other set of carbohydrates!)? A generic structure is provided below to show the transformation. Nature makes some of these too. How many chiral centers does each example have? How many stereoisomers are possible for each length of carbon? What are the absolute configurations of any chiral centers? Specify the first stereoisomer as A (then B, C, etc.) and state what each relationship is to the others. Are there any meso structures.

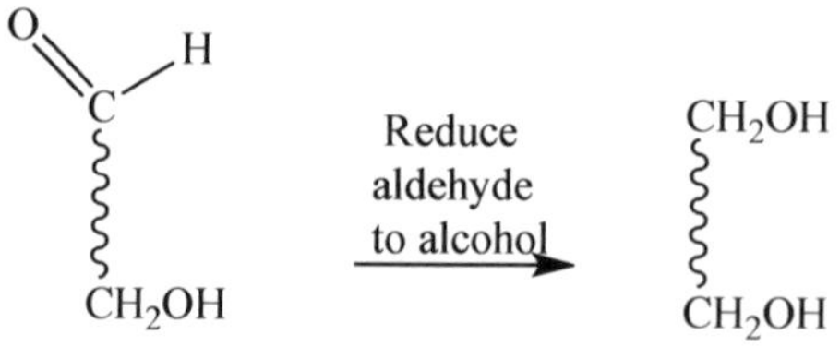

Three carbon tri-ol carbohydrate = 2 structures become only one structure, which is achiral
Four carbon tetra-ol carbohydrates = 4 structures become three stereoisomers (one meso pattern)
Five carbon penta-ol carbohydrates = 8 structures become four stereoisomers (two meso patterns, one duplication)
Six carbon hexa-ol carbohydrates = 16 structures become ten stereoisomers (two meso patterns, two duplications)

Problem 10 – How many chiral centers are found in cholesterol? How many potential stereoisomers are possible? Nature only makes one of them!

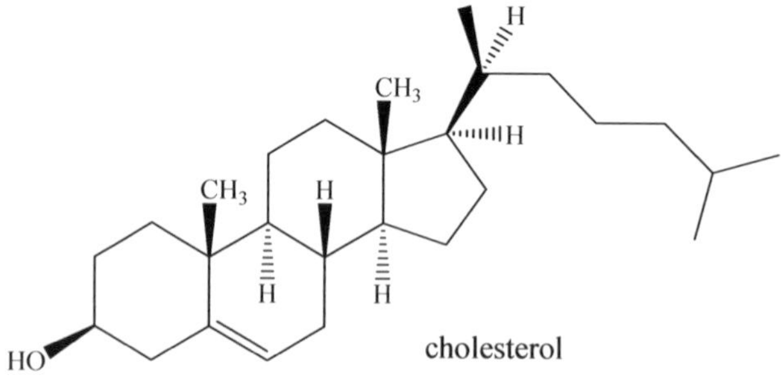

cholesterol

Chapter 7

CHAPTER 8 - ACID AND BASE CHEMISTRY

There are many good reasons for understanding the concepts of acid/base chemistry, but in organic chemistry, the concepts of Lewis acids and bases and electron movement are central. The process of showing two electron movement in reactions with curved arrows will be introduced in this topic. We will use the simple reaction of electron pair transfer (Lewis definition) and proton transfer (Bronsted definition) as our introductory reactions. These ideas are fundamental to our progress in organic chemistry. Two additional intermediates will also be introduced here: carbanions and carbocations. Resonance effects and inductive effects will be key ideas used to explain the relative stabilities of reactants, products and intermediates in these reactions. This very well could be your most important chapter in this book, because it is the start of a new way of thinking that will carry you through organic chemistry and biochemistry.

A Few Definitions of Acid/Base Chemistry

Bronsted Definition

In 1923 Johannes Bronsted (Danish, 1879-1947) developed a general theory of acids and bases based on proton transfers. Bronsted acids donate protons and Bronsted bases accept them. Four general patterns are shown below.

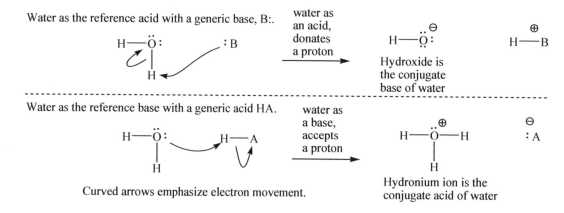

Some substances can both donate a proton (act as an acid) and accept a proton (act as a base). This is the usual focus in freshman chemistry. Water is the most common example of this behavior. Water is most commonly used as the *all purpose reference acid* with any base and the *all purpose reference base* with any acid. Such substances are referred to as amphoteric (Greek = "both"). The focus in organic chemistry is on the electrons and the curved arrows are drawn to show how the electrons change in the reaction of reactants to products. This has to become your way of thinking to succeed in organic chemistry and biochemistry.

Water as the reference acid with a generic base, B:.

water as an acid, donates a proton

Hydroxide is the conjugate base of water

Water as the reference base with a generic acid HA.

water as a base, accepts a proton

Hydronium ion is the conjugate acid of water

Curved arrows emphasize electron movement.

Lewis - Donate/ Accept electrons
Bronsted - Donate/Accept Protons

Lewis Definition

Also in 1923, Gilbert Lewis (American, 1875-1946) introduced an even more general theory of acids and bases. Lewis viewed acid/base reactions from the point of view of electron pairs about atoms, which fit in nicely with his concept of Lewis structures. Lewis' idea of electron pair donation has become a central strategy in the logic of organic chemistry and biochemistry. Lewis' definitions are not only applicable in all proton transfer reactions, they also work in a number of situations where Bronsted's definitions do not work. If no proton is transferred, then a reaction is not defined by Bronsted's definitions. Lewis' focus on electron pairs allows almost any reaction to be classified as an acid/base reaction (except one electron free radical transfers, some concerted reactions and some rearrangements).

Lewis acids: substances which accept a pair of electrons (all Bronsted examples above, plus more below)
Lewis bases: substances which donate a pair of electrons (all Bronsted examples above, plus more below)

Some "Organic" Examples where the Bronsted definitions do not work.

The first example below does show a proton transfer (Bronsted definition). Curved arrows indicate electron movement. The arrow starts at the Lewis base lone pair and points to the Lewis acid site. Additional arrows may or may not be required after the initial arrow. In subsequent examples, notice all sorts of electron acceptor atoms are shown: N, Fe, Al, C, H, and many more are possible. Lewis bases also go by the name of *nucleophile* and Lewis acids also go by the name of *electrophile*. You need to know these names!

- Lewis Base - Nucleophile
- Lewis Acid - electrophile

cyanide as a Bronsted base, the Lewis definition also works

All the reactions below show 2 electron transfers ("Lewis" acid/base reactions), but not all show proton transfer.

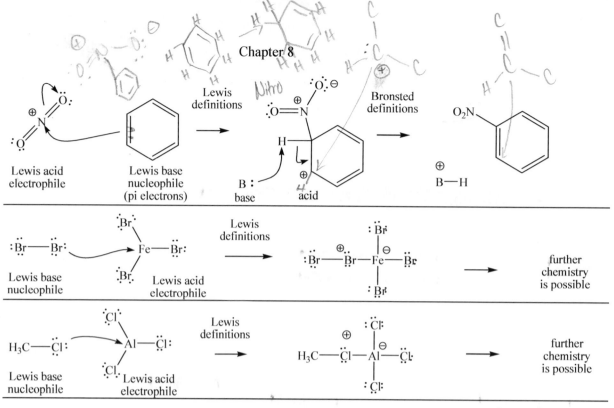

Lewis base = Nucleophile = electron pair donor (curved arrows start here)

Lewis acid = Electrophile = electron pair acceptor (curved arrows point here) } Learn these well.

Curved Arrow Formalism

Curved arrows show electron movement in the above reactions. These arrows symbolize movement of 1. the electrons that are forming new bonds and 2. the electrons that are being pushed away as old bonds are broken. We often say that the curved arrows show the flow of electrons, or show the electron movement. They are used to show the mechanisms of reactions (a chemical story for how each step of a reaction occurs).

Most of the mechanisms you write in organic chemistry (and biochemistry) will involve writing curved arrows to show how the electrons move from a nucleophile (Lewis base) to an electrophile (Lewis acid). This arrow pushing skill will be one of the most important skills for helping you understand how organic reactions work (biochemistry reactions too).

Problem 1 – Write an equation showing each of the following structures reacting as a Bronsted base, using H-A as the acid. If structures do not have any lone pairs of electrons, you will have to use the pi electrons as the base. One carbon atom of the pi bond will lose its share of the electron pair when the pi bond is broken and a new sigma bond is made using the other carbon atom and the proton. What will be the formal charge on the carbon atom losing the electrons? Hint: That carbon atom will become a carbocation. How are 'c' and 'd' related?

a. H_3C—$\ddot{O}$—H

b. (isopropyl)$_2$$\ddot{N}$:⁻

c. H_3C—C(=$\ddot{O}$:⁻)=CH_2

d. H_3C—C(=$\ddot{O}$)—CH_2⁻

e. H_3C—C(=$\ddot{O}$)—CH_3

f. H:⁻

g. H_2C=CH_2

h. (benzene ring)

Problem 2 – Write an equation showing each of the following reacting as a Bronsted acid, using B: as the base. You are going to have to leave two electrons behind and pay attention to formal charge.

a.   b.   c.   d.   e.   f.   g.

a. $H_3C-\overset{..}{\underset{..}{O}}-H$

b. $H-\overset{\overset{H}{|}}{\underset{\underset{H}{|}}{C}}-H$

c. $H-C\equiv C-H$

d. (acetone) $H_3C-\overset{\overset{\overset{..}{O}:}{\|}}{C}-CH_3$

e. (phenol) $\overset{..}{O}-H$ on benzene ring

f. $H_3C-\overset{\overset{H}{|}}{\underset{\oplus}{O}}-H$

g. $H-\overset{..}{\underset{..}{O}}:\ ^{\ominus}$

Problem 3 –There are two reasonable choices for a water molecule reacting with a t-butyl carbocation. One answer involves water acting as a Bronsted base and the other involves water acting as a Lewis base (which equation goes with each term?). Write in the necessary lone pairs of electrons, curved arrows and formal charge to show each of these possibilities. What are other terms that could be used to describe the water and carbocation reactants? Match those terms with the appropriate reactant structures. This example illustrates a common competition for electron pair donors in organic chemistry: react with a hydrogen atom or react at a carbon atom. This ambiguous competition shows up over and over in organic chemistry and is one of the reasons organic is considered difficult. In fact, one could say organic chemistry's full name is 'Organic Ambiguous Chemistry'.

t-butyl carbocation + water → partial yield (in one step) → product a + $H-O-H$

t-butyl carbocation + water → partial yield (in two steps) → (intermediate) → product b + $H-O-H$

Acid / Base Equilibria

In the examples below, A:⁻ and B: are really competing for the transferred proton in the transition state (Bronsted definition). The basic specie that is better able to donate its electron pair will win out in this competition more often. This will be the stronger base. Alternatively, we could say that the acid that is better able to give away its proton will win out (better at accepting an electron pair). This will be the stronger acid. The stronger conjugate acid will always be on the same side of the equation with the stronger conjugate base. The equilibrium will always shift towards the side of the weaker acid and weaker base. The weaker acid and weaker base will also be the thermodynamically more stable pair (lower potential energy = less reactive). This is shown in simple potential energy (PE) vs. progress of reaction (POR) diagrams.

$$B: \quad H-A \quad \rightleftharpoons \quad \overset{\oplus}{B}-H \quad :\overset{\ominus}{A}$$

stronger acid & base   weaker acid & base (more stable)

$$\overset{\delta+}{B}----H----\overset{\delta-}{A}$$

PE potential energy

stronger acid: B: H—A   TS E_a weaker base $\overset{\oplus}{B}-H$ $\overset{\ominus}{:A}$

$\Delta G = \ominus$

POR = progress of reaction

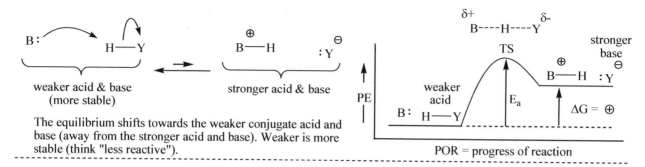

The equilibrium shifts towards the weaker conjugate acid and base (away from the stronger acid and base). Weaker is more stable (think "less reactive").

POR = progress of reaction

Quantitative Acidity

When an acid dissociates, it does this in some medium (solvent). In our world this is often water. Certainly in the earliest days of chemistry, it was almost always water because it was available. Water is the solvent of life. It flows in the rivers, lakes and oceans of our world. Our own blood, full of organic biomolecules and salts in aqueous solution, reflects this origin in its similar composition. Water often serves as both the solvent and as the base. K_{eq} values could allow us to quantify the relative ability of an acid to donate a proton (to water). However, instead, it is assumed that since water is essentially constant concentration (=55.6 M), its value can be absorbed into K_{eq} and is rewritten as K_a. Acid strength is actually measured quantitatively by referring to acidity constants, K_a values that indicate the extent of dissociation in water.

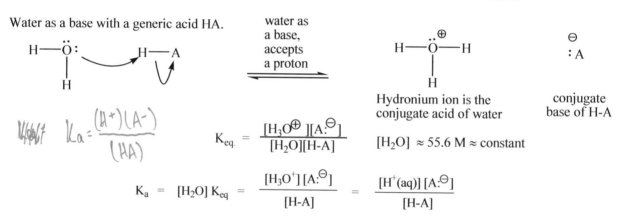

$$K_a = \frac{(H^+)(A^-)}{(HA)}$$

$$K_{eq.} = \frac{[H_3O^{\oplus}][A:^{\ominus}]}{[H_2O][H\text{-}A]}$$

$[H_2O] \approx 55.6\ M \approx$ constant

$$K_a = [H_2O]\,K_{eq} = \frac{[H_3O^+][A:^{\ominus}]}{[H\text{-}A]} = \frac{[H^+(aq)][A:^{\ominus}]}{[H\text{-}A]}$$

✱ A stronger acid (larger K_a) has a weaker base (electron pair donor) because it is a more stable conjugate base.

✱ A weaker acid (smaller K_a) has a stronger base (electron pair donor) because it is a less stable conjugate base.

Over many decades, organic chemists have learned that many organic reactions will not work at all in water, or will work a lot better if no water is present. As a result, many of the solvents used in organic chemistry are nonaqueous due to a variety of advantages. Even so, K_a values are defined with respect to water as the solvent, and acids are compared to one another based on that scale. This is what we will do.

By examining the K_a expression, we can see that large terms in the numerator will make the K_a larger and indicate a greater extent of dissociation. Thus, a stronger acid has a larger K_a and a weaker acid has a smaller K_a. We will consider strong acids as having $K_a > 1$, and can go as high as 10^{20} (but not in water). Weak acids will have a smaller $K_a < 1$, and can go as low as 10^{-50} (but not in water). These extreme numbers don't actually mean anything in water, but they do have some related meaning in other solvent systems. They are useful for comparisons of different acid strengths and different base strengths, which is how we use them.

Chapter 8

Problem 4 - Indicate whether the following compounds are stronger or weaker acids than water ($K_a = 10^{-16}$).
Write an arrow-pushing mechanisms with generic base, B:⁻.

a.

O
H\ || /H
:O—S—O:
     ||
    :O: $K_a = 10^3$

b. H—C≡C—H

$K_a = 10^{-25}$

c.

:O:
||
C
H₃C   O—H

$K_a = 10^{-5}$

d.

H H
| |
H—C—C—H
| |
H H

$K_a = 10^{-50}$

Problem 5 - Indicate whether the following bases are stronger or weaker bases than hydroxide. The K_a values
given are for the conjugate acids. You need to add a proton to evaluate each base from the point of view of its
conjugate acid and then invert your conclusion to decide "base" ability (stronger base = weaker acid). That
goes for hydroxide too (K_a of its conjugate acid = 10^{-16}. What is its conjugate acid?). Write arrow-pushing
mechanisms with generic acid, H-A.

a. H₂N:⁻

$K_a = 10^{-35}$

b. :Cl:⁻

$K_a = 10^7$

c. CH₃CH₂⁻

$K_a = 10^{-50}$

d. H₃N

$K_a = 10^{-9}$

How Do We Indicate a Proton?

Often H_2O is deleted from the $[H_3O^+]$ representation in the K_a expression. This leaves only $[H^+]$ in the K_a
expression as a symbolic representation of the acidic solvent species. This is "OK" only if you remember that
there is no such critter as H^+ in solution (H^+ (gas) > 400 kcal/mole!). A proton in solution, (H^+), is <u>always</u>
attached to a pair of electrons, somewhere. To emphasize the proton's extreme need for electrons someplace in
its valence, we explicitly write what the proton is attached to (in this case, a water molecule). We do not write
H^+ in this course!

Possible ways of indicating the hydrogen ion

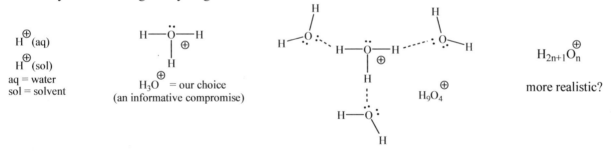

H⊕ (aq)
H⊕ (sol)
aq = water
sol = solvent

H—Ö—H
    |
    H ⊕
H₃O⊕ = our choice
(an informative compromise)

H₉O₄⊕

$H_{2n+1}O_n^{\oplus}$

more realistic?

Where is the positive charge actually located in H_3O^+? 'Formal' charge shows the positive charge on the
oxygen because the oxygen has an extra bond. However, theoretical calculations show it is really on the
hydrogen atoms. Even in H_3O^+, oxygen has a partial negative charge, though less than in hydroxide and water.

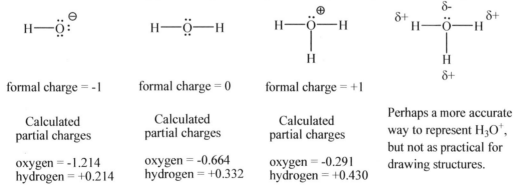

H—Ö:⁻

formal charge = -1

Calculated
partial charges

oxygen = -1.214
hydrogen = +0.214

H—Ö—H

formal charge = 0

Calculated
partial charges

oxygen = -0.664
hydrogen = +0.332

H—Ö⊕—H
    |
    H

formal charge = +1

Calculated
partial charges

oxygen = -0.291
hydrogen = +0.430

δ⁻
δ+ H—Ö—H δ+
        |
        H
       δ+

Perhaps a more accurate
way to represent H_3O^+,
but not as practical for
drawing structures.

Chapter 8

In summary, as an acid's K_a value gets larger, it has a greater extent of dissociation. The acid is better at giving its proton away. We refer to it as a stronger acid. Importantly, a stronger acid has a more stable conjugate base. A comparison of K_a values shows us that hydrochloric acid ($K_a=10^{+7}$) is a stronger acid (essentially 100% ionized in water) than ethanoic acid ($K_a= 2 \times 10^{-5}$ and less than 1% ionized in water). Therefore 1.0 M HCl has a larger percent of unprotonated chloride base (Cl$^-$, about 100% ionized in water, because it's more stable) than 1.0 M ethanoic acid has as the unprotonated ethanoate base ($CH_3CO_2^-$, less than 1% ionized in water, because it's less stable). Also, because ethanoic acid is a weaker acid than hydrochloric acid, its conjugate base, ethanoate will be a stronger base than chloride (a better electron pair donor). Understanding why this is true will provide the logic arguments that allow us to understand how organic and biochemistry works (still to come).

Problem 6 – Order the acids from strongest acid (= 1) to weakest acid in each series? What is the basis for your answers? What does that tell you about the stability of the conjugate bases? What about their basicities?

Problem 7 - Are the bases below stronger or weaker than hydroxide (K_a of conjugate acid $= 10^{-16}$)? The acidity constant provided (K_a) is that of each base's conjugate acid. Write out each conjugate acid. (Hint: Figure out the relative strengths of the conjugate acids before you predict the relative strengths of the bases.)

Problem 8 - Predict whether the products or the reactants are favored in each of the following equilibria. An equation is really a combination of two acid/base reactions (K_a), one written in the forward direction and one written in the reverse direction. Put the K_a of the forward reacting acid in the numerator (normal K_a expression) and the K_a of the reverse reacting acid in the denominator (inverted). The ratio shows the balance between the two acid/base reactions (K_{eq}). Add in curved arrows to show electron movement.

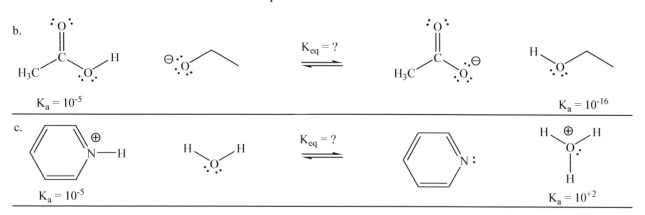

Problem 9 - Estimate whether the equilibrium in each equation below would lie to the left (reactants favored) or the right (products favored). The _acid_ is written first in each pair and the base is written second. The K_a of each conjugate acid is written in parentheses. (For bases, you have to add a proton to see what the acid looks like.) Use curved arrows to write a mechanism for the proton transfer in each part.

a.
$$H—C\equiv C—H \quad (K_a = 10^{-25})$$

$$H—C\overset{\ominus}{:} \quad (K_a = 10^{-50})$$

$H-C\equiv C:$

b.
$$H—C—N—H \quad (K_a = 10^{-37})$$

$$H_3C—C—O^{\ominus} \quad (K_a = 10^{-5})$$

c.
$$(K_a = 10^{-20})$$

$$H_3C—C—O^{\ominus} \quad (K_a = 10^{-16})$$

d.
$$(K_a = 10^{-10})$$

$$(K_a = 10^{-10})$$

- wheres the Base & wheres the Acid? (atoms)
- Whats the Charge on the Conj. Base?
  · Depends on how the Acid Starts
- What atoms Carry that Charge
↑Ka, ↑Acid strength - Is there Resonance or Inductive effect
  that helps stabilize the Conj. base

Ex. $H_2\ddot{N}^{\ominus}$ $H-C\equiv O-H \rightleftharpoons :NH_3$ $H-C\equiv C:^{\ominus}$
         $Ka_1 = 10^{-25}$ $Ka_2 = 10^{-37}$

Ka $HA \rightleftharpoons H^+ + A^-$
     $H^+ + A^- \rightleftharpoons HA$ $1/Ka$

$Keq = \dfrac{Ka_1}{Ka_2} = \dfrac{10^{-25}}{10^{-37}} = 10^{12}$

Chapter 8

Using K_a , written as pK_a (a useful complication), to estimate ΔG of a reaction,

Recall from freshman chemistry that p(something) = -log(something). One of the most important reasons for using pK_a is that it is part of the ΔG expression. A point that is almost never emphasized is that pK_a is a useful estimate of the free energy of ionization of an acid ($\Delta G°_{ionization}$). (Notice the $-\log K_a$ term in ΔG.)

$$\Delta G = -2.3RT \log K_a = 2.3RT (-\log K_a) = 2.3RT (pK_a)$$

$$\Delta G = (constant\overset{*}{)}(pK_a) = (1.4)(pK_a) \text{ kcal/mole} \approx pK_a \text{ kcal/mole}$$

$$\Delta G \approx 6 \times pK_a \text{ kJ/mole}$$

* assumes a constant temperature of 300 K

and $R = 2.0 \dfrac{\text{kcal}}{\text{mole}}$

1 kcal = 4.2 kJ

If R is the energy constant (2.0 cal/mole = 8.3 joule/mole) and we assume that room temperature is 300 K, then 2.3RT = (2.3)(2.0 cal/mole-K)(298 K) = 1.4 kcal/mole (~6 kJ/mol). Since this is organic chemistry and our approach is mostly qualitative, we can assume that 1.4 ≈ 1. This allows us to use a pK_a value as an approximate substitute for $\Delta G_{acid\ ionization}$.

Recall that if ΔG is negative (meaning that pK_a is negative), then the product (conjugate base) is favored and the acid is considered a strong acid (and has a more stable conjugate base than water). If ΔG is positive (meaning that pK_a is positive), then the reactant (conjugate acid) is favored and the acid is considered a weak acid (and a less stable conjugate base than water). We refer to the favorable reaction as downhill in energy or exergonic (ΔG = negative), while the unfavorable reaction is uphill in energy or endergonic (ΔG = positive). A figure shows this below.

Potential Energy (PE) vs. Progress of Reaction (POR) Diagrams

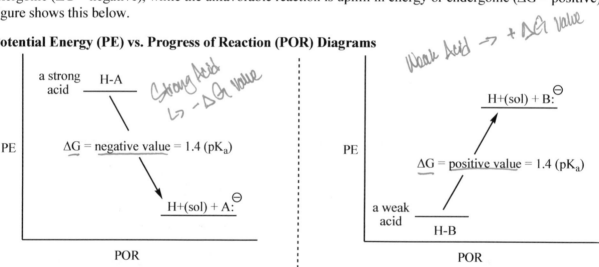

1.4 (pK_a) = ΔG = a negative number, ionization to conjugate base is favorable = strong acid, exergonic, downhill reaction

1.4 (pK_a) = ΔG = a positive number, ionization to conjugate base is not favored = weak acid, endergonic, uphill reaction

Problem 10 - What is the approximate ΔG for each reaction (in water)? What is the order of acidity in the following group (strongest acid = 1)? Write each reaction with water as the base (curved arrows, etc.).

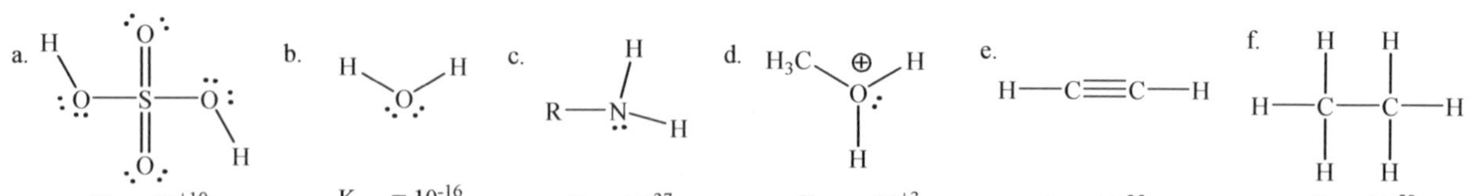

a. $K_a = 10^{+10}$

b. $K_{eq.} = 10^{-16}$

c. $K_a = 10^{-37}$

d. $K_{eq.} = 10^{+3}$

e. $K_a = 10^{-25}$

f. $K_a = 10^{-50}$

Chapter 8

Problem 11 - Draw a PE vs. POR diagram for "d" and "e" above. Show an arrow pushing mechanism using general base, B:⁻, and the conjugate acid and base for each reaction. Assume a relative starting energy value of zero as a reference point.

$$\text{d. } CH_3OH_2^{\oplus} \quad pK_a = -3 \qquad \text{e. } CH_3C{\equiv}C{-}H \quad pK_a = 25$$

A pK_a table provides us with immediate access to an acid's proton donating ability and indirectly to its conjugate base electron donating ability (and its relative stability). You can decide from the values in the pK_a table whether an acid is strong or weak and its relative acidity (or basicity) compared to other acids (or bases) in the table. Most organic acids are very weak ($pK_a > 1$ to very large), and you can evaluate approximately how large an energy input is necessary to form the conjugate base by using the pK_a. Remember, water is the reference base for all of the listed K_a's of the acids even though as a solvent for many acids, it is meaningless. You do not have to memorize pK_a/K_a values in this course. However, you do have to know how to make qualitative and quantitative comparisons in order to understand how organic and biochemistry works.

pK_a Table for a Variety of Acids – Approximately equal to $\Delta G_{acid\ ionization}$ (in kcal/mole = (1.4)x(pK_a))

Carbon Acids – There is a fair amount of uncertainty in the higher pK_a values.

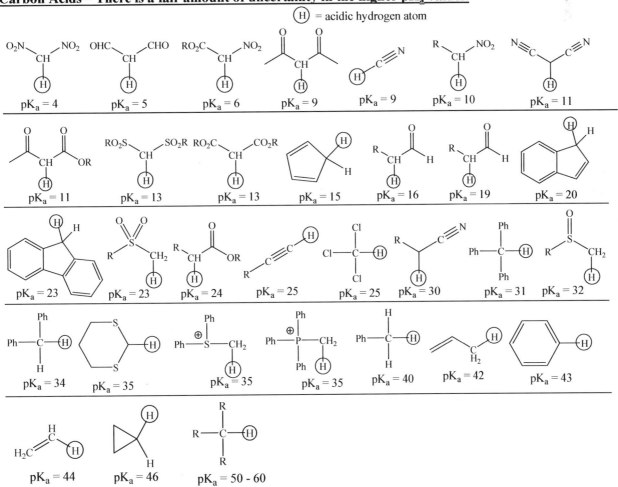

Oxygen Acids

(H) = acidic hydrogen atom

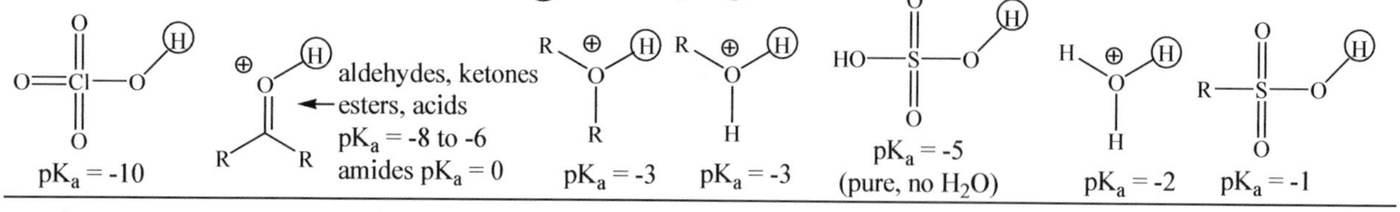

pK_a = -10

aldehydes, ketones
esters, acids
pK_a = -8 to -6
amides pK_a = 0

pK_a = -3

pK_a = -3

pK_a = -5
(pure, no H_2O)

pK_a = -2

pK_a = -1

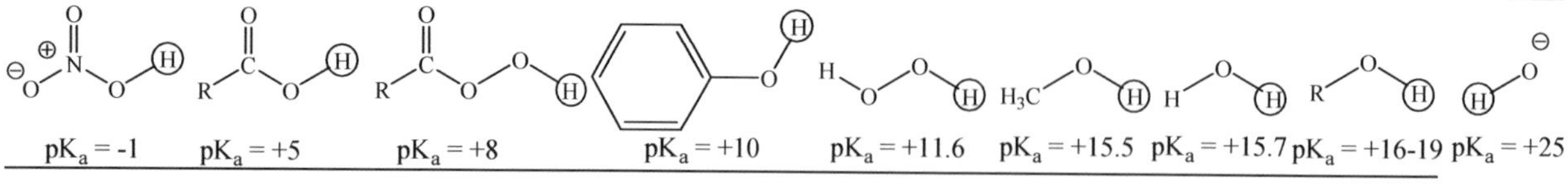

pK_a = -1   pK_a = +5   pK_a = +8   pK_a = +10   pK_a = +11.6   pK_a = +15.5   pK_a = +15.7   pK_a = +16-19   pK_a = +25

Compare the following groups.

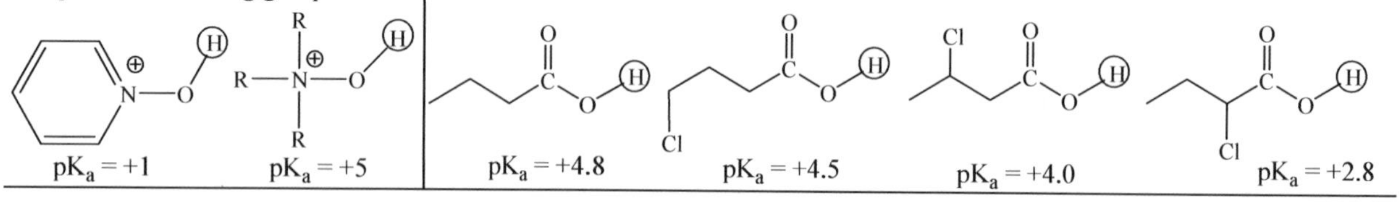

pK_a = +1   pK_a = +5   pK_a = +4.8   pK_a = +4.5   pK_a = +4.0   pK_a = +2.8

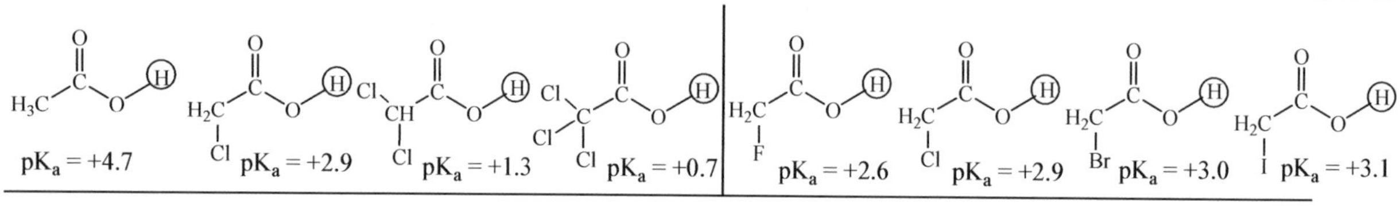

pK_a = +4.7   pK_a = +2.9   pK_a = +1.3   pK_a = +0.7   pK_a = +2.6   pK_a = +2.9   pK_a = +3.0   pK_a = +3.1

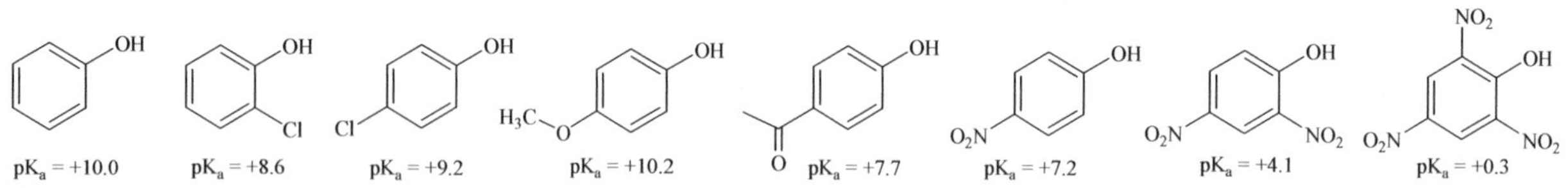

pK_a = +10.0   pK_a = +8.6   pK_a = +9.2   pK_a = +10.2   pK_a = +7.7   pK_a = +7.2   pK_a = +4.1   pK_a = +0.3

Nitrogen Acids

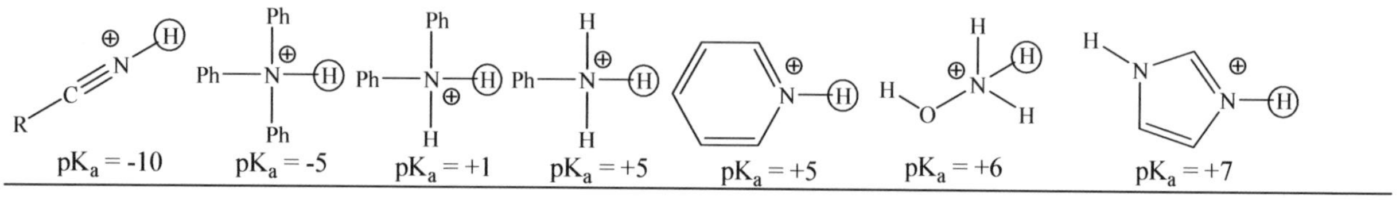

pK_a = -10   pK_a = -5   pK_a = +1   pK_a = +5   pK_a = +5   pK_a = +6   pK_a = +7

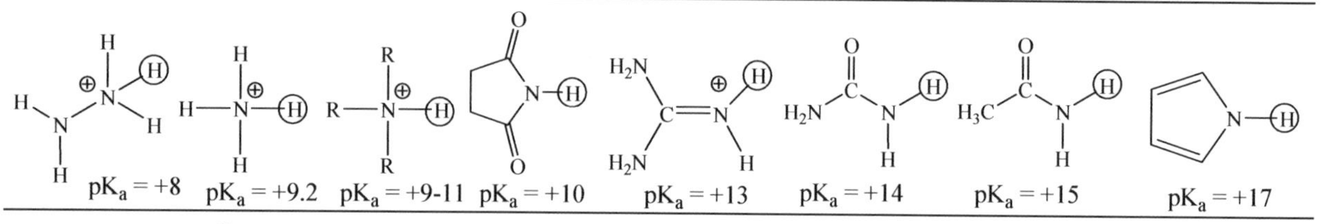

pK_a = +8   pK_a = +9.2   pK_a = +9-11   pK_a = +10   pK_a = +13   pK_a = +14   pK_a = +15   pK_a = +17

pK_a = +35   pK_a = +37

Other Miscellaneous Acids

Compare the following groups.

$R-\overset{\overset{H}{\mid}}{\underset{\oplus}{S}}-(H)$ pK$_a$ = -5	$H-\overset{(H)}{S}$ pK$_a$ = +7	$Ph-\overset{(H)}{S}$ pK$_a$ = +8	$R-\overset{(H)}{S}$ pK$_a$ = +10

$H-\overset{\overset{H}{\mid}}{\underset{\underset{H}{\mid}}{\overset{\oplus}{P}}}-(H)$ pK$_a$ = +0	$H-\overset{\overset{H}{\mid}}{\underset{\underset{H}{\mid}}{\overset{\oplus}{N}}}-(H)$ pK$_a$ = +9

$HO-\overset{\overset{H}{\mid}}{\underset{\oplus}{N}}-(H)$ pK$_a$ = +6	$H_2N-\overset{\overset{H}{\mid}}{\underset{\oplus}{N}}-(H)$ pK$_a$ = +8.1	$(H)-O-\overset{\overset{H}{\mid}}{N}-H$ pK$_a$ = +13.7

Compare the following groups. Acids with really large, negative pK$_a$'s are called "super" acids.

F$_5$SbF—(H) pK$_a$ = -20	I—(H) pK$_a$ = -10	HTe—(H) pK$_a$ = 3	O$_2$NO—(H) pK$_a$ = -1	O$_3$ClO—(H) pK$_a$ = -10
FSO$_3$—(H) pK$_a$ = -15	Br—(H) pK$_a$ = -9	HSe—(H) pK$_a$ = 4	ONO—(H) pK$_a$ = +3	O$_2$ClO—(H) pK$_a$ = -1
F$_3\overset{\ominus}{B}$F—(H)$^{\oplus}$ pK$_a$ = -15	Cl—(H) pK$_a$ = -7	HS—(H) pK$_a$ = 7	HO$_2$CO—(H) pK$_a$ = +6.4	OClO—(H) pK$_a$ = +2
O$_3$ClO—(H) pK$_a$ = -10	F—(H) pK$_a$ = +3	HO—(H) pK$_a$ = 16	$\overset{\ominus}{O_2}$CO—(H) pK$_a$ = +10.3	ClO—(H) pK$_a$ = +7.5

H$_2$PO$_4$—(H) pK$_a$ = +2.1	ClO—(H) pK$_a$ = +7.5	HO$_3$SO—(H) pK$_a$ = -3	HO$_2$SO—(H) pK$_a$ = +2	HOO—(H) pK$_a$ = +12
$\overset{\ominus}{H}$PO$_4$—(H) pK$_a$ = +7.2	BrO—(H) pK$_a$ = +8.7	$\overset{\ominus}{O_3}$SO—(H) pK$_a$ = +2	$\overset{\ominus}{O_2}$SO—(H) pK$_a$ = +7	HO—(H) pK$_a$ = +16
$\overset{-2}{PO_4}$—(H) pK$_a$ = +12.4	IO—(H) pK$_a$ = +11			

These pK$_a$ tables dramatically demonstrate how much Bronsted acids can vary in strength. The magnitude of the numbers is really beyond our comprehension. The strongest acid in the table has a K$_a$ ≈ 10^{+20}, while the weakest acid has a K$_a$ ≈ 10^{-50}. That's 70 orders of magnitude! What does 10^{70} mean? Even so, we will only use two simple arguments to rationalize the differences in acidity (…and basicity). Our two reasons for these large differences will be: 1. inductive effects (based on relative electronegativity) and 2. charge delocalization effects (usually based on resonance through 2p orbitals). We will not emphasize steric effects, hydrogen bonding or solvation effects, which can also modify relative acidities, often greatly.

Another way of looking at acidity (selected examples, pK$_a$ varies over 70 powers of 10 !)

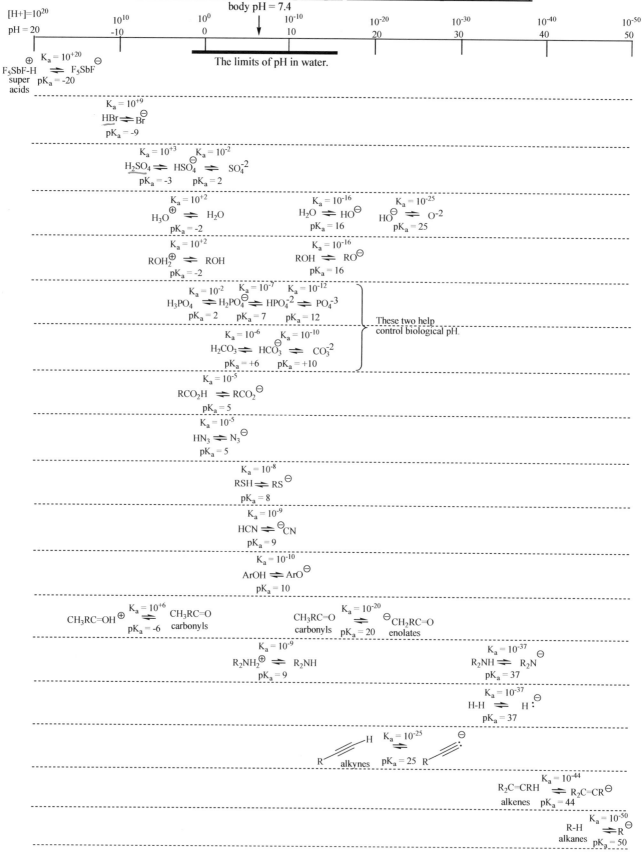

Chapter 8

Electronegativity or Inductive Effects

The first argument we will use to explain differences in acidity seems intuitively straightforward. However when resonance effects are also present, it sometimes misleads us. We begin with four potential proton donors lacking resonance. The second row hydrides of carbon (CH_4), nitrogen (NH_3), oxygen (H_2O), and fluorine (HF) are simple and informative. In each instance we are removing a proton (H^+) directly from a second row element (C,N,O,F). This is relevant because they are similar in size (n = 2 elements), so the charge density should also be similar in the conjugate bases. We use K_a or pK_a to compare relative acidities (and basicities). Notice that the acid with the strongest bond also happens to be the strongest acid (breaking the bond). We wouldn't expect that.

The acid w/ the strongest bond is the acid w/ the highest acidity level. Normally wouldn't expect this.

$$B: + H-\ddot{F}: \underset{pK_a}{\overset{K_a}{\rightleftharpoons}} B-H + :\ddot{F}:^{\ominus} \quad \text{Most Stable, Most electronegative}$$

$$B: + H-\ddot{O}H \underset{pK_a}{\overset{K_a}{\rightleftharpoons}} B-H + :\ddot{O}H^{\ominus}$$

$$B: + H-\ddot{N}H_2 \underset{pK_a}{\overset{K_a}{\rightleftharpoons}} B-H + :\ddot{N}H_2^{\ominus}$$

$$B: + H-CH_3 \underset{pK_a}{\overset{K_a}{\rightleftharpoons}} B-H + :CH_3^{\ominus}$$

Acid	conjugate base	pK_a	K_a	$\Delta G = 1.4 pK_a$ (kcal/mole)	Bond Energy (kcal/mole)	X	μ	ε	Z_{eff}
$H-\ddot{F}:$	$^{\ominus}:\ddot{F}:$	3	10^{-3}	4	135	4.0	1.82	84	+7
$H-\ddot{O}H$	$\overset{\ominus}{:\ddot{O}H}$	16	10^{-16}	22	119	3.5	1.84	78	+6
$H-\ddot{N}H_2$	$\overset{\ominus}{:\ddot{N}H_2}$	36	10^{-36}	50	107	3.0	1.47	22	+5
$H-CH_3$	$\overset{\ominus}{:CH_3}$	50	10^{-50}	70	105	2.5	0.0	2	+4

X = electronegativity

a measure of an atom's ability to attract electrons in chemical bonds

μ = molecular dipole moment

a measure of a molecule's polarity

ε = dielectric constant

a measure of a medium's ability to insulate charge

Even though all of the above acids are weak acids in water (meaning incomplete dissociation in water), there are enormous acidity differences between each of them. Two of the reactions are not even possible in water (NH_3 and CH_4). However, in a relative sense, pK_a values show us that HF is a stronger acid than H_2O, which is stronger than NH_3, which is stronger than CH_4. This is consistent with having a negative charge on the respective second row atoms of the conjugate bases. The most electronegative atom would be expected to be most stabilizing of the excess negative charge. From our earlier arguments about electronegativity we can argue that this results from a larger Z_{eff} nuclear charge for atoms using the same electron shell (n = 2), and therefore having similar charge density. Fluorine has a Z_{eff} of +7, while carbon has a Z_{eff} of +4 and both are approximately the same size. We might also say that fluorine is more electronegative, but this argument will occasionally fail (charge delocalization, also known as resonance, can change things).

Chapter 8

Problem 12 - Order the following bases in increasing strength below (1=strongest). Are these differences big or small? Provide an explanation for your choice.

This series clearly shows the stronger electron withdrawing power of F > O > N > C from the differences in their pK_a values. Because the acidic hydrogen is directly attached to each of the second row atoms, it is perhaps not surprising that there is such a large effect. But, what if these groups were farther away from the acidic hydrogen? We've seen previously that groups can inductively withdraw (pull) or donate (push) through the sigma bond framework, based on relative electronegativities. This is what we call an inductive effect. In organic chemistry, we say that the fluorine is inductively more electron withdrawing than oxygen, nitrogen and carbon. Some examples are shown below. In each of these examples the negative charge is on the oxygen atom, which is attached to a comparison atom. Unfortunately, HOF does not exist. Also, we use water as a reference compound to observe what effect each of the halogen atoms has on the acidity of the OH bond. What do they tell us?

Problem 13 - Order the substituents below, from most (=1) to least electron withdrawing, according to their apparent inductive effect based on the pK_a for the given acids (of the O-H bond). Provide a possible explanation for the relative order of acidities. They appear similar to above, but on a much reduced scale.

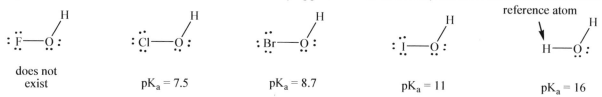

→ Inductive effects represent an electron pair tug-of-war for electrons in chemical bonds. Competition for or repulsion of electrons in sigma bonds will pull or push them in one direction or another, based mainly on the electronegativity of atoms or groups in the vicinity of the site of interest (solvation is also very important, but we are ignoring this). Hydrogen is our relative reference point, and we compare all substituent groups to what a hydrogen does. In the examples below, you can see from the higher relative pK_a that carbon groups (an alkyl group, example b) have a weaker pull (stronger push?) than hydrogen. Alkyl groups are considered inductively electron donating relative to hydrogen. Donating electron density will stabilize a positive or partial positive center and destabilize a negative or partial negative center. In the opposite sense, electron withdrawal will stabilize a negative center and destabilize a positive center. Inductive effects help explain a lot of observations in organic chemistry and biochemistry. You need to get this to succeed in these areas.

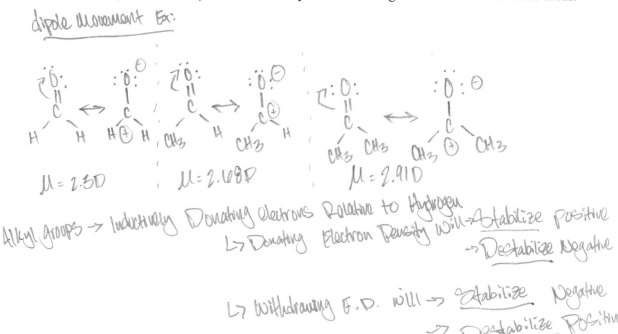

Chapter 8

Problem 14 –

a. Using the data below, decide if the given substituents on the CH_2's in each carboxylic acid are electron withdrawing (stabilizing the carboxylate anion) or electron donating (destabilizing the carboxylate anion) *relative to a hydrogen atom*. Order the substituents from most inductively electron withdrawing (= 1) to least inductively electron withdrawing (…or actually electron donating). What is the relationship between a and i?

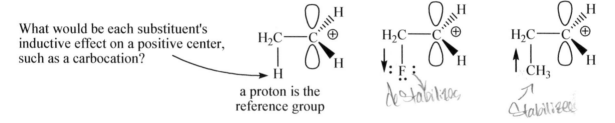

a $pK_a = 5.70$	b $pK_a = 4.87$	c reference atom $pK_a = 4.74$
d $pK_a = 4.31$	e $pK_a = 3.83$	f $pK_a = 3.57$
g $pK_a = 3.18$	h $pK_a = 2.90$	i $pK_a = 2.85$
j $pK_a = 2.85$	k $pK_a = 2.59$	l $pK_a = 2.47$
m $pK_a = 2.35$	n $pK_a = 1.68$	

b. Are inductive withdrawing effects always stabilizing? What kind of inductive effect (stabilizing or destabilizing) would be expected with fluorine (F-) versus methyl (CH_3-), above, for a positive reaction center such as the carbocation below? (Reference X = "H")

What would be each substituent's inductive effect on a positive center, such as a carbocation?

a proton is the reference group

destabilizes

stabilizes

Problem 15 – a. Order the substituents below, from most (=1) to least electron withdrawing, according to their apparent inductive effect based on the pK_a for the given acids (of the O-H bond). Are any of the substituent groups electron donating relative to the hydrogen atom (the reference atom)? Provide a possible explanation for the relative order of acidities.

H_3C—$\ddot{O}$: H
$pK_a = 16$

H_2N—$\ddot{O}$: H
$pK_a = 14$

$H\ddot{O}$—$\ddot{O}$: H
$pK_a = 12$

H_3C—C—$\ddot{O}$: H (H_3C, H_3C)
$pK_a = 19$

reference atom
H—$\ddot{O}$: H
$pK_a = 16$

b. What atom loses the proton in each molecule? Why?

$pK_a = 14$

$pK_a = 8$

Chapter 8

Problem 16 - Are the indicated pK_a's consistent with what you would expect? Explain why.

a

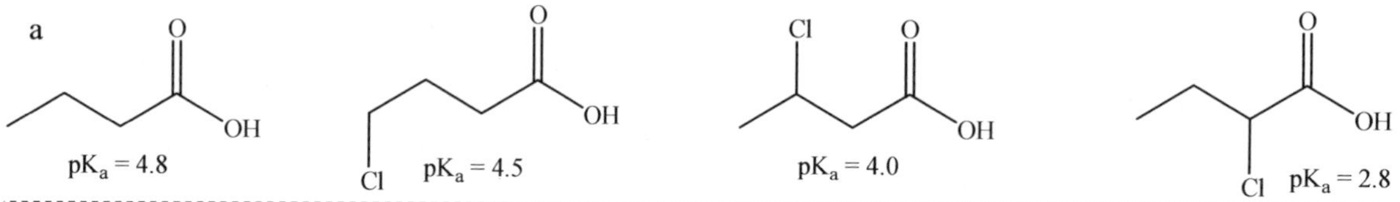
pKa = 4.8
pKa = 4.5
pKa = 4.0
pKa = 2.8

b

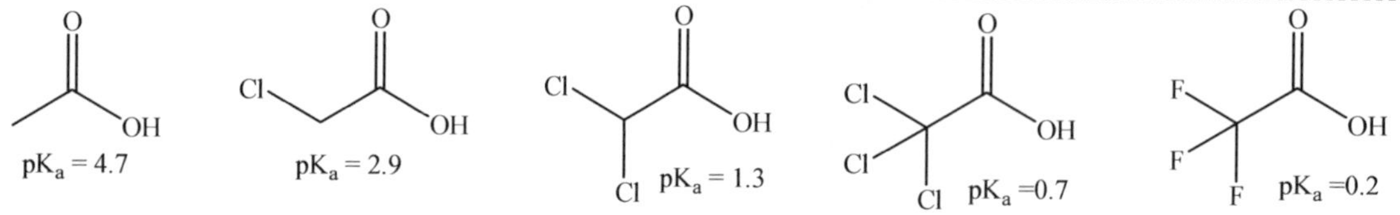
pKa = 4.7
pKa = 2.9
pKa = 1.3
pKa =0.7
pKa =0.2

Problem 17 - a. Order the carbanions below from most (=1) to least stable. (R represents an alkyl substituent.). Explain your order.

b. Order the carbocations below from most (=1) to least stable. (R represents an alkyl substituent.). Explain your order.

Problem 18 - Rationalize the following series of acidities. Using the strongest acid, write out an acid/base equation with curved arrows to show electron flow, lone pairs of electrons and formal charge using B:$^-$ as the base.

a.

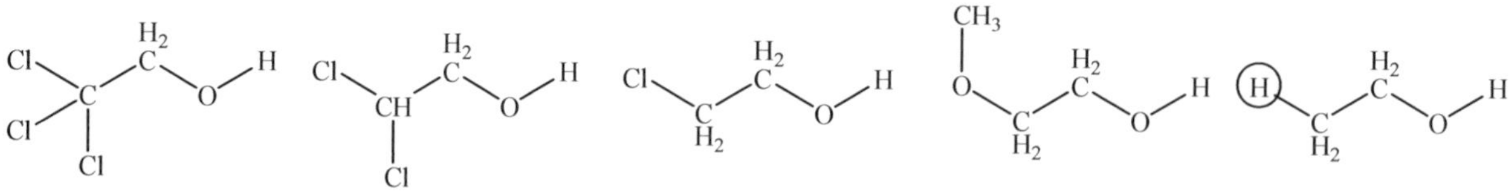

pK_a = 12 13 14.3 14.8 15.9

b.

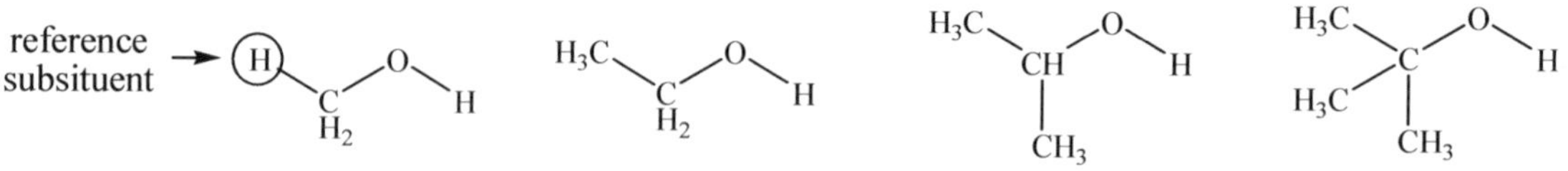
reference
subsituent

pK_a = 15.5 15.9 17.1 19.2

Chapter 8

Effect of hybridization on acidity of C-H hydrocarbon acids (Do carbon atoms have different electronegativities?)

You might think that all hydrocarbon C-H bonds would have similar acidities. This is far from the truth. If we compare just the three hydrocarbon compounds shown below we find a difference in pK_a values of approximately 25 where each pK_a unit represents approximately a power of 10. The K_a values in these examples span a range of 10^{25} (25 orders of magnitude = 10,000,000,000,000,000,000,000,000)! (That's 10 septillion.)

Acid	Conjugate Base	K_a	pK_a	$\Delta G = 1.4(pK_a)$ (kcal/mole)	Hybridization of carbon atom	%s	%p	Bond Energy
ethane	H_3C-CH_2	10^{-50}	50	70	sp^3	25	75	98
ethene	$H_2C=CH$	10^{-44}	44	62	sp^2	33	67	110
ethyne	$H-C\equiv C:$	10^{-25}	25	35	sp	50	50	124

$\ominus$: B would have to be very strong for reactions like these to work.

The s will hold the electrons tighter

G / P
P P P
25 / 75
VS

G - P
50 / 50

All three molecules are extremely weak acids, consisting of only nonpolar C-C and C-H bonds (according to our electronegativity rules). Upon first inspection, the large differences in acidity seem puzzling; since they are all C-H acids. Hybridization is the most obvious difference among the three types of carbons (we have an alkane (sp^3), an alkene (sp^2) and an alkyne (sp). Clearly, the sp hybridized carbon of ethyne is the strongest acid. It must be that it has the most stable conjugate base carbanion. But why is this the case?

In these three examples, all of the carbon atoms are approximately the same size and the charge is localized in a single hybrid orbital. The presence of pi bonds in ethene and ethyne may trick you into thinking there is resonance in their conjugate bases, but this is not the case since the lone pair is in a hybrid orbital perpendicular to the pi systems. (Resonance is our next topic.) If resonance does not explain the differences in stability of the conjugate bases, the answer must lie in differences in electronegativity of the differently hybridized carbon atoms.

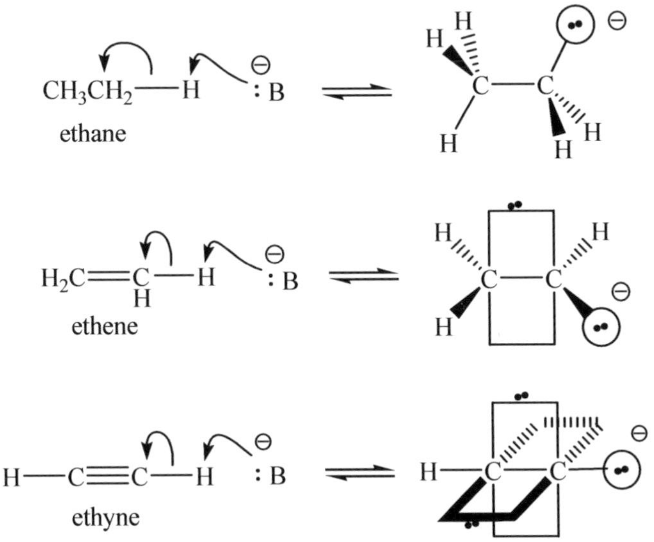

Lone pair electrons are more stable due to the greater percent s character which allows them to get closer to the nucleus and feel a greater opposite charge.

Our atomic model tells us that the 2s orbital electrons are held more tightly than the 2p orbital electrons, and on average, are closer to the +6 carbon nucleus (Z_{eff} = +4). The stronger hold of the 2s orbital for electrons is similar to what we associate with greater electronegativity among elements in the same row of the periodic table, (i.e. C < N < O < F) as Z_{eff} increases across a row.

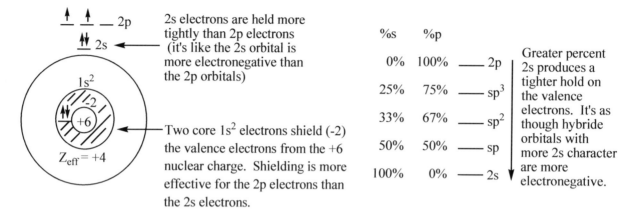

The sp orbital is 50% s, while the sp^2 orbital is 33% s, and the sp^3 orbital is 25% s. This may seem like a small difference, but the consequences are clearly profound. Ethyne (acetylene) gives up its proton better than ethene (ethylene) by a K_a factor of 10^{19} (that's 10,000,000,000,000,000,000) and ethene does so better than ethane by a K_a factor of 10^6 (that's 1,000,000). Here we argue that the sp carbanion is more stabilized by the more electronegative sp orbital. This hybridization effect is a general observation that applies to other atoms as well (the hybridization of nitrogen or oxygen also affects their relative acidities when protonated).

It's time for some more problems. Don't forget, basicity is the flip side of acidity. A stronger acid is often the result of a more stable base and a more stable base is less basic.

Problem 19 – Rationalize the following differences in pK$_a$ values.

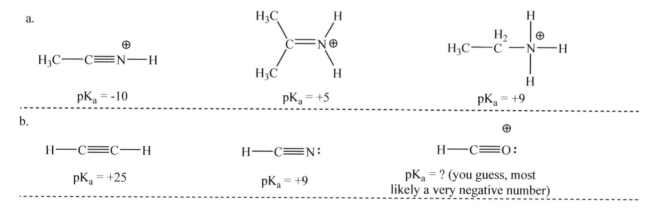

Chapter 8

Problem 20 – Propose an explanation for the following differences in acidities.

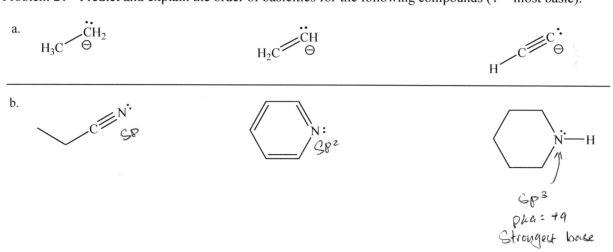

a.

pK$_a$ = 4.8   pK$_a$ = 4.4   pK$_a$ = 2.6

b.

pK$_a$ = 16.1   pK$_a$ = 15.5   pK$_a$ = 13.6

Problem 21 - Predict and explain the order of basicities for the following compounds (1 = most basic).

a.

b.

Charge Delocalization as a Stabilizing Factor in Acid/Base Chemistry (and other places too)

    A second argument often used in acid/base chemistry invokes electron delocalization in the conjugate base, when possible. A more stable conjugate base forms more easily, with lower charge density, making for a stronger conjugate acid. A simple and informative trend is present in the halogen column of the periodic table. We suddenly encounter a problem with our inductive effect argument above (electronegativity?). It seems like HF should be the strongest acid, but is that what we find? Something different is going on.

Based on electronegativity, which do you think should be the strongest acid? Is it what we expect? Why not? (See discussion.)

Acid	Conjugate Base	r_x (A) $1A = 10^{-8}cm$	$V = 4/3\pi r^3$ (volumn)	extra electron density	Z_{eff}
:F—H :B	:F: ⊖	1.36	10.5 A³	11%	+7
:Cl—H :B	:Cl: ⊖	1.81	24.8 A³	6%	+7
:Br—H :B	:Br: ⊖	1.96	31.5 A³	3%	+7
:I—H :B	:I: ⊖	2.16	42.1 A³	2%	+7

Acid	K_a	pK_a	$\Delta G = 1.4(pK_a)$ (kcal/mole)	Bond Energy (kcal/mole)	X	μ	ε	valence shell	% increase in e- density	radius	Volume element
HF	10^{-3}	3	4	135	4.0	1.82	84	n = 2	11%	$r_F = 1.36A$	$V_F = 10.5A^3$
HCl	10^7	-7	-10	103	3.2	1.08	5	n = 3	6%	$r_{Cl} = 1.81A$	$V_{Cl} = 24.8A^3$
HBr	10^9	-9	-13	88	2.7	0.82	4	n = 4	3%	$r_{Br} = 1.96A$	$V_{Br} = 31.5A^3$
HI	10^{10}	-10	-14	71	2.2	0.44	3	n = 5	2%	$r_F = 2.16A$	$V_I = 42.1A^3$

X = electronegativity	μ = molecular dipole moment	ε = dielectric constant
a measure of an atom's ability to attract electrons in chemical bonds	a measure of a molecule's polarity	a measure of a medium's ability to insulate charge

    Fluorine is obviously the most electronegative element, but it is just as clear from the data that HF is the weakest Bronsted acid in the halogen family. Our Z_{eff} argument won't work here because all of the elements have a $Z_{eff.} = +7$. What does change for each of these atoms is the number of electron shells and the total number of electrons in which the -1 charge is dispersed in the conjugate base. The extra -1 on fluorine (n=2) increases the electron density by 11% and is compressed into the smallest volume element ($r_F = 1.36A$ / $V_F = 10.5A^3$). The extra -1 on Cl⁻ (n=3), Br⁻ (n=4), and I⁻ (n=5) increases the electron density by 6%, 3%, and 2%, respectively. The volume element is larger in each case as well ($r_{Cl} = 1.81A$ / $V_{Cl} = 24.8A^3$; $r_{Br} = 1.96A$ / $V_{Br} = 31.5A^3$; $r_I = 2.16A$ / $V_I = 42.1A^3$). These two factors produce a lower charge density as we go down the halogen column. In other words, there is less energy expense for the -1 charge in each succeeding conjugate base, as we allow the charge to disperse in a larger volume element with a similar Z_{eff} (lower electron-electron repulsion). We say that the electron density is more delocalized in iodide than in bromide than in chloride and finally than in fluoride.

Chapter 8

Problem 22 - Propose an explanation for the following pK$_a$'s.

a

H₂O pK$_a$ = 16	H₂S pK$_a$ = 7	H₂Se pK$_a$ = 4	H₂Te pK$_a$ = 3

b

H₃C–O–H   pK$_a$ = 16   H₃C–S–H   pK$_a$ = 10

Problem 23 - Which is the stronger acid in each part?   Explain your choice.

a.   H–S–H   or   H–Cl

b.   H–N$^{\oplus}$H₃   or   H–P$^{\oplus}$H₃

c.   H–CH₃   or   H–SiH₃

Problem 24 - Which is the stronger base in each part (better electron pair donor)?   Explain your choice.

a.   H–O$^{\ominus}$ or   H–S$^{\ominus}$

b.   H₃C–N$^{\ominus}$H   or   H₃C–O$^{\ominus}$

c.   :Cl$^{\ominus}$:   or   :I$^{\ominus}$:

Resonance = Charge Delocalization in C, N, O structures

   How do electron delocalization arguments show up in organic chemistry? Most atoms used in organic chemistry are similar in size, e.g. C, N and O. These are the atoms that will hold the electrons of a conjugate base and bear the negative formal charge that is often present when a proton is lost from an organic molecule. Is it possible that charge could also be spread out into a larger volume by using only carbon, nitrogen and oxygen atoms?

   Your mind is hopefully racing through our earlier chapters and quickly zeroing in on...**resonance!** Electron density can be delocalized through consecutive, parallel 2p orbitals. Lone pairs in 2p orbitals, pi bonds, empty 2p orbitals and free radicals all have the potential to do this.

   When adjacent to one another, a lone pair can spread into a pi bond, if the sp^3 atom rehybridizes to sp^2 and allows that lone pair orbital to become a 2p orbital. When this is possible, the delocalization of electron density is substantial. You can probably see the next necessary step. We will push the excess electron density of the anionic site toward (into) a pi bond. This is especially good when the end atom is oxygen or nitrogen.

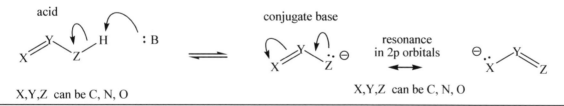

X,Y,Z can be C, N, O   X,Y,Z can be C, N, O

In the same \{ z-eff The larger the electron cloud, the less stable
The smaller → the more stable

Chapter 8

Problem 25 – Explain the relative pK$_a$'s in each part. Show all relevant structures in your explanation (e.g. resonance structures). Write an equation for each example using a general base, B: Include curved arrows to show electron movement, include formal charge and all lone pairs. In part c, can you tell which conjugate acid is more stable from their pK$_a$'s? What is the relationship between the conjugate bases of those two structures?

a

pK$_a$ = 50   pK$_a$ = 43

b

pK$_a$ = 50   pK$_a$ = 20   pK$_a$ = 9   pK$_a$ = 5

c

pK$_a$ = 12   pK$_a$ = 16

Combination Effects - Inductive and Resonance Together

Analyze each example in the following series of organic acids and explain the relative differences in acidity (vertically and horizontally).

C

pK$_a$ = 50   pK$_a$ = 40   pK$_a$ = 20

N

pK$_a$ = 37   pK$_a$ = 28   pK$_a$ = 15

O

pK$_a$ = 16   pK$_a$ = 10   pK$_a$ = 5

S

pK$_a$ = 8   pK$_a$ = 6.5   pK$_a$ = 3.1

First, all of these acids are weak acids (K$_a$ < 1, so pK$_a$ is positive), so all of the conjugate bases are higher in potential energy than their conjugate acids. The pK$_a$ values show us the order of acidities, with the lowest values indicating the strongest acids (most stable conjugate bases).

Placing the negative charge on a more electronegative atom is clearly better for the stability of the conjugate base (O > N > C). Oxygen is also more electronegative than sulfur, but sulfur is larger and the negative charge is more delocalized on sulfur (both atoms have Z$_{eff}$ = +6). Adding an adjacent aromatic ring allows three additional resonance structures which makes each acid stronger by many powers of ten than the saturated examples. The advantage is greatest where the starting acid is the weakest. Three additional

inductive effects & delocalization
↑
Reasons for strong Acids

↑ Charge delocalization from larger atoms

Chapter 8

resonance structures (all carbon) does not help the sulfur nearly as much as it helps carbon. Adding an adjacent carbonyl (C=O) helps even more than the aromatic ring because the extra atom carrying negative charge is oxygen. Again, this helps most where the starting acid is the weakest (carbon) and least where the starting acid was strongest (sulfur).

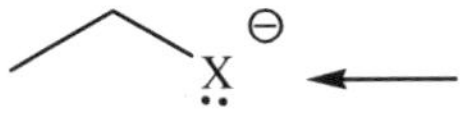

No resonance delocalization possible, charge is completely localized on a single atom.

aromatic resonance structures - 3 additional resonance structures, with negative charge on carbon atoms

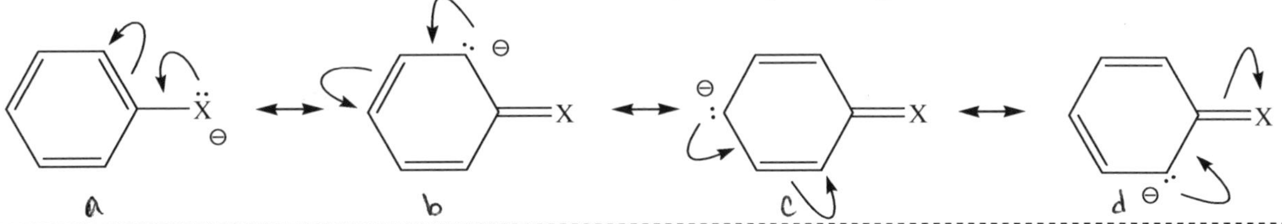

carbonyl structures - all show that the negative charge is partially localized on oxygen and one other atom.

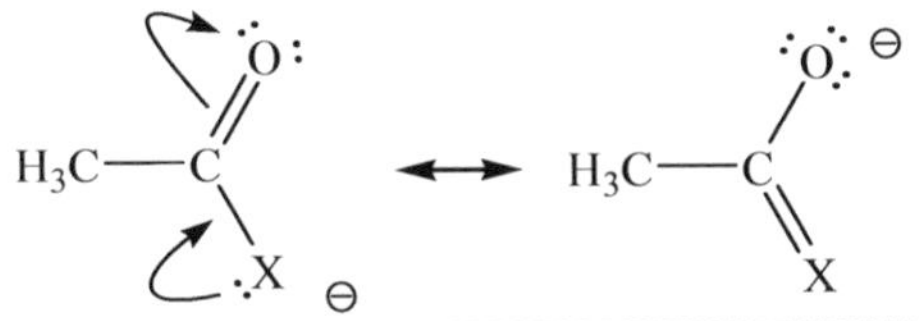

There is also an inductive withdrawing effect in the carboxylic acid from the carbonyl group (C=O). Theoretical calculations suggest that this might be account for about half the increase in acidity in this example.

Overall, there is a smaller uphill energy change in forming a more stabilized conjugate base and thus a greater extent of reaction (a larger K_a and lower pK_a). These simple arguments will carry us far in our study of organic chemistry. What we need now is lots of practice using these arguments.

Problem 26 – Which is the weaker base in each pair below. Explain your answers. Match each pK_a with its acid. *weaker base must be more stable* *Strong Acid - more stable*
more   *↳ weaker base*

a

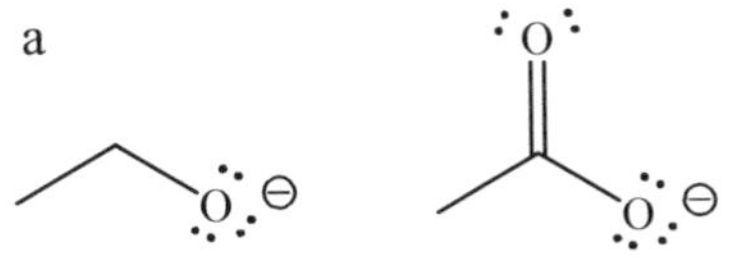

pK_a's = 5, 16

b

pK_a's = 9, 25

c

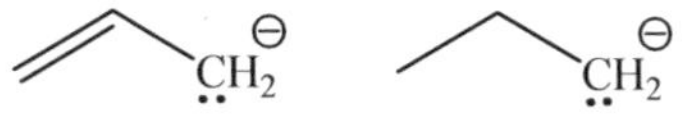

pK_a's = 43, 50

Problem 27 – Identify the most acidic hydrogen in each of the following compounds. Explain your choices. (Consult your pK_a table, if necessary.)

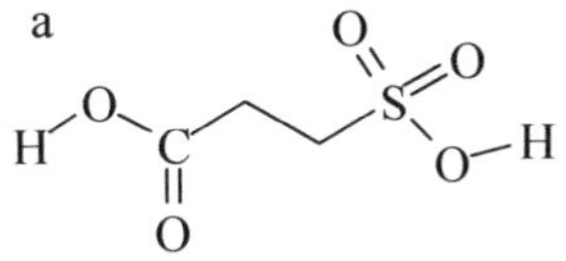

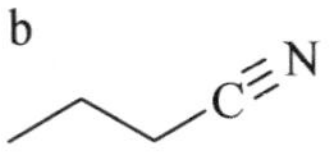

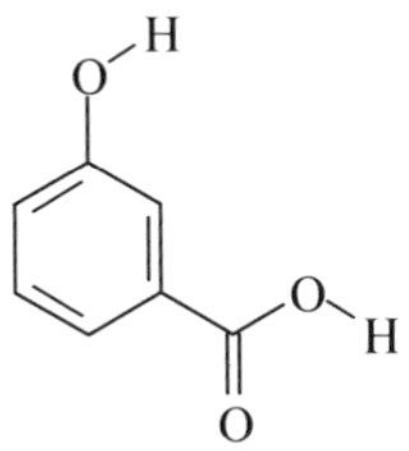

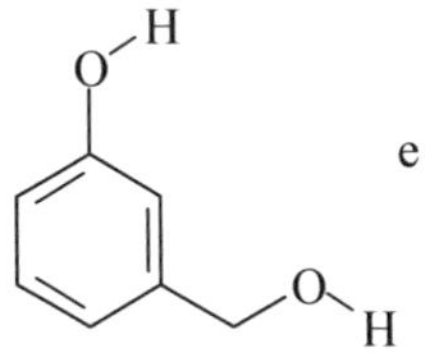

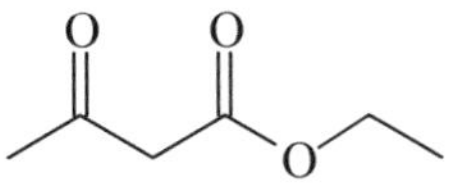

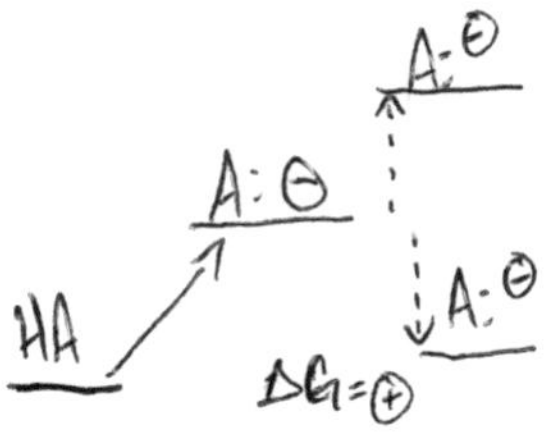

Chapter 8

Important O-H, N-H and C-H Bond Acidities in Organic Chemistry

One of the major requirements in organic synthesis is formation of new bonds (C-O, C-N, C-Br and especially C-C). There are a variety of methods developed to achieve this goal, but one of the most important strategies is where one atom donates two electrons (as a nucleophile) and a carbon accepts them (as an electrophile). The Lewis acid/base definitions fit in nicely with these reactions (electron pair donation and acceptance), and this aspect is emphasized throughout this book. Important oxygen donors will be hydroxide, alkoxides, carboxylates (all as anions = strong donors) and neutral water, alcohols and carboxylic acids (all as neutral molecules = weak donors). Important nitrogen donors will be azide, N_3^-, as a strong nucleophile and sodium amides, Na+ / R_2N^-, as strong bases. Most of our examples of strong carbon nucleophiles are carbanions. Any method that generates a carbanion is an important tool for organic chemistry. One common approach, consistent with the topic of this chapter, is where a carbanion is formed via a proton transfer reaction. Several very common possibilities are shown below in a general form in the next problem, where you can get more practice pushing curved arrows. Very strong bases are often necessary. Some common ones that we will use include, lithium diisopropyl amide = LDA ($NaNR_2$, have to make this one), sodium hydride (NaH, given), and n-butyl lithium (LiC_4H_9, given). Sodium hydroxide, NaOH, is available whenever we need it, but we have to make sodium alkoxides (NaOR, from alcohols and sodium hydride).

Problem 28 – Supply the necessary curved arrows, lone pairs of electrons and/or formal charge to show how the first step of each reaction proceeds. All of these reactions are all simple proton transfer reactions generating an anion (reactions e-j generate carbanions). Generally, there is some stabilizing feature that allows a carbanion to form via acid/base chemistry, such as inductive and/or resonance effects. In working the problem below, show any important resonance structures or identify the inductive effect that makes the reaction possible.

a.

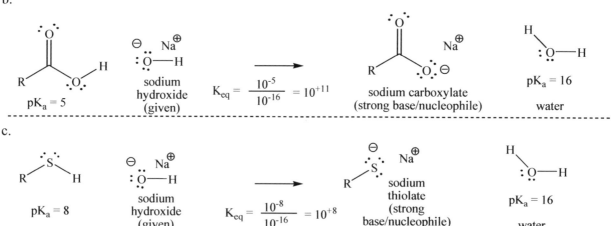

d.

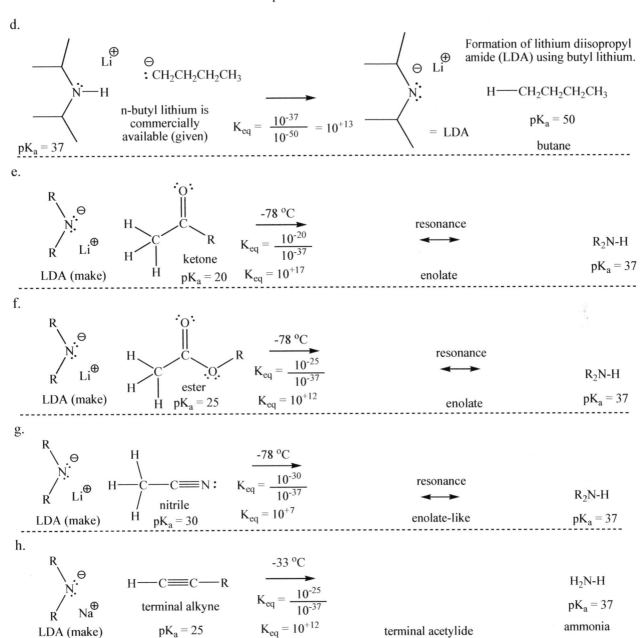

Formation of lithium diisopropyl amide (LDA) using butyl lithium.

n-butyl lithium is commercially available (given)

$K_{eq} = \dfrac{10^{-37}}{10^{-50}} = 10^{+13}$

pK$_a$ = 37

= LDA

H——CH$_2$CH$_2$CH$_2$CH$_3$

pK$_a$ = 50

butane

e.

LDA (make)

ketone
pK$_a$ = 20

-78 °C

$K_{eq} = \dfrac{10^{-20}}{10^{-37}}$

$K_{eq} = 10^{+17}$

resonance

enolate

R$_2$N-H

pK$_a$ = 37

f.

LDA (make)

ester
pK$_a$ = 25

-78 °C

$K_{eq} = \dfrac{10^{-25}}{10^{-37}}$

$K_{eq} = 10^{+12}$

resonance

enolate

R$_2$N-H

pK$_a$ = 37

g.

LDA (make)

nitrile
pK$_a$ = 30

-78 °C

$K_{eq} = \dfrac{10^{-30}}{10^{-37}}$

$K_{eq} = 10^{+7}$

resonance

enolate-like

R$_2$N-H

pK$_a$ = 37

h.

LDA (make)

H——C≡C——R
terminal alkyne
pK$_a$ = 25

-33 °C

$K_{eq} = \dfrac{10^{-25}}{10^{-37}}$

$K_{eq} = 10^{+12}$

terminal acetylide

H$_2$N-H

pK$_a$ = 37

ammonia

Chapter 8

Oxygen and Nitrogen Atoms, and a Few C=C Pi Bonds as Bases (and nucleophiles) in Organic Chemistry

Many organic reactions are run in strong acid. We will limit our strong acids to mostly sulfuric acid (H_2SO_4), when water is the solvent (the real acid will be H_3O^+), and toluenesulfonic acid ($CH_3C_6H_4SO_3H$ = TsOH), when an organic solvent is used. Occasionally we will also use other acids, e.g. HI, HBr, HCl, HNO_3, ROH_2^+ or H_3O^+, etc. If we are merely indicating an acidic environment we will sometimes use a symbolic acid, H-A, to emphasize the proton transfer step. Almost always H-A is a very strong acid, like H_2SO_4, If there is only one heteroatom (O or N), your choice is an easy one – use a lone pair of electrons on the heteroatom as a base to react with the acid. If there are two heteroatoms at a carbonyl bond (:X-C=O), always use a carbonyl lone pair electrons, in our course. This will produce a stable resonance delocalized intermediate. That won't happen if you protonate the "X" part.

Problem 29 –
a. Is protonation of an amide more likely on the oxygen atom or the nitrogen atom. Show the reaction for both and examine any possible resonance structures for clues that explain the difference.

b. Predict the more basic oxygen in a carboxylic acid. Show both oxygen atoms acting as a base and explain your choice (just like the amide in part a).

c. Would you expect the amide C=O or the acid C=O to be more basic? Explain.

Whenever an atom with a lone pair is next to a pi bond (of any kind) it will share its electrons with the pi bond, making it more electron rich and a better electron pair donor.

Use the C=O group, not the X part.

X = O or N
(in our course)

resonance stabilized positive charge

carboxylic acid   ester   H 1°, 2° 3° amide

This is even true when "X" is next to a C=C pi bond. If you protonate X, the positive charge is only on X, but if you protonate the C=C pi bond (at the carbon atom not bonded to X) you get resonance structures with positive charge on carbon and X. This is usually the better choice.

Chapter 8

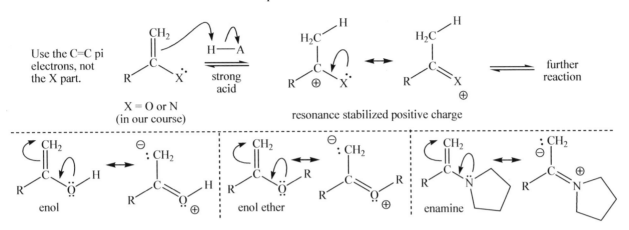

Use the C=C pi electrons, not the X part.

X = O or N (in our course)

resonance stabilized positive charge

strong acid

further reaction

enol

enol ether

enamine

In strong acid, almost every example (but not quite all) can lead to a carbocation structure or a resonance contributor that looks like a carbocation. We see carbocations quite a bit in organic chemistry, but no matter how they are generated, carbocations will always do one of three things: 1. add an electron pair donor to the carbon (add a nucleophile), 2. lose a hydrogen atom from an adjacent atom and form a pi bond (also called a beta atom) or 3. rearrange to a different carbocation that is similar to or more stable than the original carbocation (not usually observed when resonance stabilized). Every time you see a carbocation your mantra should be: "add a nucleophile or lose a beta hydrogen atom or rearrange to a similar or more stable carbocation". Learn the mantra now, because you are going to need it a lot as you study organic chemistry.

There are many ways to form carbocations.

Carbocations generally do one of three things.

C_α

1. Add a nucleophile at C_α, top or bottom

2. Lose a beta hydrogen atom (elimination)

3. Rearrange, if similar or more stable R^+ can form (except if resonance stabilized)

Problem 30

(The power of a proton in organic chemistry makes many reactions possible that otherwise would not occur.) – Many functional groups we study are shown below. They all can react with strong acid in a similar first step (H-A, below). Supply curved arrows to show the proton transfers in the acid/base reactions and additional electron movement, when indicated. (Remember, curved arrows show what forms "next".) The transfer of a proton from a strong acid generally increases the potential energy from a unreactive neutral molecule to a highly reactive, protonated cation. Even though a positive formal charge is written on the heteroatom, carbon also carries much of the positive charge. Often this can be shown with a resonance structure. Placing positive charge on carbon will almost always lead to one of the three possible outcomes just mentioned above: 1. add nucleophile, 2. lose a beta hydrogen atom or 3. rearrange. As you learn organic chemistry (and biochemistry) note the common themes that repeat themselves again and again. Some reactions below show a final product and others show possible outcomes from these initial steps at the end. Some of the reactions are covered in greater detail in the topics that follow. Think of H-A as a strong acid like HBr or H_2SO_4.

a.

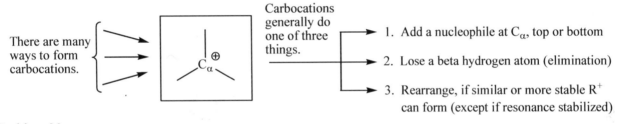

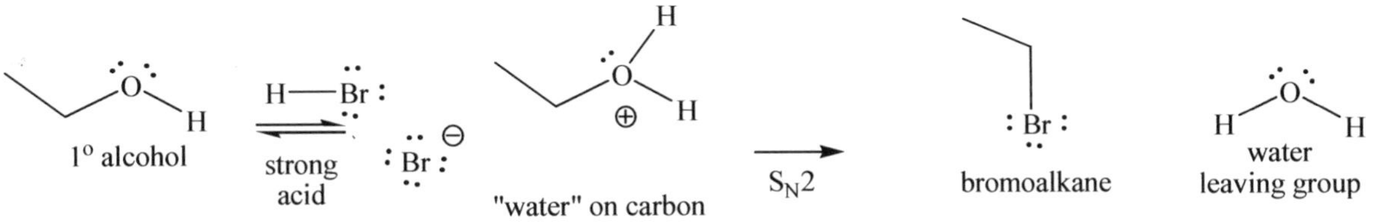

$1°$ alcohol

strong acid

"water" on carbon

S_N2

bromoalkane

water leaving group

no carbocation (R^+) forms if R = Me, $1°$ carbon atom, Br pushes off water from the backside (S_N2)

b.

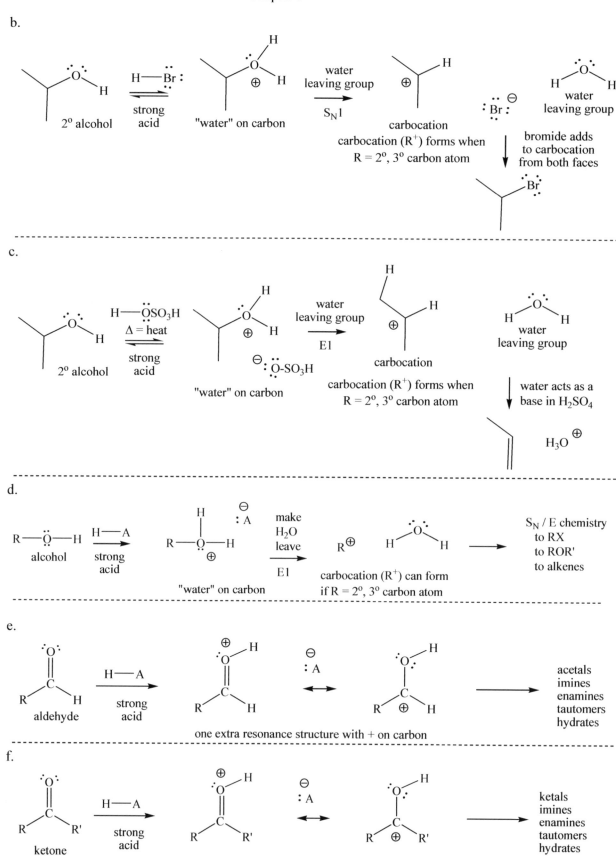

c.

d.

e.

f.

Chapter 8

g.

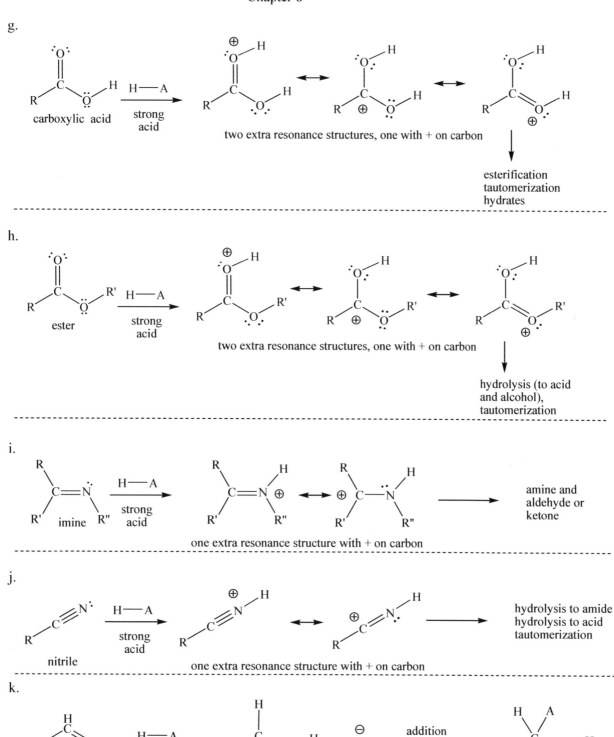

two extra resonance structures, one with + on carbon

carboxylic acid

esterification
tautomerization
hydrates

h.

ester

two extra resonance structures, one with + on carbon

hydrolysis (to acid
and alcohol),
tautomerization

i.

imine

one extra resonance structure with + on carbon

amine and
aldehyde or
ketone

j.

nitrile

one extra resonance structure with + on carbon

hydrolysis to amide
hydrolysis to acid
tautomerization

k.

alkene

proton adds to
right carbon of
pi bond

more stable 2° carbocation (R⁺) forms
on left carbon atom (sp²), instead of a
less stable 1° carbocation on the right.

addition
reactions

l.

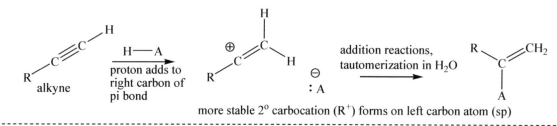

more stable 2° carbocation (R$^+$) forms on left carbon atom (sp)

m.

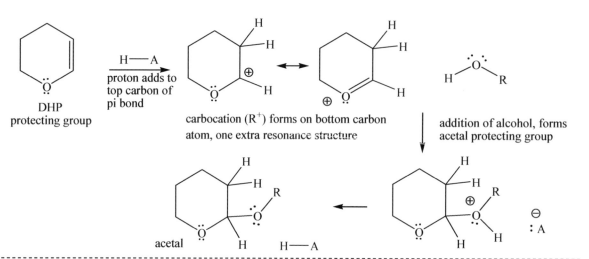

n.

o.

[reaction scheme showing aromatic benzene reacting with D—A to form carbocation resonance structures and substitution product]

deuterium adds to any carbon of a pi bond

carbocation (R⁺) forms on opposite carbon of pi bond, two extra resonance structures

substitution reactions

H—A

aromatic

We will return to some of these reactions when we study the chemistry of organic reactions, including our next topic. At this point we are mainly interested in understanding acid/base proton transfers, curved arrow pushing, formal charge, recognizing resonance structures and using the logic arguments of inductive effects and resonance effects to explain relative stabilities of acids and bases. If you can do these things, you are well on your way to understanding organic chemistry and biochemistry.

Bonus Problem

Which nitrogen atom is most basic?

a 1
b 2
c 3

Which compound's conjugate acid has the lower pK_a? Explain. (pK_a values are 5 and 10)

1   2

Chapter 8

Main Topics

→ Resonance Effects
→ Inductive Effects
→ Steric Effects

Chapter 9 – S_N and E Chemistry

This current chapter continues the theme of the previous acid/base chapter. As much as possible, try to focus on the similarities: electron pair donation and acceptance, which leads to bond making and bond breaking of chemical reactions. Because our time to study organic is short, our presentation of mechanisms is somewhat simplistic and the number of reactions covered is abbreviated, by necessity. There is a slight sacrifice in accuracy, but it reduces much of the complicated subtlety and ambiguity that makes organic chemistry so difficult for beginning students. Our approach still leads to mostly correct understanding and predictions for the outcome of the reactions under study. Even in trying to keep it simple, there are a lot of details to learn and keep track of. The bottom line is that replacing an H with a Br in an alkane is a complete game changer. There is an explosion of possibilities. We will only look at a few of them.

Four new mechanisms to learn: S_N2 vs. E2 and S_N1 vs. E1

S = substitution = a leaving group (X) is lost from a carbon atom (R) and replaced by nucleophile (Nu:)

will have a ⊖ N = nucleophilic = nucleophiles {Nu:) donate two electrons in a manner similar to bases (B:)

E = elimination = two vicinal groups (adjacent) disappear from the skeleton and are replaced by a pi bond

1 = unimolecular kinetics = only one concentration term appears in the rate law expression, Rate = k[RX]

2 = bimolecular kinetics = two concentration terms appear in the rate law expression, Rate = k[RX] [Nu: or B:]

| S_N2 competes with E2 | Reactant favored ΔG is |
| S_N1 competes with E1 | Product Favored ΔG is |

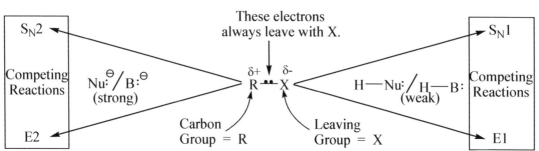

strong ($S_N2/E2$)
or
weak ($S_N1/$ E1)

Nu: / B: = is an electron pair donor to carbon (= nucleophile) or
to hydrogen (= base). It can be strong (S_N2/E2) or weak (S_N1/E1).

R = methyl, primary, secondary,
tertiary, allylic, benzylic

X = -Cl, -Br, -I, -OSO$_2$R (possible leaving groups in neutral, basic or acidic solutions)

X = -OH$_2$$^{\oplus}$ (only possible in acidic solutions)

We will limit ourselves to the RBr compounds using C1–C6 carbons and four additional special patterns. These will include methyl, primary, secondary, tertiary, neopentyl, allyl, benzyl, phenyl and vinyl. You need to be able to recognize these patterns. These RBr compounds are electron pair acceptors and are called electrophiles. They will accept electrons at the C_α carbon (S_N reactions) or a C_β proton (E reactions). A "*" marks a chiral center.

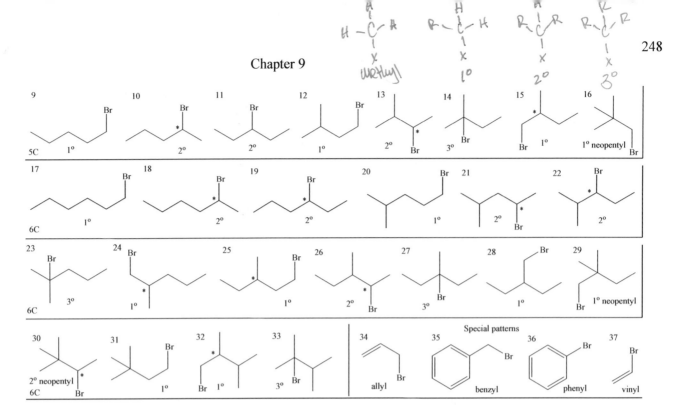

To keep life simpler, we will group our electron pair donors into two groups. Those with negative charge we will call strong nucleophile/bases (S_N2 and E2 reactions) and if they are neutral, we will call them weak nucleophile/bases (S_N1 and E1 reactions). Neutral nitrogen, sulfur and phosphorous can be good nucleophiles. We will use diphenylsulfide and triphenylphosphine as special, strong nucleophiles in our course.

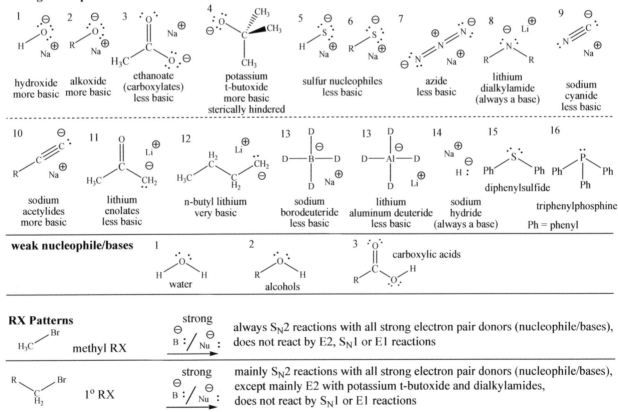

strong nucleophile/bases for this book

1 hydroxide more basic
2 alkoxide more basic
3 ethanoate (carboxylates) less basic
4 potassium t-butoxide more basic sterically hindered
5, 6 sulfur nucleophiles less basic
7 azide less basic
8 lithium dialkylamide (always a base)
9 sodium cyanide less basic

10 sodium acetylides more basic
11 lithium enolates less basic
12 n-butyl lithium very basic
13 sodium borodeuteride less basic
13 lithium aluminum deuteride less basic
14 sodium hydride (always a base)
15 diphenylsulfide
16 triphenylphosphine Ph = phenyl

weak nucleophile/bases

1 water
2 alcohols
3 carboxylic acids

RX Patterns

methyl RX — strong B : / Nu : → always S_N2 reactions with all strong electron pair donors (nucleophile/bases), does not react by E2, S_N1 or E1 reactions

1° RX — strong B : / Nu : → mainly S_N2 reactions with all strong electron pair donors (nucleophile/bases), except mainly E2 with potassium t-butoxide and dialkylamides, does not react by S_N1 or E1 reactions

Chapter 9

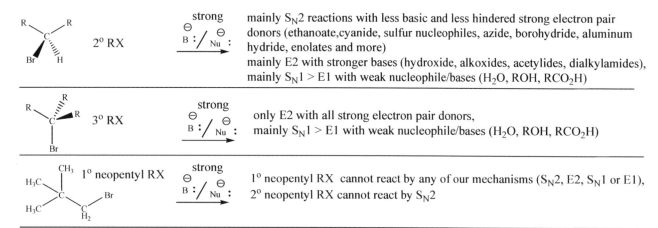

2° RX	strong	mainly S$_N$2 reactions with less basic and less hindered strong electron pair donors (ethanoate, cyanide, sulfur nucleophiles, azide, borohydride, aluminum hydride, enolates and more)	

mainly E2 with stronger bases (hydroxide, alkoxides, acetylides, dialkylamides), mainly S$_N$1 > E1 with weak nucleophile/bases (H$_2$O, ROH, RCO$_2$H)

3° RX — strong — only E2 with all strong electron pair donors,
mainly S$_N$1 > E1 with weak nucleophile/bases (H$_2$O, ROH, RCO$_2$H)

1° neopentyl RX — strong — 1° neopentyl RX cannot react by any of our mechanisms (S$_N$2, E2, S$_N$1 or E1),
2° neopentyl RX cannot react by S$_N$2

Initially, let's compare our Bronsted acid/base reactions with an expanded Lewis acid/base perspective. We can compare water (H$_2$O) and hydroxide (HO$^-$) H-Br and CH$_3$-Br. Because H-Br is such a strong acid, we can't tell the difference between water and hydroxide as bases. However, very similar looking reactions with bromomethane clearly shows that hydroxide is a much more powerful electron pair donor than water (HO$^-$ >> H$_2$O). Since oxygen is reacting at carbon instead of a proton we use the term 'nucleophile' instead of base and electrophile instead of acid. Lewis acid (electron pair acceptor) and Lewis base (electron pair donor) works for all four reactions.

Can't really tell which is stronger b/c HBr is strong & dissociates

What is similar? *Polly.*

What is different?

weak base ... acid ... 100% ...

strong base ... acid ... 100% ...

using Acetic Acid is better since its weaker

weak nucleophile ... electrophile ... poor reaction → very stable leaving group ... very slow reaction due to weak nucleophile (we say, no reaction)

strong nucleophile ... electrophile ... 100% → very stable leaving group ... fast reaction due to strong nucleophile and a very stable leaving group S$_N$2 reaction

Your intuition probably suggested to you that the second reaction looks better because of hydroxide's negative charge, and it being a much stronger electron pair donor, and it is. Both of the reactions with bromomethane are shown as S$_N$2 reactions (see terms on first page) even though only the reaction with hydroxide occurs at a reasonable rate. S$_N$2 reactions occur with the key bonds forming and breaking in a single transforming step, without any intermediate structures. Such reactions are referred to as *concerted* reactions. The "2" in S$_N$2 indicates bimolecular kinetics, meaning two participants are present in the transition state complex (slow step). These sorts of reactions require a very strong push from the electron pair donor. In fact, the first example, using water, is so slow that we designate "no reaction" as its result.

In order for the new bond to form at the same time as the old bond is breaking, the approach of the strong nucleophile always has to occur from the back side of the sp^3 carbon atom. This forces an inversion of configuration at the carbon with the leaving group, which leaves on the front side. We call this the C$_\alpha$ carbon. Any non-hydrogen atoms attached to C$_\alpha$ are referred to as beta atoms (usually C$_\beta$ for us), which will be important in the competing E2 reaction (below). Also, if the C_α carbon is chiral, the S_N2 reaction usually results in inversion of configuration.

If there are C_β atoms, S_N2 reactions compete with another reaction in consuming the R-X compound. This second reaction is called an E2 reaction. It also is a one-step, concerted reaction and forms pi bonds. E2 reactions are not possible at bromomethane, because there is only one carbon and we need two carbon atoms to make a pi bond. However, if we add one more carbon atom, e.g. bromoethane, we can consider the competition between S_N2 and E2. As in S_N2 reactions we will focus on strong electron pair donors in E2 reactions (for us, that means a negative charge). C_α is always part of the reaction, but C_β also has to be considered when elimination reactions are occurring. Because parallel overlap of orbitals is required for concerted pi bond formation in E2 reactions the C_β-H and C_α-Br have to be parallel (usually anti). C_β-H and C_α-Br are also parallel in the "syn" conformation, but there is less than 1% in that conformation because it is an eclipsed conformation, so we don't usually consider that possibility. Both S_N2 and E2 reactions have very specific stereochemical requirements for reactions to occur.

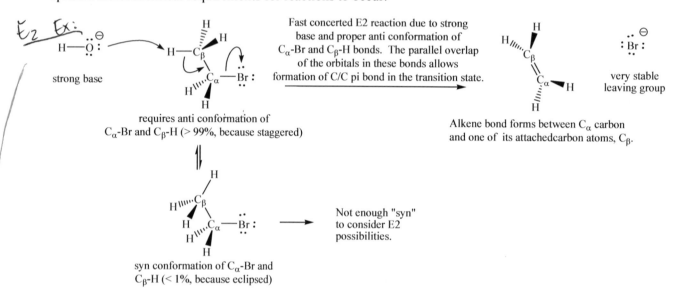

Fast concerted E2 reaction due to strong base and proper anti conformation of C_α-Br and C_β-H bonds. The parallel overlap of the orbitals in these bonds allows formation of C/C pi bond in the transition state.

strong base

requires anti conformation of C_α-Br and C_β-H (> 99%, because staggered)

very stable leaving group

Alkene bond forms between C_α carbon and one of its attached carbon atoms, C_β.

Not enough "syn" to consider E2 possibilities.

syn conformation of C_α-Br and C_β-H (< 1%, because eclipsed)

Summary of S_N2 and E2 Competition – One Step Reactions

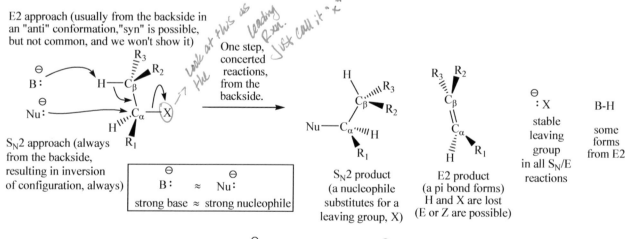

E2 approach (usually from the backside in an "anti" conformation,"syn" is possible, but not common, and we won't show it)

One step, concerted reactions, from the backside.

S_N2 approach (always from the backside, resulting in inversion of configuration, always)

B: $^\ominus$ ≈ Nu: $^\ominus$

strong base ≈ strong nucleophile

S_N2 product (a nucleophile substitutes for a leaving group, X)

E2 product (a pi bond forms) H and X are lost (E or Z are possible)

: X stable leaving group in all S_N/E reactions

B-H some forms from E2

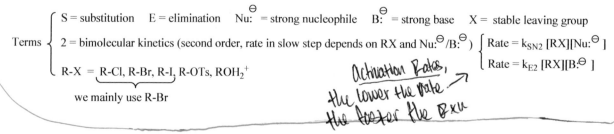

Terms
- S = substitution   E = elimination   Nu: $^\ominus$ = strong nucleophile   B: $^\ominus$ = strong base   X = stable leaving group
- 2 = bimolecular kinetics (second order, rate in slow step depends on RX and Nu: $^\ominus$/B: $^\ominus$)
- R-X = R-Cl, R-Br, R-I, R-OTs, ROH$_2^+$

we mainly use R-Br

Rate = k_{SN2} [RX][Nu: $^\ominus$]
Rate = k_{E2} [RX][B: $^\ominus$]

S$_N$2 Potential Energy vs. Progress of Reaction Diagram (= concerted, energy picture, very similar to E2)

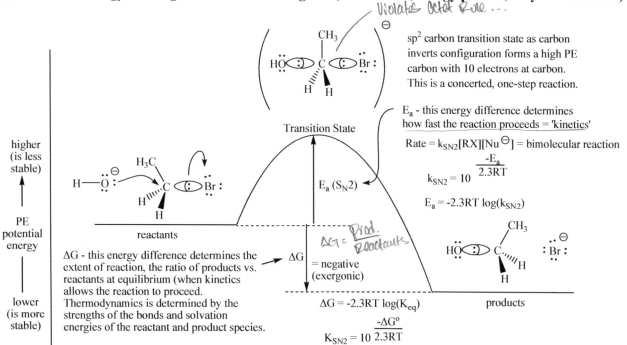

Violates Octet Rule ...

sp^2 carbon transition state as carbon inverts configuration forms a high PE carbon with 10 electrons at carbon. This is a concerted, one-step reaction.

Transition State

E$_a$ - this energy difference determines how fast the reaction proceeds = 'kinetics'

Rate = k$_{SN2}$[RX][Nu$^\ominus$] = bimolecular reaction

$$k_{SN2} = 10^{\frac{-E_a}{2.3RT}}$$

$$E_a = -2.3RT \log(k_{SN2})$$

E$_a$ (S$_N$2)

higher (is less stable)

PE potential energy

lower (is more stable)

reactants

$\Delta G = \frac{Prod.}{Reactants}$

ΔG - this energy difference determines the extent of reaction, the ratio of products vs. reactants at equilibrium (when kinetics allows the reaction to proceed. Thermodynamics is determined by the strengths of the bonds and solvation energies of the reactant and product species.

ΔG = negative (exergonic)

products

$$\Delta G = -2.3RT \log(K_{eq})$$

$$K_{SN2} = 10^{\frac{-\Delta G^\circ}{2.3RT}}$$

POR = progress of reaction - shows how PE changes as reaction proceeds

E2 Potential Energy vs. Progress of Reaction Diagram (concerted), energy picture looks very similar to S$_N$2

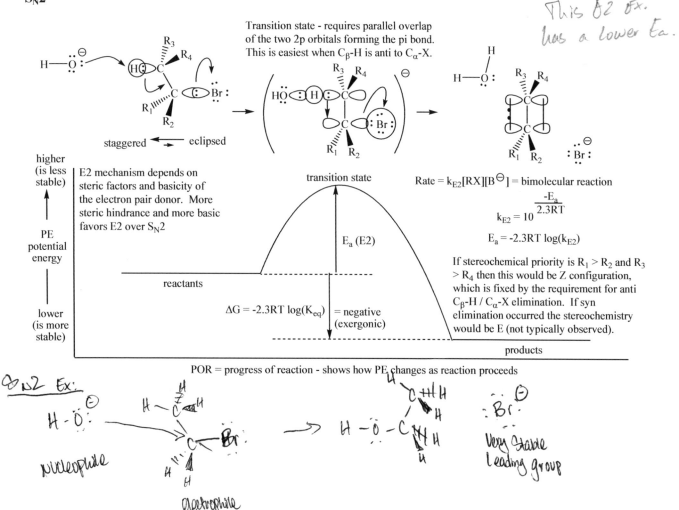

This E2 Ex. has a Lower Ea.

Transition state - requires parallel overlap of the two 2p orbitals forming the pi bond. This is easiest when C$_\beta$-H is anti to C$_\alpha$-X.

staggered ⇌ eclipsed

transition state

higher (is less stable)

E2 mechanism depends on steric factors and basicity of the electron pair donor. More steric hindrance and more basic favors E2 over S$_N$2

PE potential energy

lower (is more stable)

reactants

E$_a$ (E2)

Rate = k$_{E2}$[RX][B$^\ominus$] = bimolecular reaction

$$k_{E2} = 10^{\frac{-E_a}{2.3RT}}$$

$$E_a = -2.3RT \log(k_{E2})$$

If stereochemical priority is R$_1$ > R$_2$ and R$_3$ > R$_4$ then this would be Z configuration, which is fixed by the requirement for anti C$_\beta$-H / C$_\alpha$-X elimination. If syn elimination occurred the stereochemistry would be E (not typically observed).

$$\Delta G = -2.3RT \log(K_{eq})$$ = negative (exergonic)

products

POR = progress of reaction - shows how PE changes as reaction proceeds

S$_N$2 Ex:

H-O:$^\ominus$

nucleophile

electrophile

Very Stable Leaving group

Relative rates of S$_N$2 reactions with different substitution patterns at C$_\alpha$ and C$_\beta$ of the R-X structure

Substitution of nonhydrogen groups at C$_\alpha$ and C$_\beta$ (relative to the carbon with X) slows the rate of S$_N$2 reactions because the substituent groups block the backside approach of the nucleophile to the C$_\alpha$-X carbon, thus increasing the activation energy by increasing the congestion in the pentavalent transition state. We call this a **_steric effect_**. Both C$_\alpha$ and C$_\beta$ substitution patterns show their most dramatic decreases in rate of reaction when all possible carbon positions are switched from hydrogen to carbon (or other large group). In our course, as long as there is at least one hydrogen atom at the C$_\alpha$ and C$_\beta$ positions, there remains a possible path of approach by the nucleophile to the backside of the carbon. However, as steric effects increase, S$_N$2 slows and E2 becomes more competitive.

Relative Rates of S$_N$2 Reactions – steric hinderance at the C$_\alpha$ (C-alpha) carbon slows down the rate of S$_N$2 reactions

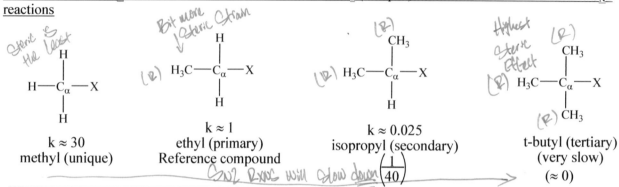

k ≈ 30	k ≈ 1	k ≈ 0.025	
methyl (unique)	ethyl (primary)	isopropyl (secondary)	t-butyl (tertiary)
	Reference compound		(very slow)
			(≈ 0)

---

methyl RX

Methyl has three easy paths of backside approach by the nucleophile. It is the least sterically hindered carbon in S$_N$2 reactions, but it is unique. E2 is not possible at methyl.

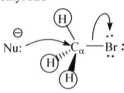

primary RX

Primary substitution allows two easy paths of backside approach by the nucleophile. It is the least sterically hindered 'general' substitution pattern for S$_N$2 reactions. Usually S$_N$2 > E2 at primary RX, except when sterically hindered, very basic t-butoxide is used, then E2 > S$_N$2.

secondary RX

Secondary substitution allows one easy path of backside approach by the nucleophile. It reacts the slowest of the possible S$_N$2 substitution patterns. Secondary RX is the most complicated pattern because either S$_N$2 or E2 can be the major outcome. Other factors have to be considered (steric size and basicity of the electron pair donor).

---

tertiary RX

Tertiary substitution has no easy path of approach by the nucleophile from the backside. We do not propose any S$_N$2 reaction at tertiary RX centers. E2 is usually the observed reaction here.

tertiary RX

When the C$_\alpha$ carbon is completely substituted the nucleophile cannot get close enough to make a bond with the C$_\alpha$ carbon. Even C$_\beta$-H sigma bonds block the nucleophile's approach to C$_\alpha$.

Chapter 9

Relative rates of S_N2 reactions with different substitution patterns at C_β (C-beta) of the R-X structure

Steric hindrance at the C_β carbon also slows down the rate in S_N2 reactions and completely shuts down S_N2 reactions when any C_β position is completely substituted. Such a pattern is called a neopentyl RBr compound, even if there are more than 5 carbon atoms present.

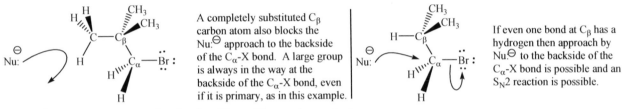

All of these structures are primary R-X compounds at C_α, but substituted differently at C_β.

A completely substituted C_β carbon atom also blocks the Nu:⊖ approach to the backside of the C_α-X bond. A large group is always in the way at the backside of the C_α-X bond, even if it is primary, as in this example.

If even one bond at C_β has a hydrogen then approach by Nu⊖ to the backside of the C_α-X bond is possible and an S_N2 reaction is possible.

Four other patterns we need to be familiar with are shown below.

Allyl and benzyl are exceptionally good S_N2 patterns because of charge delocalization in the transition state.

Phenyl and vinyl react poorly by 3 of our mechanisms, S_N2, S_N1 and E1. We say 'no reaction' in our course. (E2 reactions are possible, but we won't show it.)

Walden Inversion = special name for inversion of configuration (R → S or S → R) If a chiral center is present at C_α and priorities stay in the same relative order, the absolute configuration of the chiral center will change in an S_N2 reaction.

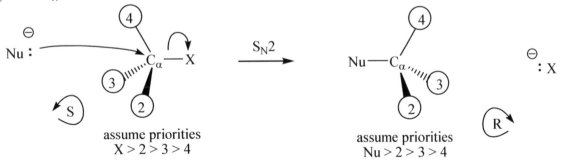

If X = #1 priority and the other groups are #2, #3 and #4 above, then S absolute configuration typically inverts to R (and vice versa). Close approach from the backside is necessary for early bond formation which helps push X away (bond cleavage). This observation assumes there is no relative change in the order of priorities of the groups at the chiral center (X = #1 priority and Nu = #1 priority).

S_N2 Inversion of Configuration in Rings

An unsubstituted cyclic RX compound with no other substituents does not allow us to detect inversion of configuration (versus retention of configuration or racemization of configuration). However, if the RX ring were substituted with another group that did not interfere with the S_N2 reaction, it would allow the observation of backside attack by the nucleophile. A very simple methyl substituent in a cyclohexane ring would allow us to see which side the nucleophilic attack occurs from. We can observe if the opposite side or the same side...or both sides are attacked by comparing the initial relative position of the leaving group to the final relative position of the nucleophile.

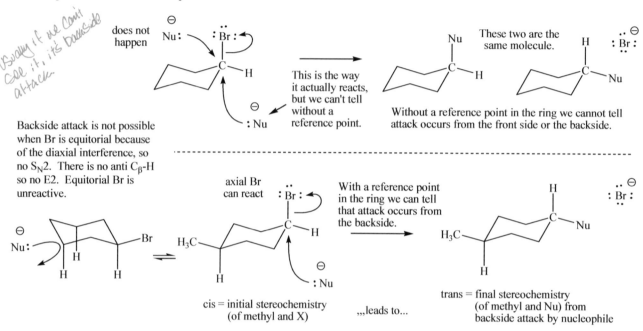

Problem 1 - How can you tell whether the S_N2 reaction occurs with front side attack (retention), backside attack (inversion) or front and backside attack (racemization)? Use both of these examples to explain your answer. Follow the curved arrow formalism to show electron movement.

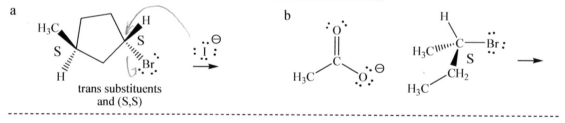

Problem 2 - In each of the following pairs of nucleophiles, one is a much better nucleophile than its closely related partner. Propose a possible explanation.

a relative rated ≈ 250 / 1

quinuclidine   triethylamine

b

tetrahydrofuran   diethyl ether
        (THF)

c

methoxide

t-butoxide

Chapter 9

So far we have mostly been using Nu: ⁻ to show strong nucleophiles. The examples and problems below show some useful strong nucleophiles that we want to learn and use in our course. They all have a negative charge. Remember, any sort of steric hindrance (in RX or in Nu: ⁻) slows down S_N2 reactions. As S_N2 reactions get slower (steric effects), E2 gets more competitive. At methyl, CH_3X, only S_N2 is possible. S_N2 also is the dominant reaction at primary RCH_2X centers, except when C_β is completely substituted (no S_N2 at neopentyl centers), or the Nu:⁻ is sterically bulky and basic, where E2 dominates (our only example is potassium t-butoxide). At secondary R_2CHX centers, S_N2 and E2 are both competitive. In our course S_N2 is the usual winner when the nucleophile is not bulky and not excessively basic (acetate, $CH_3CO_2^-$; makes esters, cyanide, NC^-; makes nitriles, azide, N_3^-, makes 1° amines after reduction with $LiAlH_4$, sulfur nucleophiles, makes thiols and sulfides and a few others). E2 dominates at secondary R_2CHX centers when the electron pair donors are strongly basic and/or sterically bulky (e.g. hydroxide, HO^-; alkoxides, RO^-; and terminal acetylides, RCC^-). Tertiary R_3CX centers are too sterically hindered for S_N2 reactions and we only propose E2 reactions when strong electron pair donors are used (negative charge in our course). These simplistic rules allow you to make reasonably unambiguous predictions. They are not 100% true, but they work well for our goals of learning the usual patterns of reactions. There is one S_N2 reaction below where the reaction does not occur at carbon, but rather at nitrogen (azides are reduced to 1° amines with $LiAlH_4$). For us, this will be an essential method for making primary amines from azido compounds. Nitrogen gas is the very good leaving group and lithium aluminum hydride provides nucleophile hydride.

Examples

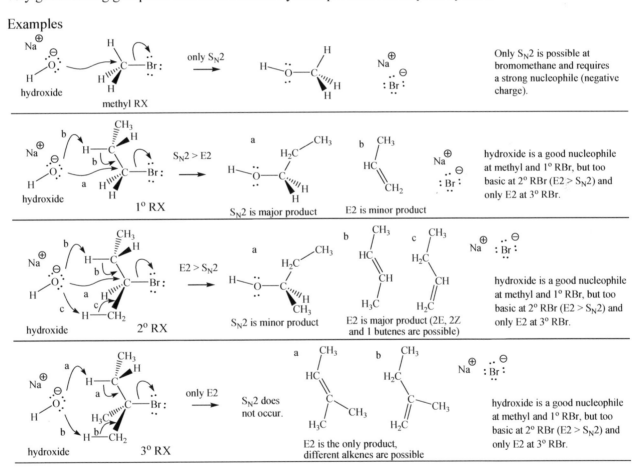

Chapter 9

Problem 3 - Fill in the necessary details to show how each reaction works. Specify as only S_N2, $S_N2 > E2$, $E2 > S_N2$ or only E2. Cover the right side to test yourself.

ethoxide
(alkoxides)   methyl RX

Only S_N2 is possible at bromomethane and requires a strong nucleophile (negative charge).

ethoxide
(alkoxides)   1° RX

alkoxides are good nucleophiles at methyl and 1° RBr, but too basic at 2° RBr (E2 > S_N2) and only E2 at 3° RBr.

ethoxide
(alkoxides)   2° RX

alkoxides are good nucleophiles at methyl and 1° RBr, but too basic at 2° RBr (E2 > S_N2) and only E2 at 3° RBr.

ethoxide
(alkoxides)   3° RX

alkoxides are good nucleophiles at methyl and 1° RBr, but too basic at 2° RBr (E2 > S_N2) and only E2 at 3° RBr.

monohydrogen
sulfide   methyl RX

Only S_N2 is possible at bromomethane and requires a strong nucleophile (negative charge).

monohydrogen
sulfide   1° RX

monohydrogen sulfide is a good nucleophile at methyl, 1° and 2° RBr, but only E2 at 3° RBr.

monohydrogen
sulfide   2° RX

monohydrogen sulfide is a good nucleophile at methyl, 1° and 2° RBr, but only E2 at 3° RBr.

monohydrogen
sulfide   3° RX

monohydrogen sulfide is a good nucleophile at methyl, 1° and 2° RBr, but only E2 at 3° RBr.

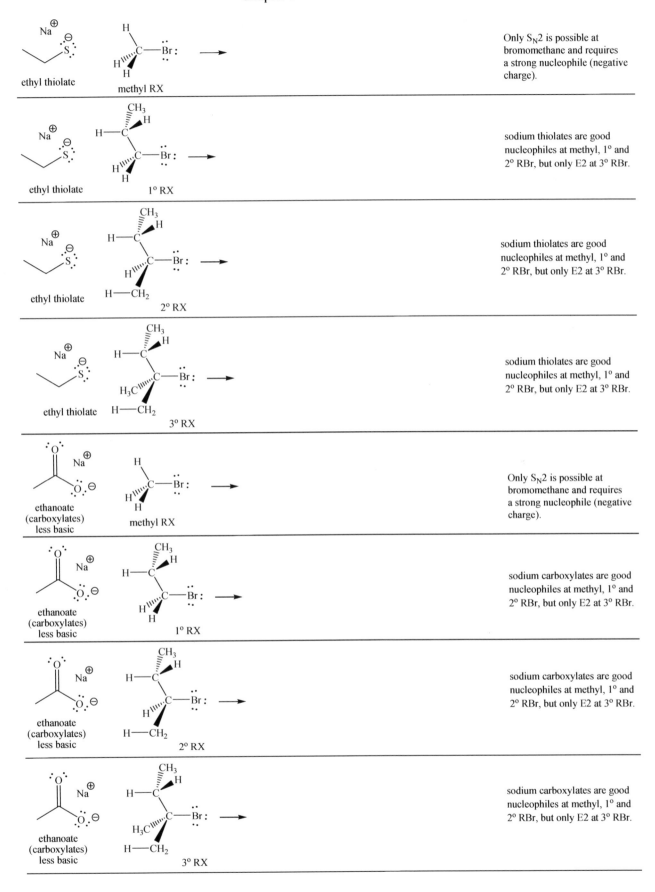

Na⊕ :S̈: ethyl thiolate   methyl RX   →   Only S_N2 is possible at bromomethane and requires a strong nucleophile (negative charge).

sodium thiolates are good nucleophiles at methyl, 1° and 2° RBr, but only E2 at 3° RBr.

sodium thiolates are good nucleophiles at methyl, 1° and 2° RBr, but only E2 at 3° RBr.

sodium thiolates are good nucleophiles at methyl, 1° and 2° RBr, but only E2 at 3° RBr.

Only S_N2 is possible at bromomethane and requires a strong nucleophile (negative charge).

sodium carboxylates are good nucleophiles at methyl, 1° and 2° RBr, but only E2 at 3° RBr.

sodium carboxylates are good nucleophiles at methyl, 1° and 2° RBr, but only E2 at 3° RBr.

sodium carboxylates are good nucleophiles at methyl, 1° and 2° RBr, but only E2 at 3° RBr.

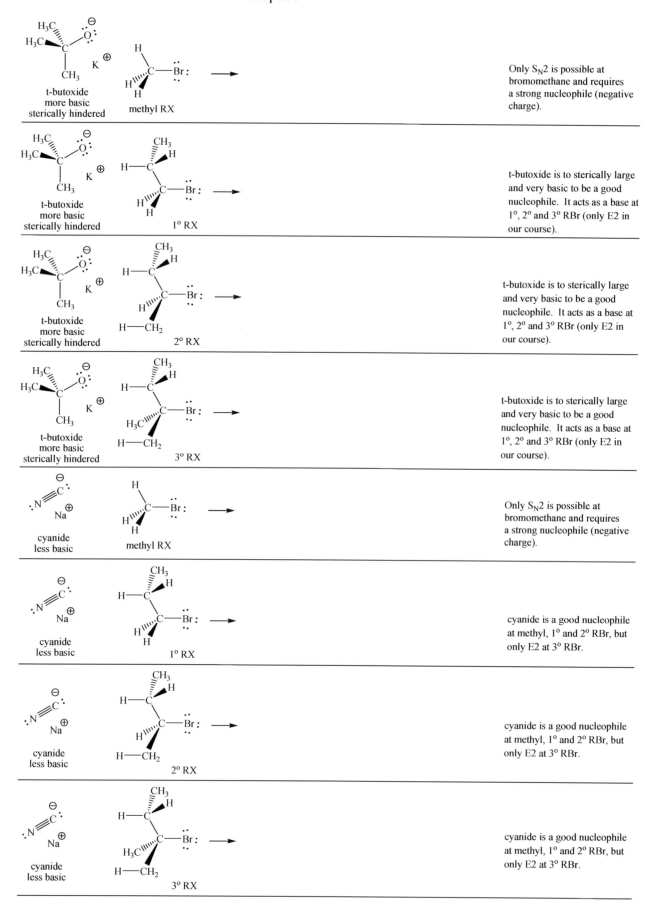

Only S_N2 is possible at bromomethane and requires a strong nucleophile (negative charge).

t-butoxide is to sterically large and very basic to be a good nucleophile. It acts as a base at 1°, 2° and 3° RBr (only E2 in our course).

t-butoxide is to sterically large and very basic to be a good nucleophile. It acts as a base at 1°, 2° and 3° RBr (only E2 in our course).

t-butoxide is to sterically large and very basic to be a good nucleophile. It acts as a base at 1°, 2° and 3° RBr (only E2 in our course).

Only S_N2 is possible at bromomethane and requires a strong nucleophile (negative charge).

cyanide is a good nucleophile at methyl, 1° and 2° RBr, but only E2 at 3° RBr.

cyanide is a good nucleophile at methyl, 1° and 2° RBr, but only E2 at 3° RBr.

cyanide is a good nucleophile at methyl, 1° and 2° RBr, but only E2 at 3° RBr.

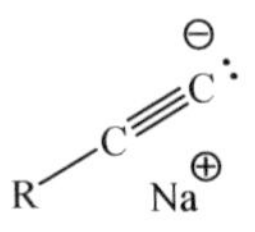

Only S_N2 is possible at bromomethane and requires a strong nucleophile (negative charge).

acetylides more basic

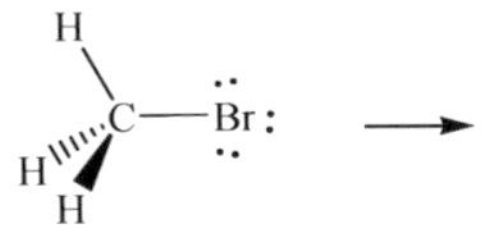

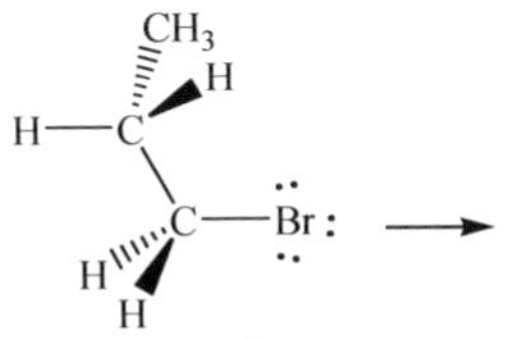

methyl RX

acetylides are good nucleophiles at methyl and 1° RBr, but too basic at 2° RBr (E2 > S_N2) and only E2 at 3° RBr.

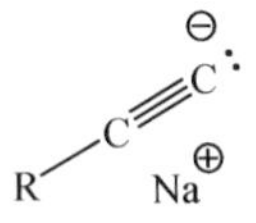

acetylides more basic

1° RX

acetylides are good nucleophiles at methyl and 1° RBr, but too basic at 2° RBr (E2 > S_N2) and only E2 at 3° RBr.

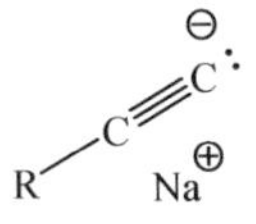

acetylides more basic

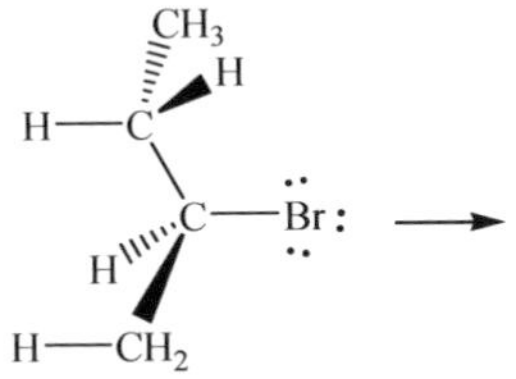

2° RX

acetylides are good nucleophiles at methyl and 1° RBr, but too basic at 2° RBr (E2 > S_N2) and only E2 at 3° RBr.

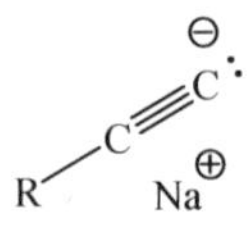

acetylides more basic

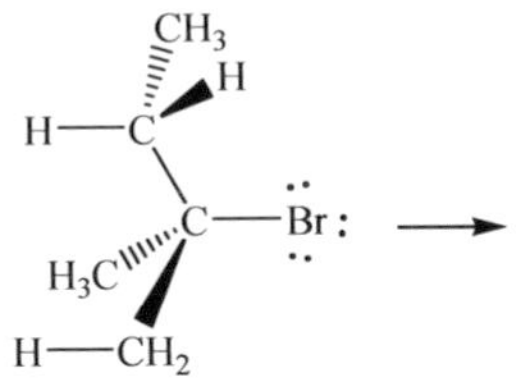

3° RX

Only S_N2 is possible at bromomethane and requires a strong nucleophile (negative charge). Azides are reacted in a second S_N2 reaction at nitrogen using lithium aluminum hydride to make 1° amines.

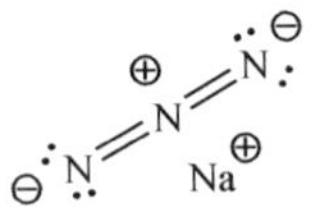

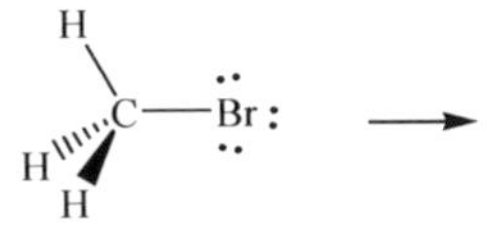

azide less basic

methyl RX

Azide is a good nucleophile at methyl, 1° and 2° RBr, but only E2 at 3° RBr. Azides are reacted in a second S_N2 reaction at nitrogen using lithium aluminum hydride to make 1° amines..

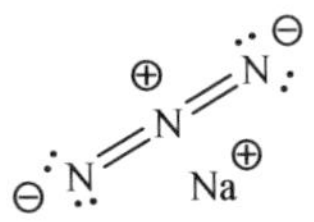

azide less basic

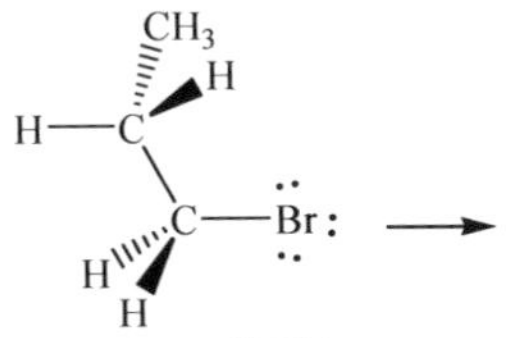

1° RX

Azide is a good nucleophile at methyl, 1° and 2° RBr, but only E2 at 3° RBr. Azides are reacted in a second S_N2 reaction at nitrogen using lithium aluminum hydride to make 1° amines.

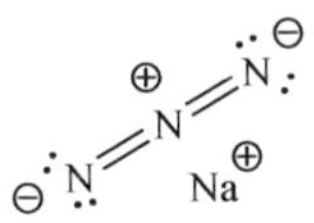

azide less basic

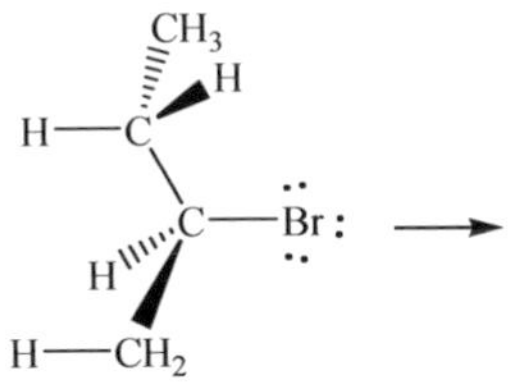

2° RX

Azide is a good nucleophile at methyl, 1° and 2° RBr, but only E2 at 3° RBr. Azides are reacted in a second S_N2 reaction at nitrogen using lithium aluminum hydride to make 1° amines.

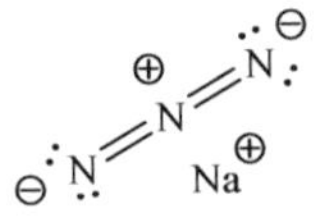

azide less basic

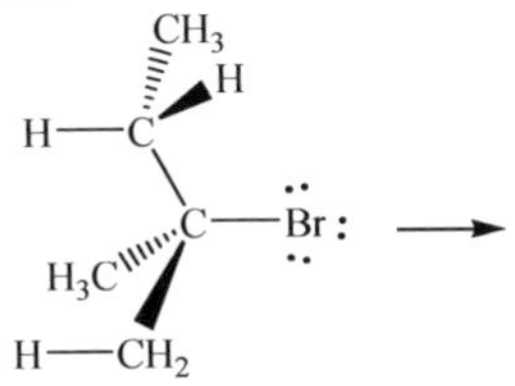

3° RX

Example of S$_N$2 reaction at nitrogen in azides to form primary amines (after acidic workup)

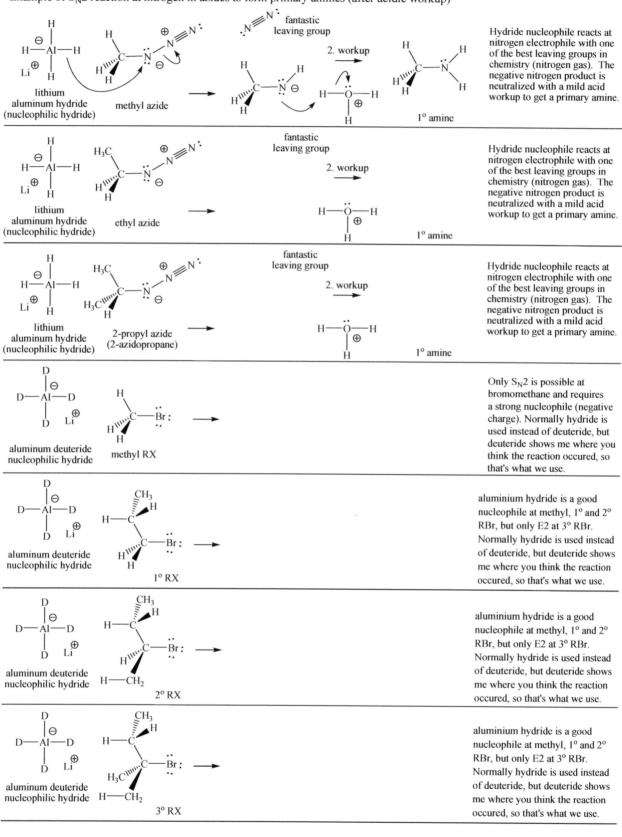

Hydride nucleophile reacts at nitrogen electrophile with one of the best leaving groups in chemistry (nitrogen gas). The negative nitrogen product is neutralized with a mild acid workup to get a primary amine.

Hydride nucleophile reacts at nitrogen electrophile with one of the best leaving groups in chemistry (nitrogen gas). The negative nitrogen product is neutralized with a mild acid workup to get a primary amine.

Hydride nucleophile reacts at nitrogen electrophile with one of the best leaving groups in chemistry (nitrogen gas). The negative nitrogen product is neutralized with a mild acid workup to get a primary amine.

Only S$_N$2 is possible at bromomethane and requires a strong nucleophile (negative charge). Normally hydride is used instead of deuteride, but deuteride shows me where you think the reaction occured, so that's what we use.

aluminium hydride is a good nucleophile at methyl, 1° and 2° RBr, but only E2 at 3° RBr. Normally hydride is used instead of deuteride, but deuteride shows me where you think the reaction occured, so that's what we use.

aluminium hydride is a good nucleophile at methyl, 1° and 2° RBr, but only E2 at 3° RBr. Normally hydride is used instead of deuteride, but deuteride shows me where you think the reaction occured, so that's what we use.

aluminium hydride is a good nucleophile at methyl, 1° and 2° RBr, but only E2 at 3° RBr. Normally hydride is used instead of deuteride, but deuteride shows me where you think the reaction occured, so that's what we use.

Only S_N2 is possible at bromomethane and requires a strong nucleophile (negative charge). There are many variations of this enolate reaction, but this is the only one we will use.

enolates less basic · methyl RX

enolates are good nucleophiles at methyl, 1° and 2° RBr, but only E2 at 3° RBr. There are many variations of this enolate reaction, but this is the only one we will use.

enolates less basic · 1° RX

enolates are good nucleophiles at methyl, 1° and 2° RBr, but only E2 at 3° RBr. There are many variations of this enolate reaction, but this is the only one we will use.

enolates less basic · 2° RX

enolates are good nucleophiles at methyl, 1° and 2° RBr, but only E2 at 3° RBr. There are many variations of this enolate reaction, but this is the only one we will use.

enolates less basic · 3° RX

The bromoalkane list on pages 247-248 provides many additional examples where you can check your ability to predict S_N2 or E2 reactions. You should also be able to write a mechanism for any of those reactions with the above strong base/nucleophiles.

There was a bit of a problem above, which you probably never even noticed. If we were trying to make a secondary alcohol from a secondary RBr using hydroxide, our main product is E2 alkene, not the alcohol. Hydroxide is too basic for the S_N2 reaction at secondary RBr centers. Situations like this always force us to do more work than we would like. In this case we have to use an indirect strategy to get a 2° ROH from 2°RBr using a weaker oxygen base than hydroxide. Ethanoate (acetate) is less basic and a better nucleophile (because of resonance), which increases the yield of S_N2 over E2 reaction. There is still a problem in that the product of the reaction is an ester (not an alcohol). However, the desired oxygen is attached to the secondary carbon. We just need to get rid of the 'ester' part.

Some alcohol forms, (S_N2) but mostly we get alkene product (from E2 reaction).

We get an ester, but an alcohol is what we want.

This is going to force us to learn a new reaction (acyl substitution), but it's not so bad because this is one reaction we will use over and over later in the book. How is the ester converted to the alcohol? There are several ways we could do this, but the one we'll use is base hydrolysis of an ester.

Ester hydrolysis in aqueous base (acyl substitution). Where are the electrophilic centers for hydroxide to attack an ester? We'll there are a few ways hydroxide could react, so welcome to "Organic Ambiguous Chemistry." There are two δ+ carbon centers, one being the carbon have one bond to oxygen (the alkoxy carbon) and one having three bonds to oxygen (the acyl carbon). It seems clear that having three bonds with oxygen would make that carbon more δ+, and it is the C=O that gets attacked faster. (Also, we are ignoring an enolate possibility.) More steps required overall from RBr, but better yields are obtained.

S_N2 attack seems logical, but it is too slow to compete with attack at the acyl carbon, which likely has a larger δ+ because of 3 bonds to oxygen.

B, C and E all have a negative charge on oxygen and are of comprable stability. Hydroxide attack on the C=O leads to an alkoxy anion (C), which can push back and eliminate a similar looking alkoxy anion (E) and a carboxylic acid. Everything to this point is in equilibrium. The conjugate base of the acid is a much more stable anion (because of resonance), so the acid gives away its proton to form the desired alcohol product and the most stable carboxylate (F), which ends the reaction because of its greater stability. Overall, this is called acyl substitution.

most stable anion   our desired secondary alcohol

Problem 4 – Hydrolysis of ester in aqueous base provides an alcohol synthesis and carboxylate anion. Write a complete mechanism and show the products?

Clarification of "Hydride" electron pair donors (basic versus nucleophilic)

Basic hydride

In our course, sodium hydride (NaH) and potassium hydride (KH) are *always* basic (= electron pair donation by hydride to a proton), *never* a nucleophile. The conjugate acid of hydride is hydrogen gas, H_2 (with a $pK_a = 37$, H_2 can hardly be considered an acid).

Sodium hydride and potassium hydride (KH)

$H:^{\ominus}$ $Na^{\oplus}$ $H:^{\ominus}$ $K^{\oplus}$

Aldrich, 2012
$39 / 100 grams
60% oil dispersion
MW = 24.0 g/mol

Aldrich, 2012
$128 / 75 grams
30% oil dispersion
MW = 40.1 g/mol

Problem 5 – Write an arrow pushing mechanism for each of the following reactions.

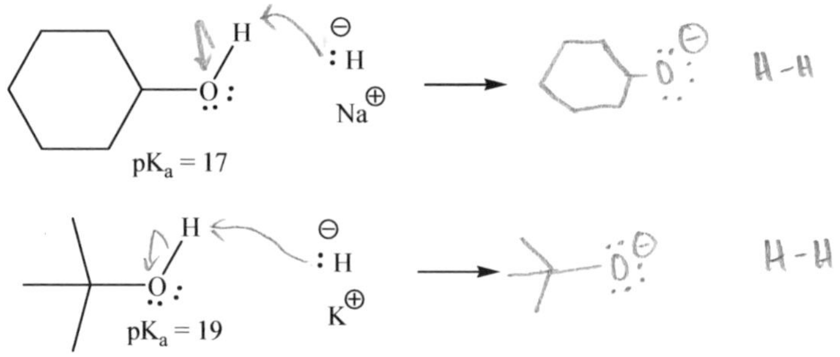

	pK_a
ROH	16-19
H-H	37

Nucleophilic hydride – Formation of C-H bonds using nucleophilic lithium aluminum hydride and sodium borohydride.

In our course, sodium borohydride ($NaBH_4$) and lithium aluminum hydride ($LiAlH_4$ = LAH) are inorganic salts containing *nucleophilic hydride*, very unusual nucleophiles. Both reagents supply nucleophilic hydride that can displace X in S_N2-like reactions with RX compounds. Sodium borohydride and lithium aluminum hydride are used in many other reactions. They also react with carbonyl compounds (C=O) and epoxides, though we won't look at those reactions until later. We will often use the deuterium version of borohydride and aluminum hydride so we can identify where a reaction occurred. In reality, this is not very common because of the expense. But, for purposes of probing your understanding of S_N2 reactions, using deuterium shows if you understand what is happening. The deuterium isotope of hydrogen reacts similarly to the proton isotope, but there are experimental methods which allow us to observe where a reaction took place, such as nuclear magnetic resonance (NMR) and mass spectrometry (MS). Neither of these is covered in this book.

Sodium borohydride and lithium aluminum hydride (LAH) - 4 equivalents of hydride per anion

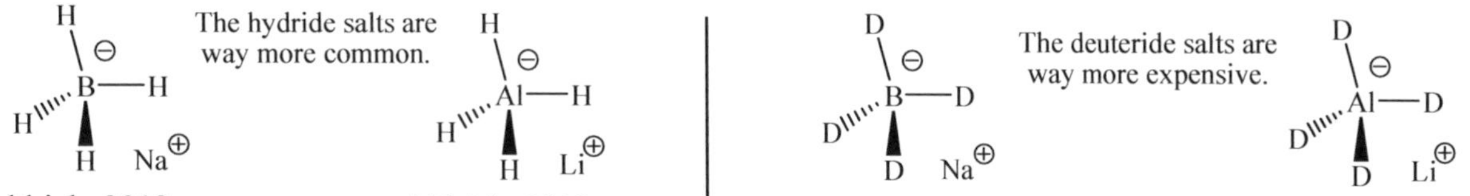

Aldrich, 2012	Aldrich, 2012	Aldrich, 2012	Aldrich, 2012
\$312 / 100 grams	\$120 / 100 grams	\$254 / 5 grams	\$125 / 5 grams
2.6 moles	2.6 moles	0.12 moles	0.12 moles
MW = 37.8 g/mol	MW = 38 g/mol	MW = 41.8 g/mol	MW = 42 g/mol

All these reagents will undergo S_N2 reactions at RX centers. Because there is a greater difference in electronegativity between aluminum and hydrogen than boron and hydrogen, LAH is dangerously more reactive than sodium borohydride.

Problem 6 – Write an arrow pushing mechanism for the following reaction.

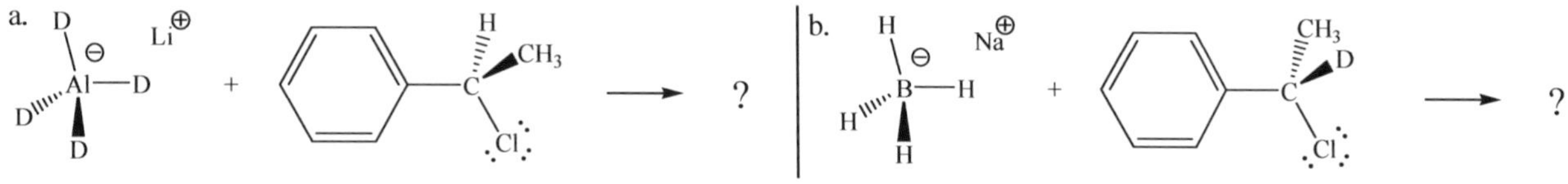

Problem 7 – It is hard to tell where the hydride was introduced since there are usually so many other hydrogens in organic molecules. Where could have X have been in the reactant molecule? There are no obvious clues. Which position(s) for X would likely be more reactive with the hydride reagent? Could we tell where X was if we used $LiAlD_4$? Note, S_N2 reactions do not occur at sp^2 carbon atoms.

Where was "X"?   LAH

Problem 8 – Predict the major products (S_N2 or E2) and only show the S_N2 product, if formed. We will develop the E2 reaction next.

a.

b.

c.

d.

e.

f.

g.

h.

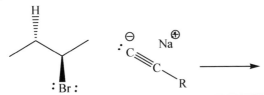

i.

j.

k.

l.

m.

n.

o.

p.

q.

r.

E2 Reactions Compete with S_N2 Reactions

As we have seen, E2 reactions also occur at the backside relative to the leaving group, but at C_β-H instead of C_α-X. The C_β-H proton has to be anti to the C_α-X to allow for parallel overlap of the 2p orbitals that form the new pi bond, lowering the transition state energy, allowing for a faster reaction. This allows elimination to occur in a concerted manner. The syn conformation also has parallel overlap of 2p orbitals, but is present in less than 1% due to an eclipsed conformation (staggered conformations > 99%).

Keeping Track of the C_β Hydrogens in E2 reactions

·At least one anti C_β position, with a hydrogen atom, is necessary for an E2 reaction to occur using a strong nucleophile/base. In more complicated systems there may be several different types of hydrogen atoms on different C_β positions. In E2 reactions there can be anywhere from one to three C_β carbons, and each C_β carbon can have zero to three hydrogen atoms.

zero C_β positions
unique methyl,
only S_N2 possible

one C_β position
1° RX, usually S_N2 > E2
(except with t-butoxide)

two C_β positions
(2° RX, S_N2 / E2
both competitive)

three C_β positions
(3° RX, only E2)

With proper rotations, each C_β-H may potentially be able to assume an anti conformation necessary for an E2 reaction to occur. This may not always be possible, such as a cyclohexane structure with an equatorial C_α-X gauche to a vicinal C_β-H. There are a lot of details to keep track of and you must be systematic in your approach to consider all possibilities. Using one of these two perspectives may help your analysis of E2 reactions.

vertical perspective
C_β-H / C_α-X

Either sketch will work for every possibility above, IF you fill in the blank positions correctly.

horizontal perspective
C_β-H / C_α-X

Chapter 9

Stability of Pi Bonds

In elimination reactions, more substituted alkenes are normally formed in greater amounts. Greater substitution of carbon groups in place of hydrogen atoms at alkene carbons (and alkyne carbon atoms too) produces a more stable alkene (lower potential energy). Alkene substitution patterns are shown below. There are three types of disubstituted alkenes and their relative stabilities are usually as follows: geminal ≈ cis < trans.

Relative stabilities of substitued alkenes.

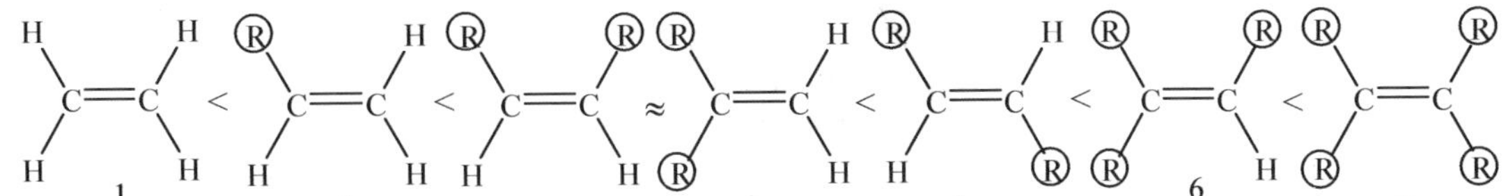

1	2	3	4	5	6	7
"unsubstituted" (unique)	"mono"	"di-cis"	"di-geminal"	"di-trans"	"tri"	"tetra"

The more substituted alkenes are usually more stable than less substitued alkenes. Substitution here, means switch an Ⓡ group for a hydrogen atom at one of the four bonding positions of the alkene.

1 = unsubstituted alkene (ethene is unique)
2 = monosubstituted alkene
3 = cis disubstituted alkene
4 = geminal disubstituted alkene
5 = trans disubstituted alkene
6 = trisubstituted alkene
7 = tetrasubstituted alkene

Saytzeff's rule: More stable alkenes tend to form faster (because of lower E_a) in dehydrohalogenation reactions (E2 and E1). They tend to be the major alkene product, though typically a little of every alkene product possible is obtained.

Possible explanations for greater stability with greater substitution of the pi bond

A fairly simple-minded explanation (the one we will use) for the relative alkene stabilities is provided by considering the greater electronegativity of an sp^2 orbital over an sp^3 orbital (remember the acid/base topic). An alkyl group (R→) inductively donates electron density better than a simple hydrogen. From the point of view of a more electronegative sp^2 alkene carbon, it is better to be connected to an electron donating alkyl carbon group than a simple hydrogen atom. The more R groups at the four sp^2 positions of a double bond, the better. However, be aware that other features, such as steric effects or resonance effects, can reverse expected orders of stability.

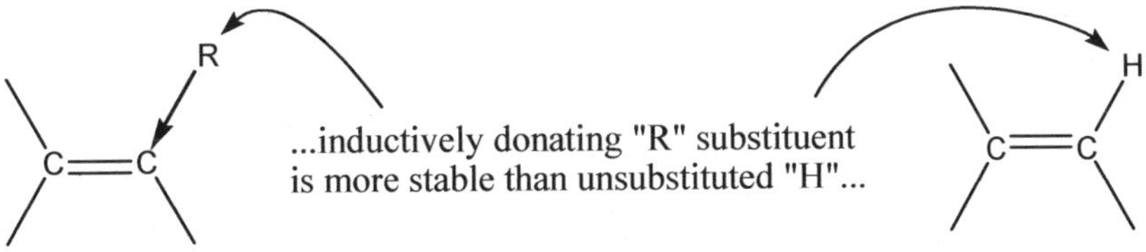

...inductively donating "R" substituent is more stable than unsubstituted "H"...

Problem 9 – How many total hydrogen atoms are on C_β carbons in the given RX compound? How many different types of hydrogen atoms are on C_β carbons (a little tricky)? How many different products are possible? Hint - Be careful of the simple CH_2. The two hydrogen atoms appear equivalent, but E/Z (cis/trans) possibilities are often present. Which elimination products would you expect to be the major and minor products. How would an absolute configuration of C_α as R, compare to C_α as S? What about $C_{\beta1}$ as R versus S?

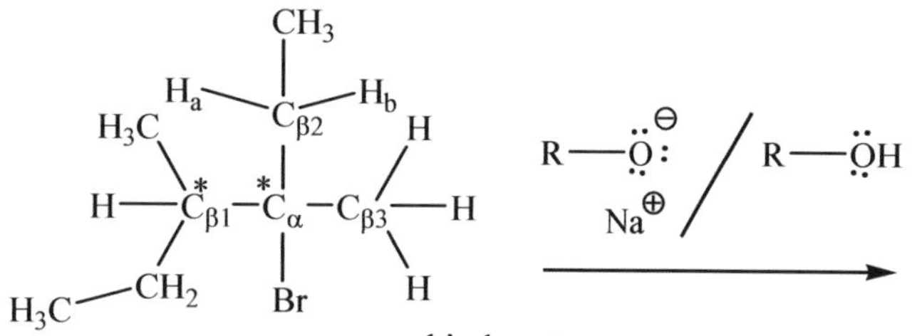

$$R\!-\!\ddot{O}\!:^{\ominus} \Big/ R\!-\!\ddot{O}H$$
$$Na^{\oplus}$$

? Compare

(3S,4R)-3-bromo-3,4-dimethylheptane

(3S,4S)-3-bromo-3,4-dimethylheptane

(3R,4R)-3-bromo-3,4-dimethylheptane

(3R,4S)-3-bromo-3,4-dimethylheptane

* = chiral center

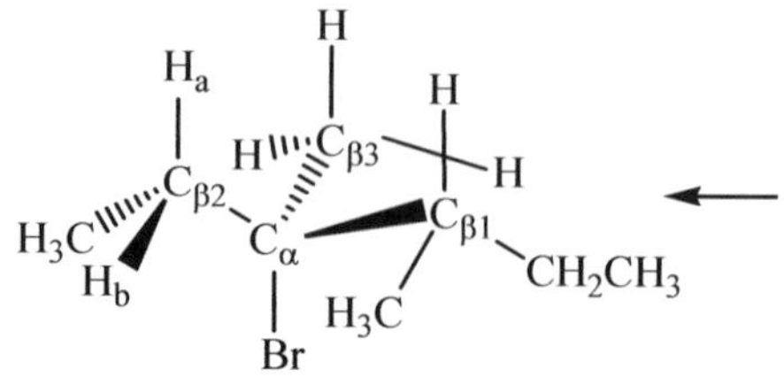

How many chiral centers are there? Which one of the listed possibilities is shown? Are H_a and H_b equivalent or different at $C_{\beta2}$? Predict the products.

Problem 10 – What is the expected order of stabilities for the alkenes below? Provide an explanation for any unexpected deviations from our general rule for alkene stabilities above. A more negative $\Delta H_{formation}$ means lower potential energy and a more stable isomer.

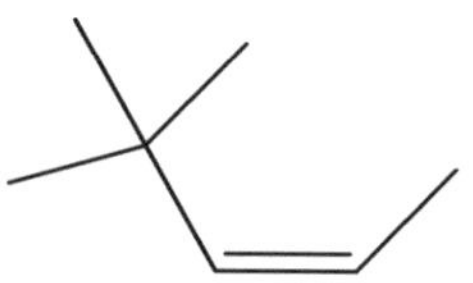

$\Delta H^o{}_f = -18.7$ kcal/mole

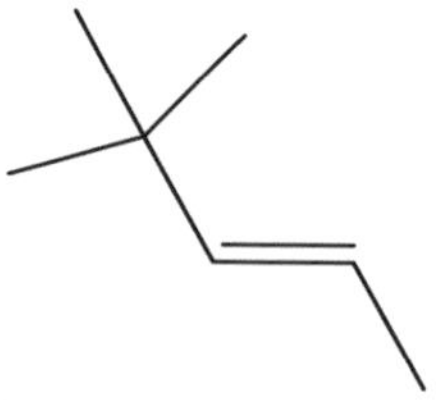

$\Delta H^o{}_f = -22.7$ kcal/mole

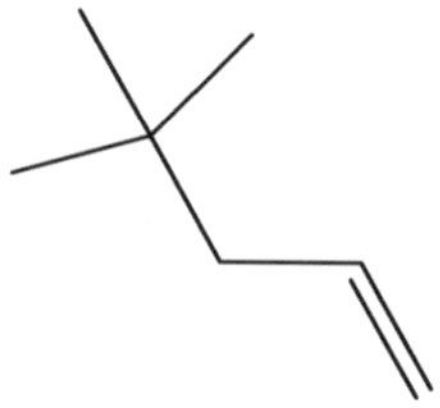

$\Delta H^o{}_f = -19.97$ kcal/mole

Problem 11 – Using the most stable alkyne as a reference point, the second most stable alkyne is 4.6 kcal/mole less stable and the least stable alkyne is 4.8 kcal/mole less stable than the second most stable alkyne. Order the alkynes in relative stability (1 = most stable) and provide a possible explanation.

$$H\!-\!C\!\equiv\!C\!-\!H \qquad\qquad R\!-\!C\!\equiv\!C\!-\!H \qquad\qquad R\!-\!C\!\equiv\!C\!-\!R$$

"R" = a simple alkyl group

Problem 12 – One of the following reactions produces over 90% S_N2 product and one of them produces about 85% E2 product in contrast to our general rules (ambiguity is organic chemistry's middle name). Match these results with the correct reaction and explain why they are different.

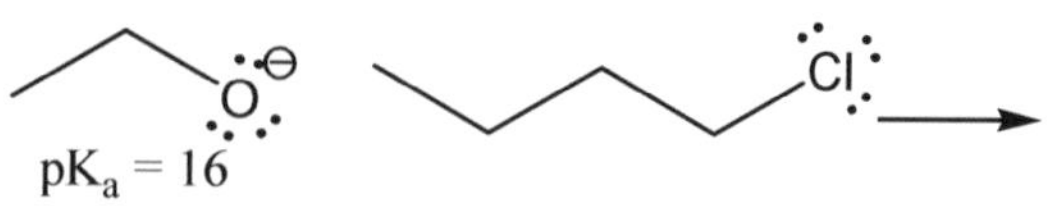

$pK_a = 16$

...versus...

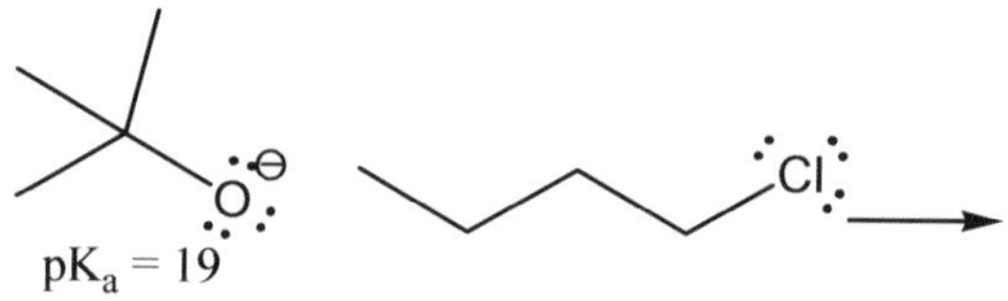

$pK_a = 19$

Chapter 9

Problem 13 - Propose an explanation for the following table of data. Write out the expected products and state by which mechanism they formed. Nu: $^-$/B: $^-$ = $CH_3CO_2^-$ (a weak base, but good nucleophile).

		percent substitution	percent elimination
	$H_3C-\overset{H_2}{C}-Br$	100 %	0 %
	$H_3C-\overset{H}{\underset{CH_3}{C}}-Br$	100 %	0 %
	$\overset{H_3C}{\underset{H_3C}{>}}CH-\overset{H}{\underset{CH_3}{C}}-Br$	11 %	89 %
	$H_3C-\overset{CH_3}{\underset{CH_3}{C}}-Br$	0 %	100 %

Problem 14 - A stronger base (as measured by a higher pK_a of its conjugate acid) tends to produce more relative amounts of E2 compared to S_N2, relative to a second (weaker) base/nucleophile. Greater substitution at C_α and C_β also increases the proportion of E2 product, because the greater steric hindrance which slows down the competing S_N2 reaction. Use this information to make predictions about which set of conditions in each part would produce relatively more elimination product. Briefly, explain your reasoning. Write out all expected elimination products. Are there any examples below where one reaction (S_N2 or E2) would completely dominate?

a.

...versus...

b.

$pK_a(RCO_2H) = 5$ ...versus... $pK_a(ROH) = 16$

c.

$pK_a(RCCH) = 25$ ...versus... $pK_a(NCH) = 9$

d.

...versus...

e.

...versus...

f.

...versus...

Problem 15 - (2R,3S)-2-bromo-3-deuteriobutane when reacted with potassium ethoxide produces
cis-2-butene having deuterium and trans-2-butene not having deuterium. The diastereomer
(2R,3R)-2-bromo-3-deuteriobutane under the same conditions produces cis-2-butene not having deuterium
and trans-2-butene having deuterium present. 1-butene is also formed. Explain these observations by
drawing the correct 3D structures, rotating to the proper conformation for elimination and showing an arrow
pushing mechanism leading to the observed products. (Protium = H and deuterium = D; H and D are
isotopes.) The chemistry of protium (H) and deuterium (D) is similar, but we can tell them apart
experimentally (NMR). S$_N$2 is not considered in this problem.

Chapter 9

Problem 16– Sodium dialkylamides are extremely basic and can be sterically bulky (a nitrogen equivalent of t-butoxide). They are often used in reactions with dibromoalkanes, RBr$_2$, to induce a double reaction (notice there are two leaving groups). We show three different dibromoalkanes below. What reaction is expected and what would be the products after workup? Remember the pK$_a$ of a terminal alkyne is about 25 and the pK$_a$ of a regular amine is about 37 (that's why a final workup step is necessary).

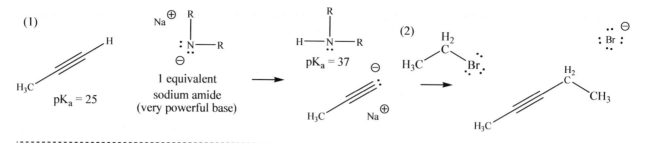

Problem 17 – Propose a reaction using our strong nucleophile/base reagents with one of our C1-C6 R-Br compounds (on pp. 247-248) to make each of the following structures.

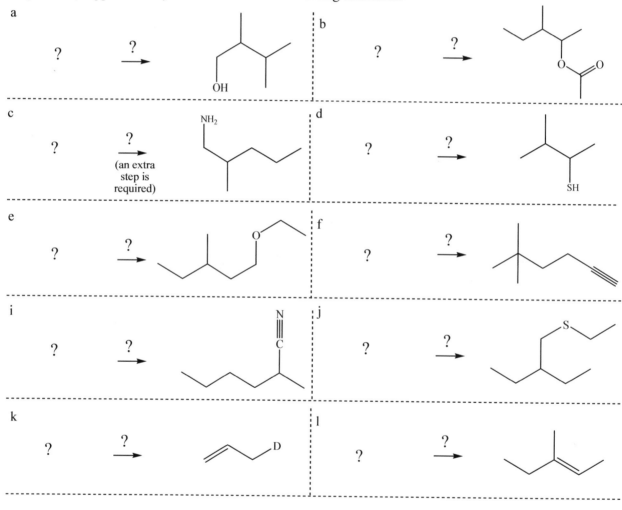

Chapter 9

Problem 18 – What are the possible products of the following reactions? What is the major product(s) and what is the minor product(s)? There are 55 possible combinations.

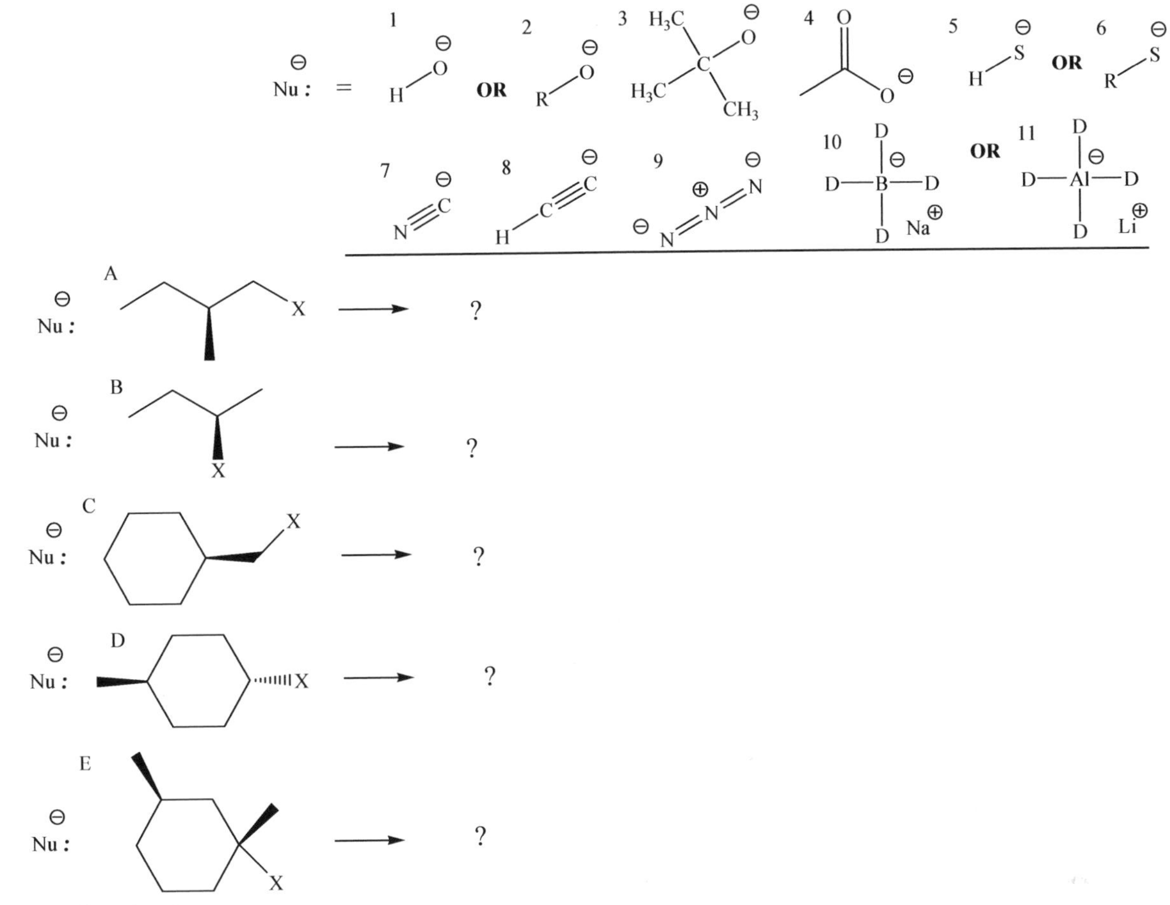

$X = Cl^{\ominus}, Br^{\ominus}, I^{\ominus}$ ...a good leaving group

S_N1 and E1 Competition – Multistep Reactions Arising From Carbocation Chemistry

$S_N2/E2$ and $S_N1/E1$ reactions look very similar overall, but there are some key differences. In $S_N2/E2$ reactions the nucleophile/base is a strong electron pair donor (negative charge in our course), which is why they participate in the slow step of the reaction and force a concerted, one-step reaction. In $S_N1/E1$ reactions the nucleophile/base is a weak electron pair donor (stable, neutral molecules: H_2O, ROH and RCO_2H for us) and that's why they don't participate in the slow step of the reaction, which is ionization of the C_α-X bond. This leads to differences in reaction mechanisms, which show up in the rate law expression (kinetics is bimolecular = 2 or unimolecular = 1). The rate of "1" reactions only depends on how fast RX forms a carbocation. The unimolecular reactions lead to possible side reactions from carbocation rearrangements. You need to carefully look at the reaction conditions to decide what mechanisms are possible. You cut you choices in half when you decide that the electron pair donor is strong (negatively charged = $S_N2/E2$) or weak (neutral = $S_N1/E1$).

H$_3$C—O—H weak nucleophil/base (neutral in our course)	+	:X: X = Cl, Br, I	$S_N1/E1$ reactions ⟶	:O—CH$_3$ S_N1 product (major) Rate = k_{SN1} [RX] E1 product (minor) Rate = k_{E1} [RX]
H$_3$C—O:$^\ominus$ strong nucleophil/base (negatively charged in our course)		:X: X = Cl, Br, I	$S_N2/E2$ reactions ⟶	:O—CH$_3$ S_N2 product (minor) Rate = k_{SN2} [RX][CH$_3$O:$^\ominus$] E2 product (major) Rate = k_{E2} [RX][CH$_3$O:$^\ominus$]

S_N1 and E1 reactions begin exactly the same way, with ionization of the C_α-X bond. Ion formation is energetically expensive and it seems strange to show a bond breaking for no reason. Protic solvents help by being able to solvate both cations and anions. Also, inductive and/or resonance effects are needed to allow carbocation formation. The initial carbocation intermediate typically follows one of three common paths: 1. addition of a weak nucleophile at carbon and/or 2. loss of a beta hydrogen to form a pi bond and/or 3. rearrangement to a similar or more stable carbocation. Rearrangement merely delays the ultimate reaction products: addition of a nucleophile or loss of a beta hydrogen atom to forma pi bond.

S_N1 and E1 reactions are multistep reactions.

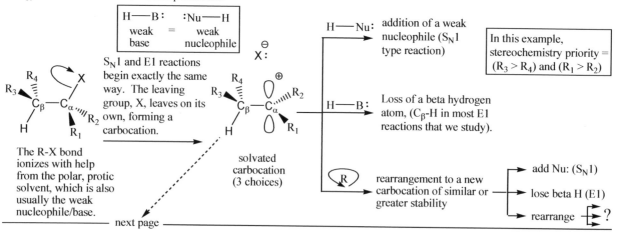

— next page —

Chapter 9

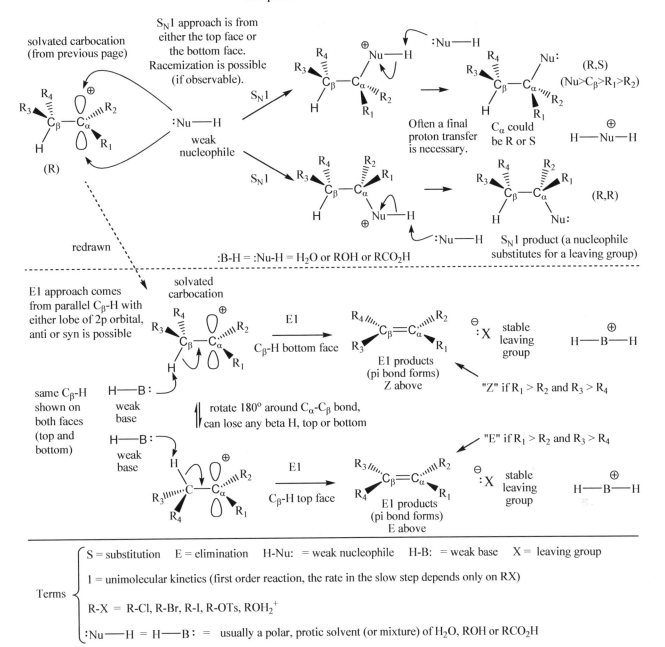

solvated carbocation (from previous page)

(R)

redrawn

S_N1 approach is from either the top face or the bottom face. Racemization is possible (if observable).

:Nu—H weak nucleophile

S_N1

Often a final proton transfer is necessary.

C_α could be R or S

(R,S) (Nu>C_β>R_1>R_2)

S_N1

(R,R)

:Nu—H

S_N1 product (a nucleophile substitutes for a leaving group)

:B-H = :Nu-H = H_2O or ROH or RCO_2H

E1 approach comes from parallel C_β-H with either lobe of 2p orbital, anti or syn is possible

solvated carbocation

E1
C_β-H bottom face

E1 products (pi bond forms) Z above

$\ominus$:X stable leaving group

H—B—H

"Z" if $R_1 > R_2$ and $R_3 > R_4$

same C_β-H shown on both faces (top and bottom)

H—B: weak base

H—B: weak base

rotate 180° around C_α-C_β bond, can lose any beta H, top or bottom

"E" if $R_1 > R_2$ and $R_3 > R_4$

E1
C_β-H top face

E1 products (pi bond forms) E above

$\ominus$:X stable leaving group

H—B—H

Terms
- S = substitution   E = elimination   H-Nu: = weak nucleophile   H-B: = weak base   X = leaving group
- 1 = unimolecular kinetics (first order reaction, the rate in the slow step depends only on RX)
- R-X = R-Cl, R-Br, R-I, R-OTs, ROH_2^+
- :Nu—H = H—B: = usually a polar, protic solvent (or mixture) of H_2O, ROH or RCO_2H

R-X Substitution Pattern and rates of S_N1 reactions

S_N1 (and E1) relative reactivities of R-X compounds:

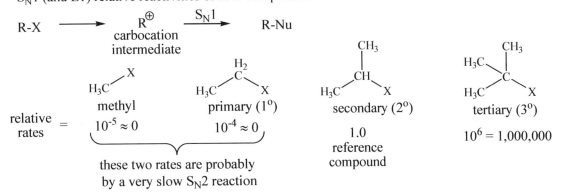

R-X $\longrightarrow$ $R^{\oplus}$ carbocation intermediate $\xrightarrow{S_N1}$ R-Nu

| methyl | primary (1°) | secondary (2°) | tertiary (3°) |

relative rates =

methyl $10^{-5} \approx 0$

primary (1°) $10^{-4} \approx 0$

these two rates are probably by a very slow S_N2 reaction

secondary (2°) 1.0 reference compound

tertiary (3°) $10^6 = 1,000,000$

The order of stability at the electron deficient carbocation carbon is methyl << primary << secondary < tertiary. This is consistent with our understanding of inductive electron donating ability of alkyl groups compared to hydrogen. R groups (alkyl groups) are electron donating (an inductive effect). We observed, previously, that this helps alkene stability and makes it harder to form an anionic conjugate base in acid/base chemistry. A carbocation is extremely electron deficient (the opposite of a carbanion) and, as such, is very electronegative. Extra electron donation to a carbocation center proves very helpful. This can occur through an inductive effect or a resonance effect.

Inductive effects are proposed to occur via polarizations of sigma bonds in the organic skeleton, helping (or hurting) a center of reactivity. We can represent these in a carbocation center, simplistically, as shown below. Hyperconjugation is an alternative explanation to rationalize how extra electron density can be donated to the electron deficient carbocation carbon. This can be considered as sigma resonance.

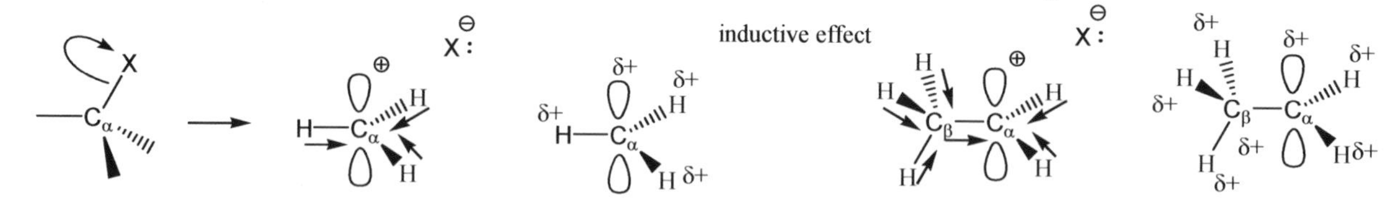

inductive effect

Sigma electrons are pulled toward the carbocation carbon. Part of the δ+ is distributed on to the hydrogen atoms, but not typically shown with formal charge.

Additional sigma bonds of alkyl substituent(s) allow further polarizations of electrons from more bonds (inductive donating effect), which spreads out δ+ charge through sigma bond polarizations and helps stabilize the electron deficient carbocation carbon.

hyperconjugation in a carbocation

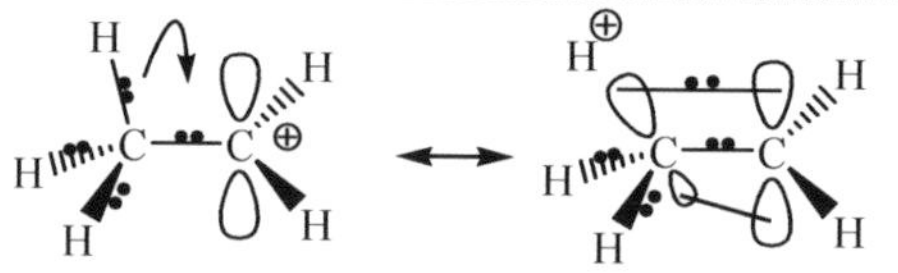

two electron delocalization

Resonance effects also make carbocations more stable. Allylic and benzylic RX compounds are very reactive in S_N1/E1 reactions (and S_N2 reactions). With resonance, there is actually full "pi electron" donation from an adjacent pi bond, instead of the weak inductive effect just mentioned above. An adjacent CC pi bond tends to produce greater stabilization of a carbocation than a single alkyl substituent. Resonance donation from lone pairs of heteroatoms (nitrogen and oxygen) can also be strongly stabilizing for carbocations. Such intermediates are very common in organic chemistry.

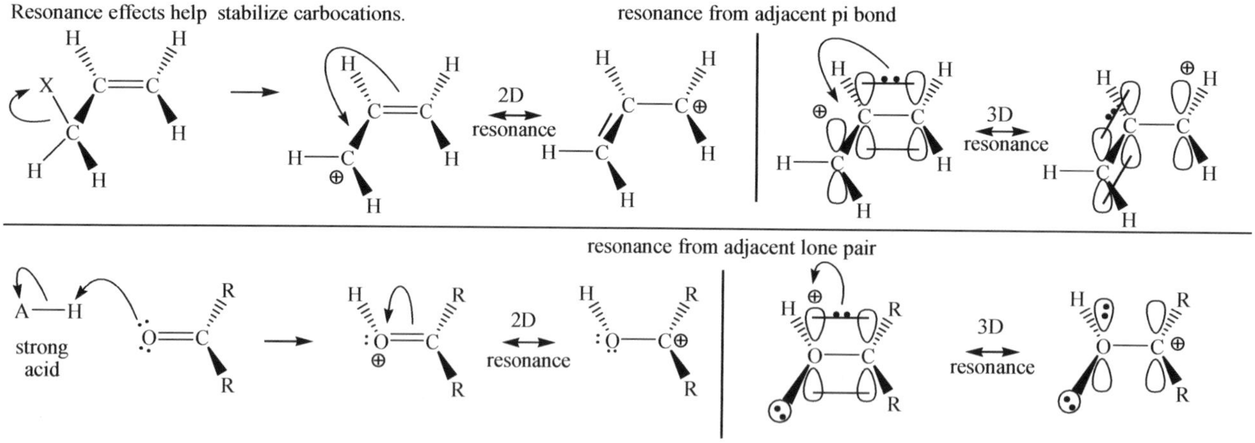

Resonance effects help stabilize carbocations.

resonance from adjacent pi bond

2D resonance

3D resonance

resonance from adjacent lone pair

strong acid

2D resonance

3D resonance

How much do these effects help a carbocation center? The following gas phase data below show the differences in carbocation stability are enormous. In fact, differences are so large that we will almost never propose methyl or primary carbocation possibilities as reaction pathways in our course. We will consider these two patterns (CH_3-X and RCH_2-X) as unreactive in S_N1 and E1 chemistry, and that should make your life a little bit easier.

All of the relative energy values (Δ) are in kcal/mole versus a primary carbocation. A positive value is less stable and a negative value is more stable relative to the reference primary carbocation.

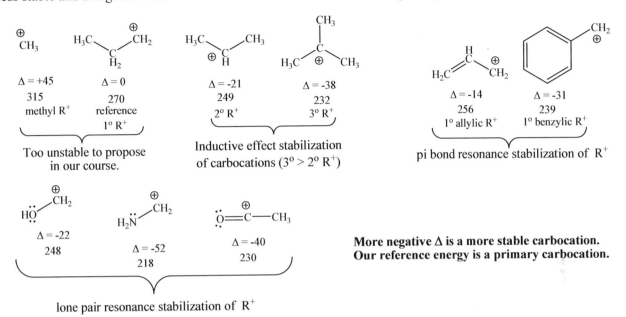

Too unstable to propose in our course.

Inductive effect stabilization of carbocations (3° > 2° R⁺)

pi bond resonance stabilization of R⁺

More negative Δ is a more stable carbocation.
Our reference energy is a primary carbocation.

lone pair resonance stabilization of R⁺

Problem 19 – The bond energy depends on charge effects in the anions too. Can you explain the differences in bond energies below? (Hint: Where is the charge more delocalized? Think back to our acid/base topic.)

$$H_3C - \underset{\underset{CH_3}{|}}{\overset{\overset{CH_3}{|}}{C}} - X$$

X =	Gas Phase bond energies
Cl	+157
Br	+149
I	+ 140

Weak Nucleophile/Bases are used in S_N1/E1 Reactions

We emphasize the term weak here because if the Nu: were strong (negative charge), the reactions observed would be S_N2/E2. Weak, for us, is represented by a neutral molecule with a pair of electrons. For us, this will be a polar solvent molecule such as water (HOH), an alcohol (ROH) or a liquid carboxylic acid (RCO_2H). All of these are protic solvents, which are reasonably good at solvating both cations and anions. In every case, there is an extra hydrogen atom on the oxygen atom of a solvent molecule that must be removed in a final acid/base reaction to produce the neutral organic product. This protonated cationic intermediate results from the addition of water (H_2O, forms alcohols) or an alcohol (ROH, forms ethers) or a liquid carboxylic acid (RCO_2H, forms esters).

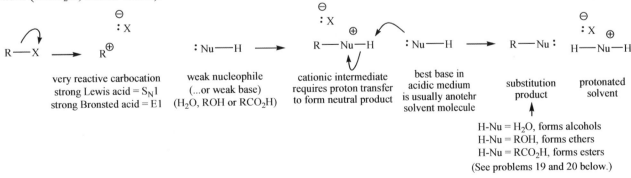

H-Nu = H_2O, forms alcohols
H-Nu = ROH, forms ethers
H-Nu = RCO_2H, forms esters
(See problems 19 and 20 below.)

We will view the attack on an sp^2 carbocation as equally accessible from either face (top or bottom).

achiral carbocation carbon

If all three attached groups at a carbocation carbon are different from one another and the attacking nucleophile, then a 50/50 racemic mixture of enantiomers will form (in our course). If one of the "R" branches has a chiral center, then diastereomers will be formed.

new stereogenic center forms a racemic mixture of enantiomers (assuming no other chiral centers exist in the molecule)

(dl) (+/-)

R/S assumes priorities Nu > R1 > R2 > R3

In our course we will view carbocations in S$_N$1 reactions, simplistically, as fully formed sp^2 centers, equally approachable from both faces (not always true).

Problem 20 – Draw in all of the mechanistic steps in an S$_N$1 reaction of 2R-bromobutane with a. water, b. methanol and c. ethanoic acid. Add in necessary details (3D stereochemistry, curved arrows, lone pairs, formal charge). What are the final products?

Donate the carbonyl (C=O) electrons to the carbocation. That way there is resonance.

S$_N$1 products generally outcompete E1 products, (always in our course using RBr compounds, S$_N$1 > E1). The only exception (presented later) will be when alcohols (ROH) are mixed with concentrated sulfuric acid at high temperatures to form alkenes (E1 products).

Keeping Track of the C$_\beta$ Hydrogens in E1 Reactions

E1 products arise from the same carbocation intermediate formed in S$_N$1 reactions. Just as in E2 reactions, we have to examine each type of C$_\beta$-H. In more complicated R-X molecules there may be several different types of hydrogen atoms on C$_\beta$ positions. After all, there can be anywhere from one to three C$_\beta$ carbons with zero to three hydrogen atoms. We will only consider secondary and tertiary RX compounds below, since methyl and primary carbocations do not form (in our course). That still leaves a lot of possibilities. There are usually more possible E1 products than E2 products because a proton can be lost from either face of a carbocation due to rotations about any C$_\beta$-C$_\alpha$ bond (as opposed to only 'anti' in E2 reactions). That means every place E/Z is possible, we will propose both are formed.

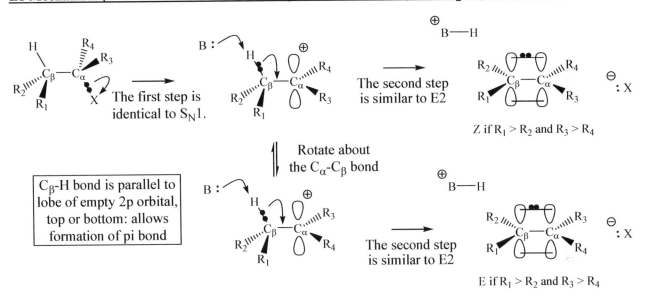

2° R-X has six C_β positions

3° R-X has nine C_β positions

H_3C—X   methyl

$\underset{R}{\overset{H_2}{C}} X$ primary

$\xrightarrow{H—Nu:}$ No Reaction in our course.

HNu: = H_2O, ROH, RCO_2H (weak, in our course)

E1 Mechanism (unimolecular kinetics) loss of proton from any adjacent C_β-H position, top or bottom

The first step is identical to S_N1.

The second step is similar to E2

Z if $R_1 > R_2$ and $R_3 > R_4$

Rotate about the C_α-C_β bond

C_β-H bond is parallel to lobe of empty 2p orbital, top or bottom: allows formation of pi bond

The second step is similar to E2

E if $R_1 > R_2$ and $R_3 > R_4$

Problem 21 – Reconsider problem 15 and draw in all of the mechanistic steps and show all possible products of the E1 reaction of 2R-bromobutane with any of the weak bases, water, methanol or ethanoic acid (use water in your work). Add in necessary details (3D stereochemistry, curved arrows, lone pairs, formal charge).

show E1 mechanism and products

+

Rearrangements of Carbocations

The high reactivity (low stability) of carbocations forces some very quick choices to try and stabilize the situation. The carbocation needs two electrons to complete its octet (in a hurry!). There are three ways it typically does this. We have studied the two ultimate pathways above, S_N1 and E1 reactions.

Rearrangements are a third possibility. Rearrangements are a temporary solution for an unstable carbocation. They transfer the unstable carbocation site to a new position having a similar energy or, better yet, to a site where the positive charge is more stable. If such possibilities exist, this will very likely be one of the observed reaction pathways. However, even with a rearrangement a carbocation will not gain the two needed electrons. The electron deficiency is merely moved to a new position. This process can occur a number of times before a carbocation encounters its ultimate fates, discussed above, S_N1 and E1. We will only consider the simplest of rearrangements.

Chapter 9

Carbocations have very large potential energy differences. These differences provide a large driving force to form more stable carbocations from less stable carbocations, in the range of ≈15-20 kcal/mole in the gas phase for 1° to 2° and 2° to 3° choices. Rearrangements are a competitive pathway in any reaction where a carbocation is formed. A relatively simple example illustrates the necessity to be systematic in your approach to determine all of the varied possibilities. Consider the migration of every group on a C_β position, whether H or C. To keep our choices simpler (than they really are) we will only consider rearrangements of 2° to 3° and 3° to 3° carbocations.

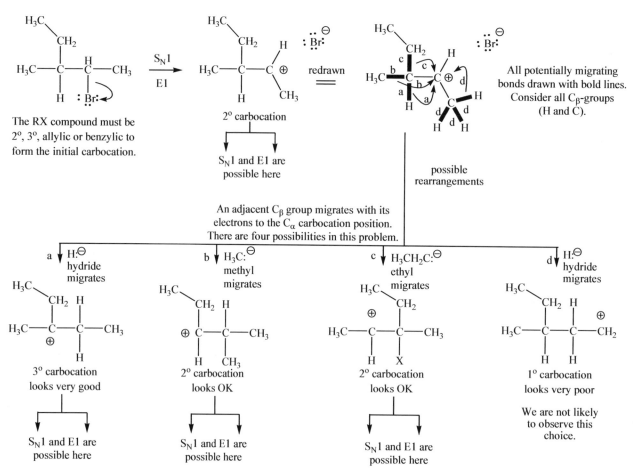

Problem 22 – What are the likely S_N1 and E1 products of the initial carbocation and the rearrangement carbocations from "a"? Assume water is the nucleophile.

The most competitive rearrangement above will be to form the more stable tertiary carbocation from the initially formed secondary carbocation. It is also likely that at least some of the initially formed carbocation will react by the S_N1 and E1 choices. However, those may be minor products when a much more stable carbocation can form by rearrangement. In the end, S_N1 and E1 possibilities are the ultimate fates of even the most stable carbocation that can form. Our goal at this point is to understand how simple rearrangements occur and what S_N1 and E1 products are possible.

All groups on any C_β carbon can potentially migrate to the adjacent carbocation carbon (also called a 1,2 shifts), if a similar or more stable carbocation can form. If hydrogen with its two electrons is the group migrating, the rearrangement is called a 1,2 hydride shift. If a carbon group migrates with its two electrons, the rearrangement if called a 1,2 alkyl shift. Hydride and alkyl shifts can occur from further away than a C_β position or even between two positions in completely different molecules. However, these we will not emphasize such possibilities in our course.

Chapter 9

Transition state of a carbocation rearrangement

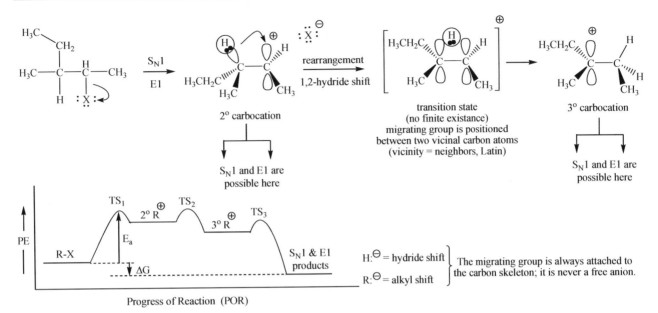

2° carbocation

S_N1 and E1 are possible here

transition state
(no finite existance)
migrating group is positioned
between two vicinal carbon atoms
(vicinity = neighbors, Latin)

3° carbocation

S_N1 and E1 are possible here

H:$^{\ominus}$ = hydride shift
R:$^{\ominus}$ = alkyl shift
} The migrating group is always attached to the carbon skeleton; it is never a free anion.

Two main rules will help guide you in evaluating possible rearrangements.

1. Rearrangements usually occur so that the migrating groups moves from a C_β atom to the C_α atom (the carbocation center). These are the 1,2 hydride or 1,2 alkyl shifts mentioned above. The C_β atom that gives up the migrating group becomes the new electron deficient carbocation center, often because it is a more stable carbocation site.

2. If a 1,2 shift of a hydrogen atom or an alkyl group can form a similar or more stable carbocation, then such a rearrangement is likely to be competitive with other reaction choices (S_N1 and E1). When *interpreting* a reaction mechanism involving rearrangements, you may have to consider both equal (3° → 3° R$^+$) and more stable (2° → 3° R$^+$) carbocation possibilities. However in this book when you are asked to *predict* what might be possible, you usually only need to consider more stable carbocation possibilities (2° → 3° R$^+$).

Problem 23 – Write out your own mechanism for all reasonable products from the given R-X compound in water (2-halo-3-methylbutane).

a

Predict the product from the best carbocation.

b

Predict the product from the best carbocation.

c

Rationalize product formation.

Problem 24 – Lanosterol is the first steroid skeletal structure on the way to cholesterol and other steroids in our bodies. It is happening inside you, even as you read these words. It is formed in a spectacular cyclization of protonated squalene oxide. The initially formed 3° carbocation rearranges 4 times before it undergoes an E1 reaction to form lanosterol. Add in the arrows and formal charge to show the rearrangements and the final E1 reaction.

Chapter 9

Summary of features to examine to decide what mechanism is in operation, suggested strategy

1. Look at the reaction conditions first. A strong base/nucleophile favors the bimolecular processes (S_N2, E2). We will simplistically view strong electron pair donation as coming from anions of all types leading to S_N2 (always backside attack) and E2 reactions (always "anti" C_β-H/C_α-X bond orientations, for us). A weak base/nucleophile favors the unimolecular processes (S_N1, E1). Weak electron pair donors will typically be neutral solvent molecules, usually water (H_2O), alcohols (ROH), or simple carboxylic acids that are liquid (RCO_2H, methanoic = formic acid or ethanoic = acetic acid), leading to S_N1 and E1 reactions (usually $S_N1 >$ E1). See more details in the chart, just below.

2. Next, look to the reactant structures (CH_3X, 1°RX, 2°RX, 3°RX, allylic, benzylic), (X = good leaving group)

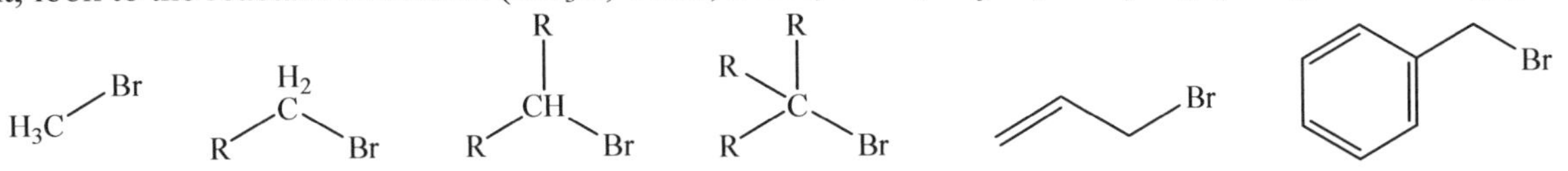

methyl (Me)   primary (1°)   secondary (2°)   tertiary (3°)   allyllic   benzylic

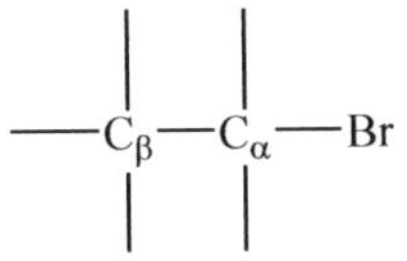

a. **CH_3X (methyl)** =

    only reacts in S_N2 reactions with strong nucleophiles (negative charge in our course) no other reactions

b. **1°RX (primary)** =

    i. bimolecular conditions – mainly S_N2, some E2, exception = mainly E2 with potassium t-butoxide

    ii. no unimolecular reactions (S_N1 / E1) due to the high energy of 1°R⁺.

c. **2°RX (secondary)** =

    i. bimolecular conditions - significant amounts of both S_N2 and E2 observed. In our course we will assume mostly S_N2, except for stronger bases and bulky bases, (see chart below) where E2 becomes the major product.

    ii. unimolecular conditions - generally S_N1 more than E1, except E1 is major with $ROH/H_2SO_4/\Delta$. (see chart below)

d. **3°RX (tertiary)** =

    i. bimolecular condition - only E2, most stable alkene is the major product, 3° RX is too sterically hindered for S_N2 reactions to compete.

    ii. unimolecular conditions - generally $S_N1 >$ E1, except E1 is major with $ROH/H_2SO_4/\Delta$.

e. **Unreactive R-X patterns (in our course) include: 1° neopentyl RX, vinyl RX and phenyl RX**

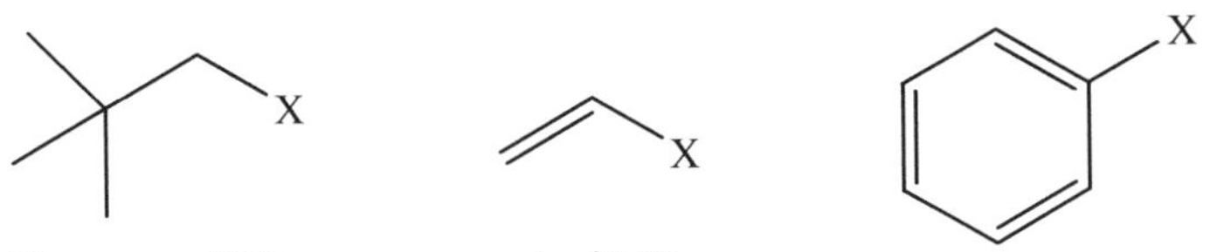

1° neopentyl RX   vinyl RX   phenyl RX

No reaction for any of these in our course.

Extending the R-X Spectrum: Making Alcohols (ROH) into Reactive S_N and E Reactants

Our general R-X pattern can be expanded by making an alcohol "OH" into a good leaving group. We will do this several ways. The first way is very simple. We will use a very strong HX acid to protonate the alcohol OH, making it into a good leaving group, H_2O. The second way is also simple. We will use an alcohol with PX_3 (X = Cl or Br). S_N2 attack on phosphorous by the alcohol oxygen displaces a halide from phosphorous, which then attacks the carbon with the oxygen-phosphorous linkage becoming the leaving group in either a second S_N2 reaction or an S_N1 reaction. A third reaction is needed at secondary alcohols where rearrangement is a problem in strong acid. A fourth reaction will be E1 elimination to alkenes using sulfuric acid and heat. Right now the alcohols we use come from our limited source of RBr compounds presented earlier, so these reactions might seem rather pointless. However, there are many commercially available alcohols that can be used as starting structures and there are a large number of reactions known to produce alcohol structures from other functional groups, some of which we will learn. The alcohol group is a central group in synthetic methodology. The following examples are the first of many to come.

strong acids (HI, HBr, HCl) have....   ...very stable conjugate bases, which make very good leaving groups = $I^{\ominus}$, $Br^{\ominus}$, $Cl^{\ominus}$.

H——Base $\longrightarrow$ $H^{\oplus}$ (solvated)   + $\ominus$:Base

A select number of acid K_a/pK_a values are provided below. The top five acids are all strong acids having very stable conjugate bases (good leaving groups). The last example is the weak acid water, which means its conjugate base, hydroxide, is not so stable (a relatively poor leaving group).

Conjugated Acid	K_a	pK_a	Conjugated Base (Leaving Group)
H——I	10^{10}	-10	$^{\ominus}$I very good
H——Br	10^{9}	-9	$^{\ominus}$Br very good
H——Cl	10^{7}	-7	$^{\ominus}$Cl very good
H——O——S——R (O, O double bonds)	10^{3}	-3	$^{\ominus}$O——S——R very good
H——$\overset{\oplus}{O}H_2$	10^{2}	-2	H_2O very good
H——OH	10^{-16}	+16	$^{\ominus}$O-H poor

Protonated Alcohols - Water as a Good Leaving Group

Strong protic acids are used to extensively protonate the alcohol OH. When the alcohol OH is protonated, the leaving group is water, not hydroxide. Water's conjugate acid is H_3O^+, ($pK_a = -2$). If substitution is the desired goal, then the strong halide acids are normally used, HCl, HBr or HI. If elimination is the desired goal, then concentrated sulfuric acid (H_2SO_4) and/or concentrated phosphoric acid (H_3PO_4) is used at an elevated temperature (Δ) to distill away the alkene product.

Using the hydrohalic acids (HCl, HBr or HI), very polar, strongly acidic conditions encourage S_N1 reactions, and these are assumed to be operating at all tertiary and secondary alcohol (ROH) centers. Rearrangements are possible under these conditions, and must be considered in every reaction. When methyl or primary alcohols are used, the large energy expense of carbocation formation prevents the escape of water on its own. The H$_2$O at a primary ROH$_2^+$ is assumed to be pushed off by the conjugate halide base partner of the strong acid used to form a methyl or primary alkyl halide without rearrangement (S$_N$2). Example mechanisms are shown below.

a. 1°, 2° and 3° ROH reacted with HX acids (HCl, HBr, HI) - usually S$_N$ chemistry

i. methanol (S$_N$2 emphasized, no rearrangement)

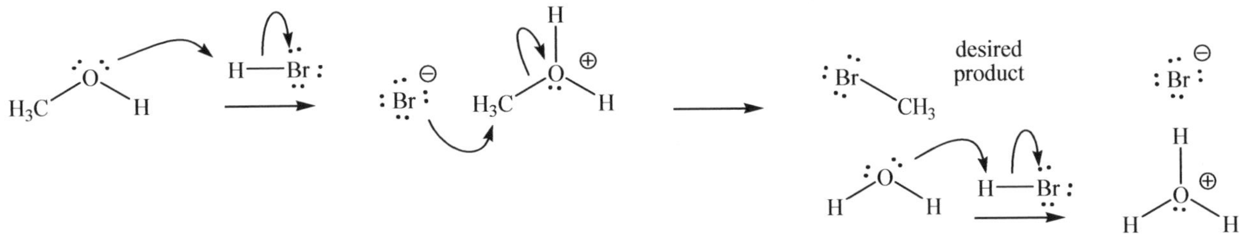

ii. primary alcohols (S$_N$2 emphasized, no rearrangement, similar at methanol)

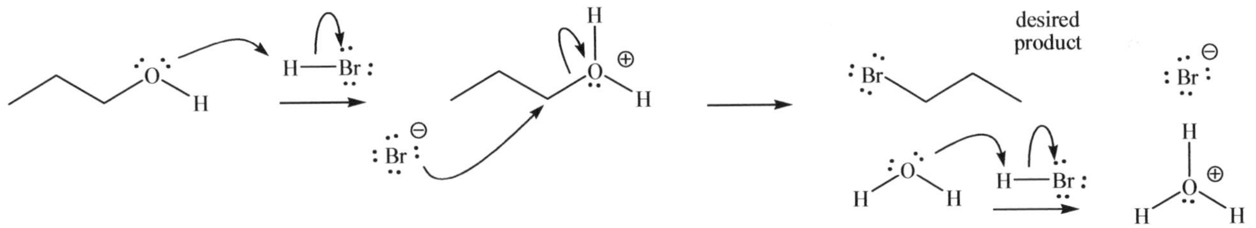

iii. secondary alcohols (S$_N$1 emphasized, rearrangements possible)

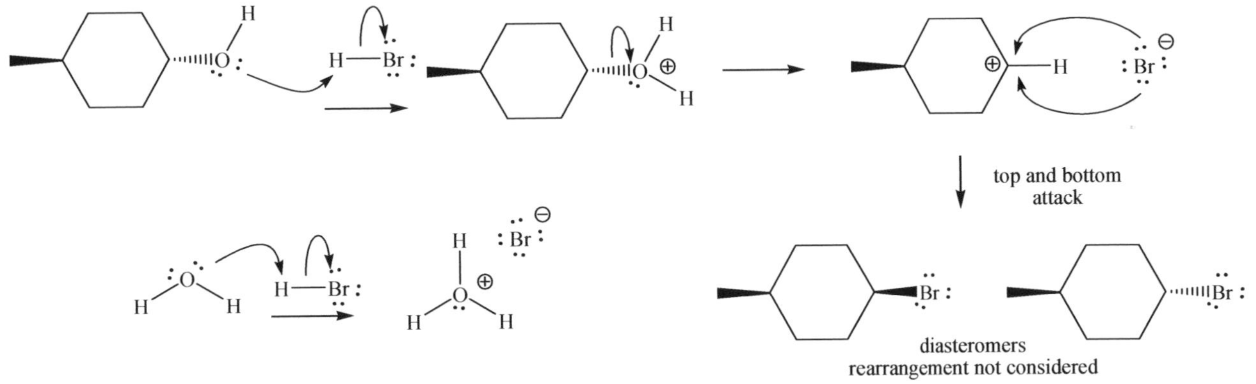

iv. tertiary alcohols (S$_N$1 emphasized, rearrangements possible)

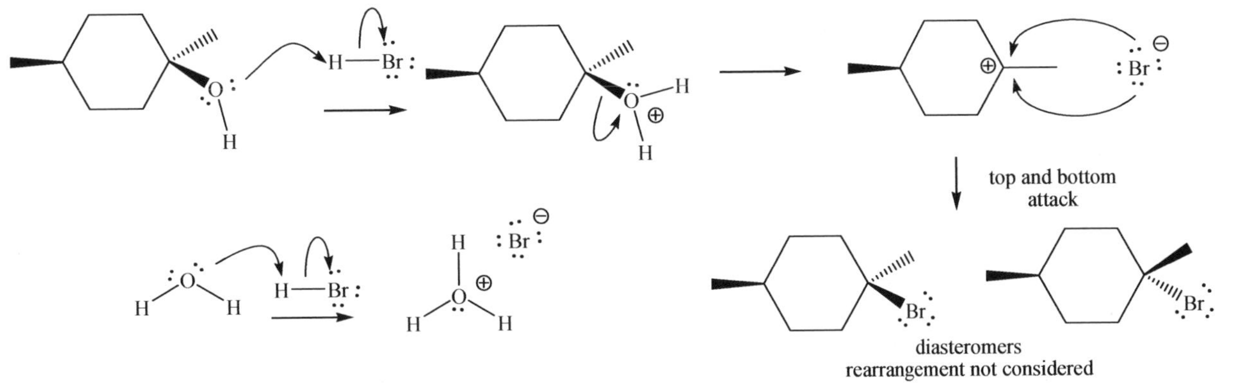

Chapter 9

b. 1°, 2° and 3° ROH reacted with PX₃ (PCl₃ and PBr₃) - usually S_N chemistry

Alcohol reactions with PX$_3$ (we use PBr$_3$) are assumed to begin with S$_N$2 attack of the alcohol oxygen on phosphorous with a bromide leaving group. The bromide then becomes the nucleophile. If the alcohol is methyl or primary, bromide attacks the alcohol carbon (backside) in a second S$_N$2 reaction, displacing the HOPBr$_2$ leaving group. If the alcohol is secondary or tertiary, we assume the reaction is S$_N$1, with possible rearrangement. The HOPBr$_2$ molecule can react two additional times.

i. methanol (S$_N$2 emphasized, no rearrangement)

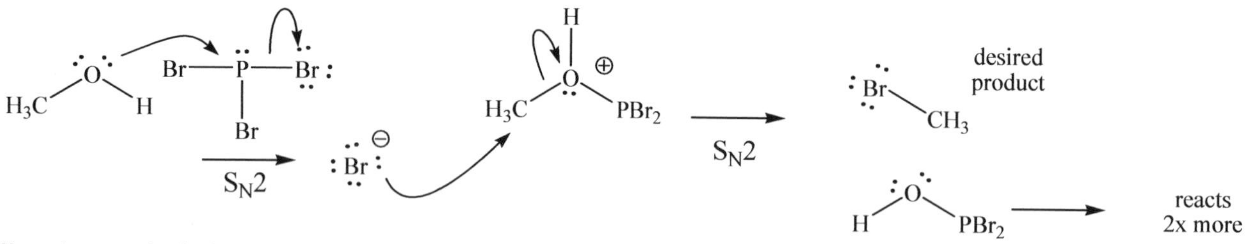

ii. primary alcohols (S$_N$2 emphasized, no rearrangement, similar at methanol)

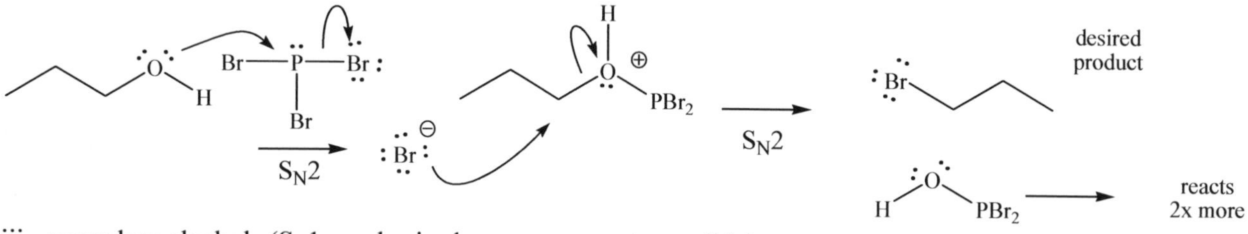

iii. secondary alcohols (S$_N$1 emphasized, rearrangements possible)

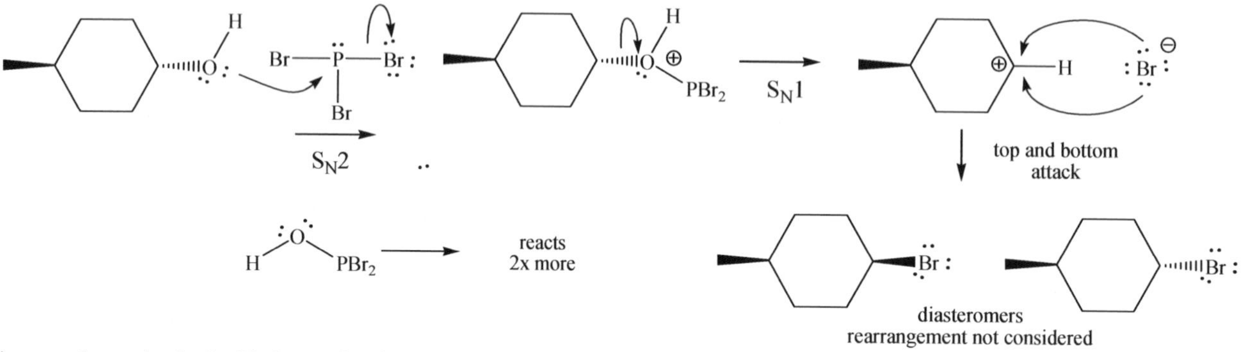

iv. tertiary alcohols (S$_N$1 emphasized, rearrangements possible)

286

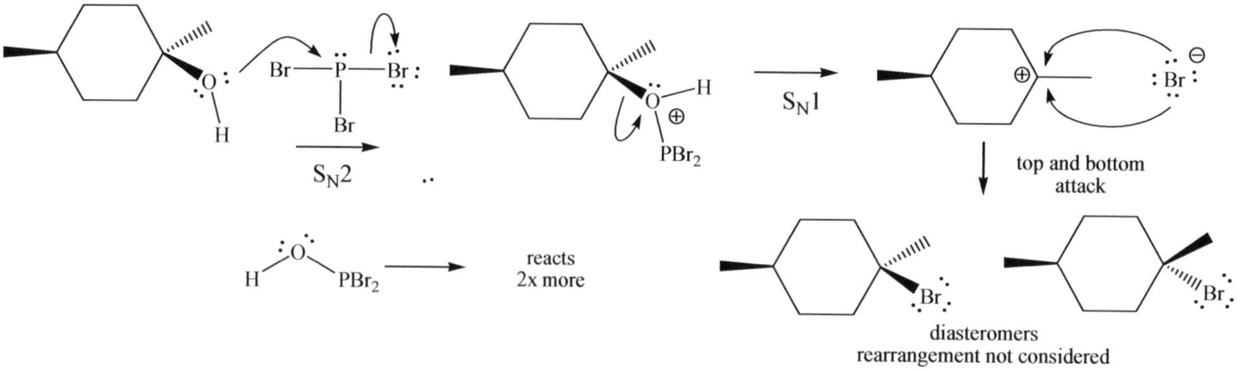

Chapter 9

Problem 25 – Write a mechanism for each of the following reactions. Rearrangements are a potential problem.

a

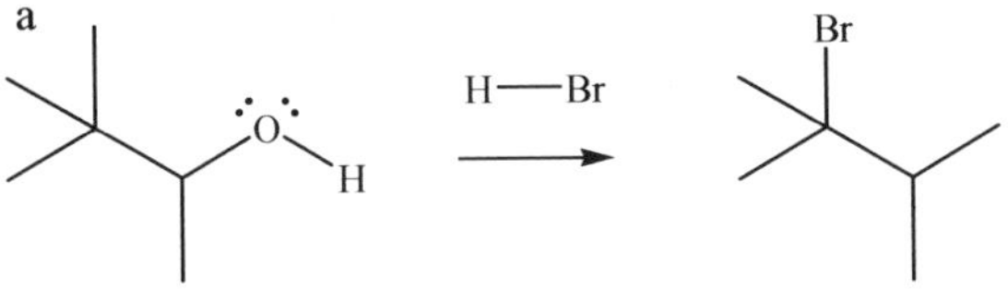

H—Br

b

Br—P—Br
|
Br

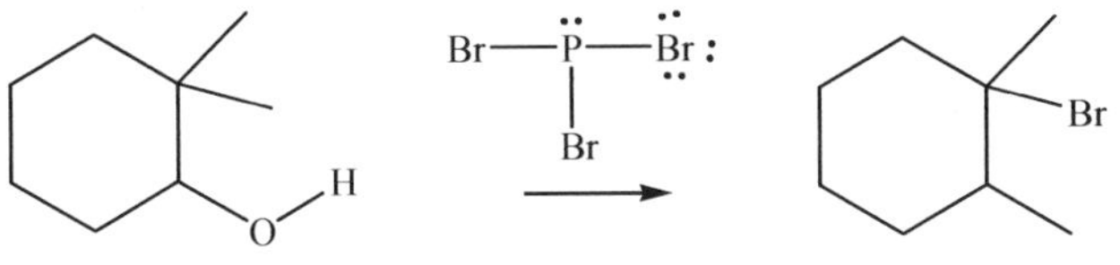

c. Make alcohols into sulfonate esters (good leaving groups) and displace with bromide (S$_N$2)

It would be nice to not consider this reaction, but we still have a problem in one particular situation, and that is converting a secondary alcohol that can rearrange into an RBr compound that does not rearrange. In this transformation, we will make the poor "OH" leaving group of an alcohol (ROH) into a good leaving group that is a sulfonate ester (R-OSO$_2$Ar). The new leaving group is the conjugate base of a sulfonic acid (a very stable anion, as seen from its conjugate acid's pK$_a$ = -3). The initial reaction occurs by 'acyl' substitution at sulfur and not by reaction at the alcohol C-O bond (so there is no rearrangement). The second step uses bromide as a good nucleophile displacing the very good leaving group, tosylate (TsO^{--}) in an S$_N$2 reaction. Because this step is S$_N$2 there is no rearrangement.

Toluenesulfonic acid is a strong acid, because it has a very stable conjugate base (looks like H$_2$SO$_4$)

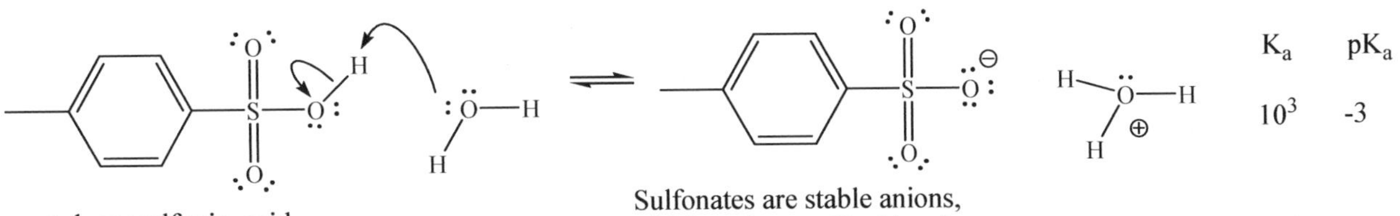

	K$_a$	pK$_a$
	10^3	-3

toluenesulfonic acid

Sulfonates are stable anions, which make excellent leaving groups when bonded to carbon.

Alcohols are converted into sulfonate esters by a different type of substitution reaction than S$_N$2 or S$_N$1. Toluenesulfonyl chloride is a sulfur acid chloride, analogous to an organic acid chloride. Both sulfur and carbon are very electron poor from being attached to electronegative oxygen and chlorine. This partial positive charge attracts the alcohol oxygen to the sulfur (or the carbon). Initially, the valency expands on sulfur (from 4 to 5) and on carbon (from 3 to 4). This is possible because sulfur uses its 3d orbitals. The intermediate collapses back to the original valency with the expulsion of the very good chloride leaving group. Generally, a tertiary amine is added to neutralize the now strongly acidic alcohol proton (we will propose pyridine). The overall result forms a sulfur ester (or a carbon ester) from an alcohol and an acid chloride. Acid chlorides are very valuable compounds that undergo many similar reactions with other nucleophiles. We will study similar reactions further in a later chapter.

Chapter 9

Formation of an inorganic sulfonate ester (mechanism = acyl-like substitution)

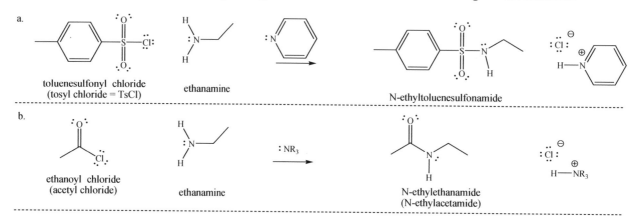

toluenesulfonyl chloride
sulfur valency = 4

alcohol

sulfur valency = 5

pyridine

acid/base

The pyridinium ion is a stable form of the otherwise very acidic proton.

Sulfonates esters have an excellent leaving groups and are useful in S$_N$2 chemistry.
sulfur valency = 4

Formation of an analogous organic alkanoate ester (mechanism = acyl substitution)

acid chloride   alcohol

carbon valency = 3

carbon valency = 4

: NR$_3$

acid base

organic ester

H—NR$_3^{\oplus}$:Cl$^{\ominus}$: discard

carbon valency = 3

Problem 26 – Write a detailed arrow-pushing mechanism for each of the following transformations.

a.

toluenesulfonyl chloride
(tosyl chloride = TsCl)

ethanamine

N-ethyltoluenesulfonamide

b.

ethanoyl chloride
(acetyl chloride)

ethanamine

: NR$_3$

N-ethylethanamide
(N-ethylacetamide)

The reaction below is the reason we need this extra method for transforming an alcohol into a bromoalkane. This is a problem for us, because of rearrangement.

HBr
S$_N$1

This is what we want, but...

...this is what we get.

rearrangement

This is how we solve our problem. Formation of 1. tosylates from ROH + TsCl/pyridine (toluenesulfonyl chloride = tosyl chloride = TsCl) and 2. NaBr, S$_N$2 chemistry is possible without rearrangements.

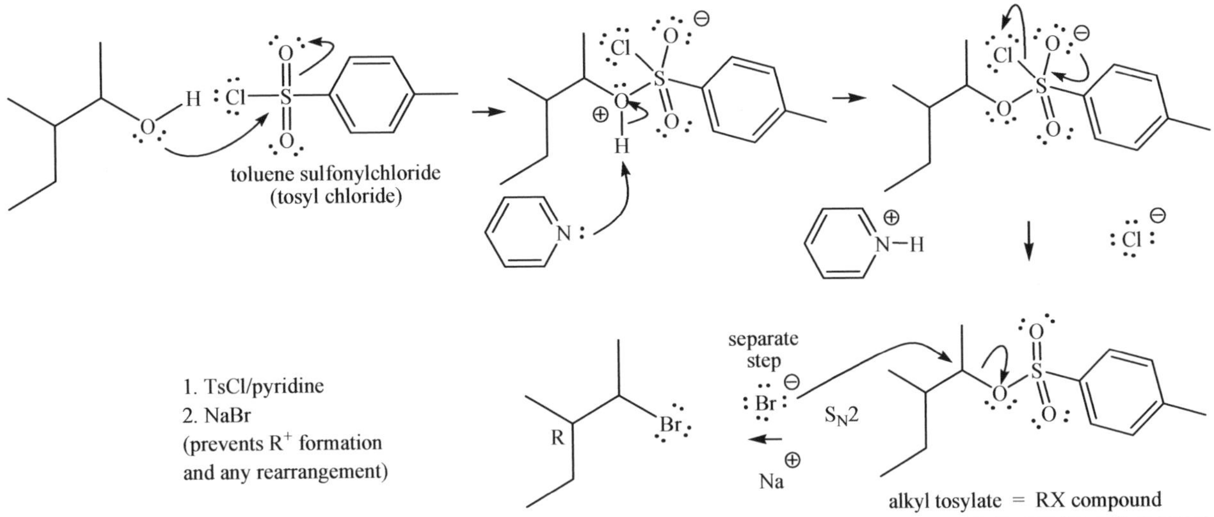

toluene sulfonylchloride
(tosyl chloride)

1. TsCl/pyridine
2. NaBr
(prevents R$^+$ formation
and any rearrangement)

separate
step

S$_N$2

alkyl tosylate = RX compound

Problem 27 – We can now make the following molecules. Propose a synthesis for each. (Tosylates formed from alcohols and tosyl chloride/pyridine via acyl substitution reaction, convert "OH" from poor leaving group into a very good leaving group, like bromide). We will only use this for secondary alcohols where rearrangement is a possibility.

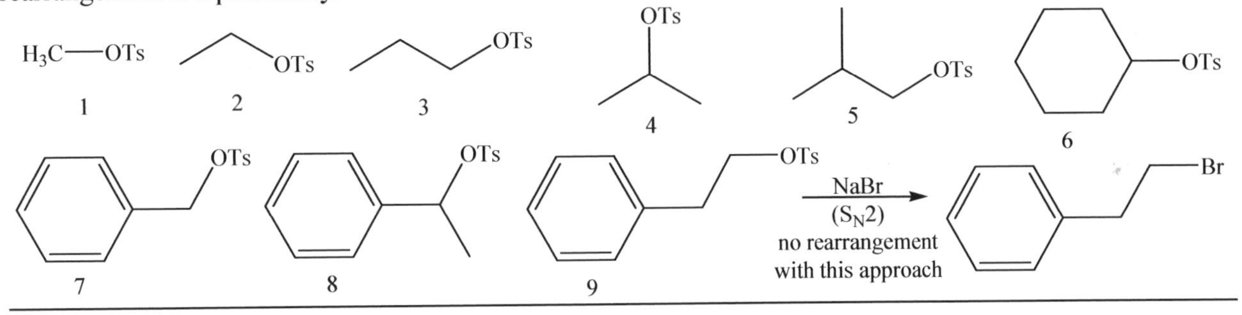

H$_3$C—OTs

1

2

3

4

5

6

7

8

9

NaBr
(S$_N$2)
no rearrangement
with this approach

Chapter 9

d. 1°, 2° and 3° ROH reacted with H$_2$SO$_4$ and high temperature - E1 chemistry

Using strongly acidic sulfuric acid, H$_2$SO$_4$, at elevated temperatures favors E1 reactions and these are assumed to be operating in all of the reactions below. Rearrangements are possible and observed.

i. primary alcohols (with high temperature, E1 is proposed, rearrangements possible)

bp = +82°C

bp = -47°C
distills away

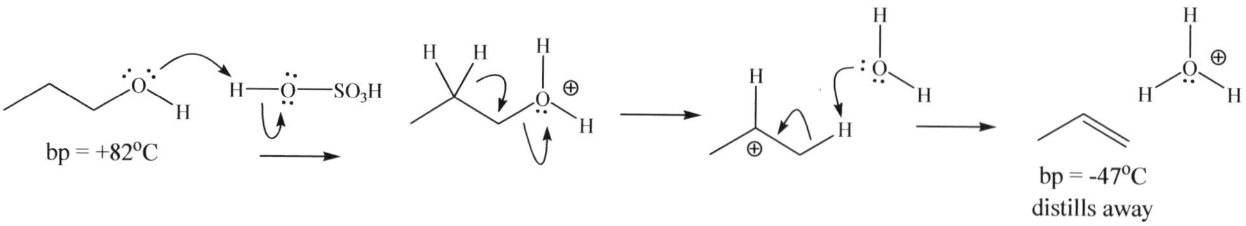

ii. secondary alcohols (E1 emphasized, rearrangements possible)

bp = +161°C

bp = +83°C
distills away

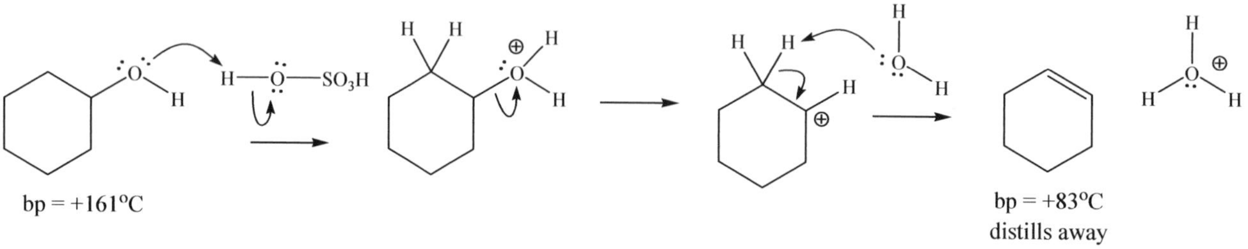

iii. tertiary alcohols (E1 emphasized, rearrangements possible)

higher boiling

lower boiling
distills away

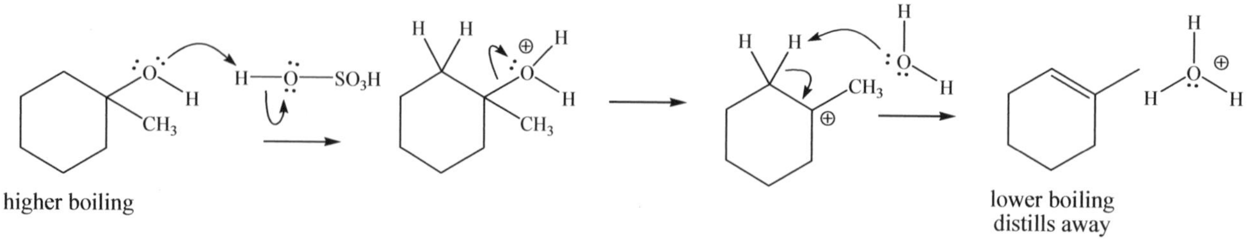

Problem 28 – Write a mechanism for the following reaction.

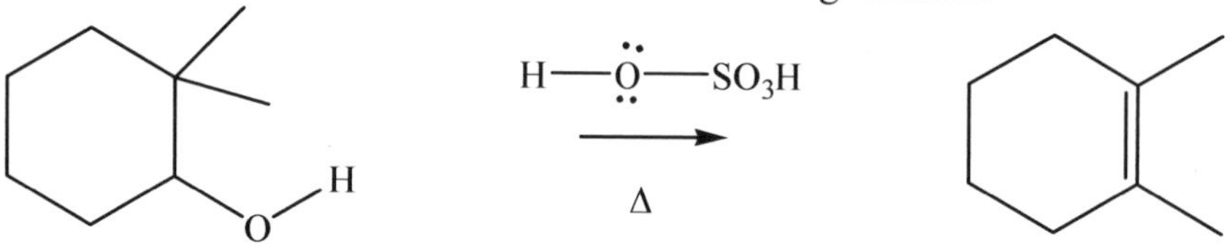

H—O—SO$_3$H

Δ

We have some, limited control to direct the alcohol functionality toward S$_N$ or E choices. The conditions to effect these different pathways are important, so you must be aware of the details mentioned above (halide acids = S$_N$ reactions and H$_2$SO$_4$/Δ = E1 reactions). Heat is a crucial aspect of the E1 reaction, since it allows the lower boiling alkene to escape from the reaction mixture by distillation, while the higher boiling alcohol or inorganic ester remains, in the reaction pot to reestablish equilibrium by forming more alkene, which distills......etc. The alkene boils at a much lower temperature than the alcohol it came from because it does not have an "OH" to form hydrogen bonds with.

Chapter 9

Examples of Boiling Point Differences Between Alcohols and Possible Alkene Products

boiling points of alcohols (°C)	boiling point of alkenes (°C)	DT_{bp}
(structure) OH (79 °C)	$H_2C{=}CH_2$ (-104 °C)	(183 °C)
(structure) OH (82 °C)	(structure) (-47 °C)	(129 °C)
(structure) OH (97 °C)	(structure) (-47 °C)	(144 °C)
(structure) OH (100 °C)	(structure) (-6 °C)	(106 °C)
	(structure) (1 °C)	(96 °C)
	(structure) (4 °C)	(99 °C)
(structure) OH (161 °C)	(structure) (83 °C)	(77 °C)

Problem 29 (R)-2-butanol retains its optical activity indefinitely in aqueous base ($^-$OH), but is rapidly converted to optically inactive 2-butanol (racemic) when in contact with dilute sulfuric acid ($H_2SO_4 + H_2O \rightarrow H_3O^+ + HSO_4{}^-$). Explain with detailed mechanisms.

RX compounds are extremely versatile in organic chemistry. We will learn additional useful methods to make them in later chapters. RX compounds provide a path to make many other organic functional groups, and there is still much more to come. As always, we have to be selective since our time is limited. We look for reactions that provide insight into how typical organic and biochemistry works.

Chapter 9

(handwritten: azides are one of the best Nu:)

Chart of S$_N$ and E Chemistry (typical patterns of reactivity in our course)

(handwritten left margin: focus on the Cα – X bond)

typical strong base nucleophiles are: (for our course)	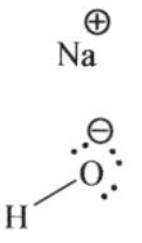 hydroxide	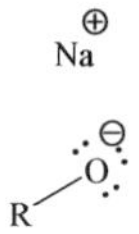 alkoxide	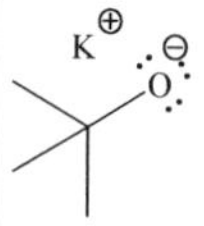 t-butoxide	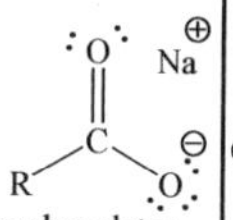 carboxylates	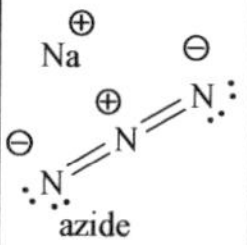 azide	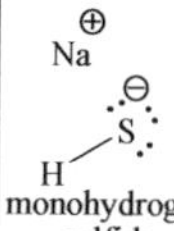 monohydrogen sulfide	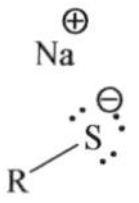 thiolate	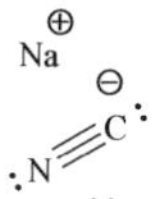 cyanide	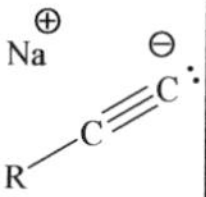 acetylides	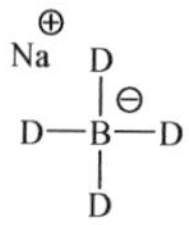 sodium borohydride
pKa	16	16	19	5	5	9		9	25	
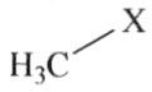 H$_3$C–X methyl	only S$_N$2	only S$_N$2	only S$_N$2	only S$_N$2	only S$_N$2	only S$_N$2	only S$_N$2	only S$_N$2	only S$_N$2	only S$_N$2
R–CH$_2$–X primary	S$_N$2 > E2	S$_N$2 > E2	E2 > S$_N$2 **exception (too basic)**	S$_N$2 > E2	S$_N$2 > E2	S$_N$2 > E2	S$_N$2 > E2	S$_N$2 > E2	S$_N$2 > E2	S$_N$2 > E2
R–CHR–X secondary	E2 > S$_N$2 **exception (too basic)**	E2 > S$_N$2 **exception (too basic)**	only E2	S$_N$2 > E2	S$_N$2 > E2	S$_N$2 > E2	S$_N$2 > E2	S$_N$2 > E2	E2 > S$_N$2 **exception (too basic)**	S$_N$2 > E2
R$_3$C–X tertiary	only E2	only E2	only E2	only E2	only E2	only E2	only E2	only E2	only E2	NA

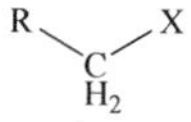

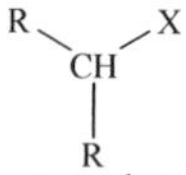

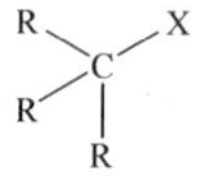

typical strong base nucleophiles are: (for our course)	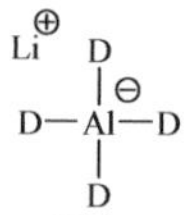 lithium aluminum hydride	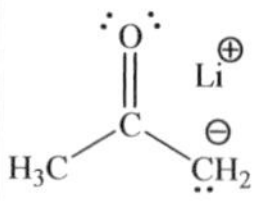 ketone enolates	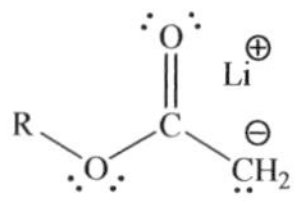 ester enolates	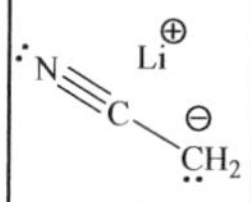 nitrile enolates	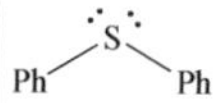 diphenyl sulfide	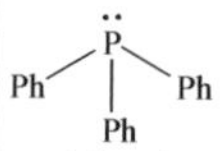 triphenyl phosphine	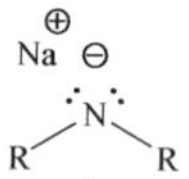 sodium dialkyl amide	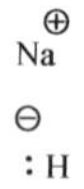 sodium hydride
pKa								
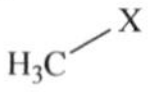 H$_3$C–X methyl	only S$_N$2	only S$_N$2	only S$_N$2	only S$_N$2	only S$_N$2	only S$_N$2	NA	NA
R–CH$_2$–X primary	S$_N$2 > E2	S$_N$2 > E2	S$_N$2 > E2	S$_N$2 > E2	S$_N$2 > E2	S$_N$2 > E2	always a base	always a base
R–CHR–X secondary	S$_N$2 > E2	S$_N$2 > E2	S$_N$2 > E2	S$_N$2 > E2	S$_N$2 > E2	S$_N$2 > E2	always a base	always a base
R$_3$C–X tertiary	only E2	only E2	only E2	only E2	only E2	only E2	always a base	always a base

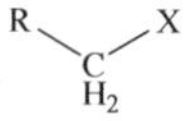

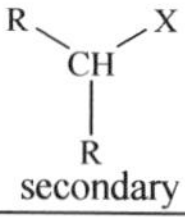

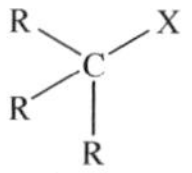

(handwritten left margin: Everything is E2, Too Sterically Hindered)

NaH, KH and NaNR$_2$ react only as bases in our course. Grignard reagents, RMgBr, do not undergo useful reactions with RBr compounds.

typical weak base nucleophiles are: (for our course)	$H-\overset{..}{\underset{..}{O}}-H$	$R-\overset{..}{\underset{..}{O}}-H$	acetic acid structure with O and $O-H$
H_3C-X methyl	no reaction	no reaction	no reaction
$R-\underset{H_2}{C}-X$ primary	no reaction	no reaction	no reaction
$R-\underset{\underset{R}{\mid}}{CH}-X$ secondary	$S_N1 > E1$ (makes alcohols)	$S_N1 > E1$ (makes ethers)	$S_N1 > E1$ (makes esters)
$R-\underset{\underset{R}{\mid}}{\overset{R}{C}}-X$ tertiary	$S_N1 > E1$ (makes alcohols)	$S_N1 > E1$ (makes ethers)	$S_N1 > E1$ (makes esters)

Discussed above

alcohol reactions in strong acid: (for our course)	$H-X$ (X = Cl, Br or I)	$X-\overset{\overset{..}{P}}{\underset{\underset{X}{\mid}}{}}-X$ phosphorous trihalide (X = Cl or Br)	H_2SO_4 Δ	1. TsCl / pyridine 2. NaBr
H_3C-OH methyl	S_N2	S_N2	not discussed	1. acyl substitution 2. S_N2r
$R-\underset{H_2}{C}-OH$ primary	S_N2	S_N2	E1	1. acyl substitution 2. S_N2r
$R-\underset{\underset{R}{\mid}}{CH}-OH$ secondary	S_N1	S_N1	E1	1. acyl substitution 2. S_N2r
$R-\underset{\underset{R}{\mid}}{\overset{R}{C}}-OH$ tertiary	S_N1	S_N1	E1	E1

Problem 30 – Suggest a mechanism for each of the following transformations. Predict major/minor products in each part.

Problem 31 – Consider all of the $C_6H_{13}Br$ compounds on pages 247-248. Do not consider stereoisomers. Use the numbers in the table. List any relevant isomers under each criteria below. (There are 17 possibilities.)

1. Isomers that can react fastest in S_N2 reactions
2. Isomers that give E2 reaction but not S_N2 with sodium methoxide
3. Isomers that react fastest in S_N1 reactions
4. Isomers that can react by all four mechanisms, S_N2, E2, S_N1 and E1 (What are the necessary conditions?)
5. Isomers that are completely unreactive with methoxide/methanol
6. Isomers that are completely unreactive with methanol, alone.

List of $C_6H_{13}X$ targets (X = Br, OH, OR, O_2CR, SH, SR, NH_2, CN, CCR, CO_2H, etc.)

Reaction Templates - sideways and vertical perspectives (either one will work)

$S_N2/E2$ (Nu: $^\ominus$ / B: $^\ominus$) always backside for S_N2 and usually anti C_α-H/C_β-X attack for E2

$S_N1/E1$ (H-Nu: / H-B:) - attack from either face of R^+ for both reactions (S_N1 and E1)

easier	**harder** (Chiral Centers)

iotopes of hydrogen
H = protium (proton)
D = deuterium
T = tritium

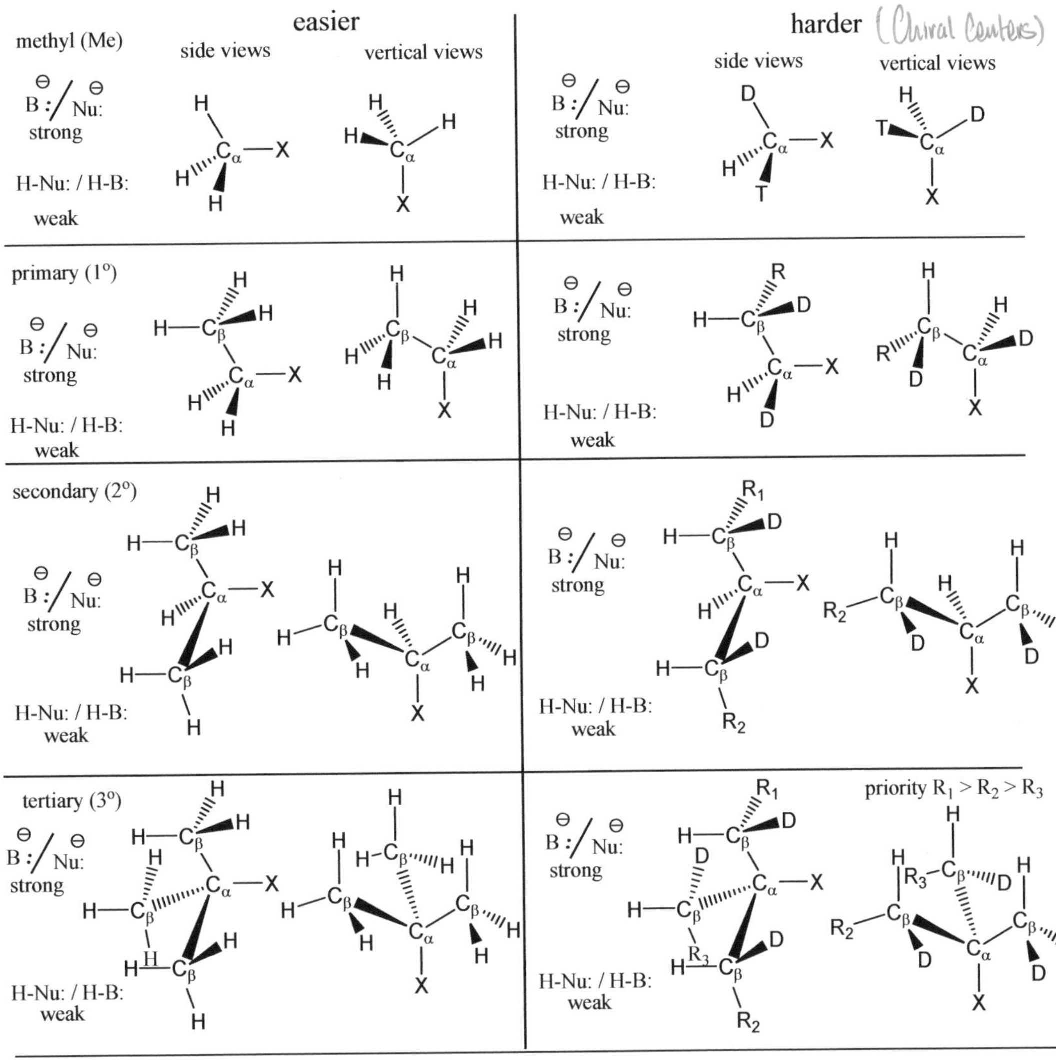

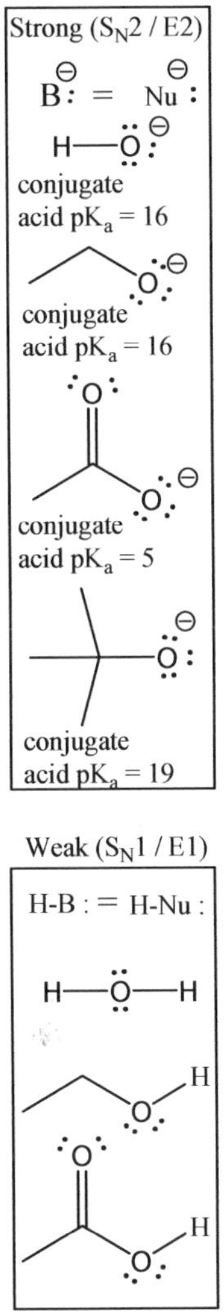

Strong (S_N2 / E2)

$B: ^\ominus = Nu: ^\ominus$

H—O: $^\ominus$
conjugate
acid $pK_a = 16$

conjugate
acid $pK_a = 16$

conjugate
acid $pK_a = 5$

conjugate
acid $pK_a = 19$

Weak (S_N1 / E1)

H-B : = H-Nu :

methyl (Me) side views vertical views

B : $^\ominus$ / Nu: $^\ominus$
strong

H-Nu: / H-B:
weak

primary (1°)

B : $^\ominus$ / Nu: $^\ominus$
strong

H-Nu: / H-B:
weak

secondary (2°)

B : $^\ominus$ / Nu: $^\ominus$
strong

H-Nu: / H-B:
weak

tertiary (3°)

B : $^\ominus$ / Nu: $^\ominus$
strong

H-Nu: / H-B:
weak

side views vertical views

B : $^\ominus$ / Nu: $^\ominus$
strong

H-Nu: / H-B:
weak

priority $R_1 > R_2 > R_3$

3-bromo-2-deuterio-4-methylhexane

R	S	S	R	R	S	R	S
R	S	R	S	S	R	R	S
R	S	R	S	R	S	S	R

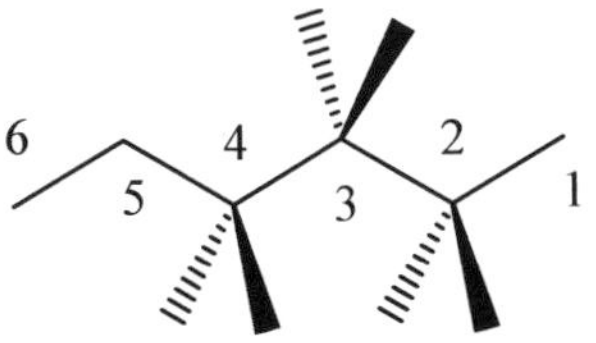

side view

vertical view

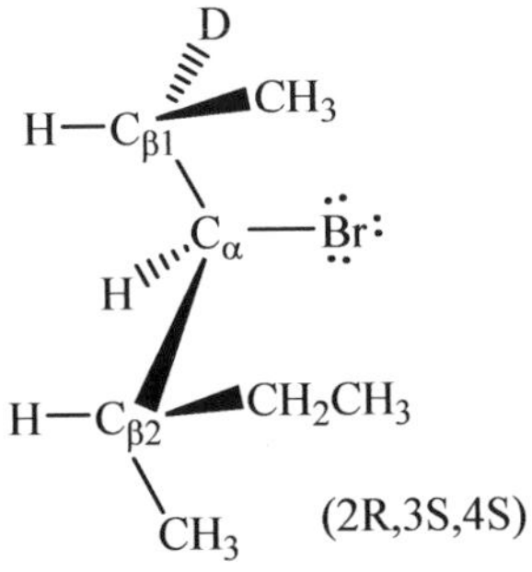

(2R,3S,4S)

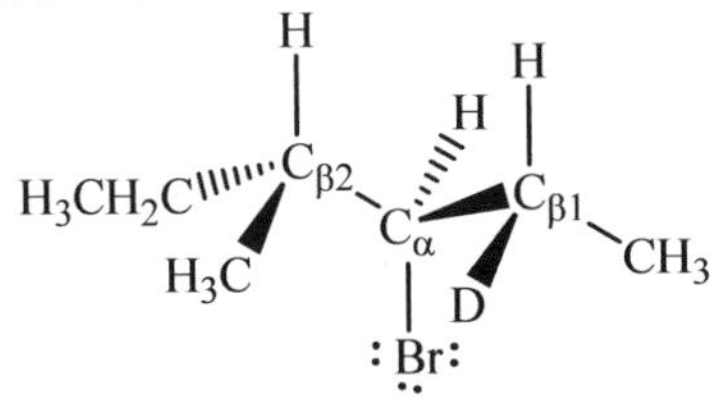

(2S,3R,4R)

6 4 2
5 3 1

template

Example Mechanisms shown below (S$_N$2 / E2).

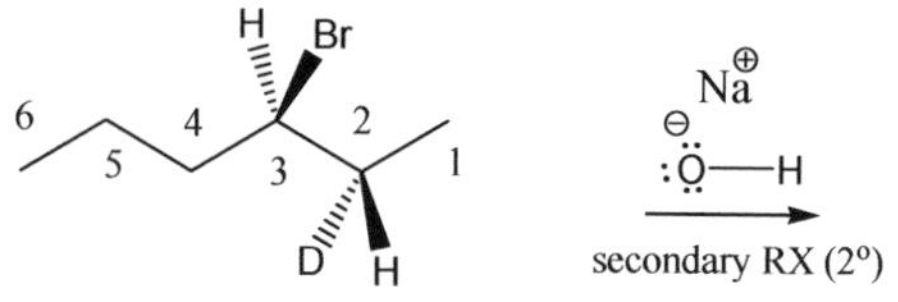

One S$_N$2 product and four E2 products.

1. strong Nu:$^\ominus$/B:$^\ominus$
2. 2° R-X

secondary RX (2°)

3R-bromo-2S-deuteriohexane

four possible E2 products

only one S$_N$2 product

One S$_N$2 product and three E2 products.

1. strong Nu:$^\ominus$/B:$^\ominus$
2. 2° R-X

secondary RX (2°)

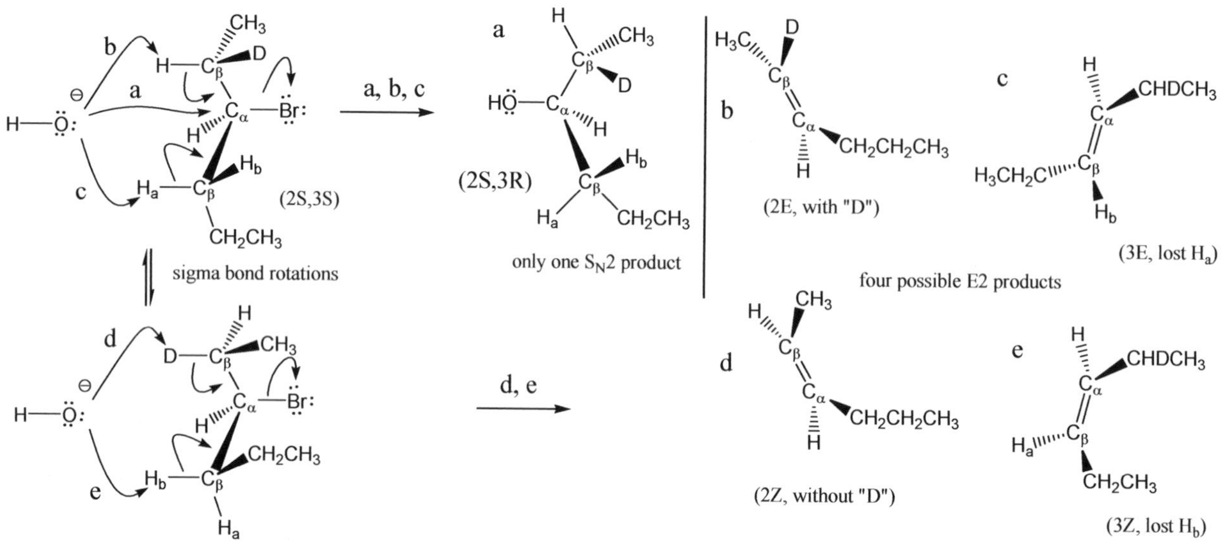

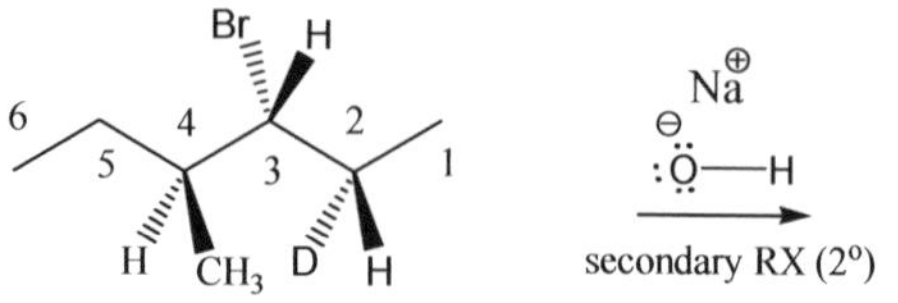

3R-bromo-2S-deuterio-4S-methylhexane   3 chiral centers = 8 possible stereoisomers

3 possible E2 products

only one S$_N$2 product

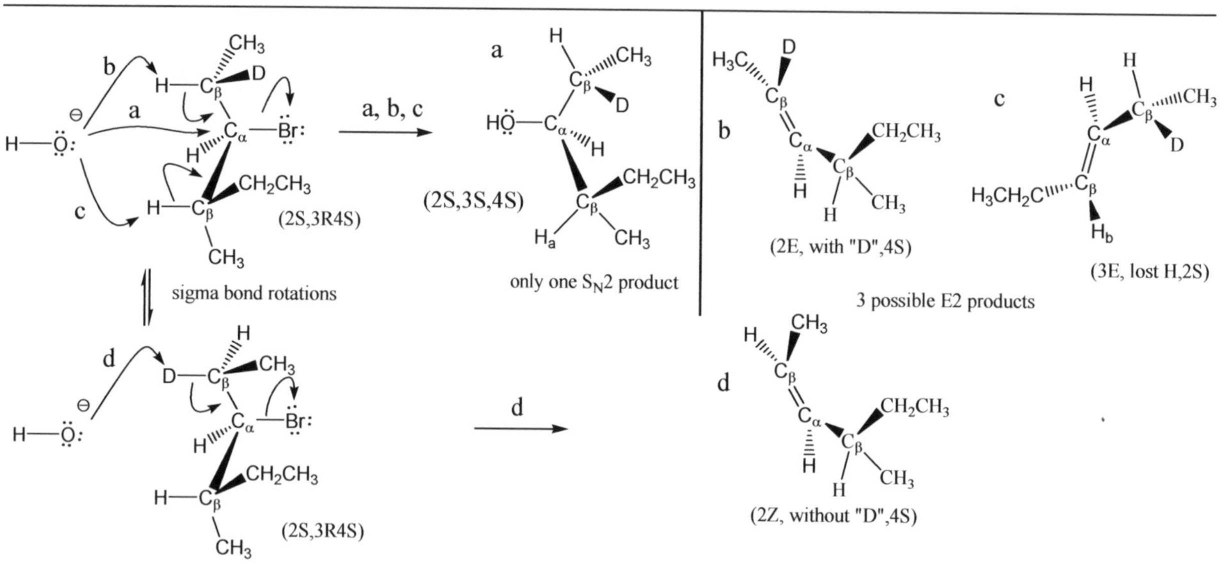

Chapter 9

Example Mechanisms shown below (S$_N$1 / E1) Only show reactions from the best carbocation.

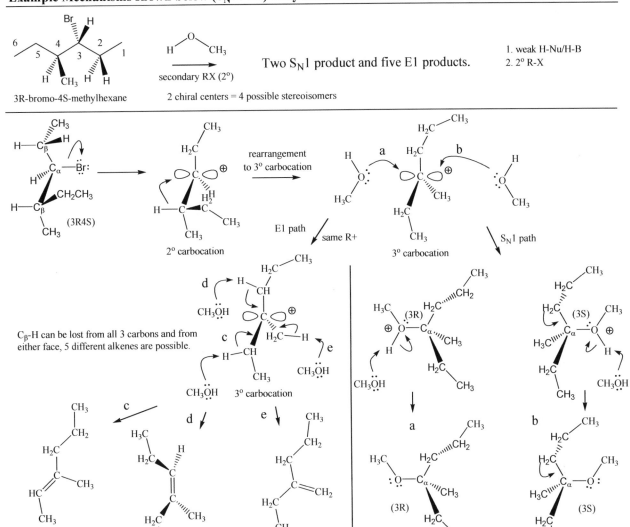

3R-bromo-4S-methylhexane

secondary RX (2°)

Two S$_N$1 product and five E1 products.

2 chiral centers = 4 possible stereoisomers

1. weak H-Nu/H-B
2. 2° R-X

rearrangement
to 3° carbocation

2° carbocation

3° carbocation

E1 path

same R+

S$_N$1 path

C$_\beta$-H can be lost from all 3 carbons and from either face, 5 different alkenes are possible.

3° carbocation

E and Z
3-methyl-2-hexene

E and Z
3-methyl-3-hexene

2-ethyl-1-pentene

two S$_N$2 product
(enantiomers)

Chapter 9

How does nature do it (S$_N$2 reactions)?

if you blaze before class

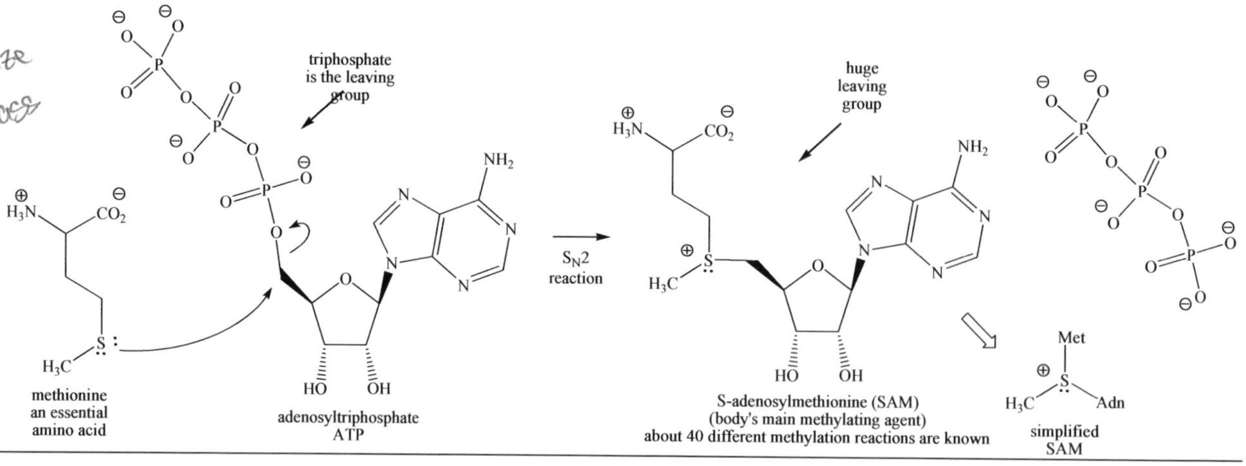

triphosphate is the leaving group

methionine
an essential
amino acid

adenosyltriphosphate
ATP

S$_N$2 reaction

huge leaving group

S-adenosylmethionine (SAM)
(body's main methylating agent)
about 40 different methylation reactions are known

simplified
SAM

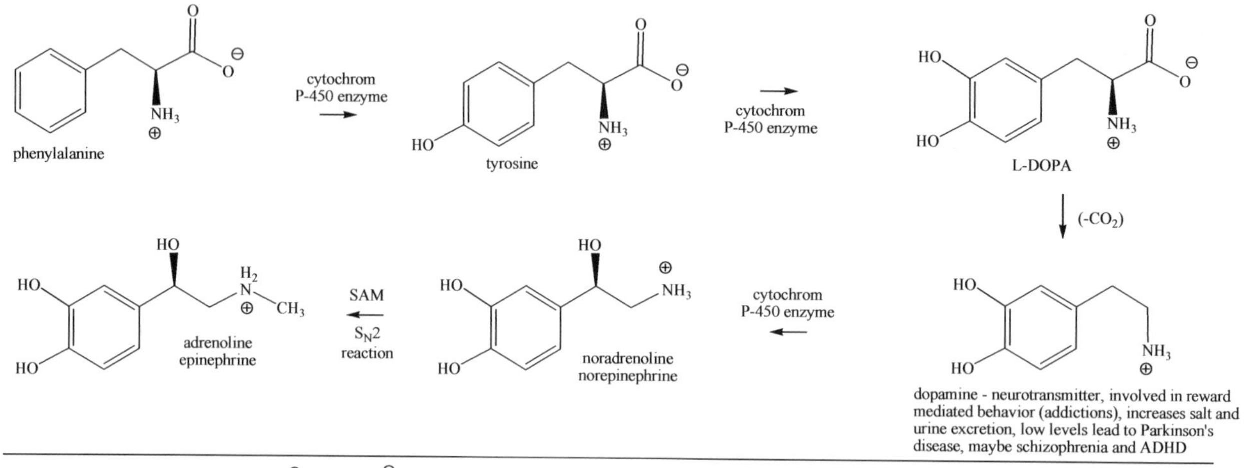

phenylalanine

cytochrom
P-450 enzyme

tyrosine

cytochrom
P-450 enzyme

L-DOPA

(-CO$_2$)

dopamine - neurotransmitter, involved in reward
mediated behavior (addictions), increases salt and
urine excretion, low levels lead to Parkinson's
disease, maybe schizophrenia and ADHD

adrenoline
epinephrine

SAM
S$_N$2
reaction

noradrenoline
norepinephrine

cytochrom
P-450 enzyme

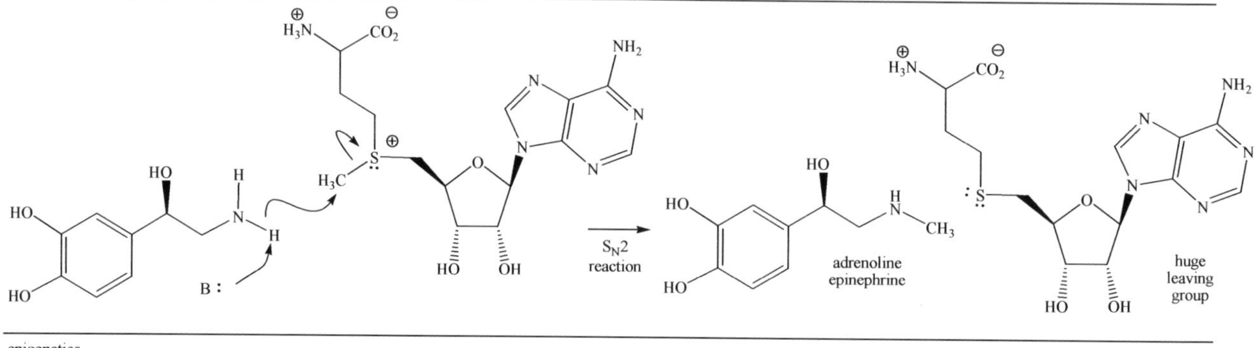

S$_N$2
reaction

adrenoline
epinephrine

huge
leaving
group

---

epigenetics

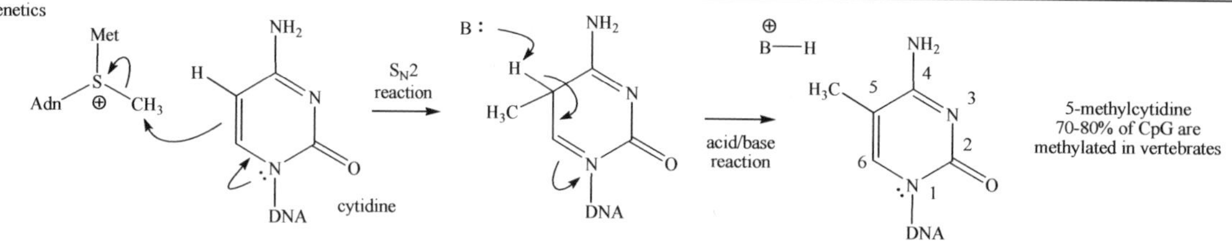

S$_N$2
reaction

acid/base
reaction

cytidine

DNA

5-methylcytidine
70-80% of CpG are
methylated in vertebrates

Chapter 9

Chapter 10 – Free Radcal Chemistry

   We will only emphasize two types of free radical reactions in this chapter: 1. halogen substitution reactions at sp^3 C-H bonds and 2. free radical addition of H-Br to alkene pi bonds (C=C). Both of these reactions follow a chain reaction sequence of 1. initiation, 2. propagation (2 steps) and 3. termination. Also important for both of these reactions is that they are regioselective and mainly react at predictable locations in an organic molecule.

   As *Newbies* to organic, the only drawback to this reaction is that the intermediates are free radicals, and the steps of the mechanisms involve one electron transfers. This is in contrast to the rest of the reactions presented in this book, which are polar reactions, involving two electron transfers (nucleophiles and electrophiles). The mechanism arrows in this chapter are half headed arrows, representing one electron movement, unlike the full headed arrows representing two electron movement.

   Free radical substitution at sp^3 C-H looks as follows (hυ = light, Δ = heat). More substituted carbon atoms tend to have weaker C-H bonds and react faster with the generated halogen free radicals. Bromine is especially selective for 3° C-H > 2° C-H > 1° C-H > methyl C-H, so we will only emphasize molecular bromine. Chlorine reacts in a similar manner, but is less selective. Fluorine reactions are too energetic (possible explosion) and iodine reactions are too sluggish to be practical.

   Free radical addition reactions follow a similar reaction sequence, 1. initiation, 2. propagation (2 steps) and 3. termination, however the overall result is that a bromine atom adds to the "less" substituted carbon and a hydrogen atom adds to the "more" substituted carbon atom of a C=C. This regioselectivity is referred to as "anti-Markovinkov addition to an alkene."

Typical Steps in Free Radical Substitution Mechanisms at sp^3 C-H bonds

   Initiation and termination occur at similar low rates (sometimes referred to as steady state conditions), while the propagation steps occur 100-1000s of times per initiation and termination steps. We will discuss each step of the mechanism in more detail later. A bare-bones mechanism is shown, just below. Bond energies are very important in deciding what will occur in each step of the mechanism, as weaker bonds react faster. We will use a "typical" bond energy table to evaluate each step of our mechanism. Notice that these are one electron transfers, so be careful to use half-headed arrows in free radical mechanisms to show one electron movement (also called fish-hook arrows). The mechanism is not difficult to learn, but it does require some practice to keep the details straight.

Chapter 10

1. **Initiation**: cleavage of weakest bond using light or heat (the halogen bond)

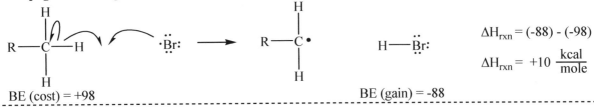

$\Delta H_{rxn} = +46 \dfrac{kcal}{mole}$

- -

2a. **Propagation**: halogen atom abstracts a hydrogen atom from a C-H bond, forming a carbon free radical

$\Delta H_{rxn} = (-88) - (-98)$

$\Delta H_{rxn} = +10 \dfrac{kcal}{mole}$

BE (cost) = +98   BE (gain) = -88

- -

2b. **Propagation**: carbon free radical abstracts a halogen atom from a halogen molecule, forming a halogen atom

start over

$\Delta H_{rxn} = (-68) - (-46)$

$\Delta H_{rxn} = -22 \dfrac{kcal}{mole}$

BE (cost) = +46   BE (gain) = -68

$\Delta H_{2a + 2b} = -12 \dfrac{kcal}{mole}$

- -

3. **termination**: combination of two reactive free radicals into one stable bond, ending two chain reaction sequences

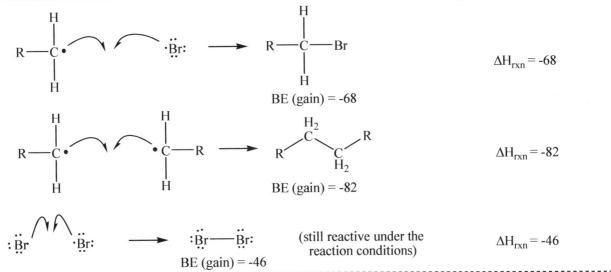

BE (gain) = -68

$\Delta H_{rxn} = -68$

BE (gain) = -82

$\Delta H_{rxn} = -82$

BE (gain) = -46

(still reactive under the reaction conditions)

$\Delta H_{rxn} = -46$

- -

    The sequence of steps depends mainly on the relative bond energies involved in each step, with the lowest energy choices followed in each step. Below is a table listing the most common types of bonds considered in these free radical reactions.

Bond Energy Table (We will use this table to decide the thermodynamics of each step of the reactions.)

Typical Substitution Pattern Bond Energy Table (X-Y = bond) $\left(\begin{array}{l}\text{Values in parentheses}\\\text{are estimated by me.}\end{array}\right)$ $\Delta H = \oplus$ (bond breaking)

$\Delta H = \ominus$ (bond making)

X ⇓ \ Y ⇒	H-	Me-	Et-	i-Pr	t-Bu	Ph	F-	Cl-	Br-	I-
CH_3- methyl	105	90	86	86	84	102	110	85	71	57
CH_3CH_2- primary	98	86	82	81	79	98	108	81	68	53
$(CH_3)_2CH-$ secondary	95	86	81	79	76	96	106	81	68	54
$(CH_3)_3C-$ tertiary	93	84	79	76	71	93	110	81	67	52
$CH_2=CHCH_2-$ allyl	86	74	70	70	67	(87)	(97)	68	54	41
$C_6H_5CH_2-$ benzyl	88	76	72	71	70	90	(100)	72	58	48
$CH_2=CH-$ vinyl	110	100	96	95	90	103	(120)	90	78	(62)
C_6H_5- phenyl	111	102	97	96	93	115	126	96	80	65
H- hydrogen	104	105	98	95	93	111	136	103	88	71

These are the types of bonds you will need to consider in our discussions below.

↓

As reactants

$:\ddot{Br}\!\!-\!\!\ddot{Br}:$

$-\overset{|}{\underset{|}{C}}\!\!-\!\!H$

$-\overset{|}{\underset{|}{C}}\!\!-\!\!\overset{|}{\underset{|}{C}}-$

As products

$H\!\!-\!\!\ddot{Br}:$

$-\overset{|}{\underset{|}{C}}\!\!-\!\!\ddot{Br}$

Molecular Halogen Bond Energies (kcal/mole)

F—F	Cl—Cl	Br—Br	I—I
37	58	46	36

Free Radical Substitution reactions at sp³ C-H

The energy required to break the molecular halogen bond comes from a photon (hν) of sufficient energy or heating (Δ) to a temperature with sufficient energy or the combination of both.

We can show bond breakage with simple sigma/sigma-star (σ/σ*) molecular orbitals. The input of energy from a photon with matching energy or very high thermal energy can excite one of the sigma bonding electrons into a sigma-star antibonding molecular orbital (MO). The bond order drops from one to zero and allows the formerly bonded atoms to separate in a very high energy state.

Sigma Bond Representation

antibonding MO   σ*____

bonding MO   σ ⇅____

hν or Δ
energy input excites electron
to antibonding orbital

σ* ↓____

σ ↑____

bond order = $\dfrac{(2)-(0)}{2} = 1$

bond order = $\dfrac{(1)-(1)}{2} = 0$

Arrow pushing representation $:\ddot{Br}\!\!-\!\!\ddot{Br}:$ hν or Δ ⟶ $:\ddot{Br}\cdot$ $\cdot\ddot{Br}:$

Chapter 10

Problem 1 - Why don't the reaction conditions (hν and/or Δ) break C-H or C-C bonds? What are the respective bond energies of those bonds? Use the following structures to explain your answer. (Use the bond energy table.)

$$Br—Br$$

1. Initiation

Only two reactants are typically present at the start of a free radical substitution reaction: halogen molecules (usually chlorine or bromine, for us) and a structure with alkane functionality (sp^3 C-H bonds). A relatively inert solvent may also be present (CCl_4, $CHCl_3$ or CH_2Cl_2 are commonly used), but we ignore this. The initial step is bond breakage of the weakest bond. The weakest bond is easily determined from a bond energy table (a partial table is provided on the previous page). Bromine molecules have weak bonds (46 kcal/mole). The alkane reactants have stronger bonds, both sp^3 C-H bonds (ranging from 86–105 kcal/mole) and sp^3 C-C bonds (ranging from 70-90 kcal/mole). It is clear that the weakest bond is the halogen bond and it is, indeed, the first bond to react. The initiation step generates halogen free radicals (halogen atoms) via homolytic bond breakage with energy input from light or heat. This starts two free radical chain reaction sequences.

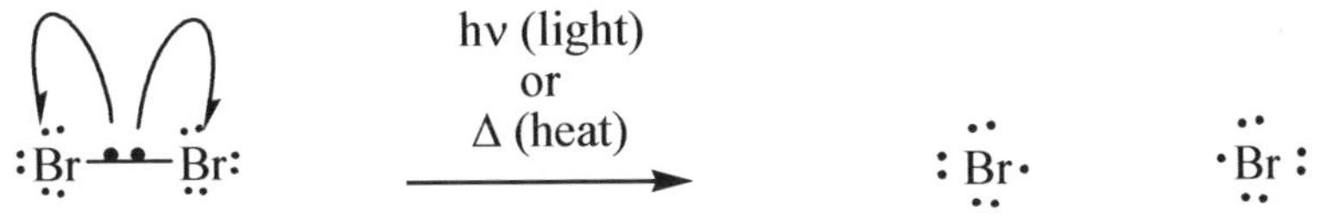

BE (cost) = -(-46) = +46

2. Propagation Steps

After the initial halogen bond is broken, the most common choice is for the two free radicals to simply reform the broken bond (overall, that would be "no reaction"). Free radicals are extremely reactive and caged in by other reactant molecules (alkane and halogen molecules).

Once in awhile the two free radicals will get separated from one another as a result of molecular collisions. When this occurs a free radical will be surrounded, mostly by the two types of reactant molecules. As a result, a free radical will have two common choices for reaction: 1. react with a halogen molecule (bromine, here) or 2. react with an alkane (ethane, here).

Because of their very low concentration and bimolecular kinetics, it is less common for two free radicals to combine with one another. When they do encounter one another, the reaction is highly favorable and terminates two free radical chain reaction sequences.

The two steps of the propagation sequence (steps 2a and 2b) generate the observed reaction products. Free radicals are normally very reactive, high energy intermediates. They need another electron for a complete octet, and generally acquire this electron by stealing an atom and one electron from two bonded atoms (sigma electrons here). The cost in energy is decided by the bond that is broken and the gain in energy decided by the bond that is formed, which allows a $\Delta H°_{rxn}$ to be calculated for each step. The overall heat of reaction ($\Delta H°_{rxn}$) will be the bond energy of the bond formed (a negative value) minus the bond energy of the bond broken (negative of a negative equals a positive value). The two step propagation sequence sustains free radicals through 100s – 1000s of steps. (We assume $\Delta G \approx \Delta H$.)

2a. There are several possible choices for the bromine atom (react with H, C or Br), but the one that leads to products in the first step of the propagation sequence (2a) is shown below (energetically most favorable). Assume a photochemical reaction run at 300 K. The temperature will affect our K_{eq} calculation.

i. A bromine atom can react with the hydrogen atom of a C-H bond in an ethane molecule. An H-Br molecule and a primary carbon free radical are formed. This is an unfavorable change, as ΔH_{rxn} is positive (endothermic). If this was the only step of propagation, the reaction would not work. Fortunately, there is a very favorable second step that completes the overall substitution reaction and that step is very favorable.

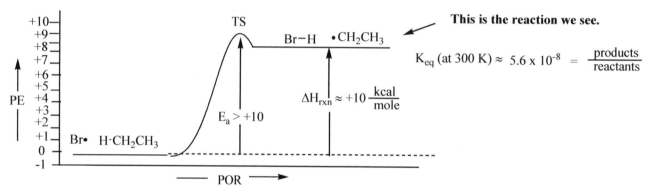

$$\Delta H_{rxn} = -88 - (-98)$$

$$\Delta H_{rxn} = +10 \frac{kcal}{mole}$$

$$K_{eq} = 10^{\frac{-\Delta G}{2.3RT}} = 10^{\frac{-10,000}{2.3(2)(300)}} = 10^{-7.25} = 5.6 \times 10^{-8} = \frac{products}{reactants}$$

PE vs. POR diagram for the observed reaction (T = 27°C = 300 K)

Reaction Coordinate for hydrogen atom abstraction from a C-H bond.

This is the reaction we see.

$$K_{eq} \text{ (at 300 K)} \approx 5.6 \times 10^{-8} = \frac{products}{reactants}$$

$$\Delta H_{rxn} \approx +10 \frac{kcal}{mole}$$

$$E_a > +10$$

2b. The second step of the propagation sequence: choices of a carbon free radical.

A carbon free radical is generated in the first step of the propagation sequence. It too needs an electron to complete its octet. It gains this by abstracting a bromine atom from Br_2, forming a C-Br bond and a bromine atom, which starts the process all over again (a chain reaction). The reaction works because of the weak Br-Br bond energy. Other choices (react with C or H) are less energetically favorable and we are ignoring them.

Chapter 10

Reaction choice of the carbon free radical

ii. React with a bromine molecule is very favorable (exothermic). We observe this one.

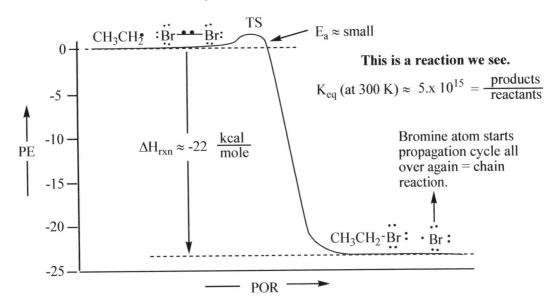

$$\Delta H_{rxn} = -68 - (-46)$$

$$\Delta H_{rxn} = -22 \ \frac{kcal}{mole}$$

$$K_{eq} = 10^{\frac{-\Delta G}{2.3RT}} = 10^{\frac{+22000}{2.3(2)(300)}} = 10^{+15.7} = 5. \times 10^{15} = \frac{products}{reactants}$$

- -

PE vs POR diagram for the most favorable reaction possibility in step 2b

Reaction Coordinate for ethyl abstraction of a bromine atom from a bromine molecule.

This is a reaction we see.

$$K_{eq} \ (at \ 300 \ K) \approx 5. \times 10^{15} = \frac{products}{reactants}$$

Bromine atom starts propagation cycle all over again = chain reaction.

$$\Delta H_{rxn} \approx -22 \ \frac{kcal}{mole}$$

$E_a \approx$ small

The second step of the propagation sequence produces a halogen free radical, which is exactly what started the first step of the propagation sequence, and the whole propagation process starts over. These two steps are repeated 100s-1000s of times for each initiation step. The free radical concentrations are so low that, statistically, it takes a relatively long time for two free radicals to find one another (from the point of view of molecular collisions). Ultimately, two free radicals will encounter one another and combine (various ways are possible), releasing a large amount of energy equal to the bond energy being formed (negative energy value). This leads to the final step of the mechanism, called termination, and shuts down two chain reactions.

3. Termination

In the termination step, two very reactive free radicals combine, stopping two chain reaction sequences. We can imagine the two free radicals combining three different ways, though disproportionation to an alkane and an alkene is another reaction possibility. Disproportionation is shown below, but not discussed. The termination reactions are balanced by the initiation step to maintain a steady state concentration of chain reactions that are running at any one time.

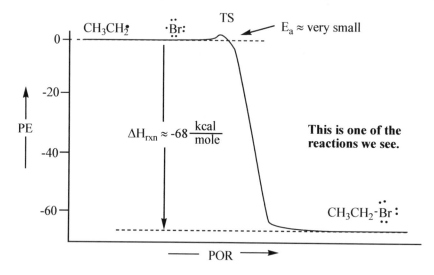

$$CH_3CH_2 \,\widehat{}\,\widehat{}\, \cdot \ddot{B}r\colon \longrightarrow CH_3CH_2\text{-}\ddot{B}r\colon \qquad\qquad \Delta H_{rxn} = BE = -68\ \frac{kcal}{mole}$$

$$CH_3CH_2 \,\widehat{}\,\widehat{}\, \cdot CH_3CH_2 \longrightarrow CH_3CH_2\text{-}CH_2CH_2 \qquad\qquad \Delta H_{rxn} = BE = -82\ \frac{kcal}{mole}$$

$$\colon\ddot{B}r \,\widehat{}\,\widehat{}\, \cdot\ddot{B}r\colon \longrightarrow \colon\ddot{B}r\text{---}\ddot{B}r\colon \text{ (still reactive)} \qquad \Delta H_{rxn} = BE = -46\ \frac{kcal}{mole}$$

disproportionation - not discussed

$$CH_3CH_2 \,\widehat{}\,\widehat{}\, H\text{-}\overset{H_2}{\underset{\cdot}{C}}\text{---}CH_2 \longrightarrow CH_3CH_2\text{-}H \quad H_2C\!\!=\!\!CH_2 \quad \Delta H_{rxn} \approx -64\frac{kcal}{mole}$$

alkane   alkene

PE vs POR diagram for the first termination example above in step 3

Reaction Coordinate for a termination reaction. The other possibilities look very similar.

CH$_3$CH$_2$ ·$\ddot{B}$r: TS $E_a \approx$ very small

$\Delta H_{rxn} \approx -68\ \frac{kcal}{mole}$

This is one of the reactions we see.

CH$_3$CH$_2$-$\ddot{B}$r:

PE (−20, −40, −60 axis)

POR

Overall Picture of the propagation sequence - PE vs POR diagram

Every time the chain reaction cycles through the above propagation steps it's going to liberate about 12 kcal/mole of energy that will show up mostly as heat to be dissipated. Because the first step in the propagation sequence for bromine is positive, the bromine atom is very selective about which hydrogen atom it reacts with (a good thing for us), and this decides what part of the alkane molecule reacts (its regioselectivity). It chooses the weakest C-H bond to react with at the fastest rate. The second step occurs very fast and completes the propagation sequence, ready to cycle through again.

These are the chain reaciton steps (2a + 2b) for bromination.

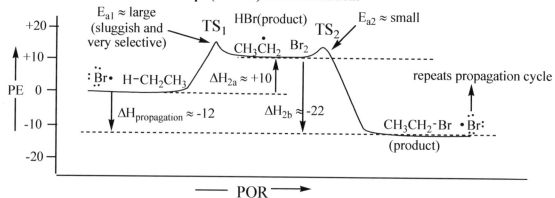

$E_{a1} \approx$ large (sluggish and very selective)   TS$_1$ HBr(product) $E_{a2} \approx$ small   TS$_2$

:$\ddot{B}$r· H-CH$_2$CH$_3$ CH$_3$$\dot{C}H_2$ Br$_2$

$\Delta H_{2a} \approx +10$

repeats propagation cycle

$\Delta H_{propagation} \approx -12$ $\Delta H_{2b} \approx -22$ CH$_3$CH$_2$-Br ·$\ddot{B}$r:

(product)

PE (+20, +10, 0, −10, −20 axis)

POR

Additional Reaction Products - Polysubstitution

As the initial product concentration begins to build up, it becomes a reactant that can be attacked by another bromine atom. This is one of the limitations of trying to use bromination as a useful synthetic reaction. Every time an additional bromine substitutes for a hydrogen atom, the yield goes down and the product is different. If this were done on an industrial scale, the various products could be fractionally distilled and separated as different commercial products to be sold. But on the small scale of a lab reaction, this would not be that useful if many different products are made. Additional possible products for ethane are shown in the flow chart below.

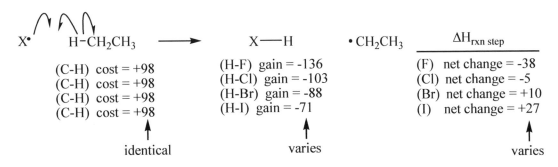

Other Halogenation Reactions

The mechanism is conceptually similar for all of the other halogen molecules. However, only chlorination and bromination are typically used. Thermodynamics will provide an explanation for the differences (by comparing bond energies). The reaction scheme below is almost identical to the mechanism discussed above, except a generic halogen molecule, X-X, is used to symbolize all of the halogen molecules (F_2, Cl_2, Br_2 and I_2).

Thermodynamics of the three sequences of alkane halogenations (kcal/mole)

1. Initiation

$$X—X \xrightarrow[\Delta \text{ (heat)}]{\substack{h\nu \text{ (light)} \\ \text{or}}} X\cdot \quad \cdot X$$

ΔH_{rxn} (bond energy) =

	F_2	Cl_2	Br_2	I_2
	+38	+58	+46	+36

All of these bonds are easy to break and the energy to do so is supplied by light or heat.

2. Propagation

a.

$$X\cdot \quad H-CH_2CH_3 \longrightarrow X—H \quad \cdot CH_2CH_3 \qquad \underline{\Delta H_{rxn \text{ step}}}$$

(C-H) cost = +98	(H-F) gain = -136	(F) net change = -38
(C-H) cost = +98	(H-Cl) gain = -103	(Cl) net change = -5
(C-H) cost = +98	(H-Br) gain = -88	(Br) net change = +10
(C-H) cost = +98	(H-I) gain = -71	(I) net change = +27

identical   varies   varies

b.

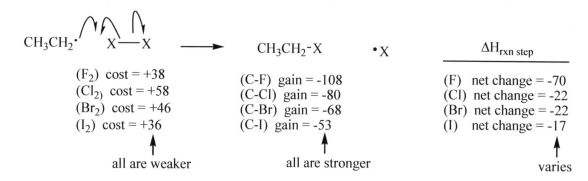

(F$_2$) cost = +38   (C-F) gain = -108   (F) net change = -70
(Cl$_2$) cost = +58   (C-Cl) gain = -80   (Cl) net change = -22
(Br$_2$) cost = +46   (C-Br) gain = -68   (Br) net change = -22
(I$_2$) cost = +36   (C-I) gain = -53   (I) net change = -17

all are weaker   all are stronger   varies

Overall ΔH for both steps (2a + 2b)

$$\underline{\Delta H_{rxn\ steps\ 2a\ +\ 2b}}$$

(F) both steps = -38 - 70 = -108 (too exothermic)
(Cl) both steps = -5 - 23 = -28 (very exothermic)
(Br) both steps = +10 - 22 = -12 (exothermic)
(I) both steps = +27 - 17 = +10 (endothermic)

3. Termination steps are all energetically favorable. The problem for free radicals is finding one another because their concentrations are so low.

$$\underline{\Delta H_{rxn\ step}}$$

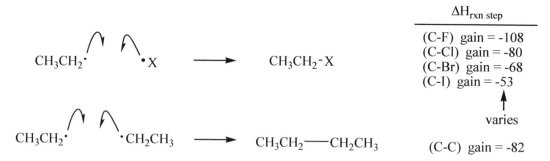

(C-F) gain = -108
(C-Cl) gain = -80
(C-Br) gain = -68
(C-I) gain = -53

varies

(C-C) gain = -82

Problem 2 - How do the results above explain the observations below? Potential Energy vs. Product of Reaction diagrams for the propagation steps are shown on the next page (PE vs POR diagrams).

a. Fluorination reactions react explosively and are totally unselective of the type of C-H bond reacted with.

b. Chlorination reactions are vigorous and only mildly selective of the type of C-H bond reacted with.

c. Bromination reactions work, but are sluggish and very selective of the type of C-H bond reacted with.

d. Iodination reactions are unfavorable and do not work with any type of C-H bond.

Chapter 10

The following PE vs. POR diagrams show the energy changes for all of the halogen (Br$_2$ works best).

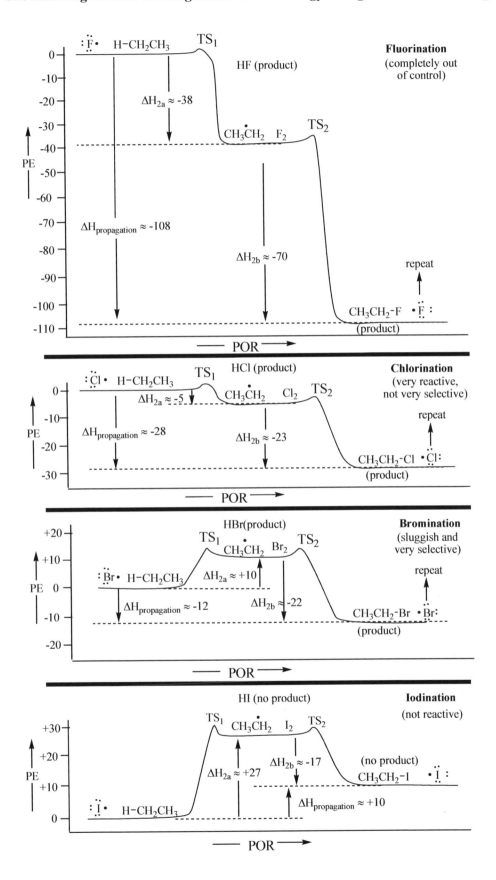

Problem 3 - Set up a table using methane, similar to the above calculations for ethane. Show each step of the free radical substitution mechanism using $X_2 = F_2$, Cl_2, Br_2, I_2. Analyze the energetics for each step. Use X_2 for your general examples to provide an arrow pushing mechanism for each step of free radical halogenation of methane. Calculate a ΔH_{rxn} for each step of your mechanisms, using the actual bond energies.

Step 1

Calculated ΔH_{Rxn}'s

X = F

X = Cl

X = Br

X = I

Step 2a

Calculated ΔH_{Rxn}'s

X = F

X = Cl

X = Br

X = I

Step 2b

Calculated ΔH_{Rxn}'s

X = F

X = Cl

X = Br

X = I

Step 3

Calculated ΔH_{Rxn}'s

X = F

X = Cl

X = Br

X = I

Free Radical Substitution Reactions of Generic Alkanes

A typical alkane could have several different types of C-H bonds that could undergo free radical substitution. Attack of any C-H bond can be viewed in a similar way. However, in the same molecule, differences in rate occur at each type of C-H because there are differences in C-H bond energies. The weaker C-H bonds will be attacked more readily (faster) and, when present, will preferentially react over a stronger C-H bond. We must also take into account how many of each type of C-H bond is present (a CH_3 has 3 positions, a CH_2 has 2 positions and a CH has 1 position, a concentration effect).

Typical hydrocarbon C-H bond energies

Type of R group	C-H bond	Homolytic Bond Energy
methyl	CH_3-H	+104
primary	CH_3CH_2-H	+98
secondary	$(CH_3)_2CH$-H	+95
tertiary	$(CH_3)_3C$-H	+92
allylic	$CH_2=CH-CH_2$-H	+86
benzylic	$C_6H_5CH_2$-H	+88
acetyl	$R-\overset{\overset{\displaystyle O}{\|\|}}{C}{\sim}H$	+86
vinyl	$CH_2=CH$-H	+110
phenyl	C_6H_5-H	+111
acetylenic	$R-C{\equiv}C-H$	+124

Lower bond energies are easier to break.

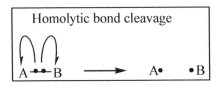

Homolytic bond cleavage

$A{-}{\bullet}{\bullet}{-}B \longrightarrow A\bullet \quad \bullet B$

Rate Law for each propagation step.

Rate = k $[R\bullet]^a[M]^b$

$k = -2.3RT \log(E_a)$
(an energy and temperature term)

$[R\bullet]^a[M]^b$ (concentration terms)

$R^\bullet$ = free radical component
M = molecular component
a = b = 1, in this reaction

C-H Bond Energies

The key mechanistic step is hydrogen atom abstraction from a C-H bond. This is where the decision is made about which C-H bond will react.

$$—\overset{\|}{\underset{\|}{C}}—H \quad \bullet X \quad \xrightarrow{\Delta H_{rxn}} \quad —\overset{/}{\underset{\backslash}{C}}\bullet \quad H—X$$

The differences in ΔH_{rxn} are a result of differences in C-H bond energies.

In our two examples below, using the same structure as an identical starting point in energy allows us to make an accurate comparison in the relative stability of different free radicals. The first comparison below shows how much more stable a secondary free radical is than a primary free radical (it shows the difference in bond energies). The second comparison shows how much more stable a tertiary free radical is than a primary free radical. Since the secondary and tertiary free radicals are both compared to a primary free radical, we can estimate how much more stable a tertiary free radical is than a secondary free radical.

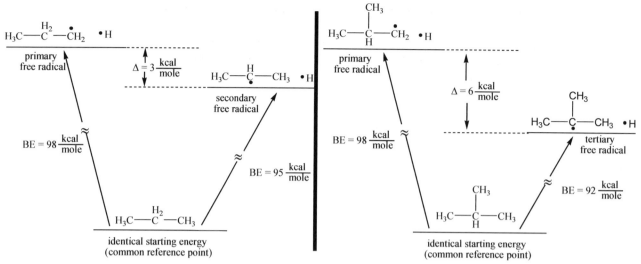

Relative stability of carbon free radicals (R•): 3° (BE ≈ 92) > 2° (BE ≈ 95) > 1° (BE ≈ 98) > CH_3 (BE = 105)

(All bond energies in kcal/mole)

Typical relative reactivity values from many experiments for chlorination and bromination show the following results.

relative rates for chlorination =	$\dfrac{\text{tertiary C-H}}{\text{secondary C-H}}$ = $\dfrac{5.1}{4.0}$
	primary C-H 1.0

relative rates for bromination =	$\dfrac{\text{tertiary C-H}}{\text{secondary C-H}}$ = $\dfrac{1600}{80}$
	primary C-H 1.0

Problem 4 - Use this data to predict the relative amounts of chlorinated product and brominated product for propane and 2-methylpropane. (Amount = [#H]x[relative rate]). Which halogen is more selective?

Problem 5 (p 11) - How many different types of hydrogen atoms are in each of the following molecules? Which type of hydrogen atom in each molecule is most reactive? How did you decide this? What is the maximum number of monosubstituted products that could form if each type of hydrogen atom was substituted in each molecule? Use the given table of relative rates (partly real, partly made up data) to show the product distribution for chlorine and bromine. We won't consider stereoisomers here. Using the most reactive hydrogen atom in the first molecule, write out a mechanism for formation of the major reaction product, using bromine as the halogen reactant. (Amount = [#H]x[relative rate]).

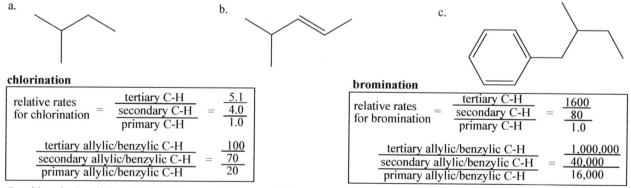

a.

b.

c.

chlorination

relative rates for chlorination =	$\dfrac{\text{tertiary C-H}}{\text{secondary C-H}}$ = $\dfrac{5.1}{4.0}$
	primary C-H 1.0
$\dfrac{\text{tertiary allylic/benzylic C-H}}{\text{secondary allylic/benzylic C-H}}$ =	$\dfrac{100}{70}$
primary allylic/benzylic C-H	20

bromination

relative rates for bromination =	$\dfrac{\text{tertiary C-H}}{\text{secondary C-H}}$ = $\dfrac{1600}{80}$
	primary C-H 1.0
$\dfrac{\text{tertiary allylic/benzylic C-H}}{\text{secondary allylic/benzylic C-H}}$ =	$\dfrac{1,000,000}{40,000}$
primary allylic/benzylic C-H	16,000

Consider vinyl and phenyl C-H to have a relative rate of 0.

Chapter 10

Inductive and Steric Effects – Possible Reasons for Different Stabilities of Free Radicals

Because a free radical has less than the desired octet of electrons, it is considered electron deficient. Anything that can donate more electron density to the deficient free radical should help stabilize it. Additional carbon groups are inductively electron donating (relative to hydrogen) through the sigma bond framework. More alkyl substituents at a free radical center can share some of their electron density with the electron deficient center via a donating inductive effect. The estimation of electron donation or withdrawal is typically made on the basis of relative electronegativities. However, care has to be taken when resonance effects are also present.

Sigma electrons are polarized closer to the electron deficient carbon atom.

Additional sigma electrons are all polarized closer to the electron deficient carbon atom. Alkyl groups are typically considered to be inductively donating.

Free Radicals

A methyl free radical only has three bonds (six additional electrons) that it can polarize towards the free radical carbon atom to help stabilize its electron deficient condition.

Each "R" group supplies many additional electrons that can be polarized towards the free radical carbon atom to help stabilize its electron deficient condition.

A similar argument was used to rationalize carbocation stabilities. The inductive effect is even more important with carbocations, since these reactive intermediates are missing two electrons.

Sigma electrons are polarized closer to the electron deficient carbon atom.

Additional sigma electrons are all polarized closer to the electron deficient carbon atom. Alkyl groups are typically considered to be inductively donating.

Carbocations

A methyl carbocation only has three bonds (six additional electrons) that it can polarize towards the carbocation atom to help stabilize its electron deficient condition.

Each "R" group supplies many additional electrons that can be polarized towards the carbocation atom to help stabilize its electron deficient condition.

The inductive effect is even more important for stabilizing a carbocation than a carbon free radical because the carbocation is more electron deficient.

A hyperconjugation effect is also used to explain these differences in stability of free radicals and carbocations. We can approximate this as sigma resonance effects.

one electron delocalization in a free radical

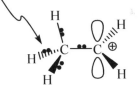

two electron delocalization in a carbocation

The differences in relative carbocation stabilities parallel the trend seen in free radicals, but are greater than the differences in free radical stabilities, as seen in relative energies (kcal/mole). Tertiary is referenced as zero in the table below and the less stable substitution patterns listed as positive energies above that value (= 0).

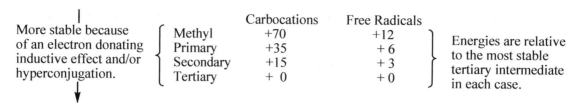

		Carbocations	Free Radicals	
More stable because of an electron donating inductive effect and/or hyperconjugation.	Methyl	+70	+12	Energies are relative to the most stable tertiary intermediate in each case.
	Primary	+35	+ 6	
	Secondary	+15	+ 3	
	Tertiary	+ 0	+ 0	

A steric explanation is also possible.

When a hydrogen atom is abstracted there is less crowding of the remaining three groups (sp³ → sp²). This will be better for 3° CH than 2° CH than 1° CH than methyl CH. There should be a smaller E_a and a faster rate with larger "R" groups.

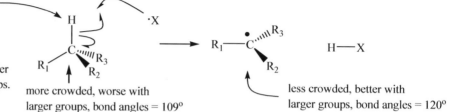

more crowded, worse with larger groups, bond angles = 109°

less crowded, better with larger groups, bond angles = 120°

Anomolous Bond Energies

There were a few anomalous C-H bond energies in the bond energy table. For the typical alkane sp³ C-H bonds discussed in this topic, the bond energies range from 92-105 kcal/mole. All of the examples outside this range have pi bonds somewhere in the structure. One group has C-H bonds that are weaker (less) than the range above and one group has C-H bonds that are stronger (more) than this range. Why do we see these deviations?

a. Resonance Stabilization – makes intermediates more stable

According to information from our bond energy table, a primary free radical is 3 kcal/mole less stable than a secondary free radical and 6 kcal/mole less stable than a tertiary free radical. Two of the primary free radicals in the bond energy table are actually more stable than tertiary free radicals. As is often the case in organic chemistry, this is because of electron delocalization, or what we call "resonance".

Type of R group	C-H bond	Homolytic Bond Energy	
primary (1°)	CH_3CH_2-H	+98	Typical primary BE
secondary (2°)	$(CH_3)_2CH$-H	+95	
tertiary (3°)	$(CH_3)_3C$-H	+92	compare...
allylic (1°)	CH_2=CH-CH_2-H	+86	more stable than expected by 12 $\frac{kcal}{mole}$
benzylic (1°)	$C_6H_5CH_2$-H	+88	more stable than expected by 10 $\frac{kcal}{mole}$

	primary	secondary	tertiary	allyl	benzyl
	$R—CH_2^{\bullet}$	$R—CH^{\bullet}$ with R below	$R—C^{\bullet}$ with R above and below	$H_2C=CH—CH_2^{\bullet}$	benzyl radical
Relative stabilities compared to the tertiary free radical =	+6	+3	0	-6	-4

Chapter 10

Both the allyl and benzyl free radicals can delocalize the unpaired electron in the 2p orbital into their adjacent pi systems. There is a slight difference in drawing the resonance structures of free radicals, in that we have to remember to use half-headed arrows to show the one electron movement. That means we use more arrows, since we need one arrow for every single electron, rather than one arrow for every pair of electrons.

2D representations

Delocalization of electron density (resonance) is always stabilizing. Each structure requires three half-headed arrows to show resonance of free radicals. Remember, arrows show how electrons move to generate the "next" structure(s). The last structure won't have any arrows.

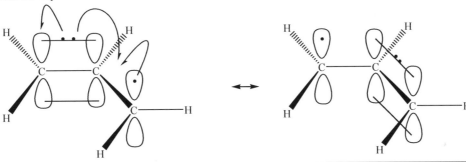

allyl **free radical** 3D representation of resonance

compare allyl **carbanion** 3D representation of resonance | compare allyl **carbocation** 3D representation of resonance

b. More electronegative hybrid orbitals make it harder to abstract a hydrogen atom

Three bond energies from the table are much higher than the others: vinyl C-H (+110), phenyl C-H (+111) and acetylenic C-H (+124). As we learned in the acid/base topic, among atoms of the same type (all carbon atoms here), an atom's relative electronegativity is dependent on the amount of 2s character in the hybrid orbital [2s (100% s) > sp (50% s) > sp^2 (33% s) > sp^3 (25% s) > 2p (0% s)]. The relative electronegativity of hybridized carbon increases with increasing percent 2s contribution: sp > sp^2 > sp^3. In homolytic C-H bond cleavage an electron is removed from one of these orbitals. This should be more difficult as the electronegativity of the carbon atom increases. That is exactly what we observe in the bond energies below. The sp C-H bond has the tightest hold on its electrons (largest bond energy = 124), followed by the sp^2 C-H bonds (= 110), which are tighter than the sp^3 C-H electrons (lowest bond energy = 95). We will never propose free radical substitution reactions at vinyl, phenyl or acetylenic C-H bonds, rather we will only show free radical attack at sp^3 C-H. Remember, the pK_a's of the C-H bonds is exactly opposite to the bond energies. An sp C-H is the most acidic and has the lowest pK_a (=25). That's because sp orbitals hold onto their electrons the tightest.

Typical sp³ C-H bond energies » 92-105 $\frac{kcal}{mole}$

sp³ carbon atom

sp² carbon free radical
(carbon rehybridizes)

H—X

This is the
reaction we
observe.

Typical sp² C-H bond energies » 110-111 $\frac{kcal}{mole}$

a. vinyl

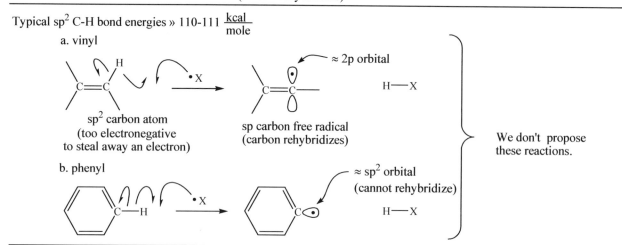

sp² carbon atom
(too electronegative
to steal away an electron)

≈ 2p orbital

sp carbon free radical
(carbon rehybridizes)

H—X

b. phenyl

≈ sp² orbital
(cannot rehybridize)

H—X

We don't propose
these reactions.

Typical sp C-H bond energies ≈ 124 $\frac{kcal}{mole}$

acetylenic C-H

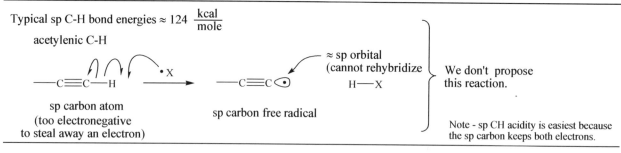

sp carbon atom
(too electronegative
to steal away an electron)

sp carbon free radical

≈ sp orbital
(cannot rehybridize

H—X

We don't propose
this reaction.

Note - sp CH acidity is easiest because
the sp carbon keeps both electrons.

Problem 6 – Predict the products and the approximate relative amounts of each (as a percent). Include stereoisomers, if present (enantiomers, diastereomers, meso, cis/trans). Assume the relative reactivities are the same as listed at the end of the problem. Write an arrow-pushing mechanism for the major product formed in part f. Calculate ΔH_{step} for each step of your mechanism.

a. $CH_3CH_2CH_2CH_2CH_3$ + Cl-Cl

3 possible products if stereoisomers are ignored

(4 different products, including positional isomers, enantiomers, diastereomers and meso compounds.)

b. $CH_3CH_2CH_2CH_2CH_3$ + Br-Br

3 possible products if stereoisomers are ignored

(4 different products, including positional isomers, enantiomers, diastereomers and meso compounds.)

c.

$$H_3C-\overset{\overset{\displaystyle CH_3}{|}}{\underset{\overset{\displaystyle |}{H}}{C}}-\overset{H_2}{C}-CH_3 \qquad Cl-Cl$$

4 possible products if stereoisomers are ignored

(6 different products, including positional isomers, enantiomers, diastereomers and meso compounds.)

d.

$$H_3C-\overset{\overset{\displaystyle CH_3}{|}}{\underset{\overset{\displaystyle |}{H}}{C}}-\overset{H_2}{C}-CH_3 \qquad Br-Br$$

4 possible products if stereoisomers are ignored

(6 different products, including positional isomers, enantiomers, diastereomers and meso compounds.)

e.

⬡—Br   Br—Br

4 possible products if stereoisomers are ignored

(9 different products, including positional isomers, enantiomers, diastereomers and meso compounds.)

f.

⬡—   Br—Br

5 possible products if stereoisomers are ignored

(12 different products, including positional isomers, enantiomers, diastereomers and meso compounds.)

relative rates for chlorination	=	$\dfrac{tertiary}{\dfrac{secondary}{primary}}$	=	$\dfrac{5.1}{\dfrac{4.0}{1.0}}$
relative rates for bromination	=	$\dfrac{tertiary}{\dfrac{secondary}{primary}}$	=	$\dfrac{1600}{\dfrac{80}{1.0}}$

Synthesis – Using organic reactions to construct new organic molecules.

We can now make several R-Br compounds from the seven hydrocarbon starting points listed below. We are still stuck when it comes to making primary R-Br (except for 2 and 6), but a possible solution will be found in our next reaction.

Available hydrocarbons in our chemical catalog.

CH₄
1   2   3   4   5   6   7

Bromo hydrocarbons that can be made from above using free radical substitution.

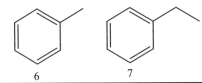

Bromo hydrocarbons that cannot be made from above using free radical substitution. These must be made from free radical addition of H-Br to alkene pi bonds.

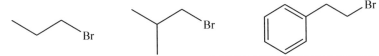

possible reactions, etc.

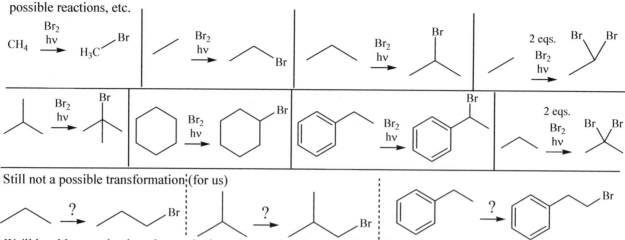

Still not a possible transformation (for us)

We'll be able to make these bromo hydrocarbons using free radical addition of HBr to alkenes.

We are not quite through with this topic. Putting a leaving group (Br) on an alkane allows many substitution possibilities (S_N2 and S_N1), with the potential to prepare a wide range of functional groups. However, it also allows an elimination possibility (E2). Let's take a moment to show how our starting alkanes can make valuable additions to our synthetic tool chest, which is about to explode with additional possibilities. From ethane we can make ethene and ethyne. From propane we can make propene and propyne, from cyclohexane we can make cyclohexene and from ethyl benzene we can make styrene (ethenylbenzene) and ethynylbenzene (phenylacetylene). When we propose making the alkynes, we will just use two equivalents of Br_2 in the reaction. The second bromine prefers to add regioselectively at the same position as the first bromine. The synthetic sequences look as follows. Remember, we use very strong, bulky bases to force an E2 reaction.

Many new reactions to learn for alkenes in a later topic.

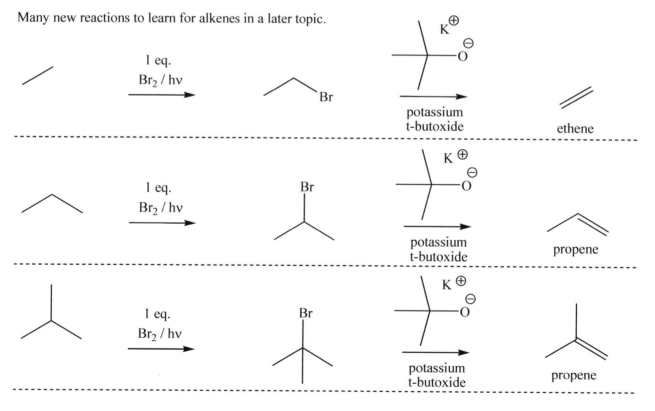

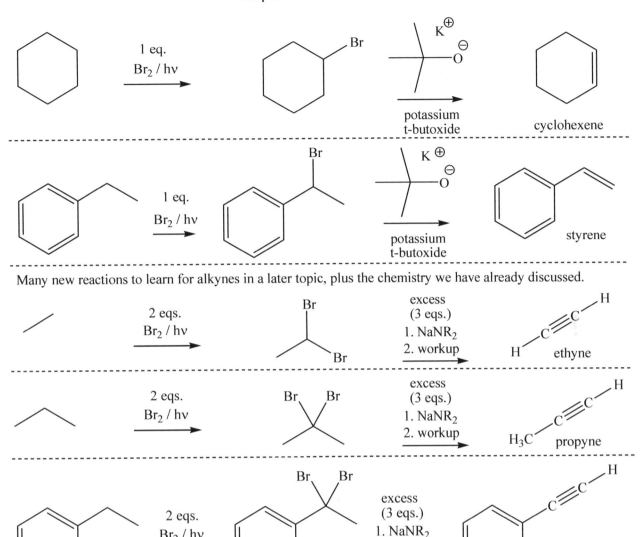

Free radical addition mechanism of H-Br to alkene pi bonds (anti-Markovnikov addition to alkenes)

Our second free radical reaction also puts a bromine leaving group on a hydrocarbon molecule. In this example H-Br adds to an alkene pi bond in the presence of a peroxide catalyst and light. We know that alkenes can be made from E2 or E1 reactions at this point in course. Just as in free radical substitution reactions, initiation is the first step. The first bond to break is the very weak peroxide –O-O- bone (bond energy are generally < 40 kcal/mole). An oxygen atom abstracts an "H" from H-Br to make a very strong O-H bond (> 110 kcal/mole). The very reactive bromine atom adds to the pi bond, in a very selective way, so that the more stable free radical forms (the more substituted free radical). We call this a regioselective choice. The newly formed carbon free radical then abstracts an H atom from H-Br to make a strong C-H bond and the observed product. That brings everything back to a bromine atom which starts the entire process over in the chain reaction. Just a catalytic amount of peroxide is used to get the reaction started. All of the arrow-pushing steps are shown in the first example (half-headed arrows). Termination remains the same, any possible combination of two free radicals.

overall reaction

1. initiation (two steps)

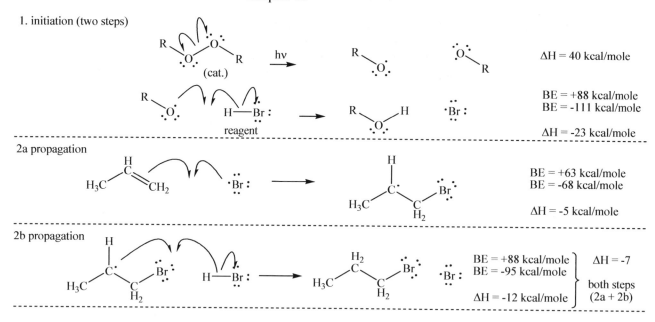

Problem 7 – Supply all missing mechanism arrows and any formal charge to complete the following free radical addition reaction.

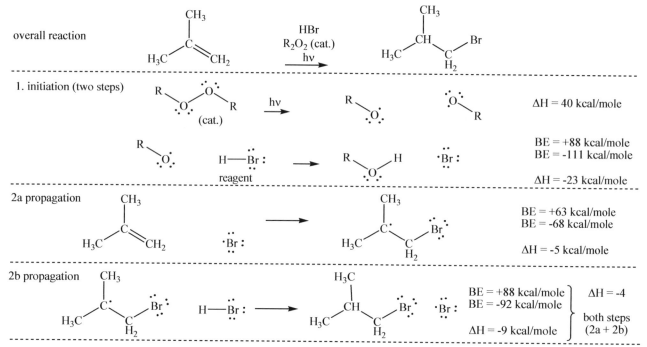

Problem 8 – Write a complete mechanism for the following free radical addition reaction.

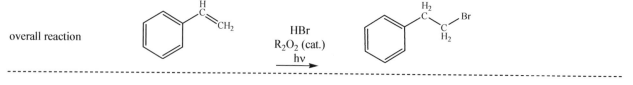

Chapter 10

For now, the structures below represent your hydrocarbon starting points to synthesize target molecules (TM) that are specified. We only study two free radical reactions in our course, but they are very important reactions because they make versatile functionalized starting molecules for synthesis of all the other functional groups studied in this course. These reactions allow us to make 13 bromohydrocarbons that we can use to study all of the other chemistry in our course.

Allowed starting structures – our main sources of carbon – 1. Free radical substitution of sp³ C-H bonds to form sp³ C-Br bonds at the weakest C-H position and 2. Anti-Markovnikov addition to alkenes makes 1° R-Br. From these two reactions we can make 13 R-Br molecules below.

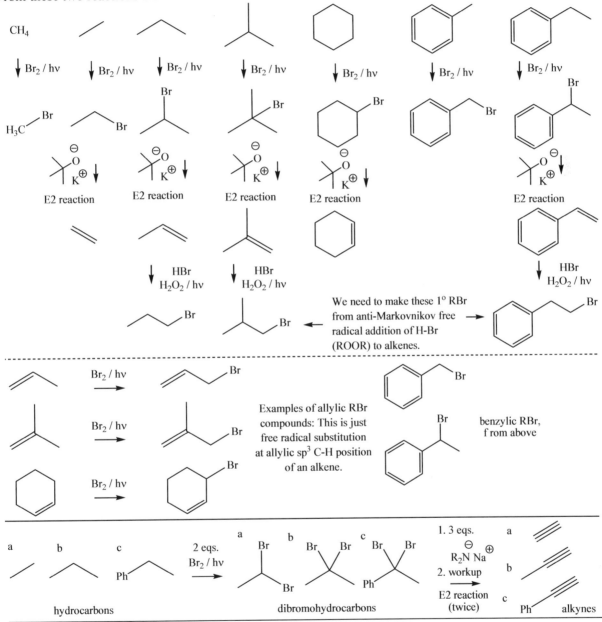

Chapter 10

Problem 9 – Propose a reasonable synthesis for the following molecules from the given starting materials.

Carbon compounds available in our chemical catalog (plus the usual inorganic reagents).

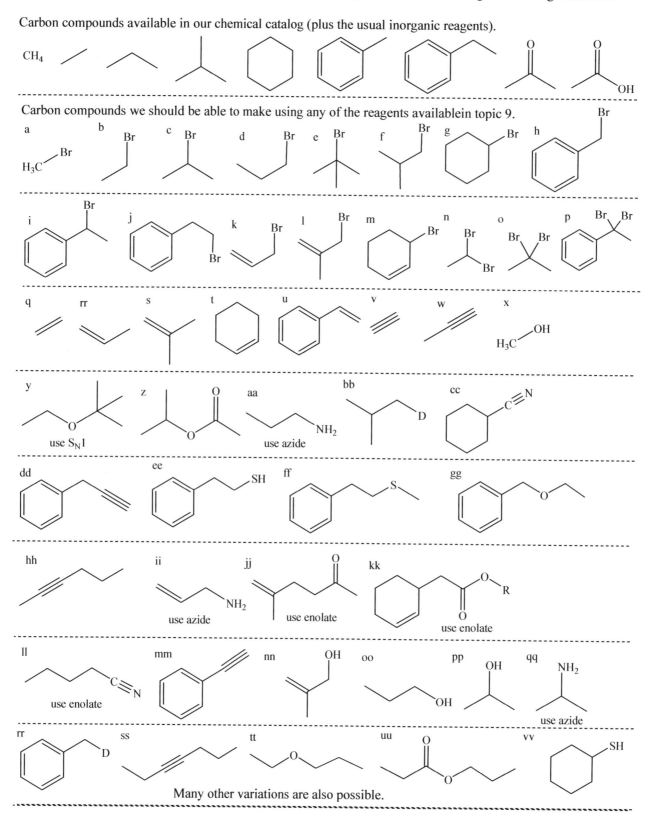

Carbon compounds we should be able to make using any of the reagents availablein topic 9.

Many other variations are also possible.

Chapter 10

Possible free radical mechanism steps for combustion of alkanes in oxygen

Combustion reactions are likely the most common reaction used by human beings. We use them when we cook, drive our cars, dry our clothes, heat our houses, heat our water and even heat our world. Combustion reactions very likely involve free radical steps as well. However, to convert hydrocarbon molecules like ethane combined with oxygen to form carbon dioxide and water must involve many complicated steps (that happen very quickly). Below is a proposed possible path between starting structures and product structures (one of many possibilities). I took a few short cuts to save paper. It is, of course, highly speculative, but tries to use the logic suggested above. Incomplete combustion products that stop before carbon dioxide and water are formed constitute air pollution (smog).

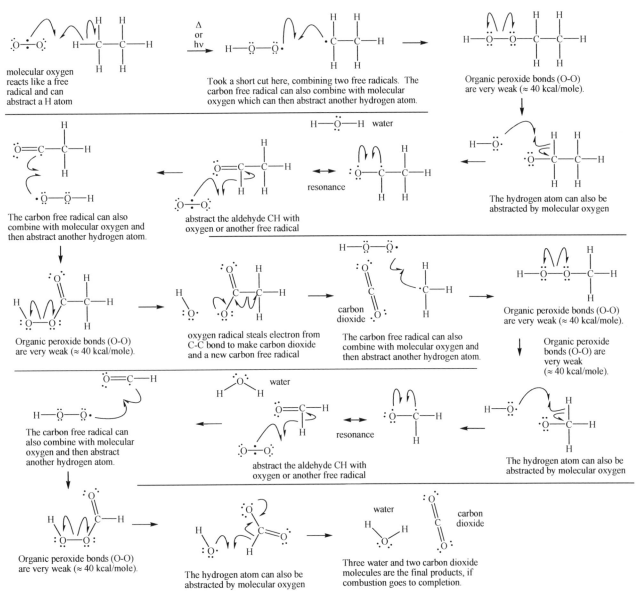

molecular oxygen reacts like a free radical and can abstract a H atom

Took a short cut here, combining two free radicals. The carbon free radical can also combine with molecular oxygen which can then abstract another hydrogen atom.

Organic peroxide bonds (O-O) are very weak (≈ 40 kcal/mole).

H—O—H water

The hydrogen atom can also be abstracted by molecular oxygen

The carbon free radical can also combine with molecular oxygen and then abstract another hydrogen atom.

abstract the aldehyde CH with oxygen or another free radical

resonance

Organic peroxide bonds (O-O) are very weak (≈ 40 kcal/mole).

oxygen radical steals electron from C-C bond to make carbon dioxide and a new carbon free radical

carbon dioxide

The carbon free radical can also combine with molecular oxygen and then abstract another hydrogen atom.

Organic peroxide bonds (O-O) are very weak (≈ 40 kcal/mole).

Organic peroxide bonds (O-O) are very weak (≈ 40 kcal/mole).

The carbon free radical can also combine with molecular oxygen and then abstract another hydrogen atom.

water

abstract the aldehyde CH with oxygen or another free radical

resonance

The hydrogen atom can also be abstracted by molecular oxygen

Organic peroxide bonds (O-O) are very weak (≈ 40 kcal/mole).

The hydrogen atom can also be abstracted by molecular oxygen

water

carbon dioxide

Three water and two carbon dioxide molecules are the final products, if combustion goes to completion.

Chapter 10

Biochemical Oxidation

How does the body do it? Your body can also attack C-H bonds with free radicals, though in a much more controlled manner. Random free radicals in the body can cause a lot of damage, so the body has built in extensive protections to minimize these sorts of reactions. Also, instead of wasting much of the energy as heat in a single gigantic explosion, the body captures a little bit of energy at a time in many small, single bond oxidation steps using highly specific enzymes. We can simplify the complicated biomolecules to take a brief look at a biochemical oxidation reaction. We do need oxygen, which we breathe with every breath, so let's start there. An oxygen molecule is picked up by an iron atom in myoglobin in the lungs, where an oxygen atom steals an electron from an iron atom bound in myoglobin. Myoglobin transports oxygen through the blood, transferring it to hemoglobin molecules in the cells of the body. The iron atoms in myoglobin and hemoglobin are supported at four valence locations inside huge prophyrin ring structures. Four porphyrin nitrogen atoms surround the iron in a square planar shape, and a sulfur atom (or nitrogen atom) from a protein molecule is connected from the bottom to keep the whole complex in the correct location on the enzyme. There are actually six bonding positions at the iron atom, so there is one open position. This is the position that will bond to oxygen to begin the oxidative attack on reactive sites in the body. The iron atom changes its oxidation state as the oxygen reacts, ranging anywhere from +2 to +4.

The simplistic schematic below shows how iron could pick up an oxygen molecule and how that oxygen molecule could attack and oxidize an sp3 C-H in the body. Oxidizing cytochrom P-450 enzymes in the liver have the main job of making external metabolites more water soluble so they can be eliminated from the body. This example clearly shows why organic chemistry is taught before biochemistry. While many of the reactions are similar, the structures in biochemistry are way more complicated. It's better to learn the essential ideas of reactions on simple molecules and then use them to understand those reactions on more complicated molecules. There is a lot of free radical chemistry going on inside you, so make sure you eat your fruits and vegetables to keep yourself well stocked up on free radical protecting anti-oxidants.

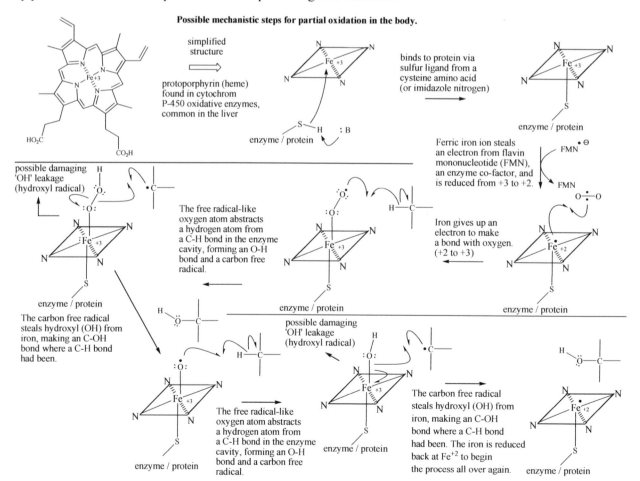

Possible mechanistic steps for partial oxidation in the body.

Chapter 10

Chapter 11 - Oxidations, Imines and Acid Derivatives

Oxidation of alcohols – E2 elimination reactions can form carbonyl functional groups (C=O), including aldehydes, ketones and carboxylic acids.

In our last chapter, we learned how to convert relatively inert CH hydrocarbons into very versatile RBr compounds that could lead to many organic functional groups. One of the most valuable of those groups is the alcohols, having only a single bond between carbon and oxygen. Increasing the number of bonds between carbon and oxygen is referred to as oxidation. Methods that do this are important and common and are usually exothermic. In this section we will focus our attention on oxidation of alcohols to aldehydes, ketones and carboxylic acids. There are many ways to do this, but we will emphasize chromium reagents, because they are reasonably simple to understand and commonly used. We will simplify the wide variety chromium reagents to CrO_3/pyridine without water present and CrO_3 with acidic water present. Our presentation is not entirely rigorous, but will provide us with insights into key aspects of such reactions. These reactions will allow us to perform some very important additional transformations that will greatly extend our synthetic possibilities. This also means the potential compounds that can be made will expand enormously. Similar oxidations are very common in nature, but are usually quite different in mechanism (using NAD^+).

The crucial details of the oxidation transformation

With many organic oxidizing reagents, these two electrons are lost from "O" by combining the oxygen with an electron poor atom (Cr, Mn, S, I and others), but immediately filled back in from the C-H bond (E2). The two protons are lost in proton transfers to a base (acid/base chemistry).

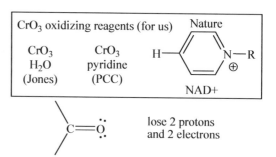

oxidation of an alcohol to a carbonyl group

lose 2 protons and 2 electrons

In nature / biochemistry, these two electrons are lost from "C", often by a hydride transfer to NAD^+ or FAD, always with base abstraction of the oxygen proton (backwards to the way we do it).

Two electrons are lost from the carbon atom to an electron poor atom and two hydrogens are lost from the alcohol functional group in either type of oxidation to form a carbonyl (C=O) functional group.

There are many organic reducing reagents that can add electrons, via a carbanion or hydride to an oxidized carbon atom. An acidic proton is often added in a workup step. Our nucleophilic hydrogen will be lithium aluminium hydride and we will use verious forms of nucleophilic carbon (cyanide, acetylides, enolates, Grignard reagents and sulfur and phosphorous ylids).

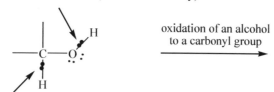

1. reduction of a carbonyl group to an alcohol

2. pick up a proton on the oxygen

gain a nucleophile (2 electrons) and a proton

In nature / biochemistry, these two electrons and a hydrogen atom are added from by a hydride transfer from NADH to form NAD^+, along with protonation of the oxygen.

Examples of $Nu:^{\ominus}$

$H_3Al{-}H$ $H_3C:^{\ominus}$ $Li^{\oplus}$ $(MgBr)^+$

(NaBH$_4$ too)

Nature

NADH

Freshman chemistry oxidation/reduction electron counting rules give all electron credit in bonds to the more electronegative atom. Using these rules, oxygen almost always is –2, unless it is bonded to another oxygen atom (peroxides) or a fluorine atom (very rare and very dangerous). Hydrogen atoms are usually +1, unless bonded to another hydrogen atom (unique = H_2, oxidation number = 0) or a metal atom (e.g. NaH, sodium hydride, H oxidation number = -1). Formal charge on the other hand, views all bonded electrons as shared evenly between bonded atoms (single, double or triple bonds).

Chapter 11

Problem 1 – What are the oxidation states of each carbon atom below? What is the formal charge of every carbon atom below?

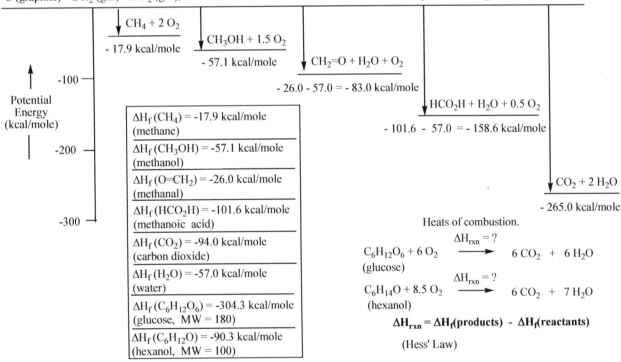

$CH_4 \longrightarrow H_3C—OH \longrightarrow$ (H, C=O, H) $\longrightarrow$ (H, C=O, HO) $\longrightarrow O=C=O$

oxidation state of carbon = ____ ?

oxidation state of carbon = ____ ?

oxidation state of carbon = ____ ?

oxidation state of carbon = ____ ?

oxidation state of carbon = ____ ?

$\pm H_2O$

$H_3C—CH_3 \longrightarrow H_3C—\overset{H_2}{\underset{*}{C}}—OH \longrightarrow H_3C—\overset{H}{\underset{*}{C}}=O \longrightarrow H_3C—\overset{OH}{\underset{*}{C}}=O \longrightarrow (HO)(HO)C=O$

oxidation state of * carbon = ____ ?

oxidation state of * carbon = ____ ?

oxidation state of * carbon = ____ ?

oxidation state of * carbon = ____ ?

oxidation state of * carbon = ____ ?

The energy content of carbon compounds changes with the number of oxygen bonds. When there are fewer oxygen bonds, the energy content is higher and when there are more oxygen bonds, the energy content is lower. If you think about fats, which are largely "alkane-like" there is minimal oxygen present and they are high energy (high calorie). On the other hand carbohydrates have an oxygen on almost every carbon and the energy content is much lower. Carbon dioxide and water are completely oxidized and their energy content is zero (from our perspective). The following table shows various energy content in some C1 organic molecules.

C (graphite) 2 H₂ (gas) 2 O₂ (gas), elements in standard state are defined to have zero potential energy = Reference point

$CH_4 + 2 O_2$

- 17.9 kcal/mole

$CH_3OH + 1.5 O_2$

- 57.1 kcal/mole

$CH_2=O + H_2O + O_2$

- 26.0 - 57.0 = - 83.0 kcal/mole

Potential Energy (kcal/mole)

-100

$HCO_2H + H_2O + 0.5 O_2$

- 101.6 - 57.0 = - 158.6 kcal/mole

-200

$\Delta H_f (CH_4) = -17.9$ kcal/mole (methane)

$\Delta H_f (CH_3OH) = -57.1$ kcal/mole (methanol)

$\Delta H_f (O=CH_2) = -26.0$ kcal/mole (methanal)

$\Delta H_f (HCO_2H) = -101.6$ kcal/mole (methanoic acid)

-300

$\Delta H_f (CO_2) = -94.0$ kcal/mole (carbon dioxide)

$\Delta H_f (H_2O) = -57.0$ kcal/mole (water)

$\Delta H_f (C_6H_{12}O_6) = -304.3$ kcal/mole (glucose, MW = 180)

$\Delta H_f (C_6H_{14}O) = -90.3$ kcal/mole (hexanol, MW = 100)

$CO_2 + 2 H_2O$

- 265.0 kcal/mole

Heats of combustion.

$\Delta H_{rxn} = ?$

$C_6H_{12}O_6 + 6 O_2 \longrightarrow 6 CO_2 + 6 H_2O$ (glucose)

$\Delta H_{rxn} = ?$

$C_6H_{14}O + 8.5 O_2 \longrightarrow 6 CO_2 + 7 H_2O$ (hexanol)

$\Delta H_{rxn} = \Delta H_f(products) - \Delta H_f(reactants)$

(Hess' Law)

Problem 2 – Which has a higher energy content per "*gram*", glucose (carbohydrates) or hexan-1-ol (model for fats)? You might also need the molecular weights to solve this problem (given). Speculate why this is the case. The necessary data is in the table above. Use the given reactions and necessary heats of formations.

Chapter 11

Oxidation of alcohols to make aldehydes, ketones and carboxylic acids

In this book we will use CrO_3 as an oxidizing reagent for alcohols. Resonance structures give an indication of just how electrophilic (electron poor) chromium is in CrO_3. Even though the formal charge changes with every structure, the oxidation state remains constant at +6.

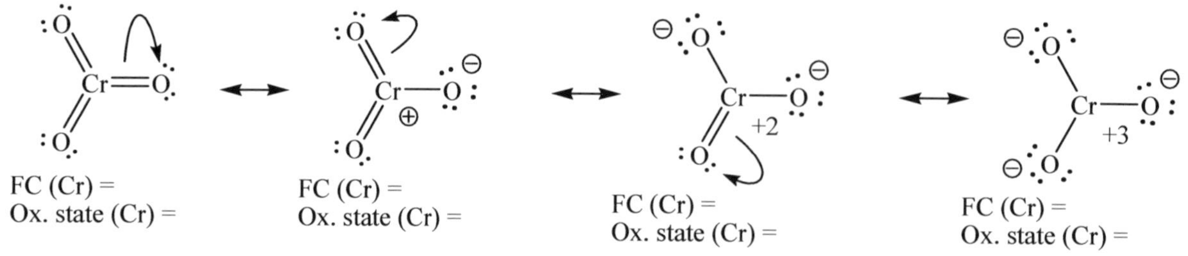

FC (Cr) =
Ox. state (Cr) =

FC (Cr) =
Ox. state (Cr) =

FC (Cr) =
Ox. state (Cr) =

FC (Cr) =
Ox. state (Cr) =

We will view the following sequence as the key steps of chromium oxidation reactions. A very electron poor Cr atom bonds with the oxygen of a methyl, primary or secondary alcohol (step 1), followed by proton transfer away from the alcohol oxygen using a base (step 2). This makes an inorganic chromium ester, turning the very electron poor chromium atom into a good leaving group from an oxygen atom. Normally, we would never show oxygen losing electrons to another atom, but the chromium is in a +6 oxidation step, and the oxygen isn't really losing electrons, since electrons are supplied from the other side where a C-H bond is broken in the concerted step of an E2 elimination (step 3). This is also the slow step of the reaction (rate determining step = R.D.S.). There are numerous variations that have a similar flow of electrons, with only the atom receiving the electrons from oxygen changed (Moffit, Swern and periodinane are common). We will emphasize chromium oxidations, and not worry about any of its limitations or variations, except for the distinction of running the reaction without water present (PCC = CrO_3/pyridine) or with water present (Jones = CrO_3/H_2O/acidic).

Problem 3 – What are the oxidation states below on the carbon atom and the chromium atom as the reaction proceeds? Which step does the oxidation/reduction occur? (PCC, B: = pyridine and Jones, B: = water)

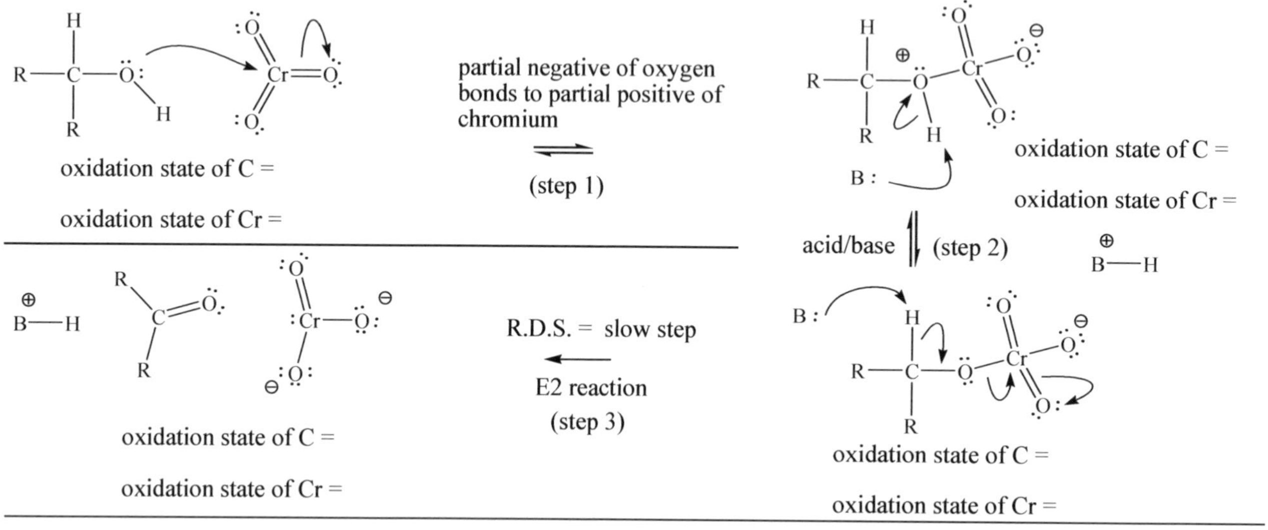

All by itself, Cr^{+6} has many variations, but we will consider them all as CrO_3 (chromic anhydride). Our only distinctions will be nonaqueous or aqueous conditions. We will consider Jones conditions to include water, acid (H_3O^+) and CrO_3. We will consider pyridinium chlorochromate (PCC = CrO_3/pyridine) to be anhydrous (does not include water). PCC is slightly acidic while a close variation, called pyridinium dichromate (PDC), is slightly basic. Such experimental details will not be a concern in this book (see below). Four chromium reagents are shown below, but we will only use the first two.

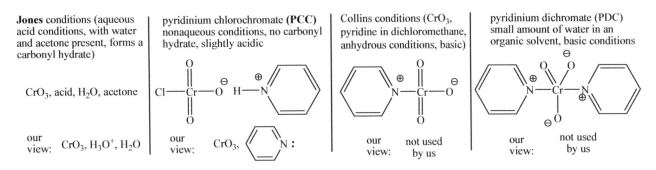

There is a lot of "recipe" chemistry in the above reagents (and several others not presented). If we were actually doing an oxidation reaction in the lab, we would want to get out our "cook book" to find a procedure as close to our reaction as possible, and hope that any minor differences would not affect the yields. To keep life simple, we will only view two versions of the above oxidizing conditions: Jones conditions (CrO_3 with water) and PCC (CrO_3 with pyridine). The crucial difference between aqueous and nonaqueous conditions is that the carbonyl group (C=O), once formed, can hydrate forming two "O-H" groups in aqueous conditions. One of these OH groups can re-esterify with a chromium atom and the oxidation can occur a second time, if there is still a hydrogen atom on the hydrated carbon atom. This means that the original alcohol carbon would have to have at least two hydrogen atoms and have been a methyl (unique) or primary alcohol (general) to begin with. A secondary alcohol can only oxidize once, regardless of aqueous (Jones) or nonaqueous conditions (PCC). Jones conditions ($CrO_3/H_2O/H_3O^+$) is going to force us to consider another reaction mechanism, an addition reaction called hydration (addition of water). It's not really that hard because it's just a combination of steps that we have already studied.

1. PCC = pyridinium chlorochromate, (CrO_3/pyridine), CrO_3 oxidations of alcohols (methyl, 1° and 2° ROH) without water. Steps are: 1. Cr=O addition, 2. acid/base and 3. E2 to form C=O (aldehydes and ketones).

primary alcohols

PCC = pyridinium chlorochromate oxidation of primary alcohol to an aldehyde (no water to hydrate the carbonyl group)

aldehydes

Chapter 11

CrO$_3$ oxidations of alcohols (methyl, 1° and 2° ROH) without water = PCC, Cr=O addition, acid/base and E2 to form C=O (aldehydes and ketones)

secondary alcohols

PCC = pyridinium chlorochromate oxidation of primary alcohol to an aldehyde (no water to hydrate the carbonyl group)

ketones

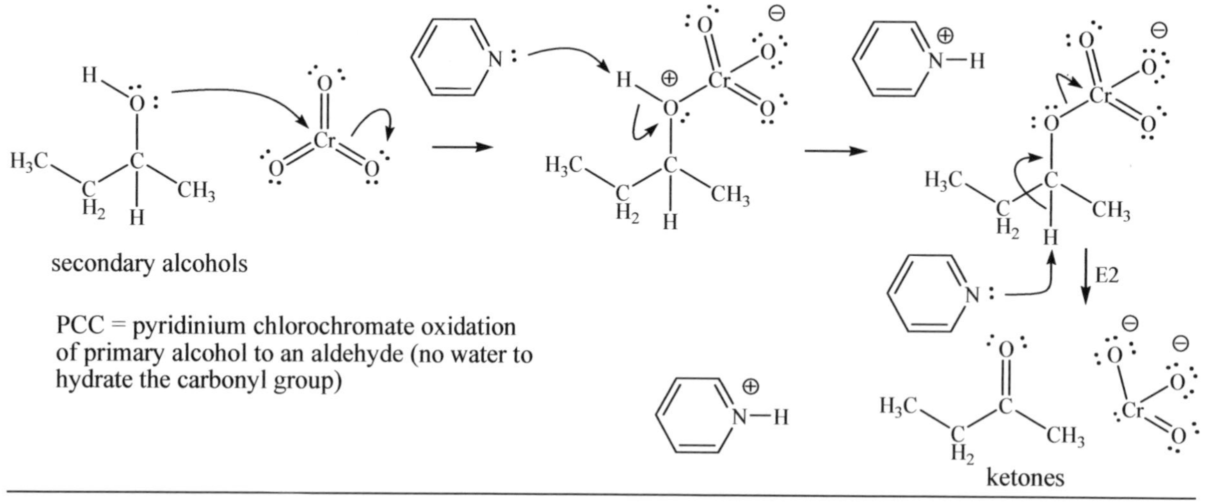

2. Jones reagent = CrO$_3$/water/acid, CrO$_3$ oxidations of alcohols (methyl, 1° and 2° ROH) with water. Steps are: 1. Cr=O addition, 2. acid/base and 3. E2 to form C=O (aldehydes and ketones) 4. hydration of C=O and repeat reactions when the starting alcohol is a 1° alcohol (forms carboxylic acids from primary alcohols and ketones from secondary alcohols).

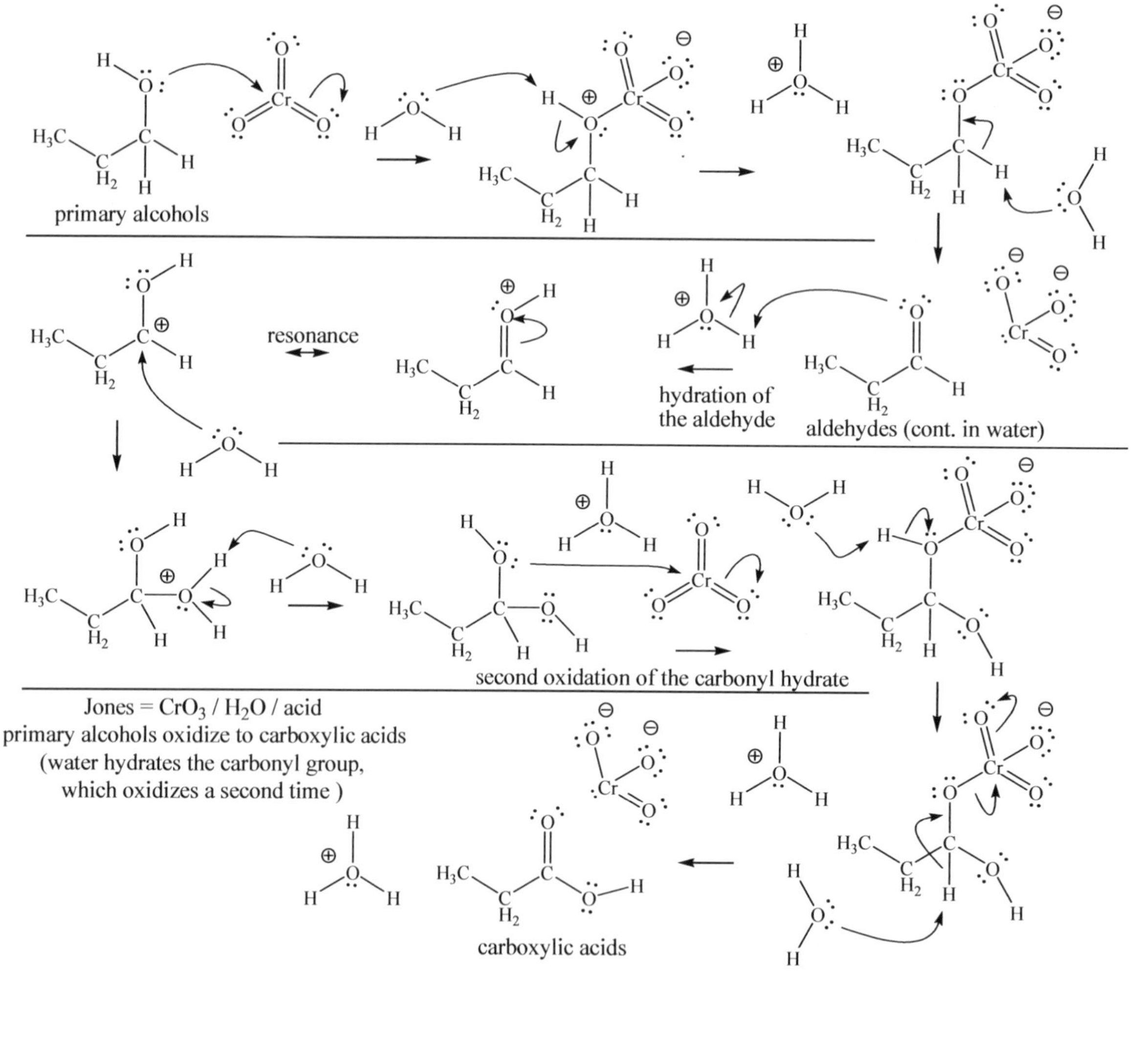

primary alcohols

resonance

hydration of the aldehyde

aldehydes (cont. in water)

second oxidation of the carbonyl hydrate

Jones = CrO$_3$ / H$_2$O / acid
primary alcohols oxidize to carboxylic acids
(water hydrates the carbonyl group,
which oxidizes a second time)

carboxylic acids

Chapter 11

Problem 4 – Supply all of the mechanistic details in the sequences below showing 1. the oxidation of a primary alcohol, 2. hydration of the carbonyl group and 3. oxidation of the carbonyl hydrate (Jones conditions).

Under aqueous conditions, a hydrate of a carbonyl group has two OH groups which allow a second oxidation, if another C-H bond is present. This is only possible if the starting carbonyl group was an aldehyde (true when starting with methyl and primary alcohols).

Your next step is to write out the above mechanism completely on your own, using the following equation.

Chapter 11

Typical oxidation possibilities are shown below for common alcohol patterns. Currently we can make the alcohols from RBr compounds using S_N chemistry, and that's a lot of alcohols. The alcohol carbon with oxygen is the key. Does it have any hydrogen atoms (is it methyl, primary or secondary)? How many hydrogen atoms does it have? Potentially all of the hydrogen atoms can be oxidized off, or only one of them, depending on the conditions you choose (Jones or PCC).

1. Primary alcohols (or methanol), without any water in the reaction mixture (PCC), oxidize only to aldehydes. No carbonyl hydrates can form without water, so there is no way to oxidize a second time. Pyridine is the base.

2. Primary alcohols (or methanol) with water in the reaction mixture can oxidize twice (Jones). Once the aldehyde is formed, it can hydrate (add H_2O) and form a chromium ester a second time, which oxidizes off a second hydrogen atom. Water is the base.

3. Secondary alcohols can only oxidize once in either aqueous or nonaqueous conditions. Either reagent produces a ketone product.

There are no additional C-H bonds at the original alcohol carbon, so there is no additional oxidation possible at the ketone carbon.

4. Tertiary alcohols can form chromium esters, but there is no hydrogen atom to eliminate at the alcohol carbon. Tertiary alcohols are unreactive with either reagent. At higher temperatures C-C bonds can be cleaved (usually making a mess).

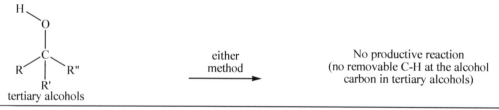

No productive reaction
(no removable C-H at the alcohol carbon in tertiary alcohols)

Chromium's big disadvantage is its toxicity. Cr^{+6} is a carcinogen, and workups of reactions can be messy, particularly on a large scale. For those reasons, chromium reagents are used less than they used to be. However, using pencil and paper reactions in exercises does not generate any waste. If you are in a class that uses a different oxidizing reagent, use that one. The mechanisms are all pretty similar. (Did you ever watch the movie "Erin Brokovich"?)

Chapter 11

Aldehydes, Ketones, Nitriles, Imines and Carboxylic Acid Reduction using Lithium Aluminum Hydride and Sodium Borohydride

We have just seen how we can take alcohols and make them into three valuable functional groups: aldehydes, ketones and carboxylic acids. What about the reverse direction? We want reactions that will allow us to do whatever is our heart's desire. So yes, there are ways to make aldehydes, ketones and carboxylic acids into alcohols. In this section we will look at hydride reduction of these carbonyl groups to alcohols. We have already used nucleophilic lithium aluminum hydride and sodium borohydride to reduce RBr compounds and the azide functional group. What we discover below is just an extension to a new class of electrophiles, compounds having a C=O, C=N and C≡N.

Aldehydes and ketones can be reduced to alcohols using lithium aluminum hydride or sodium borohydride, followed by neutralization with mild acid (called 'workup'). The acid has to be added separately after the first step is over or it will destroy the hydride reagent, but it is dumped into the same flask and called "workup". An example mechanism of each is provided below. Notice the ketone forms a secondary alcohol and the aldehyde forms a primary alcohol. The carbonyl carbon is electrophilic and the hydride is nucleophilic. Transfer of the hydride nucleophile to the carbon requires the oxygen to accept the excess electron density, forming an alkoxide. To get the neutral alcohol an acidic proton is provided in a final workup step.

Similar, but slightly more complicated mechanisms allow for the reduction of esters and carboxylic acids to primary alcohols. Since these functional groups are less reactive than aldehydes and ketones, they require the more reactive lithium aluminum hydride to make the reaction work (mechanisms on the next page). Also, because there is a potential leaving group at the carbonyl (OR for esters and OH for carboxylic acids) the reaction occurs two times. The first addition of hydride leads to acyl substitution and the second hydride leads to a carbonyl addition reaction. An example mechanism of each is provided below. In each instance the carbonyl carbon is electrophilic and the hydride is nucleophilic. Transfer of the hydride nucleophile requires the oxygen to accept the excess electron density, forming an alkoxide. To get the neutral alcohol an acidic proton is provided in a final workup step. We can now make alcohols from RBr compounds (S_N chemistry) and carbonyl chemistry (C=O + aluminum hydride or borohydride).

Reduction of an ester to a primary alcohol using lithium aluminum hydride (2x)

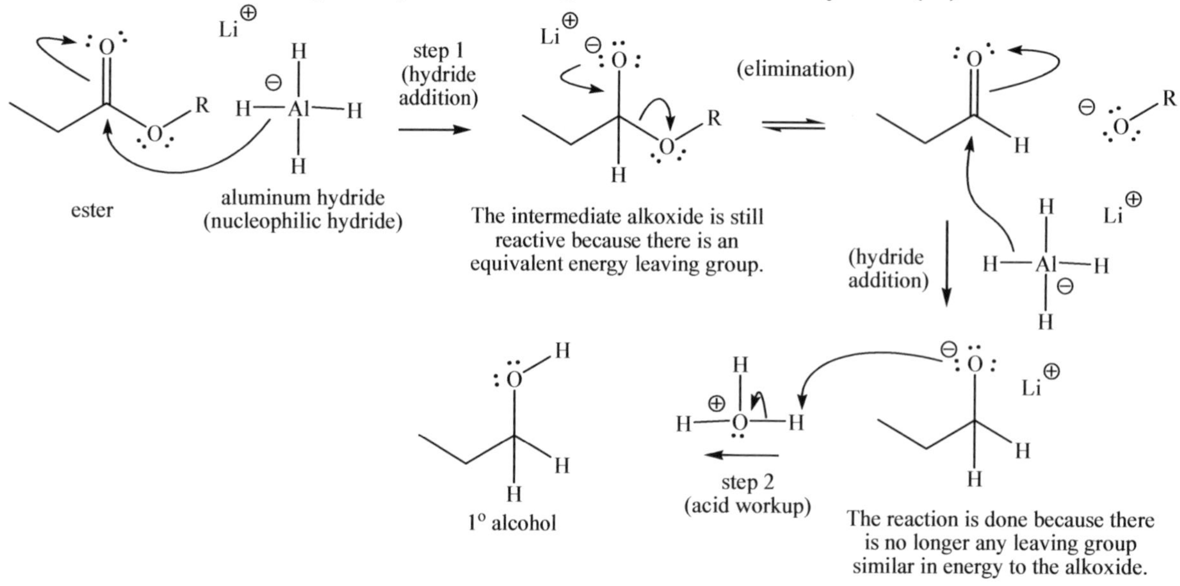

step 1 (hydride addition)

(elimination)

aluminum hydride (nucleophilic hydride)

ester

The intermediate alkoxide is still reactive because there is an equivalent energy leaving group.

(hydride addition)

1° alcohol

step 2 (acid workup)

The reaction is done because there is no longer any leaving group similar in energy to the alkoxide.

Reduction of a carboxylic acid to a primary alcohol using lithium aluminum hydride (This is a little more complicated because the first step is neutralization of the acid and the negative oxygen has to be made into a better leaving group by complexation with aluminum.

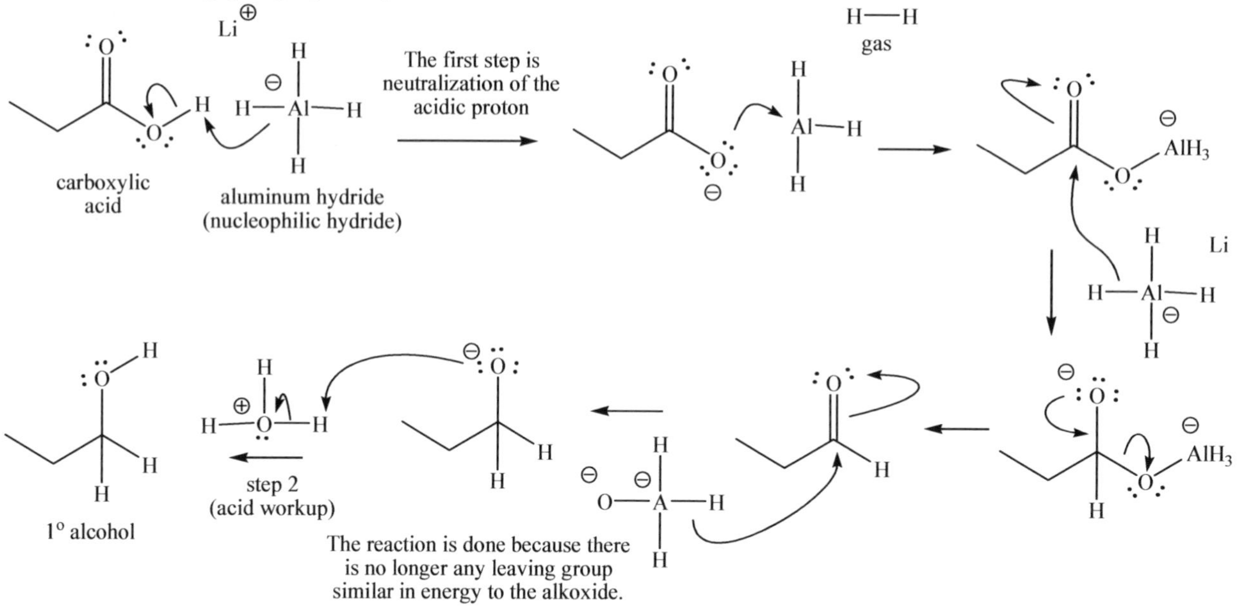

carboxylic acid

aluminum hydride (nucleophilic hydride)

The first step is neutralization of the acidic proton

H—H gas

1° alcohol

step 2 (acid workup)

The reaction is done because there is no longer any leaving group similar in energy to the alkoxide.

Nitrile Reduction with Lithium Aluminum Hydride

Nitriles share similarities with the reactions above. The carbon is electrophilic because of the pi bonds with nitrogen and the nitrogen accepts the electron density when the electrons are shifted away from the carbon. You wouldn't think that nitrogen is as good at accepting electrons as oxygen, but a nitrile nitrogen is sp hybridized which is more electronegative than an sp^2 hybridized nitrogen and is a good place to accept the excess electron density. The nitrogen complexes with the trivalent aluminum, which allows the carbon to accept a second hydride. After mild acid workup the product is a primary amine. This allows us a second method to make primary amines, in addition to the azide, LAH reactions we studied earlier (double S_N2 chemistry).

Chapter 11

LAH reduction of a nitrile to a primary amine.

Secondary (2°) amines from imines and tertiary (3°) amines from iminium ions

As we have seen, weak nucleophiles (water, H_2O, and alcohols, ROH) react with secondary and tertiary RBr compounds via $S_N1 > E1$ reactions. Methyl and primary alcohols react under strong acid conditions via S_N2 (ROH + HBr) secondary, and tertiary alcohols react under strong acid conditions via S_N1 (ROH + HBr), and via E1 reactions (ROH + H_2SO_4/Δ). Also, carbonyl hydrates form in Jones conditions (CrO_3/H_2O/acid).

We will now examine how weakly electrophilic carbonyl compounds can react with moderately good nucleophiles, primary or secondary amines (RNH_2 or R_2NH) and a catalytic amount of acid. A nucleophilic amine attacks the weakly electrophilic carbonyl carbon of an aldehyde or ketone to form and imine ($R_2C=N-R$). Water is used to shift the equilibrium, in acid, to one side or the other. By adding water (hydration) the equilibrium is shifted towards the carbonyl (C=O) and the amine. By removing water (dehydration) the equilibrium is shifted towards the imine ($R_2C=N-R$) and water. We will use sulfuric acid (H_2SO_4) when water is the solvent and we will use toluenesulfonic acid (TsOH) when nonaqueous solvent is used. You can think of TsOH as 'organic' sulfuric acid.

Use sulfuric acid in aqueous solutions and toluenesulfonic acid in organic solvents. They have similar acidities.

The pKa of sulfuric acid is listed with many values. Often a pKa of -2 is listed. This is the pKa of H_3O^+ in water, which is reasonable because sulfuric acid is a strong acid and dissociates completely to H_3O^+ in water. However, pure sulfuric acid is a much stronger acid and values as low as -10 have been reported. We will use a value of -5, but for us pure sulfuric acid just represents the strongest acid that we work with, and can protonate anything it is mixed with except an alkane.

Essentially, the order of events is to make the imine (or iminium ion): 1. amine adds to the carbonyl carbon, 2. add another proton to the oxygen atom to make a good leaving group, water, 3. remove a proton from the nitrogen atom, 4. water leaves, 5. nitrogen shares its lone pair electrons, forming a very good resonance structure and 6. a base removes any remaining proton on the nitrogen atom, if one is present (1° amines). Writing mechanisms in acid tends to be longer because there is an initial protonation step, more intermediate resonance structures, and a final deprotonation step is required.

Overall reaction: Aldehyde or Ketone Reacts with Primary NH₂ groups to form Imine + Water

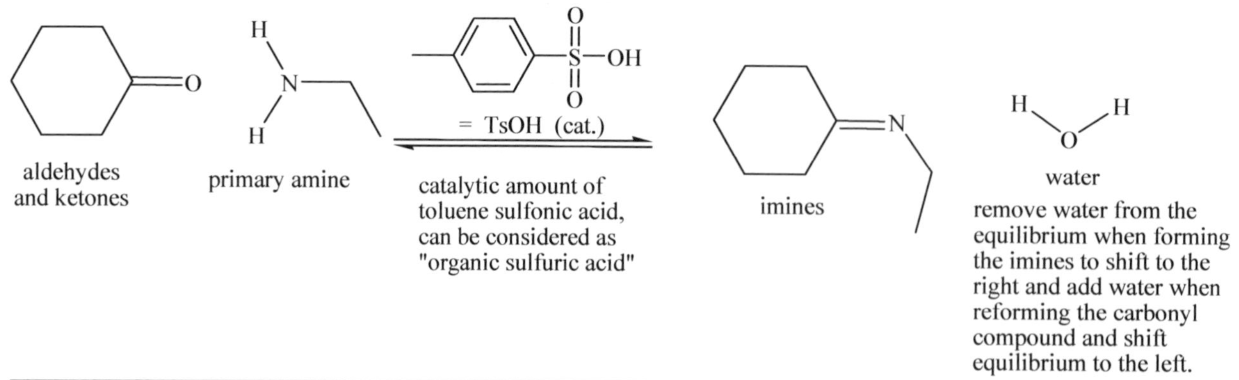

aldehydes and ketones — primary amine — = TsOH (cat.) catalytic amount of toluene sulfonic acid, can be considered as "organic sulfuric acid" — imines — water

remove water from the equilibrium when forming the imines to shift to the right and add water when reforming the carbonyl compound and shift equilibrium to the left.

Our proposed mechanism for imine formation (RNH₂ + R₂C=O, catalytic TsOH and removal of water)

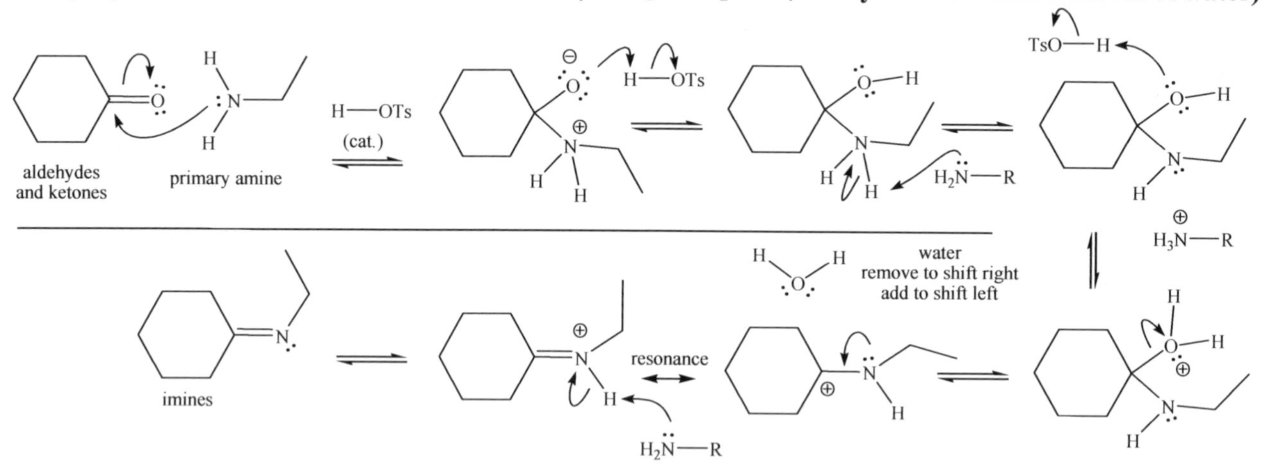

aldehydes and ketones   primary amine

H—OTs (cat.)

water remove to shift right add to shift left

resonance

imines

Imines and imminium ions can be reduced with sodium cyanoborohydride to a secondary or tertiary amines. In our course, we will use sodium borohydride to simplify writing the mechanism.

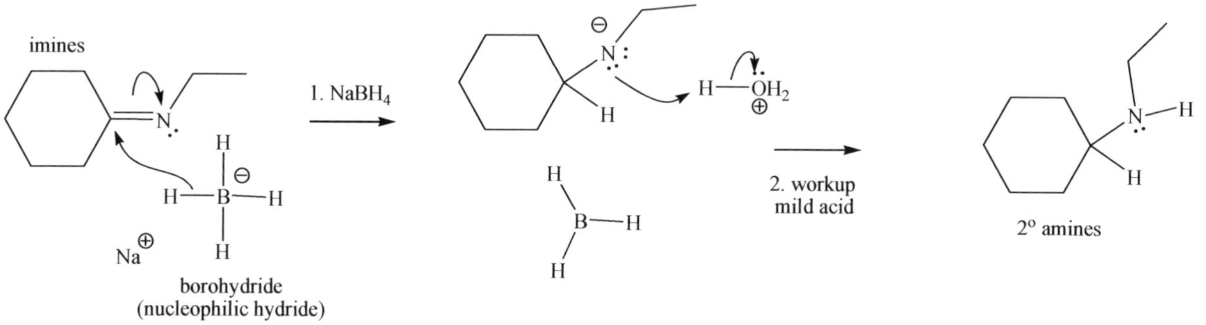

imines

1. NaBH₄

2. workup mild acid

2° amines

borohydride (nucleophilic hydride)

Chapter 11

Our proposed mechanism for iminium ion formation (R_2NH + $R_2C=O$, catalytic TsOH and removal of water) followed by reduction with LAH to a tertiary amine

Problem 5 – Propose a mechanism for the reverse reaction, hydrolysis of an imine with water to form a carbonyl compound (ketone here) and a primary amine.

If we can make primary amines, we can make secondary amines. If we can make secondary amines, we can make tertiary amides. And, if we can make amines, we can make amides (next). Yikes, it never stops!

Problem 6 – Fill in the necessary reagents to accomplish the following transformations.

Problem 7 – Propose a synthetic path to make the triethylamine.

Chapter 11

Acyl Substitution

Being able to synthesize aldehydes, ketones and carboxylic acids opens up an enormous range of additional reactions. Our time is limited, so we will be very selective to pick a few that demonstrate the essential logic of organic chemistry.

We will begin with the carboxylic acids because there is a large family of related functional groups that can be made. If we convert a carboxylic acid into an acid chloride it can be further converted into, thioesters, anhydrides, esters, amides, aldehydes, ketones and more. We will only look at four of these here. Our carboxyl functional group reaction products are shown below in order of relative reactivity. Acid chlorides are the top of the energy mountain, so all other functional group are downhill in energy (favorable transformations). The order of reactivity of the carboxyl group is the same as the order of stability of the anion leaving group, as judged by the pK$_a$ of the anion's conjugate acid.

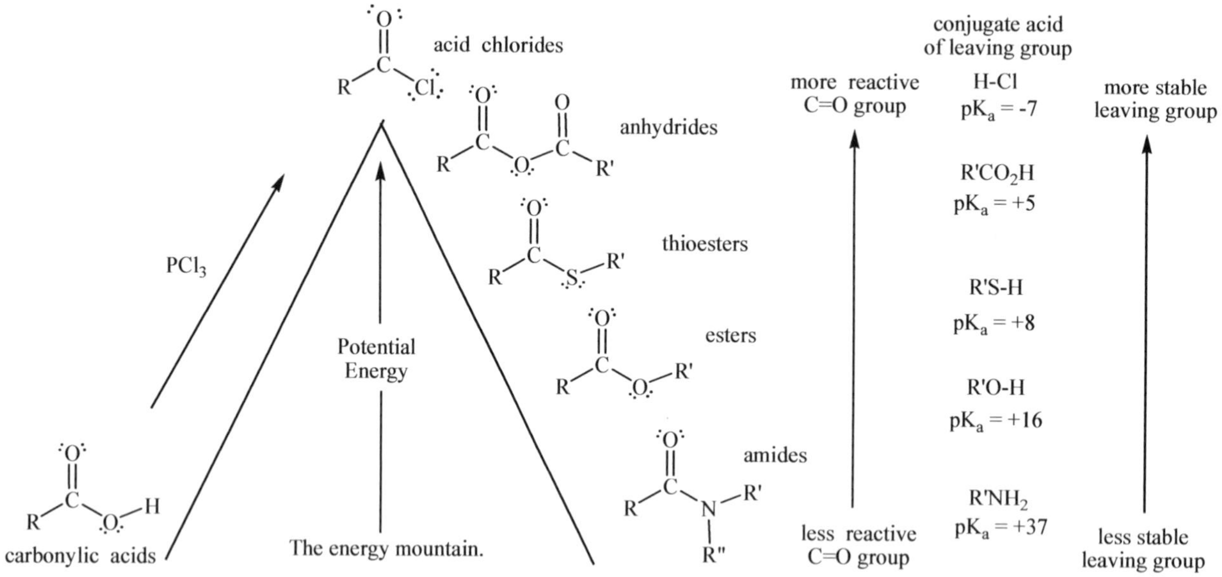

How do we get to the top of the energy mountain? There are several ways to make acid chlorides from carboxylic acids, but we will only propose one, because it follows the simple logic that we have learned so far. We will start with a carboxylic acid, which we can now make (1° ROH + Jones → RCO$_2$H), and react it with phosphorous trichloride, PCl$_3$. So, what is a possible mechanism? (We will soon have a second, very versatile method to make carboxylic acids.)

Proposed mechanism for phosphorous trichloride making an acid chloride

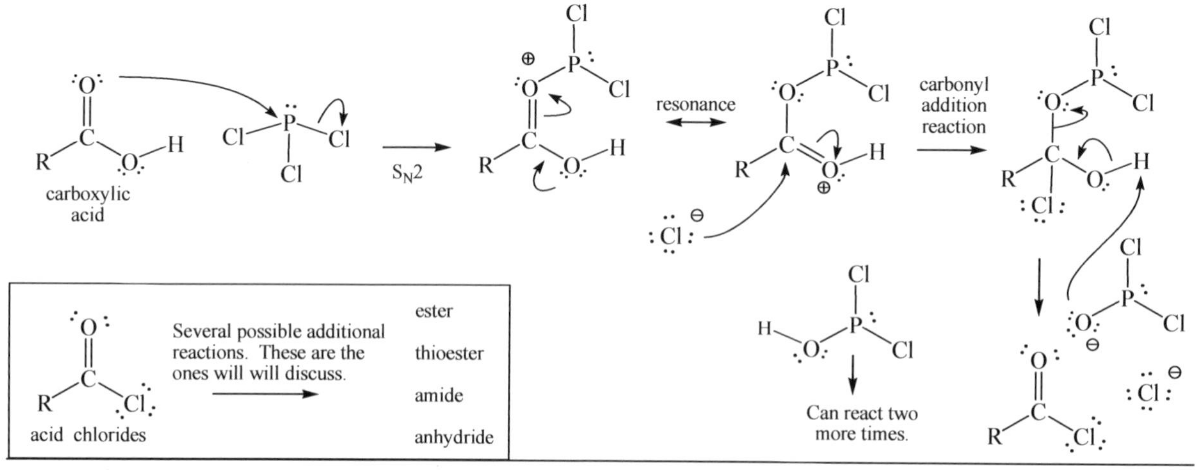

Chapter 11

Problem 8 - All the functional groups we make using acid chlorides are formed by "acyl substitution" reactions. Supply the necessary mechanistic details to complete the mechanisms (curved arrows, lone pairs and formal charge).

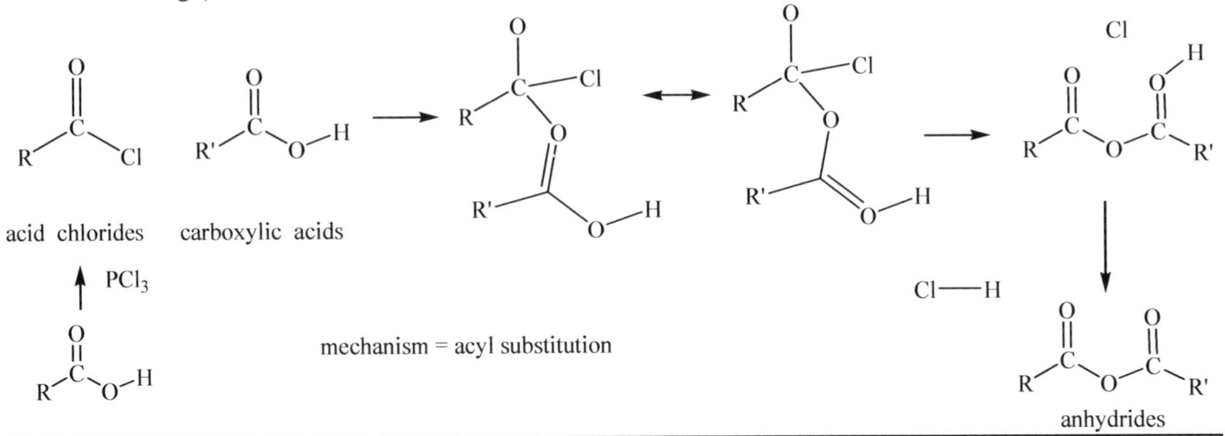

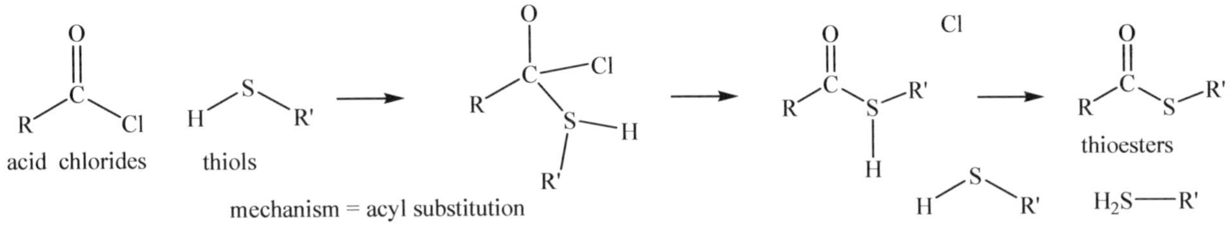

acid chlorides   carboxylic acids

mechanism = acyl substitution

anhydrides

Cl—H

acid chlorides   thiols

mechanism = acyl substitution

thioesters

H_2S—R'

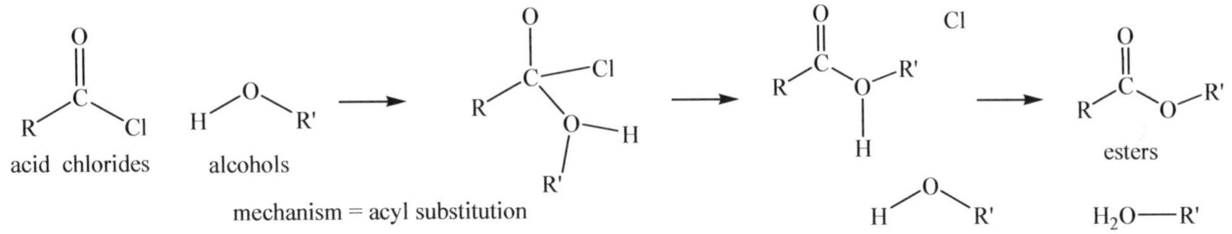

acid chlorides   alcohols

mechanism = acyl substitution

esters

H_2O—R'

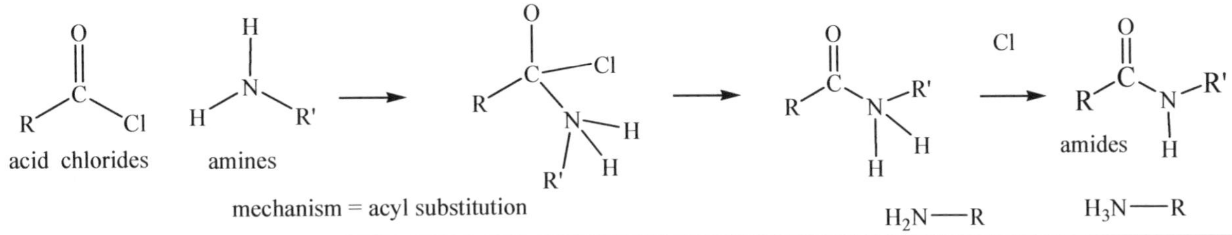

acid chlorides   amines

mechanism = acyl substitution

amides

H_2N—R H_3N—R

Acid chlorides are extremely reactive electrophiles. What would one do if it got in your eyes, nose or mouth. The water in your body would immediately react as the nucleophile, as above and form a carboxylic acid and H-Cl. H-Cl is what your stomach uses to help hydrolyze amide bonds in proteins to free up amino acids. If that happened in your eye, similar chemistry would start to break down the protein of your eyeballs. That can't be good. That's one reason you wear gloves, goggles and a lab coat when working in a lab.

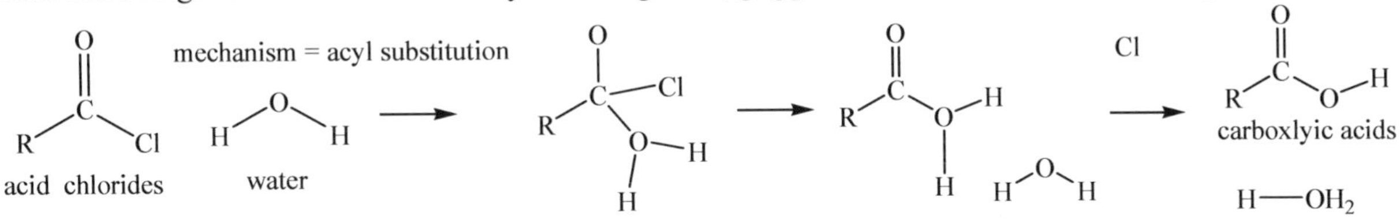

acid chlorides   water   mechanism = acyl substitution   carboxlyic acids

H—OH_2

Chapter 11

Relative reactivity of carbonyl groups

How can we use our organic logic to explain the relative reactivity of carbonyl compounds (C=O)? This is a useful exercise because it forces us to use the three main logic arguments or organic and biochemistry: inductive effects, steric effects and resonance effects. The usual order of reactivity is shown below.

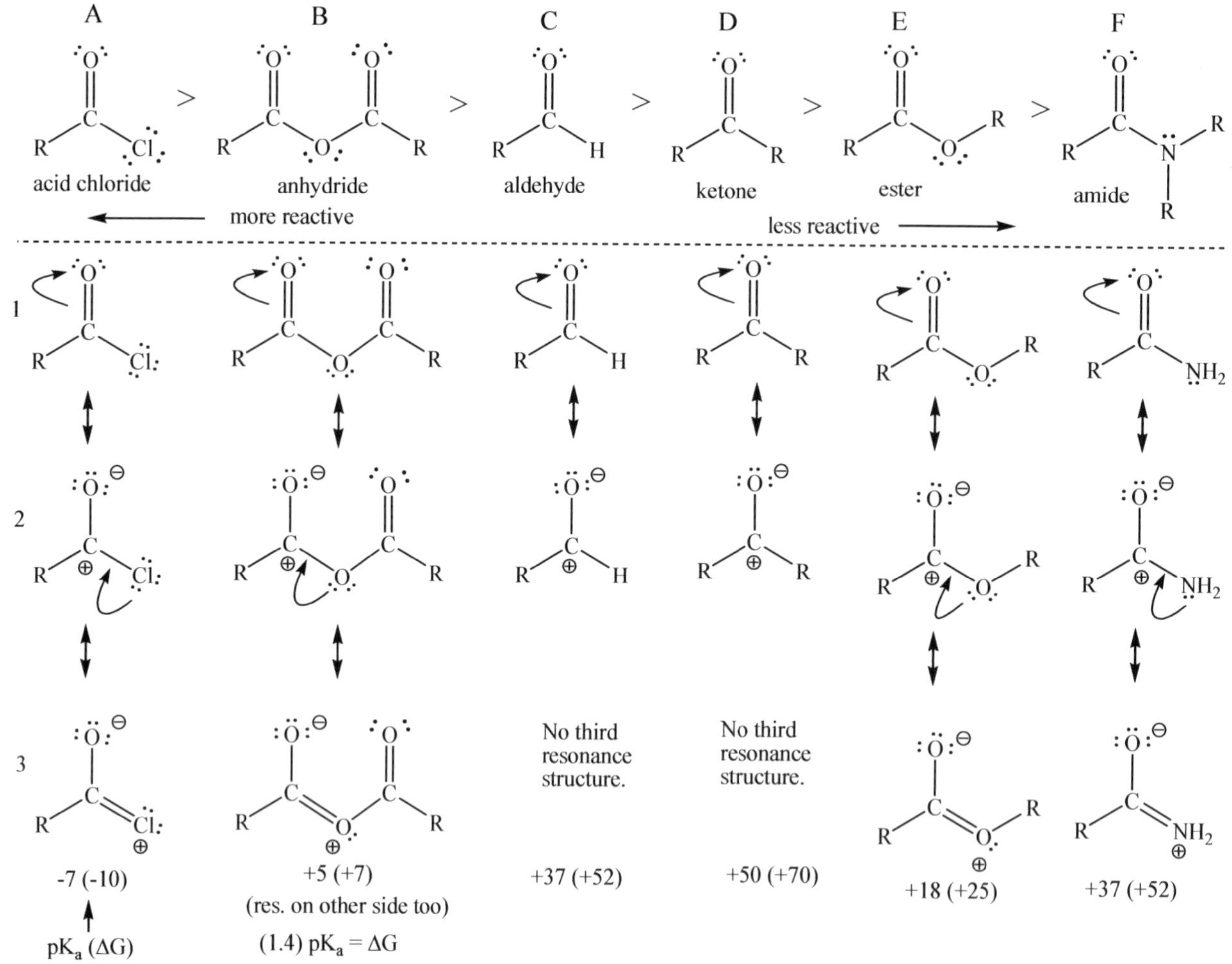

A	B	C	D	E	F
acid chloride	anhydride	aldehyde	ketone	ester	amide

⟵——— more reactive   less reactive ———⟶

| 3 | -7 (-10) | +5 (+7) (res. on other side too) | +37 (+52) | +50 (+70) | +18 (+25) | +37 (+52) |

pK_a (ΔG)   (1.4) $pK_a = \Delta G$

No third resonance structure. (C, D)

The first number below the resonance structures represents the pK_a of the leaving group (LG) part of the acyl group when it is protonated (the corresponding ΔG for acid ionization is shown in parentheses in kcal/mole). This provides a measure of how stable the leaving group is on its own. The lower the number, the more stable the leaving group is (chloride is the best leaving group and the amide anion is the worst).

Problem 9 – Write a pK_a equation for each conjugate acid group referred to above and write the ΔG of the reaction.

We study a wide range of carbonyl groups in organic chemistry, and as we have already seen, they span a wide range of reactivity. Which type of carbonyl group is more electrophilic (has greater δ+ and is better able to attract nucleophiles) depends on steric effects, inductive effects and resonance effects. Carbonyl compounds are divided on the basis of whether they usually undergo additions with nucleophiles (aldehyde and ketones) or substitutions with nucleophiles (esters, amides, carboxylic acids, anhydrides and acid chlorides). In comparing aldehydes and ketones, steric effects and inductive factors are most important. When considering carbonyl groups having an adjacent lone pair of electrons (esters, amides, acids, etc.), resonance effects must also be included. So, how do these factors affect carbonyl reactivity with nucleophiles?

Steric effects between aldehydes and ketones are fairly obvious. A ketone has carbon groups (larger) on both sides of the carbonyl group, while an aldehyde has only one carbon group and a hydrogen atom (smaller) on the other side. Not only does a larger steric effect slow attack by a nucleophile, but it makes the sp³ hybridized product more crowded in reactions of ketones than aldehydes (bond angles get tighter changing from 120° to 109°). Also important are inductive effects because they alter the amount of partial positive charge on the carbonyl carbon. A ketone has two inductively donating "R" groups, which reduces the amount of partial positive charge on the carbonyl carbon compared to an aldehyde, with only one "R" group. Both of these factors work to make ketones less reactive than aldehydes.

All carbonyl groups have the resonance shown (aldehydes, ketones, acids, esters, amides, anhydrides, acid chlorides, etc.)

Aldehydes are generally more reactive than ketones.

Ketones are generally more reactive than esters and amides.

➚ = inductive effect

1. The extra inductive effect of a ketone from its second "R" group stabilizes the carbonyl carbon and makes it less reactive (less partial positive).

2. Switching a hydrogen for a carbon group adds a larger steric effect, which further reduces the reactivity of a ketone.

3. Esters and amides are even less reactive because of a resonance effect from an adjacent lone pair of electrons (discussed below).

Problem 10 – Which ketone reacts faster with strong nucleophiles (A or B)? Explain your reasoning.

A

B

A notable difference in the carboxyl groups (having a carbonyl bond and a leaving group, Cl, O or N above) is that a third resonance structure is possible. Both oxygen (the alkoxy group of an ester) and nitrogen (the amine group of an amide) have a lone pair that can be shared with the positive carbon of the second resonance structure. This third resonance contributor is important because it has an extra bond and full octets (over the second, carbocation resonance structure). Its disadvantage with the first structure is that formal charge is created. This third resonance structure moves the positive charge off of carbon and puts it on the alkoxy oxygen in esters or the nitrogen in amides. As a result, both of these functional groups have less partial positive charge on the carbonyl carbon than aldehydes and ketones. This reduces their reactivity with nucleophiles below that of aldehydes and ketones. There is a complicating factor in that oxygen (esters) and nitrogen (amides) are more electronegative than carbon (ketones) and hydrogen (aldehydes). The inductive effect from oxygen or nitrogen is electron withdrawing relative to carbon or hydrogen. This should make the carbonyl carbon more positive and more reactive. Ambiguity strikes again! Usually, but not always, when a resonance effect opposes an inductive effect, resonance wins out. That is the case in these examples, and the third contributing resonance structure of esters and amides makes these groups less reactive towards nucleophiles, when compared to aldehydes and ketones. (However, that is not true for acid chlorides and anhydrides discussed just below. In acid halides (Cl or Br) there is poor overlap of a chlorine 3p or bromine 4p orbital with carbon 2p orbitals, so the resonance contribution is poor.

In contrast to our general rules of resonance, structure C of the acid chloride (below) is a minor contributor due to poor 3p/2p orbital overlap. Inductive effects of the chlorine dominate over weak resonance effects in this instance. This makes the electrophilic carbon of the carbonyl more reactive towards nucleophiles (greater partial positive) than aldehydes and ketones. Because of the decreased importance of structure C, structures A and B are relatively more important and the carbonyl group is more electrophilic.

Also, the pK$_a$'s of the leaving group conjugate acids suggest that Cl $^-$ is a very stable anion, and therefore an excellent leaving group.

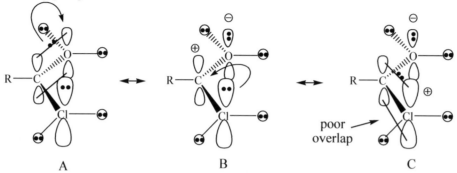

A

A is the best resonance structure because it has maximum bonds and no charge separation

B

C

poor overlap

Normally C would be the second best resonance structure, but inefficient overlap of the 2p and 3p orbitals allows the inductive withdrawal of cholorine to overpower its weak resonance effect.

In amides and esters, the third resonance structures are more important than the second resonance structures. There is good overlap of the carbon 2p orbital with the oxygen and nitrogen 2p orbitals and the third resonance structure has full octets. This is especially true for amides, where nitrogen is donating (estimated to be a 40% contributor). In these structures the resonance effect overpowers the inductive withdrawing effect of oxygen or nitrogen. Recall the difference in basicity (electron donating ability) between nitrogen and oxygen is many, many orders of magnitude [pK$_a$(H$_2$O)=16, pK$_a$(NH$_3$)=35].

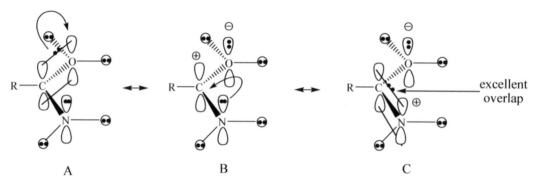

A

A is the best resonance structure because it has maximum bonds and no charge separation

B

C

excellent overlap

C is the second best resonance structure. Overlap of the nitrogen lone pair fills in the octet of the carbon in B and makes an extra bond. There is efficient overlap because both atoms are using 2p orbitals. Donation from nitrogen is much better thanfrom oxygen because it's less electronegative. This third structure is even better for the amide than it is for the ester.

Anhydrides seem out of place in reactivity, since they too have an oxygen atom that can donate electron density into the carbonyl bond (like the ester group). However, there is a carbonyl group on both sides of the middle oxygen, and the ability to share electrons by resonance is cut in half. The strong inductive effect of oxygen becomes dominant, and makes the carbonyl carbon more electrophilic and reactive than expected in comparison to the ester example. The carboxylate leaving group (in anhydrides) is also much more stable, and it is a better leaving group than the alkoxide leaving group (in esters). .

Chapter 11

The neutral resonance
structure is the best.

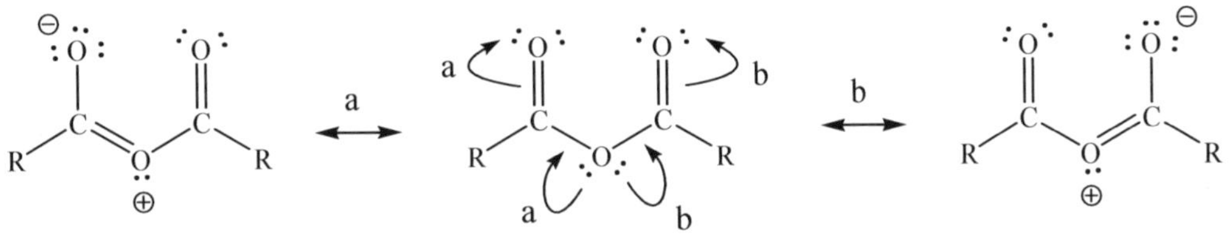

The middle oxygen can donate in two opposite directions, which weakens its donating power into both carbonyl groups by 50%. Inductive withdrawal by the middle oxygen over powers the resonance effect and makes the carbonyl carbons more partially positive and more reactivve with nucleophiles. Also a carboxylate anion leaving group (from an anhydride) is a better leaving group than an alkoxide leaving group (from an ester).

If you reread the explanations above, you will find every argument we use in organic chemistry (and biochemistry): steric effects, inductive effects and resonance effects. You can really sharpen you organic skills by learning how to explain these differences, which helps you in biochemistry too.

Problem 11 –
a. How does the reactivity of methanal compare with the reactivity of a simple aldehyde? Explain your reasoning using structures. (Hint: write out the resonance structures and evaluate the partial positive on the carbonyl carbon.)

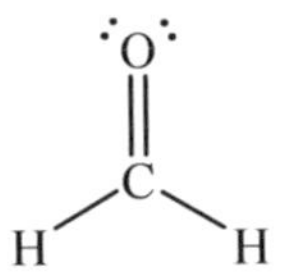

methanal

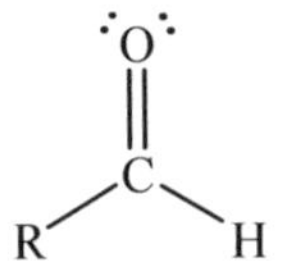

simple aldehyde

b. How does the reactivity of a methanoate ester compare with the reactivity of a regular simple ester? Explain your reasoning using structures. (Hint: write out the resonance structures and evaluate the partial positive on the carbonyl carbon.)

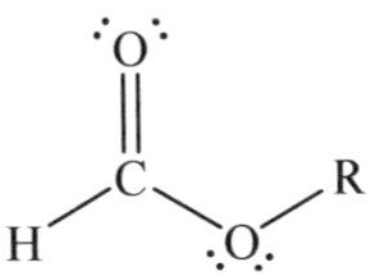

alkyl methanoate
(alkyl formate)

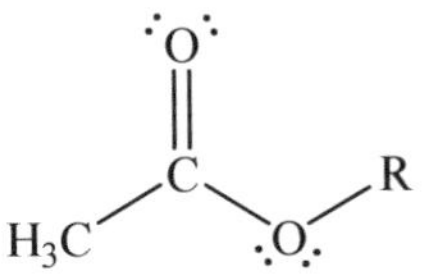

alkyl ethanoate
(alkyl acetate)

c. How does the reactivity of methanamide compare with the reactivity of a regular simple amide? Explain your reasoning using structures. (Hint: write out the resonance structures and evaluate the partial positive on the carbonyl carbon.)

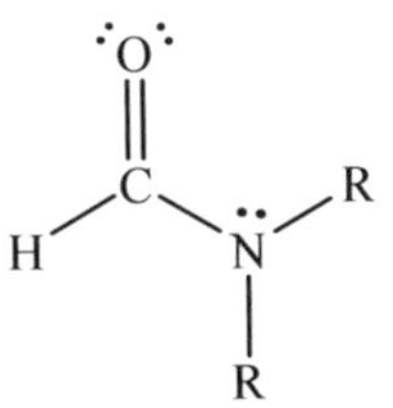

N-alkyl methanamide
(N-alkyl formamide)

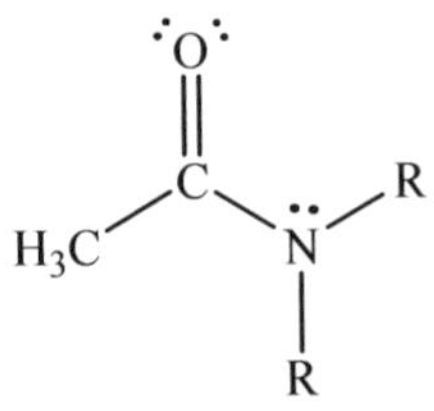

N-alkyl ethanamide
(N-alkyl acetamide)

d. Is a thioester more like an oxygen ester or an acid chloride? Explain your answer. What order of reactivity would you predict for these 3 carboxyl functional groups?

thioester   acid chloride   ester

We can now make the following molecules (and more). Not every possibility is shown. From our starting hydrocarbons, propose a synthesis for each type of functional group until you know what you are doing.

Starting sources of carbon (10 compounds)

CH_4 NaCN CO_2

----

1. Possible RBr (bromohydrocarbons) compounds from these starting hydrocarbons using free radical substitution at C-H (Br$_2$/hv) and free radical addition to C=C (ROOR/HBr/hv). When other functional groups are available we can use alcohols (HBr or PBr$_3$ or 1. TsCl/py 2. NaBr) and alkenes (electrophilic HBr), which are common starting points. Dibromoalkanes can make alkynes using double E2 reactions.

dibromoalkanes can make alkynes
(use hydrocarbon and 2 eqs of Br$_2$)

Examples

2. Possible C=C (alkenes) and C≡C (alkynes) from the above starting hydrocarbons. Currently, our only strategies are to use E2 with potassium t-butoxide and RBr compounds or E1 with an alcohol / H$_2$SO$_4$ / Δ. Also possible later is the reduction of alkynes to E alkenes (Na/NH$_3$) and Z alkenes (Lindlar's catalyst) and variations of the Wittig reaction (phosphorous ylids and aldehydes or ketones). To make alkynes we have to start with dibromoalkanes and use 3 equivalents of NaNR$_2$ (ioinic), followed by mild acidic workup.

Chapter 11

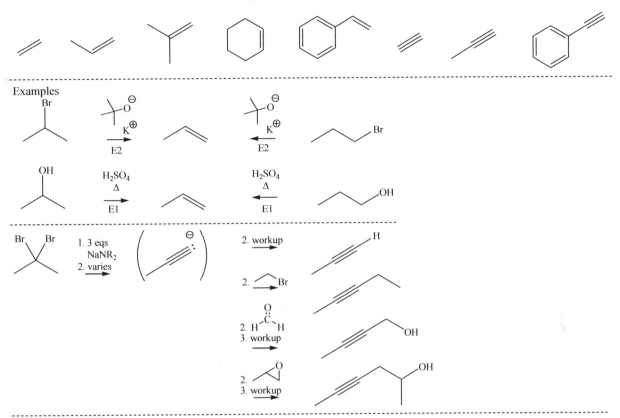

3. Possible ROH compounds (alcohols) from the above RBr compounds, so far we can use S_N2 (HO⁻) or S_N1 (H_2O), and there are many additional ways to make alcohols from alkenes, C=O compounds and epoxide compounds. Remember, hydroxide is too basic for good S_N2 at 2° RBr, so either use S_N1 if rearrangement is not a problem or two steps: 1. S_N2 with ethanoate (acetate) and 2. ester hydrolysis with NaOH.

4. Possible ROR compounds (ethers) from the above RBr compounds, so far we only use S_N2 (HO⁻) or S_N1 (H_2O), but there are other ways to make ethers, such as alkenes. Only one R group is explicitly shown. Some similar strategies to ROH.

5. Possible RSH compounds (thiols) from the above RBr compounds. We only have one approach using sodium hydrosulfide in an S_N2 reaction (HS⁻) at methyl, 1° and 2° RBr. At 3° RBr, E2 is the main product.

6. Possible RSR compounds (sulfides) from the above RBr compounds. We only have one approach using sodium thiolates (NaSR, made from the thiol and NaOH). Reactions occur as S_N2 (RS⁻) at methyl, 1° and 2° RBr. At 3° RBr, E2 is the main product. Only one R group is explicitly shown.

Chapter 11

7. Possible RNH_2 compounds (1° amines) from the above RBr compounds. We have three approaches. One uses sodium azide in an S_N2 reaction (N_3^-), followed by reduction with $LiAlH_4$ (another S_N2 reaction) and workup. Another uses sodium cyanide (NC^-) followed by reduction with $LiAlH_4$ and workup. For us, these will both work at methyl, 1° and 2° RBr. At 3° RBr, E2 is the main product. A third possible approach is to make a primary amide from an acid chloride and ammonia, then reduce it using $LiAlH_4$ and workup.

8. Possible RR'NH and RR'NR" compounds (2° and 3° amines) from 1° and 2° amines and aldehyde or ketone compounds (C=O). Mild acid is used (pH = 5) and removal of water (-H_2O) is used to form imines. The equilibrium can be shifted in the opposite direction with acidic water. A second possible approach is to make a secondary or tertiary amide from an acid chloride and a primary or secondary amine, then reduce it using $LiAlH_4$ and workup. Only one R group is explicitly shown.

9. Possible RCHO compounds (aldehydes) from the above RBr → ROH → RCHO using PCC (CrO₃/pyridine) and other approaches (terminal alkyne → 1. R₂BH, 2. H₂O₂, HO⁻).

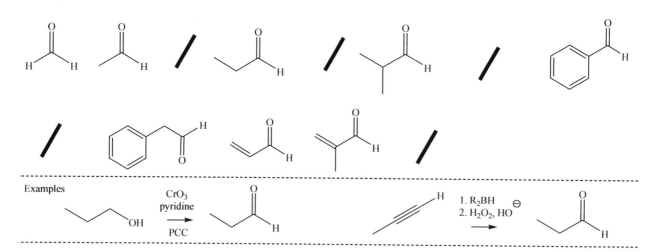

10. Possible RCOR' (ketones) compounds from the above RBr → ROH → RCOR' using PCC (CrO₃/pyridine) or Jones (CrO₃/H₂O) and acid hydrolysis of alkynes (1. HgX₂, H₂O, 2 NaBH₄).

There are other ways to make nitriles, but we do not cover them.

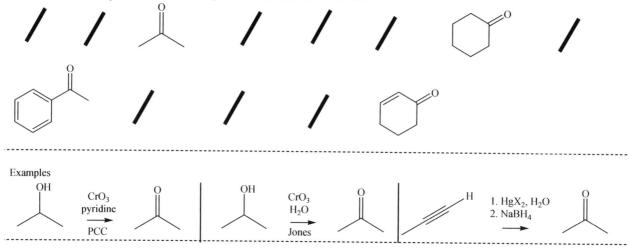

11. Possible RCO$_2$R' compounds (esters) from the above RBr compounds and other approaches. We have four approaches. One uses sodium carboxylates in an S$_N$2 reaction (RCO$_2$ $^-$) at methyl, 1° and 2° RBr. Another uses carboxylic acids in an S$_N$1 reaction with 2° and 3° RBr (possible rearrangements). Another uses the Fischer ester synthesis, combining a carboxylic acid and an alcohol with acid catalyst (TsOH) and removal of water (-H$_2$O). This is an equilibrium reaction that can be reversed by adding water and H$_2$SO$_4$. Finally, mixing an acid chloride with an alcohol favors ester formation.

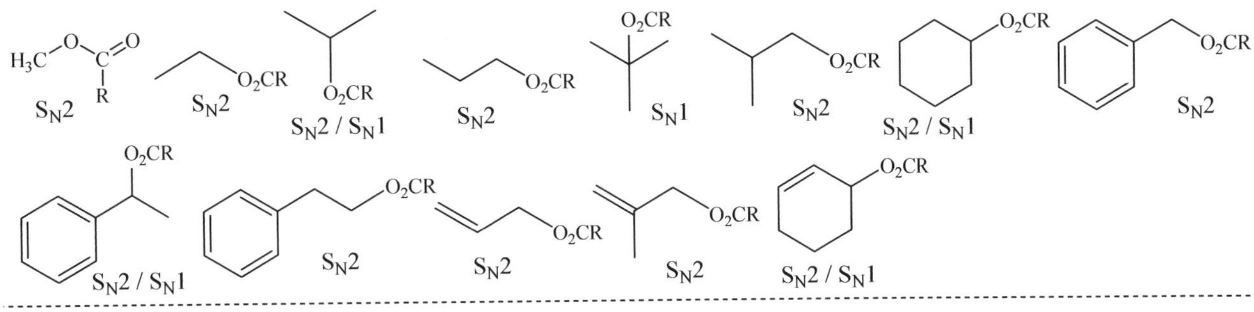

Examples

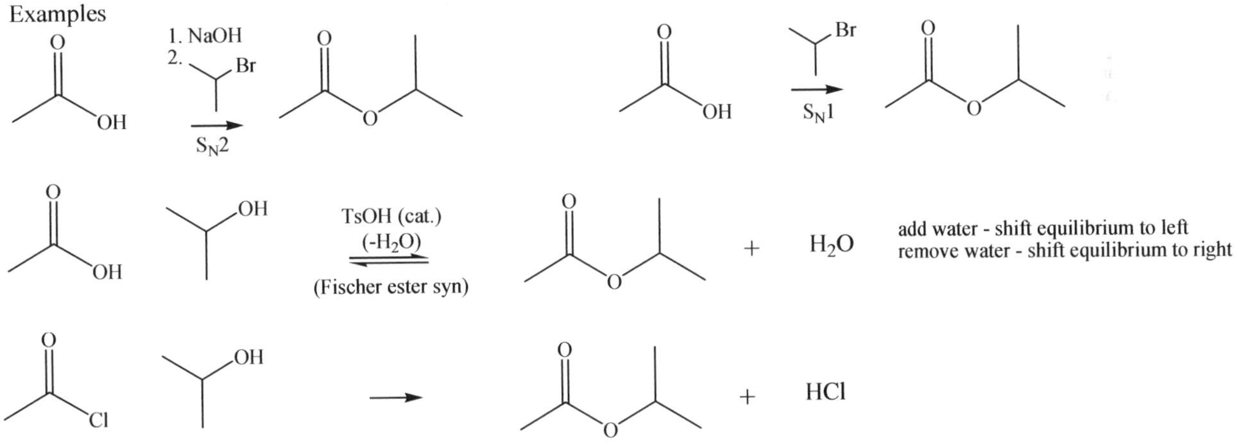

12. Possible RCN compounds (nitriles) from the above RBr compounds, we only use S$_N$2 (NC $^-$) at methyl, 1° and 2° RBr. Reaction of cyanide at 3° RBr centers only leads to E2 reactions (in our course). There are other ways to make nitriles, but we do not cover them.

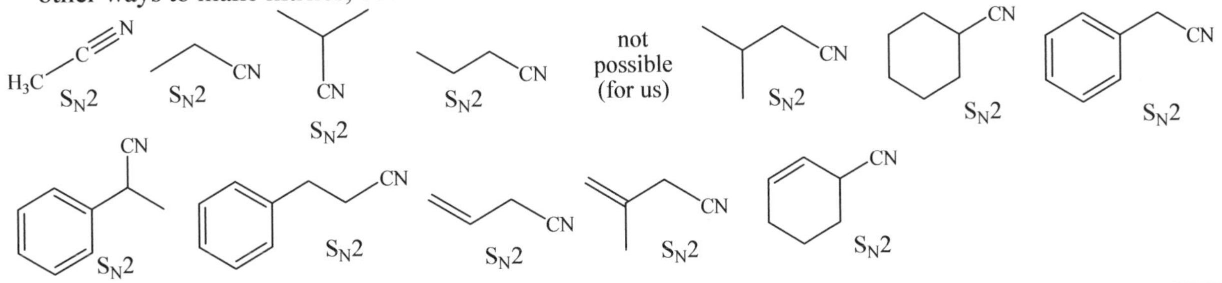

Examples

13. Three possible RC≡CR' (R' can be C or H) compounds (alkyne extensions) can be made from the above RBr compounds using a double E2 reaction from dibromoalkanes (ethyne, propyne and phenylethyne). These become starting points for additional alkynes using S_N2 (RCC⁻) reactions at methyl and 1° RBr. Reaction of acetylides at 2° and 3° RBr centers leads to E2 reactions (in our course). Acetylides will add to C=O and epoxides too. There are even other ways to make alkynes, but we do not cover them.

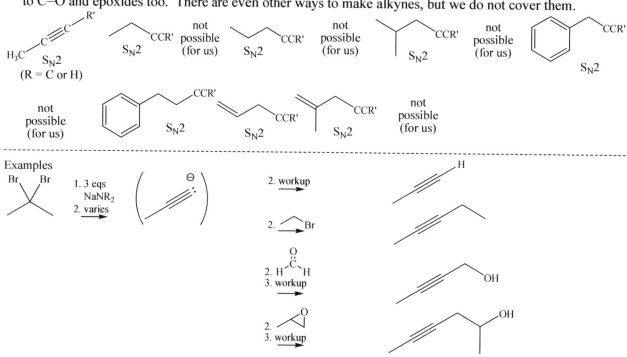

14. We will use 3 variations of enolate compounds with the above RBr compounds. These will include simple ketone enolates (propanone and cyclohexanone), ester enolates (mainly from alkyl ethanoates) and acetonitrile enolates. We propose that all of these are S_N2 reactions at methyl, 1° and 2° RBr. There is much, much more to this topic, but not in our short course.

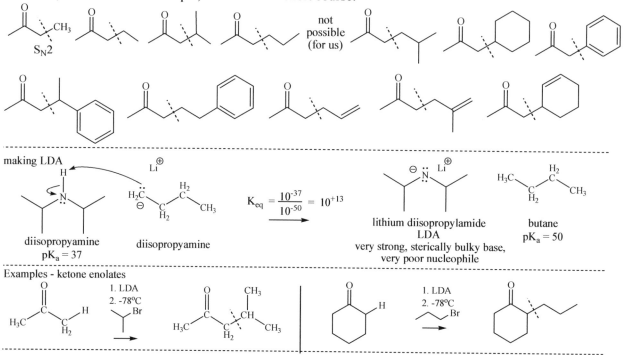

Chapter 11

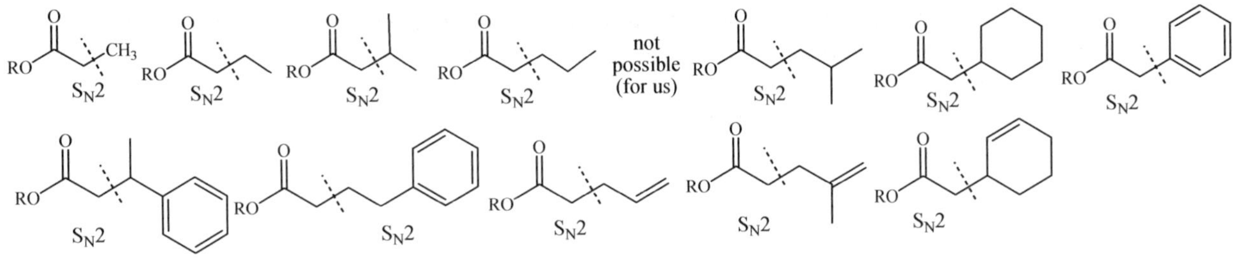
RO
CH3
SN2
not
possible
(for us)

Examples - ester enolates

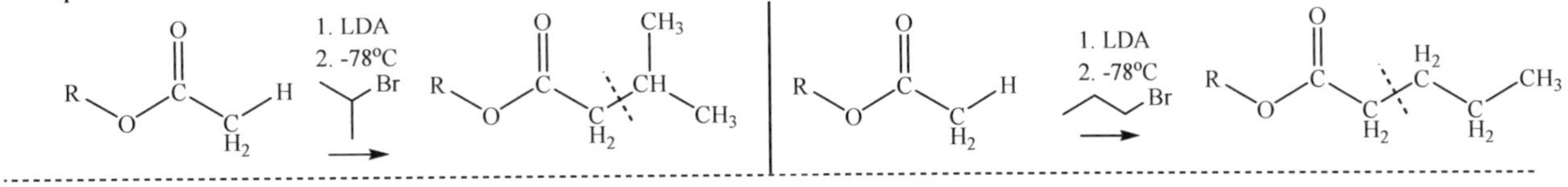
1. LDA
2. -78ºC
Br
CH3
CH
H2

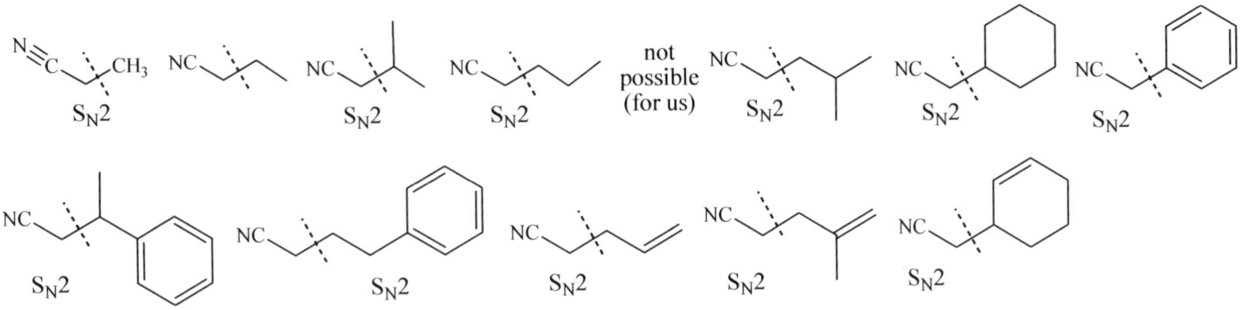
NC
SN2
not
possible
(for us)

Examples - nitrile enolates

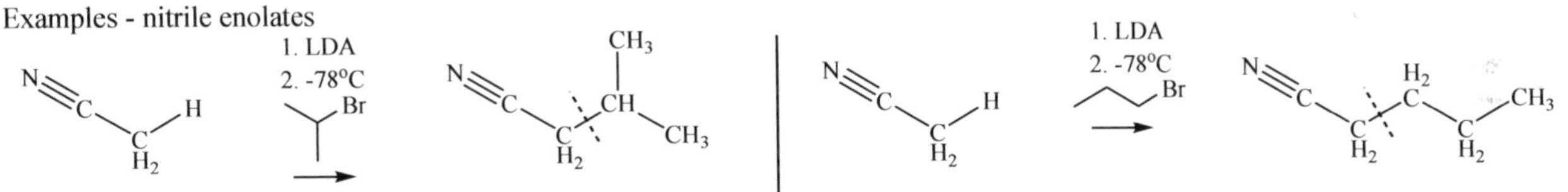
1. LDA
2. -78ºC
Br
CH3
CH
H2

15. We will use 2 variations of ylid chemistry. Sulfur salts will be made with diphenylsulfide and the above RBr compounds. Phosphorous salts will be made with triphenylphosphine and the above RBr compounds. We propose that both of these are S$_N$2 reactions at methyl, 1° and 2° RBr. Sulfur salts will make sulfur ylids when reacted with n-butyl lithium that will react with aldehydes and ketones to make epoxides. Phosphorous salts will make phosphorous ylids when reacted with n-butyl lithium that will react with aldehydes and ketones to make alkenes. We will study those reactions in our next chapter.

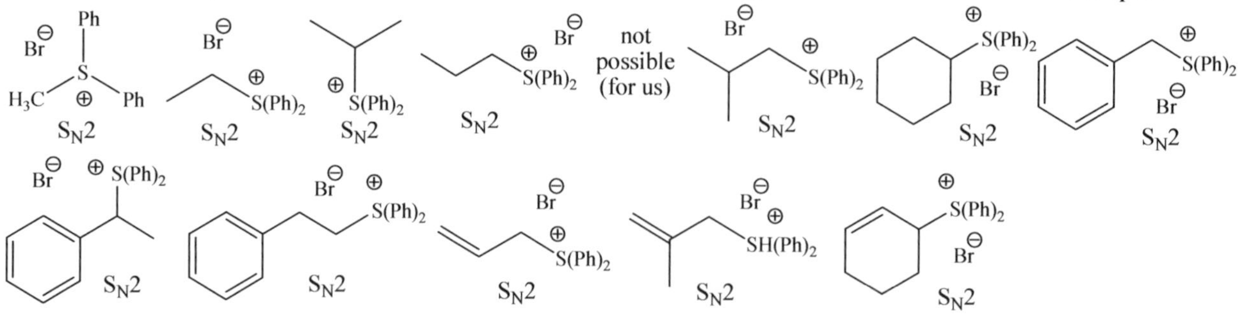

Examples

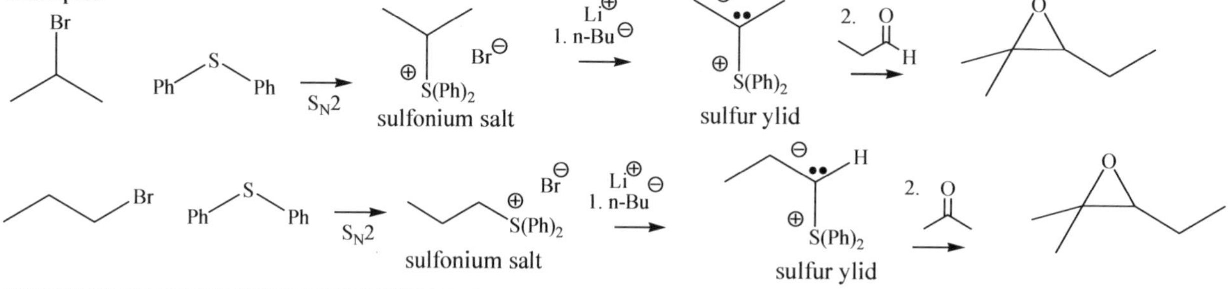

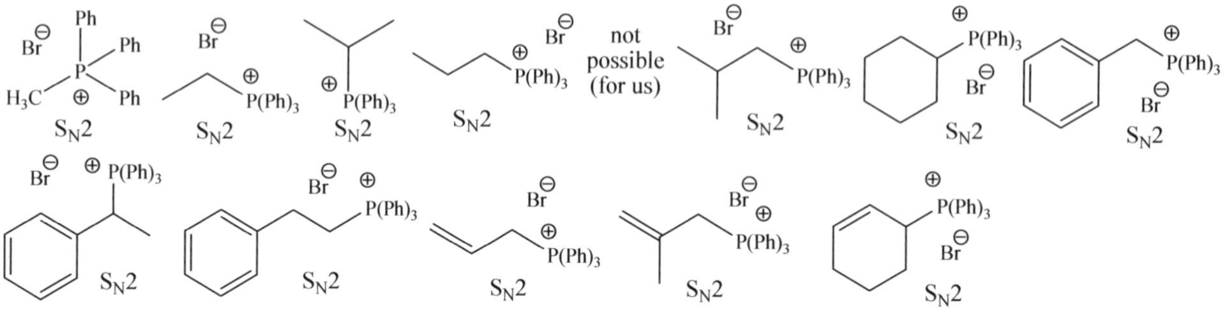

Examples

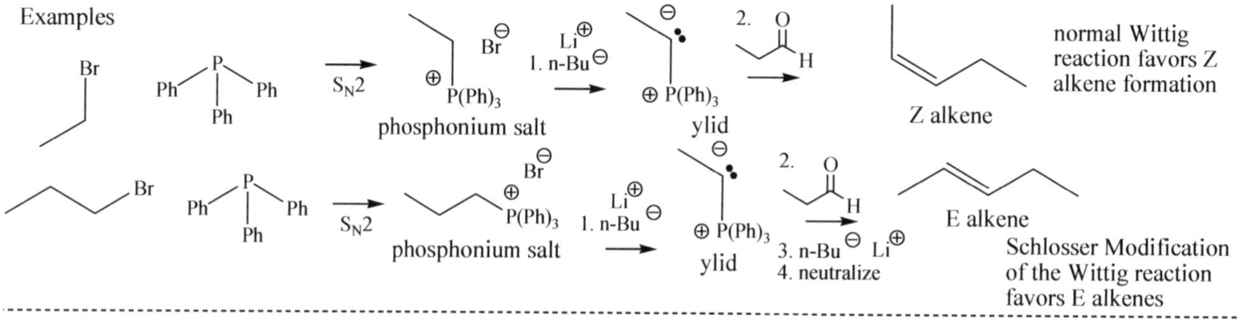

16. We will use two approaches to make carboxylic acids. One approach uses methyl and primary alcohols and Jones reagent (CrO_3/H_2O). The other approach uses RBr compounds and Mg to make a Grignard reagent, which is followed with carbon dioxide, CO_2 and acidic workup. There are several other approaches, but not in our short course.

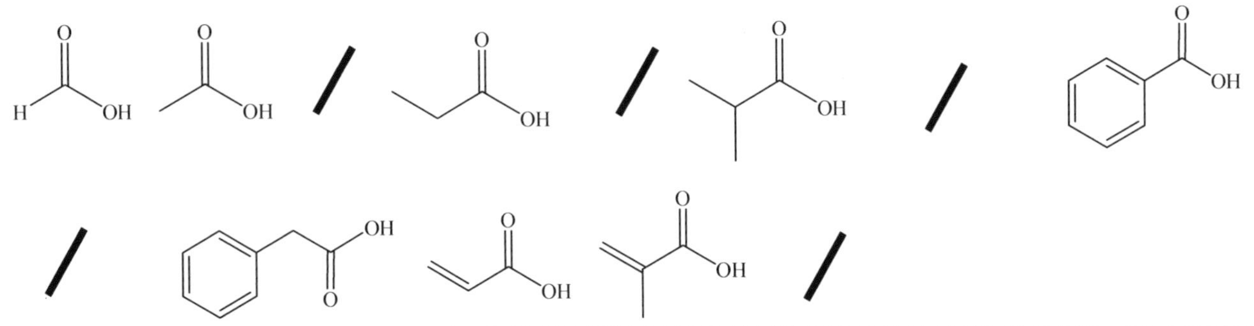

Examples

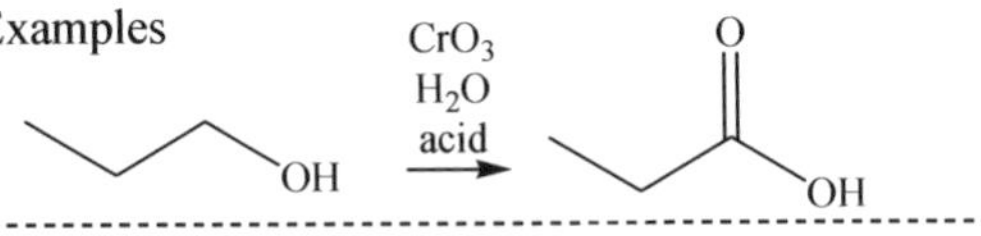
CrO3
H2O
acid

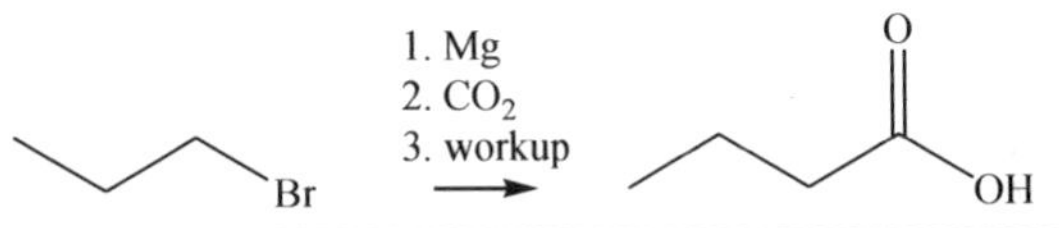
1. Mg
2. CO2
3. workup

17. We will only use one approach to make acid chlorides. We will mix carboxylic acids and phosphorous trichloride. There are other approaches, but not in our short course. Acid chlorides are versatile compounds because they can make many other functional groups.

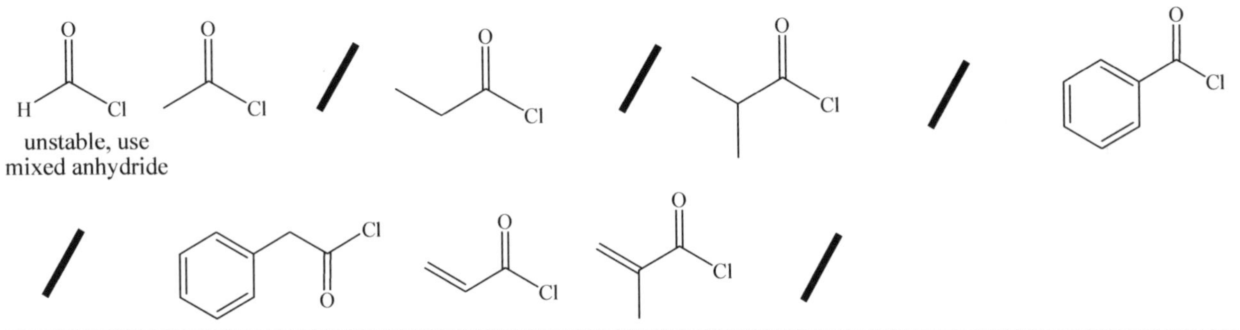
unstable, use
mixed anhydride

Examples

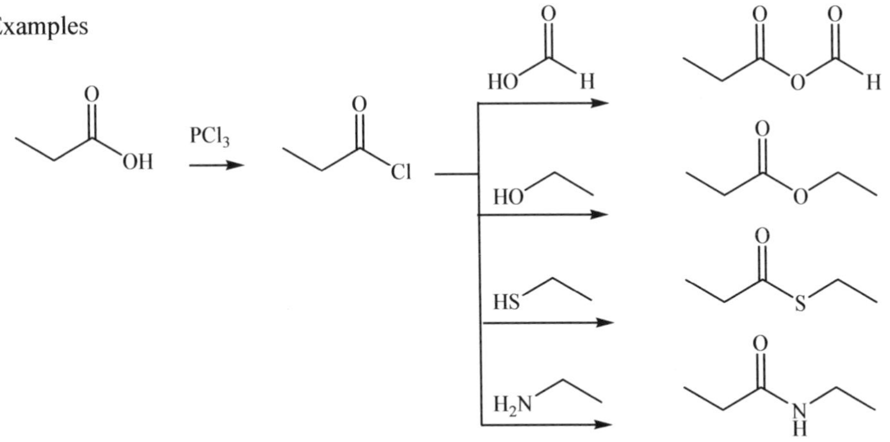
PCl3

18. We will two approaches to make amides. One approach uses nitrile hydrolysis in aqueous acid solution to make primary amides. The other approach uses acid chlorides with ammonia, primary amines or secondary amines. Since there is no methanoyl chloride, we need to use an alkanoic methanoic anhydride (mixed anhydride). There are several other approaches, but not in our short course.

19. We will only use one approach to make anhydrides. We will mix carboxylic acids and acid chlorides. There are other approaches, but not in our short course. Anhydrides are versatile compounds because they can make many other functional groups, but we mainly use the more reactive acid chlorides. The only reaction we emphasize in our course is right above, alkanoic methanoic anhydride and that is because methanoyl chloride is not stable and this gives us a 1C acyl derivative that we can make into other functional groups.

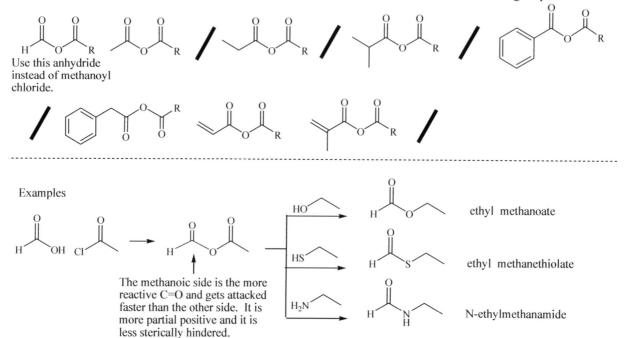

20. We will only use one approach to make thioesters. We will mix thiols and acid chlorides. They are very reactive acyl groups. The most famous of them all is a biological molecule called acetyl Co-A, which you have probably heard of in one of your biology courses.

21. We will only use one approach to make α,β-unsaturated compounds. We will mix lithium diisopropylamide with epoxides that have a C-H bond next to the epoxides. An E2-like reaction forms a C=C and an allylic alcohol. The alcohol has to be oxidized (with a CrO_3 reagent in our course). We will use PCC so we can get the aldehyde or ketone.

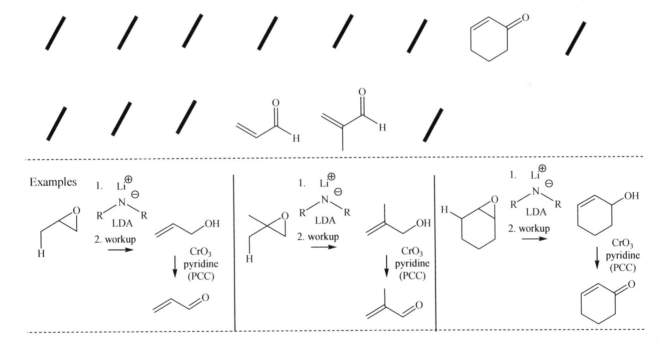

Chapter 11

22. Possible epoxides from the above RBr compounds and other approaches. Carbonyl (C=O) plus methyldiphenylsulfonium bromide (CH_3SPh_2 / Br^-). We will introduce another method in the alkene chapter (1. Br_2/H2O 2. NaOH). This will allow us to make cyclic epoxides, which the sulfur ylid approach does not allow.

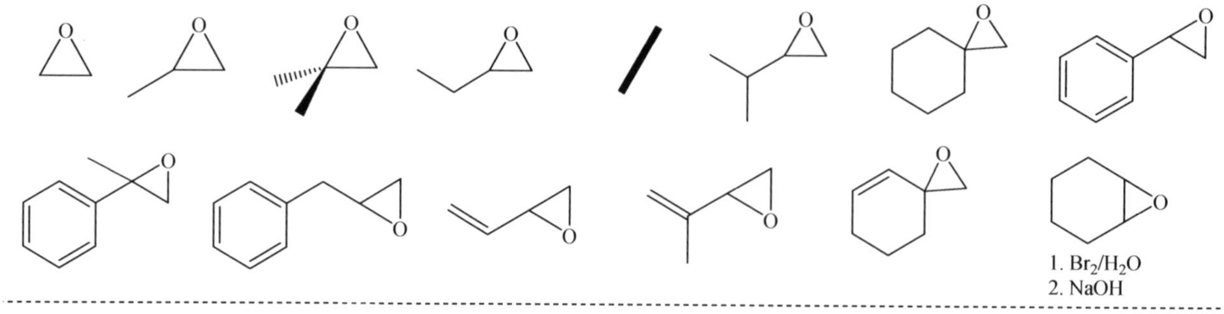

1. Br_2/H_2O
2. NaOH

Examples

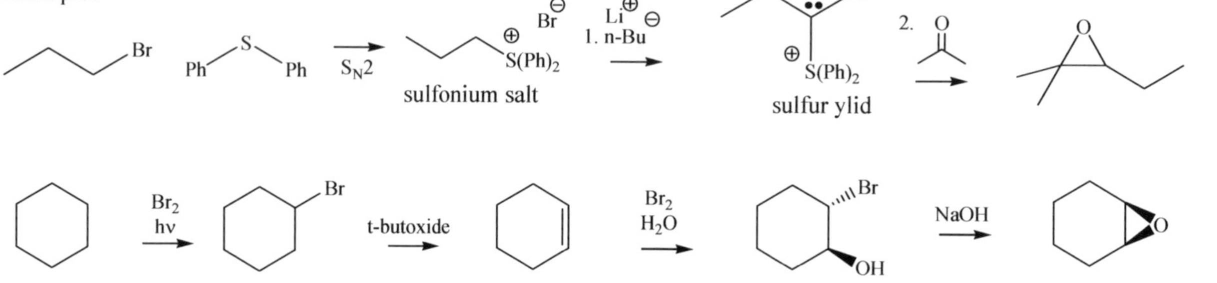

sulfonium salt

sulfur ylid

Br$_2$
hv

t-butoxide

Br$_2$
H$_2$O

NaOH

It's pretty amazing what is possible with the chemistry we have learned so far. We only have 7 hydrocarbon starting points, but the variety of compounds we can make is very impressive. And, there's more still to come. Next up, epoxides, Grignard reagents and sulfur and phosphorous ylids.

Chapter 11

Chapter 12 - Epoxides, Grignards, LDA Enolates, Sulfur Ylids and Phosphorous Ylids

Why are epoxides different?

Epoxides are ethers, but they're not typical ethers. The sp^3 hybridization suggests the atoms prefer $109°$ bond angles, but the geometry of the ring enforces $60°$ bond angles. There is a lot of angle strain that inclines the ring to open. Strain energy is estimated at ~27 kcal/mole and that is energy that is released upon rupture of one of the bonds, and that makes them special electrophiles that we can use in synthesis. Because of their large strain energy epoxides undergo many reactions where normal ethers do not react

Simple ethers are fairly unreactive when in basic and neutral conditions. They are even used as solvents in Grignard reactions (discussed after epoxides), which form very basic and nucleophilic carbanions. The two most common ether solvents are diethyl ether and tetrahydrofuran (THF). In Grignard reactions they help solubilize an organomagnesium reagent, yet they do not react with it. However, they will react in strong acid conditions when the ether oxygen is protonated.

diethyl ether	THF (tetrahydrofuran)	ethylene oxide (oxirane)

These are the most common solvents to run Grignard reactions in. They do not usually react with our strong electron pair donors.

Epoxides aren't typical ethers. Because of their large ring strain they do react with our strong electron pair donors, like Grignard reagents.

Problem 1 – THF boils at 66°C and its dipole moment is 1.63 D. The boiling point of diethyl ether is 35°C and its dipole moment is 1.15 D. The boiling point of ethylene oxide is 11°C and its dipole moment is 1.8 D. The boiling point of dimethyl ether is -24°C and its dipole moment is 1.30 D. Propose a possible explanation for the different physical properties of these similar looking ethers.

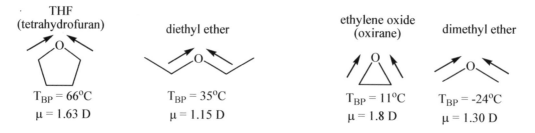

THF (tetrahydrofuran)	diethyl ether	ethylene oxide (oxirane)	dimethyl ether
$T_{BP} = 66°C$	$T_{BP} = 35°C$	$T_{BP} = 11°C$	$T_{BP} = -24°C$
$\mu = 1.63$ D	$\mu = 1.15$ D	$\mu = 1.8$ D	$\mu = 1.30$ D

Making Epoxides from Sulfur Ylids and Aldehydes or Ketones

There are several ways to make epoxides, but as usual, we will be selective in our choice of reactions. We will look at one way here and one in the alkene chapter. Here, we will use a sulfonium salt, reacted with the strongest base available to us, n-butyl lithium, to make a sulfur ylid, a neutral dipolar molecule. The sulfur ylid has a strongly nucleophilic carbanion that can attack the electrophilic carbon of an aldehyde or ketone. The negative oxygen that is formed closes back down on the adjacent carbon with the sulfonium ion and kicks out a good, neutral sulfur leaving group, thus forming the epoxides (an intramolecular S_N2 reaction). The sulfonium salts that we are proposing are easy to make from an S_N2 reaction of a sulfide with an RBr compound (C1 – C6 are examples of what is possible to us). We will limit ourselves to our usual 7 hydrocarbon starting points.

Once again we are learning a reaction that repeats our common strategy of joining nucleophile to electrophile, in this case, to make epoxides. Aldehydes and ketones are electrophiles that we are already familiar with. We will make a new nucleophilic carbon from a salt, diphenylalkylsulfonium bromide, and n-butyl lithium, using a simple acid/base reaction. There is no base stronger (in our course) than a simple alkyl carbanion like n-butyl carbanion (pK$_a$ of conjugate alkane acid ~50). The positive charge on the sulfur makes the adjacent methyl proton weakly acidic (pK$_a$ ~35). The doubly and oppositely charged sulfur conjugate base is called an ylid, and provides the nucleophilic carbon to make this reaction work. There are other ylids, e.g. phosphorous ylids, that we will use later in this chapter, and some we won't talk about. Diphenylsulfide and n-butyl lithium are available in our chemical catalog, and always will be. We will use these to make the sulfonium salts (from RBr or RI) that are used to make epoxides.

Make a sulfur ylid to react with aldehydes and ketones to make epoxides

diphenyl sulfide — RX compounds — S_N2 — diphenylalkylsulfonium bromide

This is a salt.

We can make a lot of variations of these salts. Just think of all of the RBr choices in our table that can undergo S_N2 reactions (p. 2, S_N/E topic).

diphenylalkylsulfonium bromide pK$_a$ ≈ 35

n-butyl lithium the strongest base we have

very favorable acid/base reaction to make the ylids

$$K_{eq} \approx \frac{10^{-35}}{10^{-50}} \approx 10^{+15}$$

sulfur ylids (two opposite charges in one molecule), good carbanion nucleophiles at carbonyl carbon atoms

butane pK$_a$ ≈ 50 (throw away)

The nucleophilic carbanion of the sulfur ylid attacks the electrophilic carbonyl carbon (aldehyde or ketone) and adds to the carbonyl group, something we have seen many times by now. You might not predict what happens next, but once you see it, the mechanistic step should seem logical. The strongly nucleophilic/basic alkoxide does a backside attack on the carbon bonded to positively charged sulfur, an excellent leaving group as diphenylsulfide. Transfer of negative charge on oxygen to neutral sulfur helps to compensate for ring strain in the epoxide. The result is the desired epoxide functional group and stinky diphenylsulfide (a throw away for us). Suddenly we can make a lot of epoxides. Our combinations are all of those new aldehydes and ketones, plus most of our RBr compounds that can now be turned into epoxides. That's a lot of possibilities.

We can combine all of these ⤴ with all of these. ⤵ to make epoxides

not possible (for us)

Using this approach, one of the carbons in an epoxide comes from a carbonyl compound and one comes from the sulfur ylid, which comes from R-Br. This means there will often be two approaches to the synthesis of an epoxide, as shown in the following examples.

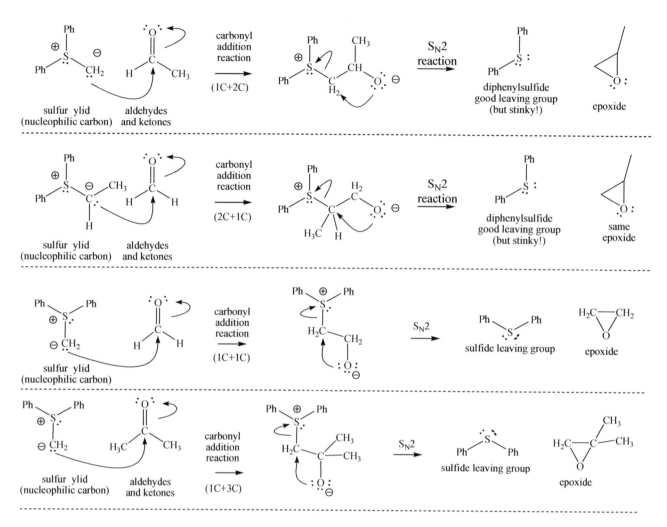

Using the single diphenylmethylsulfonium ylid we can make twelve different epoxides from twelve aldehydes and ketones made from our seven hydrocarbon starting structures.

The new bond forms here using this strategy. One carbon is the nucleophile and one carbon is the electrophile.

(continued on next page)

Chapter 12

Problem 2 – Nearly all of the epoxides in this problem can be made two ways using the sulfur ylid/carbonyl strategy. Propose a reasonable synthesis for each of the following epoxides using the sulfur ylid approach.

The new bond forms here using this strategy. One carbon is the nucleophile and one carbon is the electrophile.

Ph = phenyl

We still cannot make cycloalkene epoxide (like cyclohexene oxide), but once we study alkenes we will be able to make almost any needed epoxide.

Problem 3 –Propose a mechanism for how the bromohydrin of cyclohexene (given for now, but made from cyclohexene and Br_2/H_2O in our next topic) could make cyclohexene oxide. Consider what is missing in the product epoxide from the starting bromohydrin, and the reaction conditions that make it happen.

cyclohexene

Br$_2$
H$_2$O

Covered in the alkene topic.

cyclohexane bromohydrin
(given for now)

mild
NaOH

Propose a mechanism
for this reaction using
two mechanisms we
have previously studied.

cyclohexene oxide

We now have three classes of very useful electrophiles: bromoalkanes, carbonyl compounds and epoxides and a whole slew of strong nucleophiles and bases. There is one more powerful nucleophile we will introduce. This nucleophile won a Nobel Prize in 1912 and it is still included in every organic book used. A French chemist named Victor Grignard discovered that if magnesium metal was mixed with bromohydrocarbons, a carbanion-like reactant was formed. This was one BIG discovery!

The Grignard Reaction

Up to this point, we have avoided any mention of metals in our discussions. Metals are actually used a lot in organic chemistry, but their mechanisms are often "magical" in how they occur. Metallic elements are pretty rare in our world. Gold and silver are two that are common, however, you will never find magnesium metal naturally occurring in nature. That's because magnesium wants to give away its two valence electrons so it can assume a noble gas configuration (like neon). It's made artificially, using electrochemistry, and then we buy it. We put it in a situation where it can give away its two electrons (oxidation) to a C-Br bond (reduction). A carbanion-like nucleophile and bromide are formed and complex to the Mg^{+2} cation. This has become known as the Grignard Reaction.

Using the Grignard reaction, we have an extremely reactive (and valuable) carbanion electron pair donor (as a base to a proton or as a nucleophile to a carbon electrophile). The magnesium variation was actually

Chapter 12

developed first, in the late 1800s by a French Chemist named Victor Grignard, and won him the 1912 Noble Prize. His advisor, Biot, was working with Zn compounds but, as the story goes, they kept catching on fire and he thought Mg might make a good substitute (little did he know his student would win the Noble Prize, or he might have run the reactions himself). Magnesium and lithium can supply two electrons (2xLi or 1xMg) to reduce both the bromine and the carbon. This one electron transfer chemistry is a little different than we are used to. However, as soon as the carbanion is made, we are back to our nucleophile/electrophile chemistry (two electron transfers).

Making Carbanions from RX Compounds and Metals (Li and Mg) – A Nobel Prize winning combination

To make Grignard carbanion nucleophiles, we are again going to have to move out of our comfort zone. Our source of carbon will be R-Br compounds in combination with metals (lithium and magnesium are common, but we will only use magnesium). The table of R-Br compounds on pages 247-248 of the S_N/E topic gives just a hint of the possibilities. Neither metallic lithium nor metallic magnesium is found in nature. The pure metals are too unstable towards losing their valence electrons, so we have to buy these (lithium loses one, Li → Li$^+$ + 1e- and magnesium loses two, Mg → Mg^{+2} + 2e-). We only emphasize magnesium in our course.

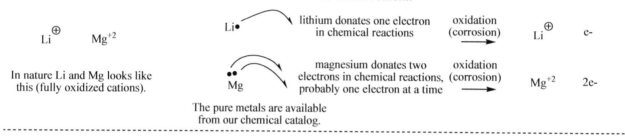

Our goal is to put the magnesium metal in a controlled situation where it vigorously gives away its electrons to a site of our choosing (the C-Br bond here). The mechanism is not going to be like anything we have studied before. We use half-headed arrows to show one electron movement, but we will simplify this to 2 electron movement below.

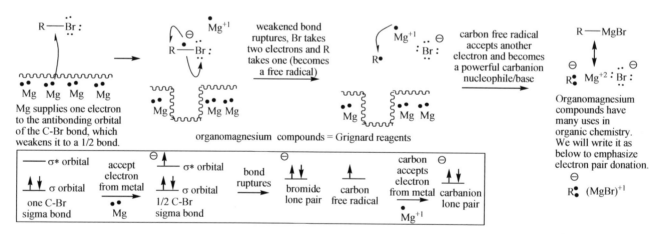

Problem 4 – A side product of magnesium organometallic reactions is coupling of two R groups (R-R). Propose a possible mechanism for how this could happen.

Chapter 12

Possible ways to write a Grignard reagent, in an ether solvent.

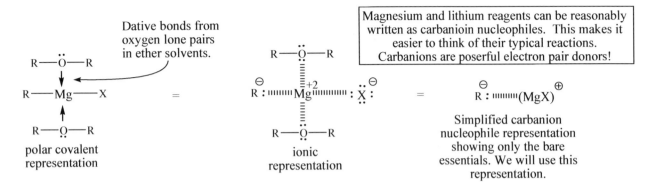

Dative bonds from oxygen lone pairs in ether solvents.

Magnesium and lithium reagents can be reasonably written as carbanioin nucleophiles. This makes it easier to think of their typical reactions. Carbanions are poserful electron pair donors!

polar covalent representation

ionic representation

Simplified carbanion nucleophile representation showing only the bare essentials. We will use this representation.

Now, we will simplify this mechanism a bit and look at it simplistically as a two electron transfer from magnesium into the C-Br bond (one electron goes to each atom), making two anions paired up with the Mg^{+2} cation, shown as follows. It is estimated that the R-Mg bond is about 50% ionic and 50% covalent. Many books write it as covalent, but we will write a carbanion structure with the Mg and Br paired together having a +1 charge, since that shows logic consistent with the other nucleophiles that we study.

We can use any R-Br in our list.

ether solvent

actually a metal complex with the ether solvent

We write this as ...

nucleophilic carbanion

This is fantastic! An almost impossible to make carbanion, by acid/base chemistry, can easily be formed from simple, available metals (Li and Mg) and R-Br compounds, of which we can make quite a few in using our allowed starting hydrocarbons and bromobenzene. However, that number is about to explode (see pages 247-248) of the S_N/E topic for a hint using $C_1 - C_6$ RBr compounds). This is why Grignard won the Nobel Prize. Your possibilities are multiplying as I write these words.

Problem 5 – Propose a reasonable synthesis to make the following organomagnesium compounds (use a single reaction arrow). Pick one of your reactions and write a simplistic mechanism for the reaction (2 electron transfer).

Some examples of simple organometallic reagents 'makeable' from our R-Br list   (lots will be possible)

Something pretty remarkable just happened. The C_α-Br carbon, that accepted electrons (as an electrophile) in all of our S_N and E reactions just turned into a powerful electron pair donor (as a nucleophile). We can choose whatever polarity we want for that alpha carbon atom by controlling the conditions under which it reacts. In chemistry this is described as "umpolung", a German word meaning "reverse polarity". If we had more time we could study many other examples of umpolung, but we don't. Let's take a look at some of the possible reactions of our new carbon nucleophiles with our carbon electrophiles.

Chapter 12

1. R-Br electrophiles – a poor result with Mg organometallics, (but cuprates R₂Cu ⁻ work well).

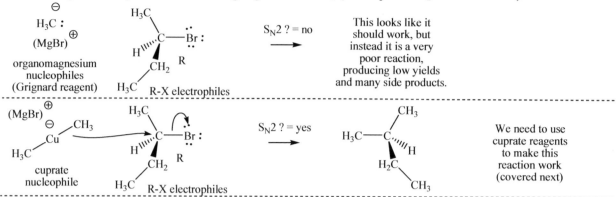

2. Aldehyde electrophiles – the carbanion nucleophile and carbonyl oxygen, as OH, are geminal groups in the product (on the same carbon atom).

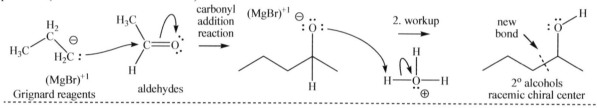

3. Ketone electrophiles– the carbanion nucleophile and carbonyl oxygen, as OH, are geminal groups in the product (on the same carbon atom).

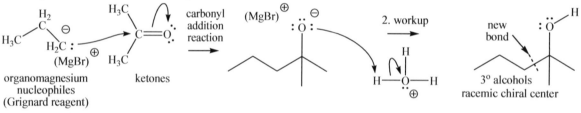

4. Carbon dioxide electrophile– the carbanion nucleophile and acid OH are geminal groups in the product (on the same carbon atom). This reaction will make a lot of new carboxylic acids.

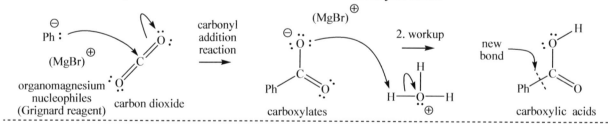

Chapter 12

5. Ester electrophiles (reacts twice) – two carbanion nucleophiles and carbonyl oxygen, as OH, are geminal groups in the product (on the same carbon atom).

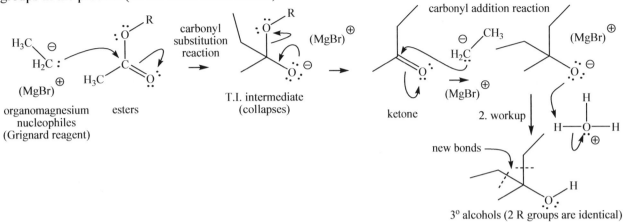

3° alcohols (2 R groups are identical)

----

6. Acid chloride electrophiles (reacts twice) – two carbanion nucleophiles and carbonyl oxygen, as OH, are geminal groups in the product (on the same carbon atom).

3° alcohols (2 R groups are identical)

----

7. Epoxide electrophiles– the carbanion nucleophile and epoxide oxygen are vicinal groups in the product (vicinal = neighbors on adjacent carbon atoms). S_N2 attack occurs at the less hindered carbon of the epoxide.

organomagnesium nucleophiles (Grignard reagent)   epoxide electrophiles   alkoxides   alcohols

----

8. Nitrile electrophiles – The nitrile nitrogen becomes the ketone oxygen in the product. One R group comes from the nitrile and the other comes from the Grignard reagent (usually two options are possible).

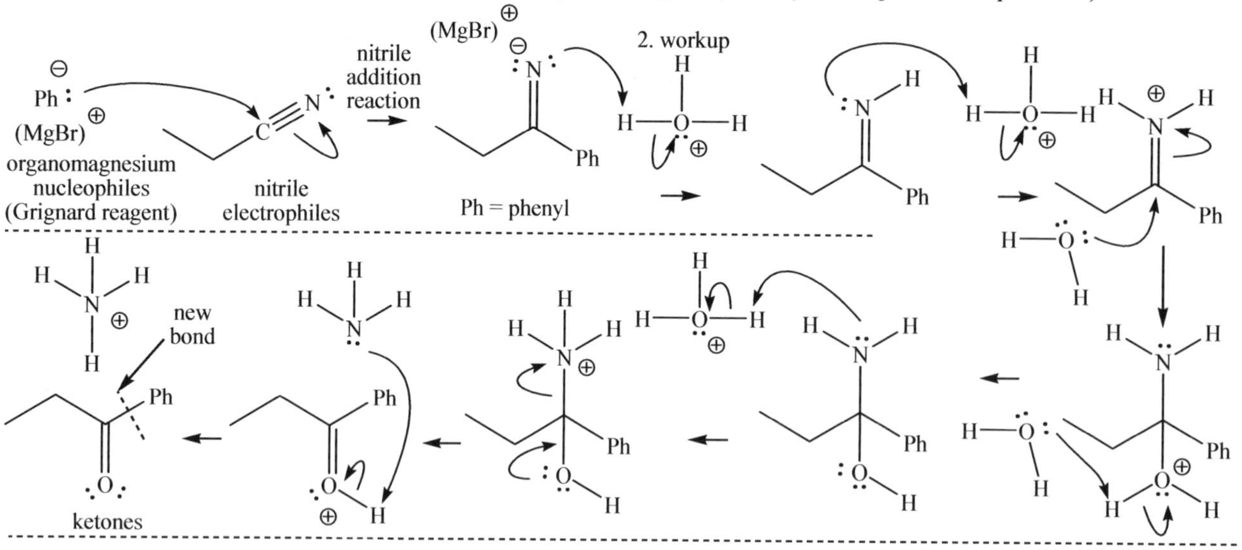

9. Acid chloride electrophiles – Cuprates substitute one time, making ketones.(discussed below)

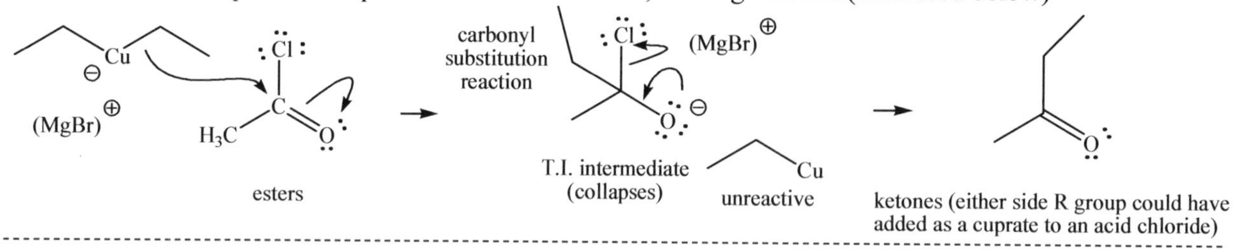

10. Acidic protons kill an organometallic reagent. Using D_3O^+ would label the carbanion position with deuterium. This also limits other functional groups that are in a molecule you are trying to make into a Grignard reagent. If there is an alcohol, thiol, carboxylic acid, amine or any other acidic functional group, then the Grignard reaction won't work.

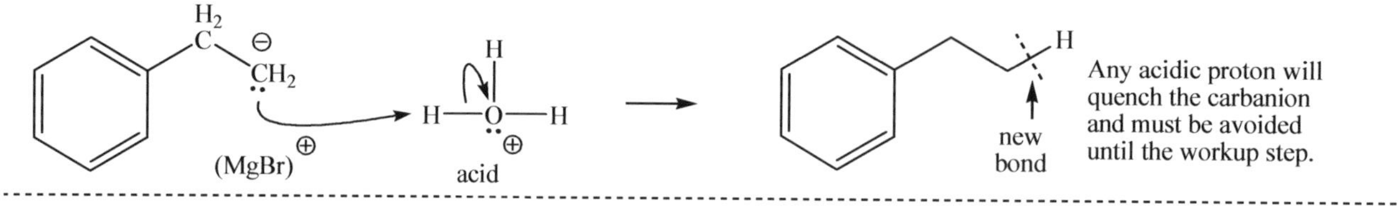

Any acidic proton will quench the carbanion and must be avoided until the workup step.

Problem 6 – Complete the mechanistic details in each of the following reactions and write the expected product. Workup means to neutralize with acid (H_3O^+).

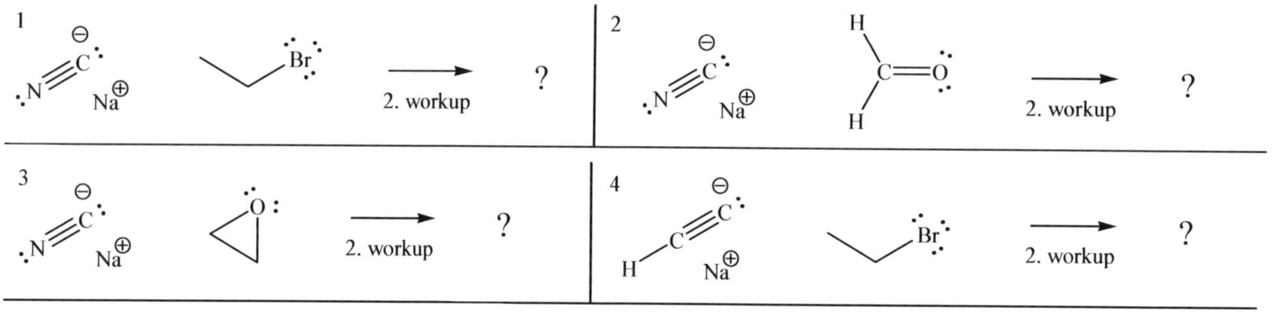

Chapter 12

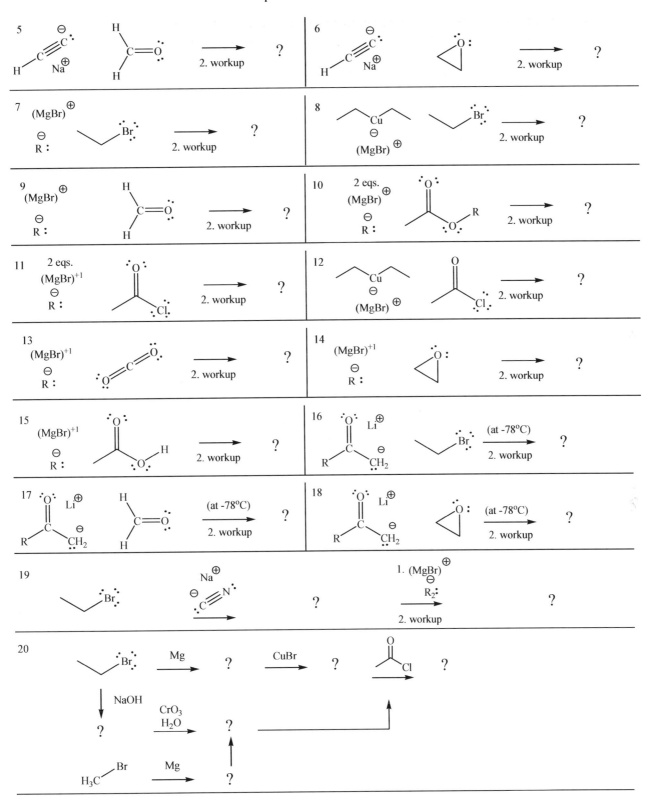

Problem 7 – Use LiAlD$_4$ with the given molecules and show expected products when followed with a workup step. Write arrow-pushing mechanisms.

Many target molecules (TM) below can be made using more than one approach. Our possibilities will expand even farther as we learn new material.

The examples below shows three possible strategies for making 3-phenylpentan-3-ol.

Any of the C-C bonds (1, 2 or 3) could be retrosynthetically cleaved back to a starting ketone and organolithium reagent. Because the two ethyl branches are identical they could also be put on in one reaction with an ester.

Problem 8 – A few examples of different strategies to the same target molecules are shown. Show short reaction sequences to make the following combinations work. No mechanisms are necessary. Show each step with the necessary reagents over each arrow.

Reaction 1 - few possible approaches (one part nucleophile and one part electrophile)

Reaction 2 – a few possible approaches (one part nucleophile and one part electrophile)

Reaction 3 – a few possible approaches (one part nucleophile and one part electrophile)

Reaction 4 – a few possible approaches (one part nucleophile and one part electrophile)

The target is a carboxylic acid ($Ph-C(=O)-O-H$).

Possible approaches:
- $Ph-Br$ with CO_2
- $Ph-CH_2-OH$ with CrO_3 / H_2O (Jones)
- $Ph-C(=O)-H$ with CrO_3 / H_2O (Jones)
- $Ph-C≡N$ with H_2SO_4 / H_2O
- $Ph-C(=O)-O-R$ with H_2SO_4 / H_2O
- $Ph-C(=O)-O-R$ with 1. $NaOH / H_2O$ 2. workup

Reaction 5 – a few possible approaches (one part nucleophile and one part electrophile)

The target is a 3° alcohol (tertiary alcohol).

Possible approaches:
- a ketone with $Br-CH_3$
- a ketone with $Br-$ (ethyl)
- an ester ($-C(=O)-O-R$) with $Br-$ (ethyl), 2 eqs.
- an epoxide with $Br-CH_3$
- a tertiary bromide (Br) with H_2O

Special Reactions of Copper Organometallic Nucleophiles (also called cuprates) showing reactions that do not work using Mg organometallics (Grignard reagents)

Cuprates provide choices not possible with lithium and magnesium reagents and are therefore valuable additions to our organometallic tool kit. Because our time and space is limited, we will only emphasize the following two differences. There are many other metals that make useful organometallic reagents, but there is not enough time to cover them in a first course of organic chemistry. Only two metals (Mg and Cu) will serve as our examples in this chapter.

Formation of the cuprate

To form a cuprate, we mix a cuprous salt, CuBr or CuI, (as opposed to cupric = CuX_2) with an organolithium or organomagnesium reagent (we only use Mg). The more electronegative copper metal (over magnesium) prefers to bond with the electron donating "carbanion" and the more electropositive magnesium prefers to associate with the more ionic bromide. This switching of metals with their bonding partners is called transmetallation. The transmetallation of copper with magnesium occurs because copper is better able to accept the powerful electron donation from the "carbanion" component (than magnesium). The more electropositive magnesium is better able (than copper) to take on an ionic role when matched with a weaker electron pair donor partner (bromide). We need two 'R' groups to complex at copper to make the cuprate reactive in the reactions we study, so a second organomagnesium equivalent is added that forms an anion complex that we call "cuprate"

X (electronegativity) = Cu (2.0), Li (1.0), Mg(1.3), C (2.5), Cl (3.2), Br (2.9), I (2.7)

$$R-X \xrightarrow{Mg} R:^{\ominus} ----- (MgBr)^{\oplus} \xrightarrow{Cu^{\oplus} :Br:^{\ominus} \ 1 \ eq.} R-Cu-R \ (MgBr)^{\oplus} \ ^{\ominus} \quad MgBr_2$$

2 eqs.

simple dialkylcuprate (there are many variations)   discarded

1. Start with an RX structure.
2. Form the organomagnesium reagent
3. Mix with cuprous halide (CuBr) in 1/2 ratio to form magnesium dialkylcuprate
4. React cuprate with the electrophile (electron pair acceptor)
5. Workup

"Organometallic clusters" bring the participants together in close proximity. The transmetallation allows both metal ions (Mg^{+2} and Cu^+) to choose a more compatible partner. Magnesium becomes more ionic and copper becomes more covalent. We will draw a line representing a bond between copper and the carbanion because it is more covalent. Even so, the 'R' group tends to act like a carbanion nucleophile so we will push a full headed arrow from that bond to the electrophiles it reacts with.

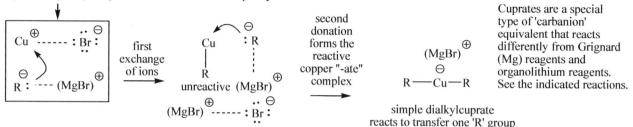

Cluster species may allow for transfer of groups. The boxed figure is a greatly simplified verw used for its simplicity.

first exchange of ions

second donation forms the reactive copper "-ate" complex

unreactive

simple dialkylcuprate reacts to transfer one 'R' group

Cuprates are a special type of 'carbanion' equivalent that reacts differently from Grignard (Mg) reagents and organolithium reagents. See the indicated reactions.

1. Cuprates react with other RX compounds (S_N2 style) in coupling reactions in good yields. Lithium and magnesium reagents tend to produce messy mixtures in low yield. There are too many side reactions that occur (E2, free radical reactions, etc) and yields are typically very low. Cuprates are commonly formed from magnesium and lithium organometallics when mixed with a cuprous salt (like CuBr). Organomagnesium reagents, in turn, come from RX compounds. Since the organometallic reagents (nucleophiles) come from RX compounds, and RX compounds are also the electrophiles, the coupling reaction can be considered a combination of two RX compounds.

General Cuprate Coupling Reaction with RX Compounds

An RX compound is made into a nucleophilic organomagnesium reagent. Because these do no couple well with other RX compounds, they are converted into cuprates, which will couple in good yields with other RX compounds.

R_1-R_2 comes from two coupled RX molecules, one donated the electrons and one accepted the electrons. In most examples in our course this reaction could proceed by either possibility. Pd offers many very popular alternative ways of coupling two carbon groups (called the Heck reaction, the Suzuki reaction and the Stille reaction), but we will not emphasize any of these reactions in this book.

Specific example of a duel approach to a target molecule.

approach 1

Two RBr compounds are coupled together. This doesn't work with Mg and Li reagents

approach 2

Two RBr compounds are coupled together. This doesn't work with Mg and Li reagents

The following compound shows how ambiguous cuprate RX couplings can be. There is no functional group clue in our TM, as is the case in most other TMs we study. Almost any carbon-carbon bond with an sp^3 carbon atom can be disconnected using a cuprate approach. This leads to six possible disconnections below. Both parts of any disconnection can trace back to an RX compound. One will have to become an

Chapter 12

organomagnesium reagent, which is then transmetallated into a cuprate with a cuprous salt. The other will be the electrophilic RX compound that reacts with the cuprate. As above, this can be viewed from two different perspectives. A few examples are provided (the others are asked in the following problem). We will look at the synthetic strategy of doing this in much more detail later in this chapter. Consult the key, if you have to and just try to understand what each step is doing. Later in this chapter we will develop strategies to synthesize more complicated skeletons. I'm assuming you can simple monofunctional functional group interconversions (FGI) using our 7 stating alkanes.

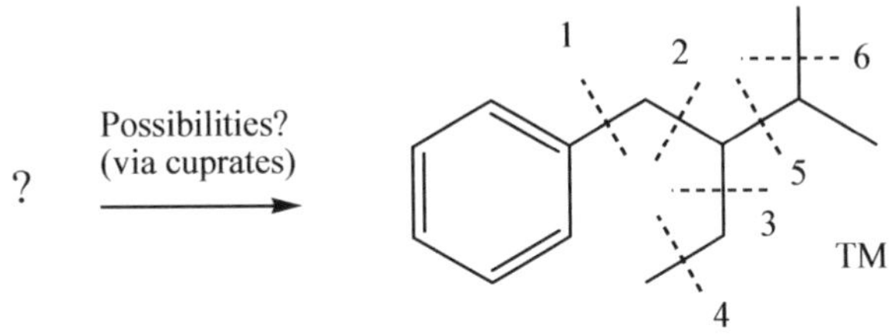

Sources of carbon for this example

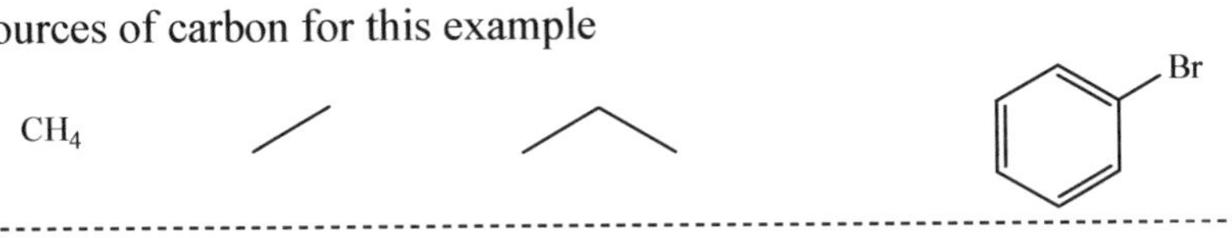

Bond 1

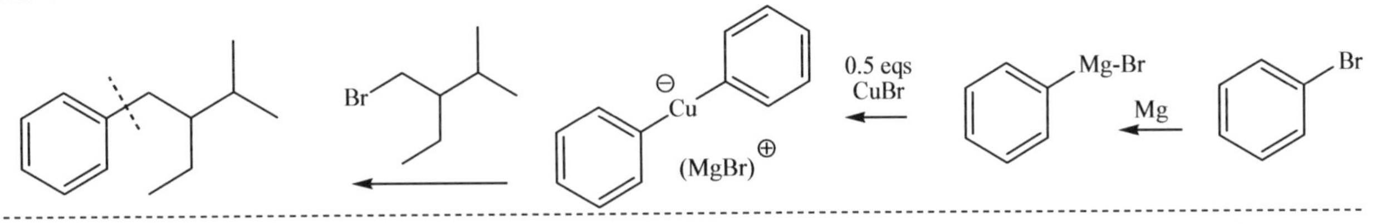

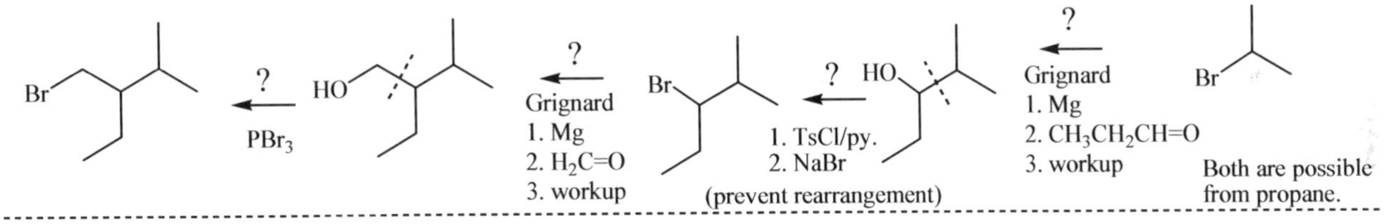

Bond 2

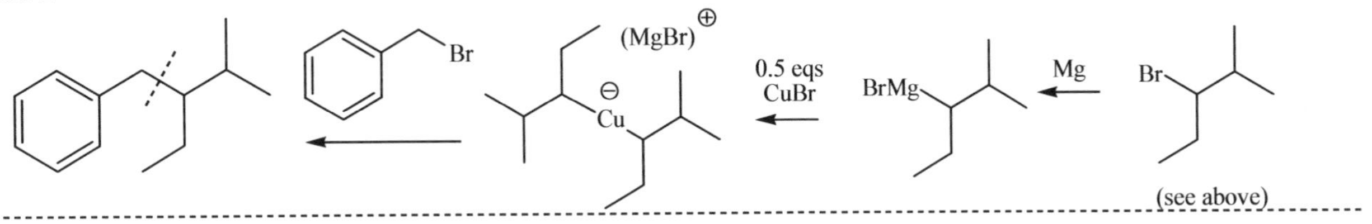

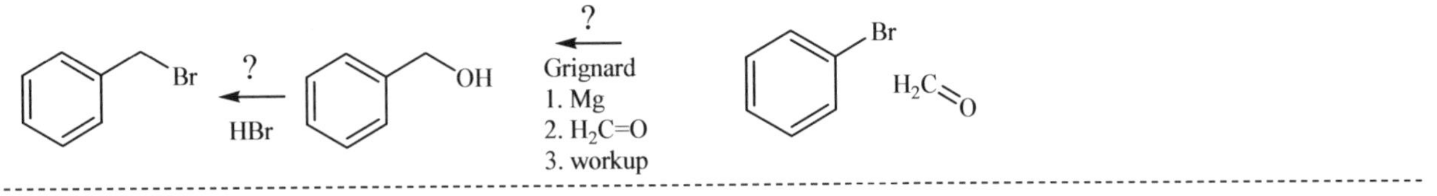

Bond 3 (a couple of ways)

Cuprate reactions (the way we propose them) waste one of the 'R' groups. Because of this the diethylcuprate would be the better approach here. There are alternative approaches that we don't discuss.

Problem 9 – Provide your own cuprate approaches for coupling at bonds 4, 5 and 6. Use the key if you have to. We will use C4 and C5 target molecules in just a bit to develop explicit synthetic strategies, using our available reactions.

2. Acid chlorides are the most reactive carbonyl group we study. Because acid chlorides are more reactive than ketones, and cuprates are less reactive than magnesium and lithium reagents, cuprates can be selectively added to acid chlorides to form ketones without any further reaction (they react just once). Grignard and lithium reagents are too reactive for this sort of selectivity and will also attack a ketone, even in the presence of the more reactive acid chloride, forming tertiary alcohols as part of a mixture of products.

This synthetic transformation (cuprate + acid chloride) provides a straightforward route to many ketones. Just separate either 'R' group to the side of the ketone carbonyl group. The side that retains the carbonyl group will come from the acid chloride and the side without the carbonyl group will come from a cuprate that came from an organomagnesium (or organolithium) compound that came from an RX compound. Often there are two logical approaches.

Generic approach to ketone target

Chapter 12

Specific approach to ketone target

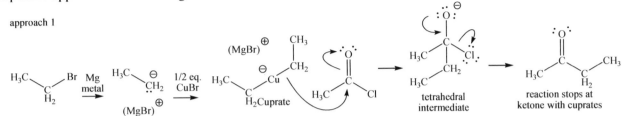

approach 1

Acid chlorides are the most reactive carbonyl group and rapidly react with cuprates to form ketones, which in turn react very sluggishly with cuprates.

tetrahedral intermediate

reaction stops at ketone with cuprates

Ketones do not react well with cuprates so the reaction stops at this point. There are two possible approaches to this ketone. The other one is shown next.

approach 2

tetrahedral intermediate

reaction stops at ketone with cuprates

magnesium and lithium reagents react twice and make tertiary alcohols (similar to esters)

2 eqs.

2. workup

RMgBr reacts twice

Tertiary alcohol products (similar to esters)

Li or Mg reagents

Chapter 12

Think about how many carboxylic acids are available to us from Grignard reagents using RBr compounds and carbon dioxide. This surpasses the number from primary alcohols reacting with Jones reagent, although there is a little overlap.

1. Mg ⟶ Can be made into cuprates.

2. CO_2
3. workup ⟶ Can be made into acid chlorides (and all sorts of other functional groups)

* Can also be made uisng primary alcohols from our 7 hydrocarbon starting structures and Jones reagent.

Many new ketones are possible, plus several compounds made from ketones (alcohols, amines, epoxides, etc.).

Chapter 12

We've learned a lot of reactions, but there are a lot of more reactions to come, so organization is crucial. You really need to remember and catalog the reactions we study as you go along. Functional group interconversions (FGI) are very important because it's part of how we get from one place to another place. If you can make bromoalkanes, you can make lots of alcohols, and then you can make aldehydes, ketones, carboxylic acids, acid chlorides, esters, amides, nitriles, alkenes, new bromoalkanes and alcohols and more. And if you can make bromoalkanes you can use all of the S_N and E chemistry we studied previously, plus organometallics (Mg and Cu). You also have to remember the limitations each reaction. For example, using HBr with a secondary alcohol that can rearrange will not make the desired product. Instead you need to know how to make a tosylate ester and follow that with an S_N2 reaction with NaBr, which does not rearrange. Overall, the steps needed to transform an alcohol that is prone to rearrangement in HBr are as follows.

1. TsCl/pyridine 2. NaBr.

This is the problem and...   This is the solution.

Review of nucleophiles and electrophiles to this point in our course.

Representative electrophiles (RBr, carbonyl groups, carbon dioxide, epoxides and nitriles).

A bromo hydrocarbons (Me, 1°, 2°, 3°) B methanal C ethanal D propanone cyclohexanone E methyl ethanoate F ethanoyl chloride

N,N-dimethyl-ethanamide G carbon dioxide H ethanenitrile (acetonitrile) I ethylene oxide J propylene oxide K isobutylene oxide L cyclohexene oxide

All of these classes of electrophiles react under two extreme conditions: very strong nucleophile/base conditions (very weak acid) and very strong acid conditions (weak nucleophiles). The ideas presented here have many close analogies with our previous topics. They are important enough to be repeated, again and again. Our first class of electrophiles were the simple bromoalkanes that introduced us to the concepts of nucleophile/electrophile and substitution and elimination. Our second class of electrophiles, carbonyl groups (C=O), is a topic so vast that we can barely scratch the surface in our short course. We include nitriles and carbon dioxide here as honorary members of the carbonyl family. Our third class of electrophiles, the epoxides, is more limited, but still very important in organic synthesis and in nature. Together, these groups share many themes with one another. A few specific compounds from these two classes will serve as examples to show typical reactions with some of the nucleophile/bases that we have used in the S_N/E topic. We will use a small, representative group of nucleophile/bases (given for now), also presented below, to show typical reactions with these electrophiles.

Representative nucleophile/bases

1. 2. 3. 4. 5. 6. 7. always a base 8. if in acid = H_2SO_4 9. if in acid = TsOH

Strong Nucleophile/Base Conditions with Aldehydes, Ketones, Carbon Dioxide and Nitriles)

Strong nucleophile/bases (negatively charged, for us) can attack the carbon of a carbonyl group. The target is obvious in a carbonyl group, because there is only one carbon bonded to oxygen. When the nucleophile attacks the carbonyl group the pi electrons merely shift over to the oxygen forming an alkoxide intermediate (similar to the "X" leaving group in RX compounds taking its electrons away). The next step depends on whether the solvent is protic (has an OH) or aprotic (does not have an OH). If the solvent is protic, the negatively charged alkoxide can pick up a proton from the solvent. If the solvent is aprotic, a final workup step is necessary to neutralize the alkoxide with mild acid. When the charge on the nucleophilic atom is of similar stability to the newly formed alkoxide (both as negative oxygen) the process is usually reversible (in equilibrium). When the charge on the nucleophilic atom is much less stable than the newly formed alkoxide, the process is usually irreversible and requires a final workup step. Generic examples of both of these possibilities are shown below.

Generic Carbonyl Addition Reaction in Protic Solvent: examples 1 and 2

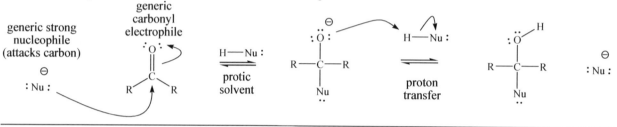

Generic Carbonyl Addition Reaction in Aprotic Solvent: examples 3, 4 and 5

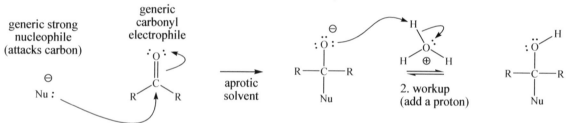

Notice that the attacking nucleophile and the carbonyl oxygen are attached to the same carbon after the reaction is over. This is an important clue that we will use later when deciding you a target molecule might be synthesized.

Problem 10 – Show the expected product and provide a mechanism for each of the following reactions.

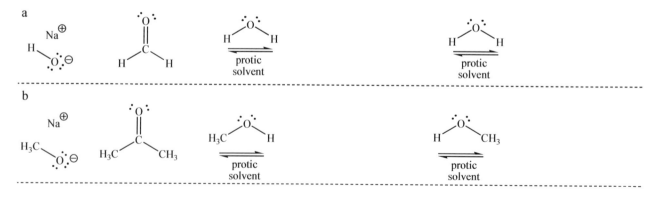

Chapter 12

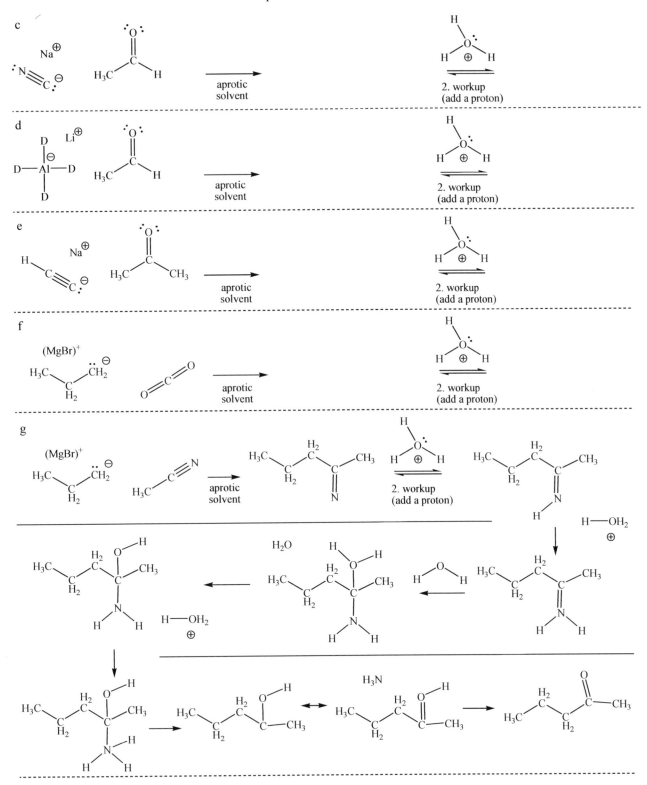

Chapter 12

h. Base hydrolysis of a nitrile to form a primary amide.

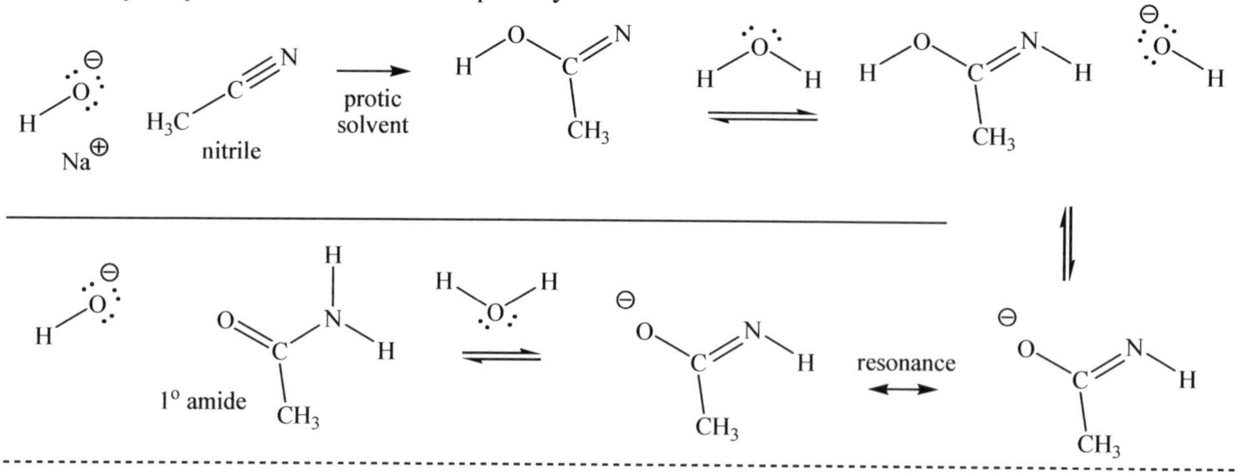

Strong Nucleophile/Base Conditions with Ester Carbonyl Compounds

Esters share the carbonyl feature (C=O) with aldehydes and ketones, but with one important difference, they have a third bond with oxygen (the –OR' group).

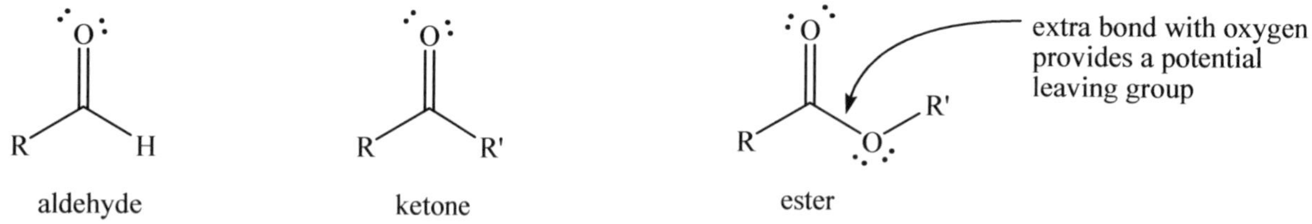

aldehyde   ketone   ester

That alkoxy group allows and an additional elimination reaction that was not possible with aldehydes and ketones. Addition followed by elimination leads to overall substitution, which is the dominant reaction path for esters (acyl substitution).

Problem 11 – Predict the expected product and provide a mechanism for each of the following reactions.

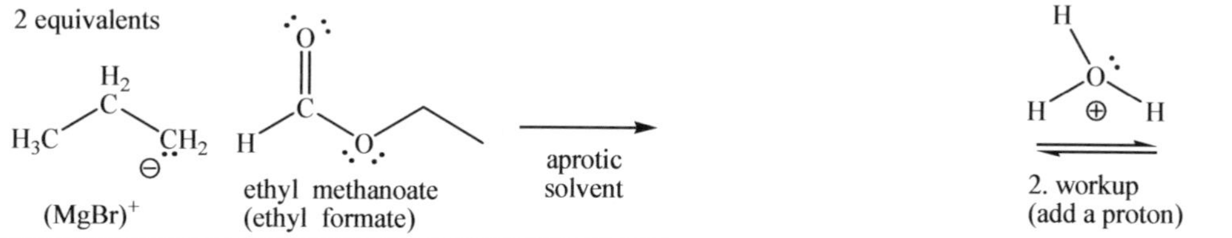

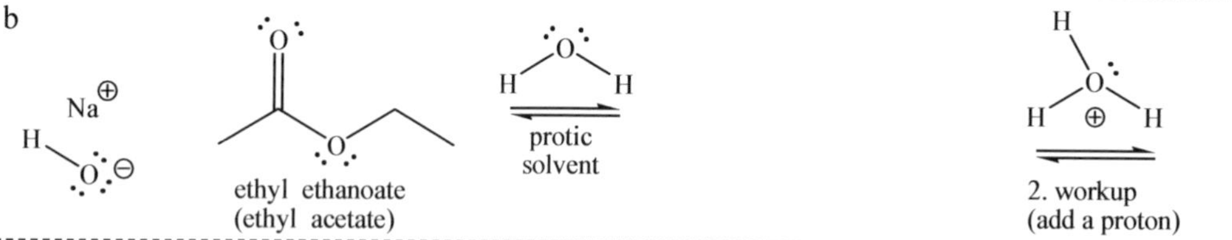

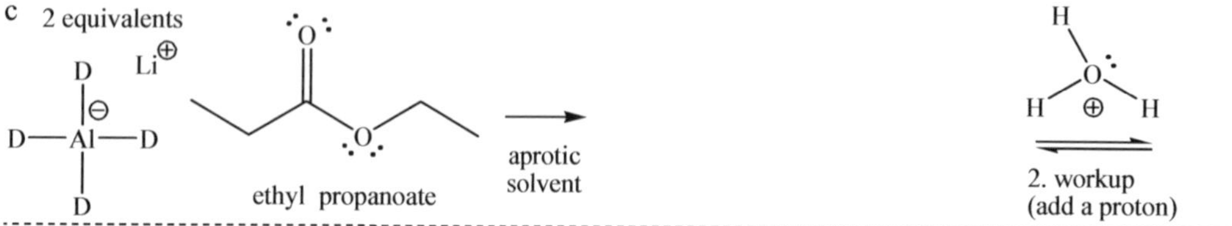

Strong Nucleophile/Base Conditions with Epoxide Compounds

Epoxides are ethers with very large ring strain (~ 27 kcal/mole). This large ring strain allows strong nucleophiles to attack one of the epoxide carbons and open the ring, with oxygen as the leaving group. A critical decision you have to make is which carbon will be the target (less obvious than carbonyl compounds). From our studies of S_N2 reactions, we expect the nucleophile will approach from the backside to the leaving group (the negatively charged oxygen) and occur at the least hindered position (less substituted carbon atom). The choices are clear in the three simple epoxide examples below. Ethylene oxide is symmetrical and it does not make any difference which carbon we attack (both are primary). In the remaining two epoxides, one side is a primary CH_2 and the other side is either secondary or tertiary. We expect that the primary CH_2 would be the target carbon atom, and it is. In epoxide (b and e) there is also a chiral center with absolute configuration "S" that does not change in the reaction since it is not the carbon that is attacked.

As in the carbonyl reactions, the next step depends on whether the solvent is protic (has an OH) or aprotic (does not have an OH). If the solvent is protic, the negatively charged alkoxide can pick up a proton from the solvent (in equilibrium). If the solvent is aprotic, a final workup step is necessary to neutralize the alkoxide with mild acid. Because of the large ring strain in epoxides, these reactions are not usually reversible. Generic examples of epoxide reactions are shown below.

Generic Epoxide Addition Reaction in Protic Solvent: examples 1 and 2

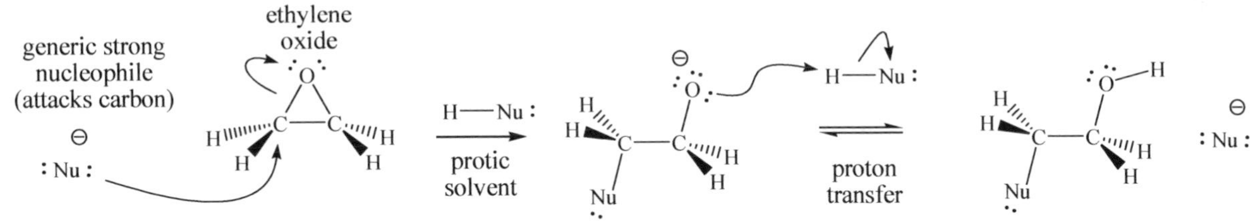

Generic Epoxide Addition Reaction in Aprotic Solvent: examples 3, 4 and 5

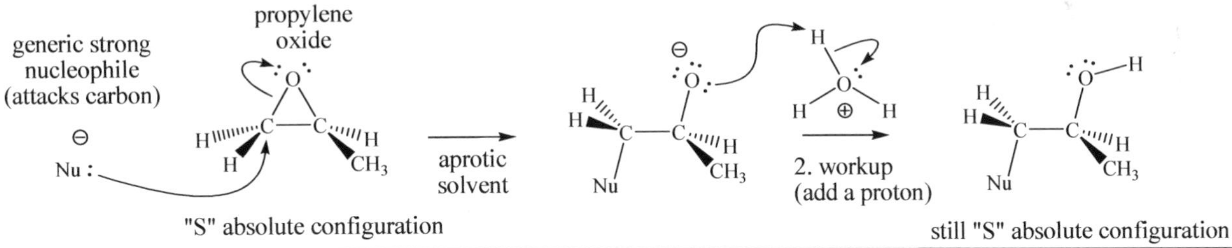

Notice in the epoxide reactions that the attacking nucleophile and the epoxide oxygen are NOT on the same carbon, but rather on vicinal (neighbor) carbon atoms. This is also an important (but different) clue that we can use later in deciding how we might synthesize a target molecule.

Problem 12 – Show the expected product and provide a mechanisms for each of the following reactions.

a

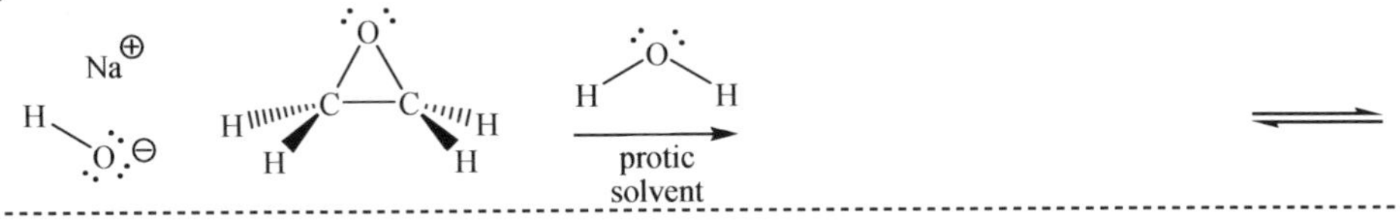

b

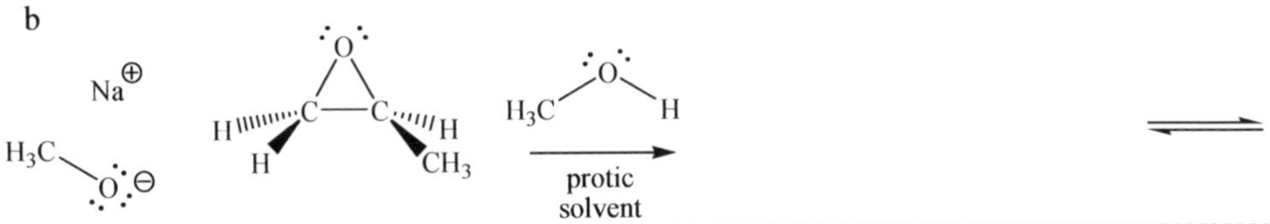

c

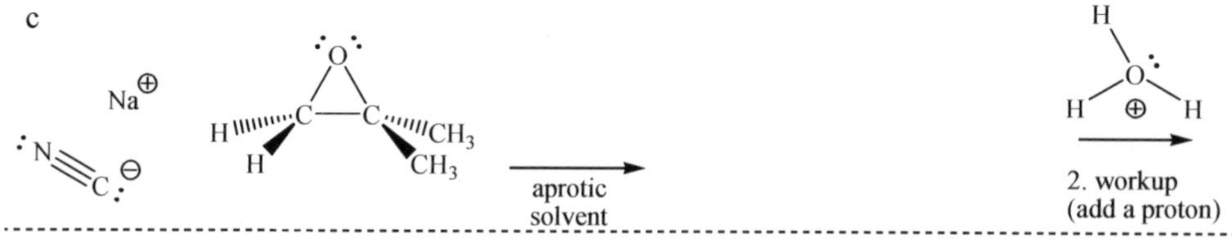

d

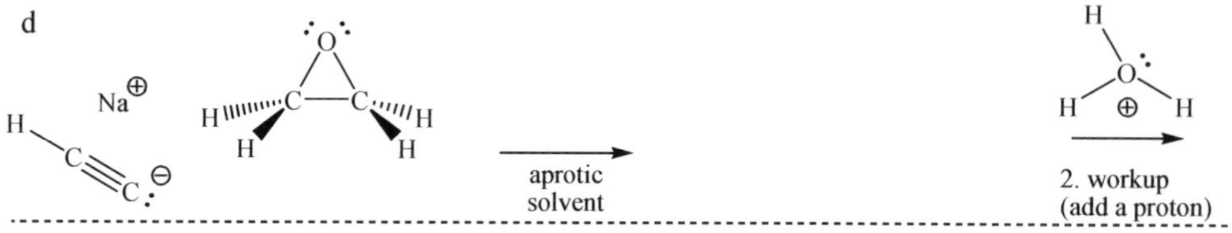

e

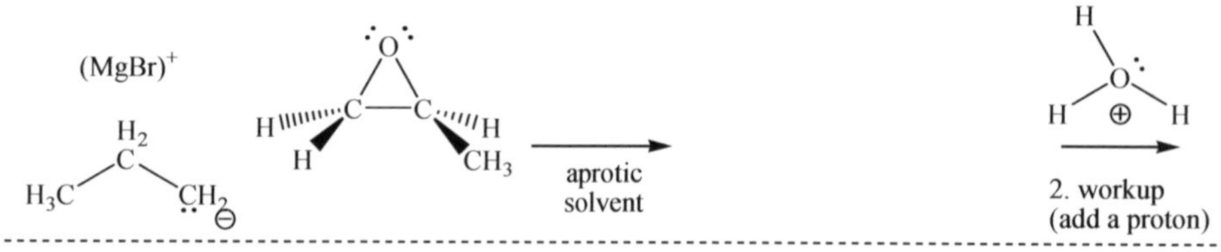

f

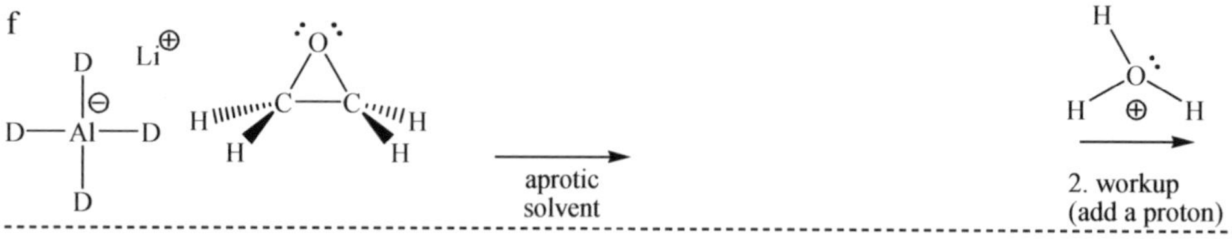

Strong Sterically Bulky Base Conditions – Attack at the Proton instead of the Carbon

Just as was the case in S_N2/E2 reactions, there is competition between attacking a carbon atom versus attacking a proton. In all of the above reactions we only emphasized attack on a carbon because that is what we mainly observe. However, if we want to see the "proton" attack as the dominant reaction, we have to adopt a similar strategy to that used in E2 reactions: use a very bulky, very basic electron pair donor. There are several possibilities, but our choice will be lithium diisopropyl amide (LDA for short). We have to make this, but it's an easy acid/base reaction from reagents that will always be available to us, in our course. As with the above information, we will repeat some of this presentation in a later topic, which is a good thing for beginning students, like yourselves. Generic examples using both of these functional group patterns are shown below.

Making Enolates = carbanion nucleophiles from carbonyl compounds (aldehydes, ketones and esters, for us) using LDA and low temperature (-78°C)

Carbonyl compounds are a very important source of carbanion nucleophiles, called enolates. We will limit ourselves to enolates of very simple ketones, aldehydes, esters and nitriles. As a class these are some of the most important variations of carbon nucleophiles in organic chemistry. However, we will only use the simplest examples possible: propanone, cyclohexanone, ethanol, alkyl ethanoates and ethanenitrile. Things get a lot more complicate when the two sides of a ketone are different or we have to consider stereochemistry of chiral centers.

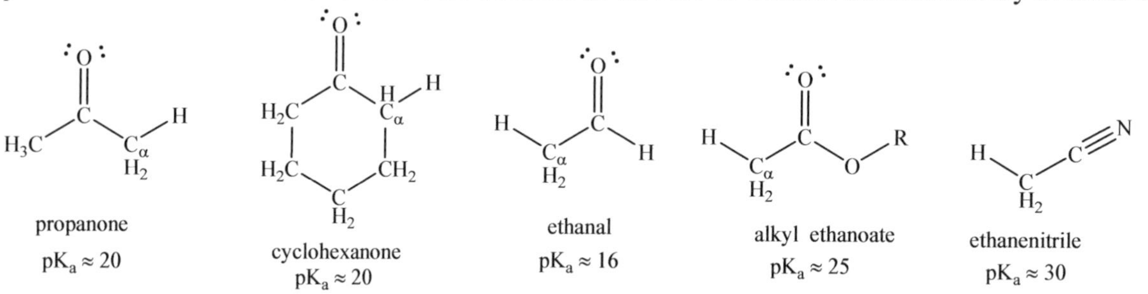

propanone
$pK_a \approx 20$

cyclohexanone
$pK_a \approx 20$

ethanal
$pK_a \approx 16$

alkyl ethanoate
$pK_a \approx 25$

ethanenitrile
$pK_a \approx 30$

The C_α-H of a carbonyl compound ($pK_a \approx 16$-25) is much more acidic than an isolated sp^3 C-H ($pK_a \approx 50$) because negative charge can be taken onto the oxygen atom via resonance. A similar explanation is used with nitriles.

Chapter 12

As neutral carbonyl compounds, none of these are carbanion nucleophiles. However, the carbonyl group is the key to creating this valuable class of carbanion enolates. If we use a very strong non-nucleophilic base (means sterically bulky) we can remove a proton at the C_α position to the carbonyl because of resonance delocalization of negative charge onto the oxygen atom. As is usually the case with very strong electron donation in organic chemistry, we have to worry about donation to a carbon (the C=O) versus donation to a proton (the C_α-H proton). To control the possible choices we pick a very sterically hindered base that has a very high pK_a for its conjugate acid. This favors reaction at the proton over reaction at the carbon atom. Our choice will be lithium diisopropyl amide (LDA), which only reacts with C_α-H of our representative carbonyl compounds (and not the carbon). We also limit thermal energy to a minimum by running these reactions at -78°C to minimize possible side reactions.

1. First, make lithium diisopropyl amide (LDA) – easy to do with available n-butyl lithium at 0°C, then lower to -78°C

$$K_{eq} = \frac{10^{-37}}{10^{-50}} = 10^{+13}$$

very favorable

n-butyl lithium
the strongest
base we have
(in the catalog)

diisopropyl amine
(in the catalog)
$pK_a = 37$

THF, 0°C

LDA
lithium diisopropyl amide
(not in the catalog)

butane
$pK_a = 50$

Using these controls (LDA and a very cold temperature) provides the following examples for preparing carbanion enolate nucleophiles. First we make LDA (above), then lower the temperature to -78°C. Next we add in our carbonyl compound and form the desired enolate. Once the enolate is formed, the electrophile is added in to make the desired carbon-carbon bond and the temperature allowed to rise. In our course, the electrophiles will be R-X compounds, other carbonyl compounds (C=O) and epoxides. A final workup step will neutralize the basic mixture with mild acid and provide our target molecules (TM).

2. React any of the above carbonyl compounds with LDA at -78°C to form an enolate. LDA is sterically large and very basic, which favors reaction at the C_α-H over attack at carbon.

propanone

LDA, -78°C
nonnucleophilic base

enolate nucleophile

resonance

E$^+$
(electrophile)

3. workup

TM = target molecule

cyclohexanone

LDA, -78°C
nonnucleophilic base

enolate nucleophile

resonance

E$^+$
(electrophile)

3. workup

TM = target molecule

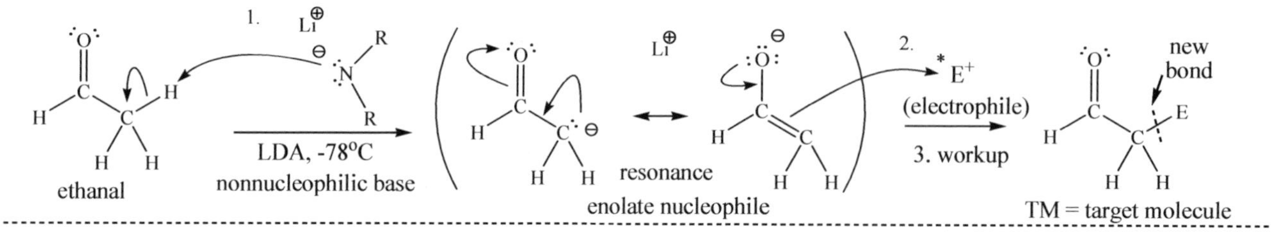

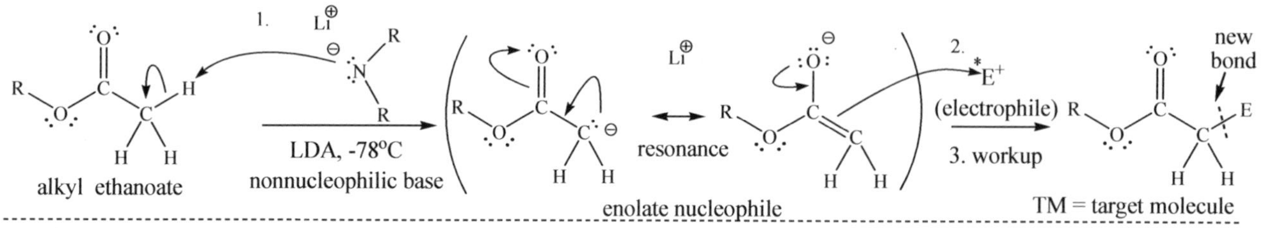

The reaction schemes for ethanal, alkyl ethanoate, and ethanal (nitrile) with LDA forming enolate nucleophiles that react with electrophiles:

Row 1 (ethanal):
ethanal → LDA, -78°C, nonnucleophilic base → enolate nucleophile (with resonance) → 2. E+ (electrophile), 3. workup → TM = target molecule (new bond to E)

Row 2 (alkyl ethanoate):
alkyl ethanoate → LDA, -78°C, nonnucleophilic base → enolate nucleophile (with resonance) → 2. E+ (electrophile), 3. workup → TM = target molecule (new bond to E)

* E+ = R-Br, other carbonyl compounds (C=O) and epoxides

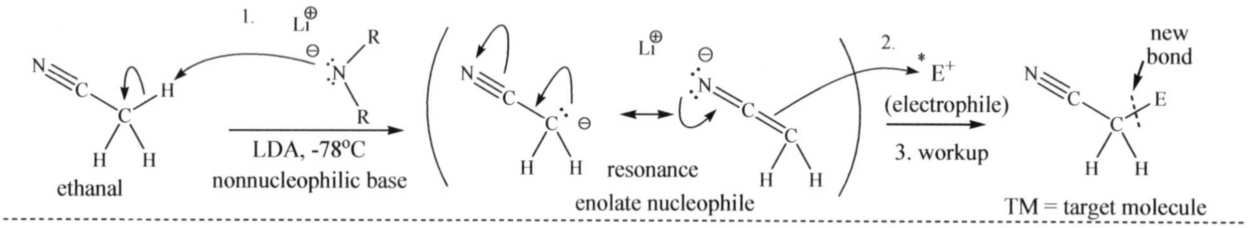

Row 3 (ethanal / nitrile):
ethanal → LDA, -78°C, nonnucleophilic base → enolate nucleophile (with resonance) → 2. E+ (electrophile), 3. workup → TM = target molecule (new bond to E)

Problem 13 – Show the enolate of propanone reactions with the following electrophiles: a. bromoethane, b. methanal and c. ethylene oxide. Assume a final workup step where necessary.

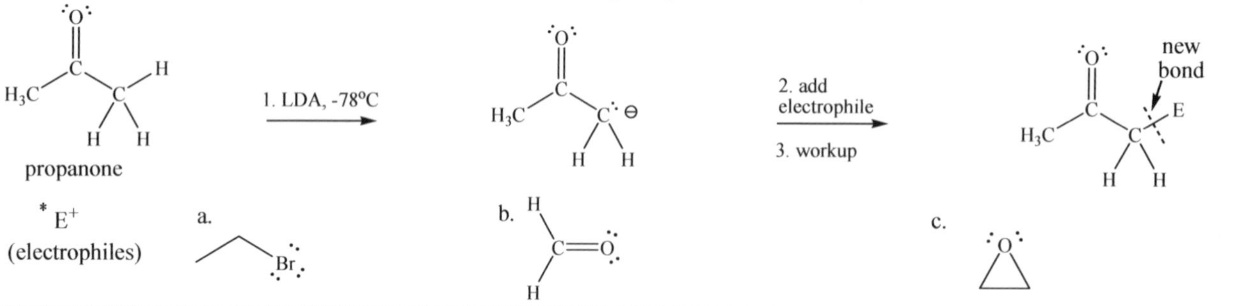

propanone → 1. LDA, -78°C → enolate → 2. add electrophile, 3. workup → product (new bond to E)

* E+ (electrophiles)
a. (bromoethane)
b. (methanal, H₂C=O)
c. (ethylene oxide, epoxide)

Making Allylic Alcohols from Epoxides using LDA

Epoxides with a side chain also react predominantly in a manner to our E2 reaction studied earlier in this topic, when a very bulky, very basic electron pair donor is used (LDA, again!). The large strain of the epoxide ring and the adequate stability of the negatively charged oxygen atom leaving group causes this reaction to occur. Two example reactions are shown below. LDA reactions require a second workup step, because any sort of acid would react with the LDA.

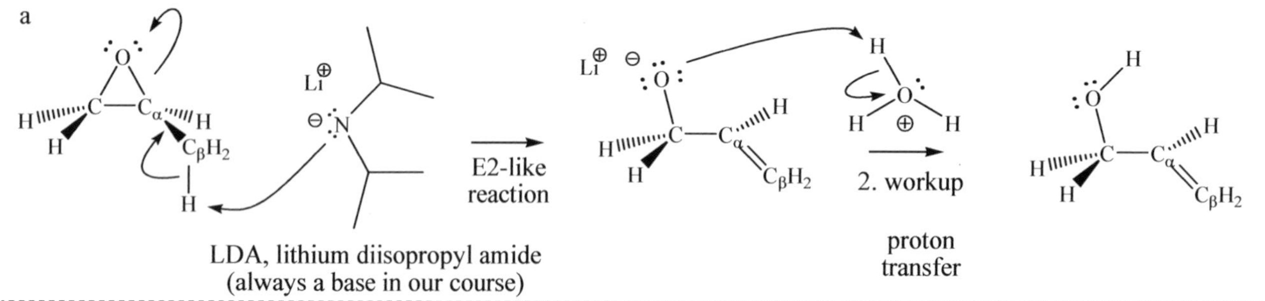

a. epoxide + LDA, lithium diisopropyl amide (always a base in our course) → E2-like reaction → alkoxide intermediate + H₃O⁺ → 2. workup, proton transfer → allylic alcohol

b

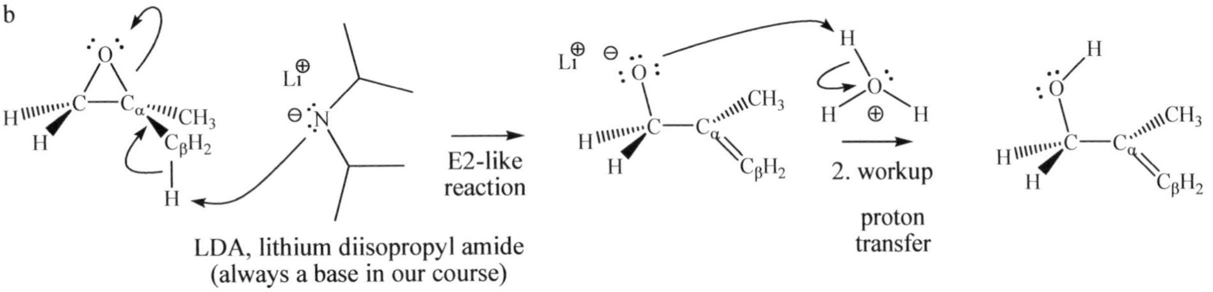

LDA, lithium diisopropyl amide
(always a base in our course)

E2-like reaction

2. workup

proton transfer

Attack at carbon is too hindered, when using LDA, so proton abstraction is observed at C_β-H position to oxygen in a reaction analogous to E2. Release of ring strain is the driving force and compensates for the negatively charged oxygen leaving group.

Problem 14 – Predict the product and show a mechanism for the following reaction of LDA with cyclohexeneoxide, followed by neutralization (workup).

cyclohexene oxide   LDA   intermediate   product

2. workup

?

Weak Nucleophile/Base (and strong acid) Conditions with Carbonyl Compounds and Epoxide Compounds

As we have seen, weak nucleophiles (water, H_2O, and alcohols, ROH) react with secondary and tertiary RX compounds via $S_N1 > E1$ reactions. Also, methyl and primary alcohols react under strong acid conditions via S_N2 secondary, and tertiary alcohols react under strong acid conditions via S_N1 (ROH with HBr) and via E1 reactions (ROH with H_2SO_4/Δ).

We will now examine how weakly electrophilic carbonyl compounds and epoxides can also become strongly electrophilic in strong acid and protic solvents. We will only use two examples (H_2O and CH_3OH) to introduce these reactions and extend the possibilities in later topics. Essentially, the order of events is reversed in strong acid conditions (1. add proton, 2. add weak nucleophile) from the strong nucleophile/base conditions (1. add strong nucleophile, 2. add proton). Because the conditions are usually strong protic acid, the solvents used are typically protic (H_2O or ROH).

Carbonyl reactions with water and alcohols usually have products and reactants in equilibrium with one another, while with epoxide reactions are irreversible, due to the large ring strain. Generic examples of these possibilities are shown below. Mechanisms in acid are often longer because there is an initial protonation step, intermediate resonance structures, and a final deprotonation step required.

One final point is that water is often used to shift an equilibrium, in acid, to one side or the other by adding (hydration) or removing it (dehydration). The acid we will use when water is the solvent will be sulfuric acid (H_2SO_4) and the acid we will use when an alcohol is the solvent (nonaqueous) will be toluenesulfonic acid (TsOH). You can think of TsOH as 'organic' sulfuric acid.

Very strong acids used in our course.

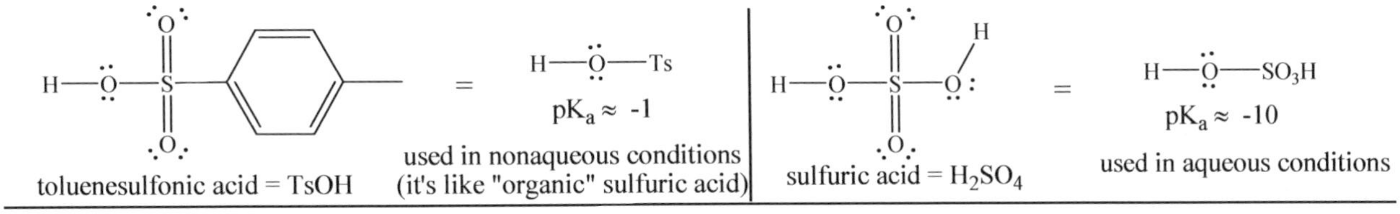

toluenesulfonic acid = TsOH   used in nonaqueous conditions (it's like "organic" sulfuric acid)   sulfuric acid = H_2SO_4 used in aqueous conditions

$H—O—Ts$
$pK_a \approx -1$

$H—O—SO_3H$
$pK_a \approx -10$

Generic Carbonyl Addition Reactions of Aldehydes and Ketones in Aqueous Acid (hydration)

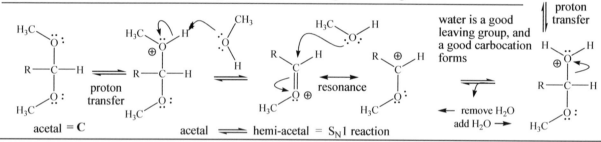

equilibrium reaction, usually favors carbonyl compounds, except when R = R' = H (methanal)

Generic Carbonyl Addition Reaction in Alcoholic Acid (forms a hemi-acetal/ketal), followed by an S_N1 reaction (forms an acetal/ketal). To form the acetal/ketal requires the removal of water to shift the equilibrium to the right. If water is added the equilibrium will shift to the left, reforming the carbonyl group. This is an equilibrium reaction that can be pushed in either direction.

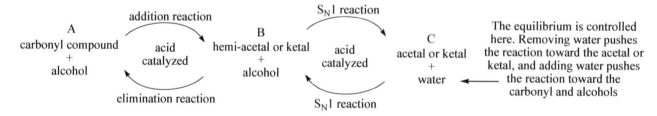

equilibrium is controlled by water, removing water shifts it to the right and adding water shifts it to the left

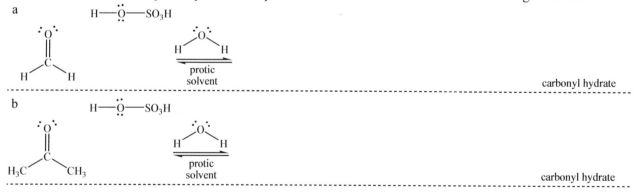

Problem 15 – Show the expected product and provide a mechanism for each of the following reactions.

a

b

Chapter 12

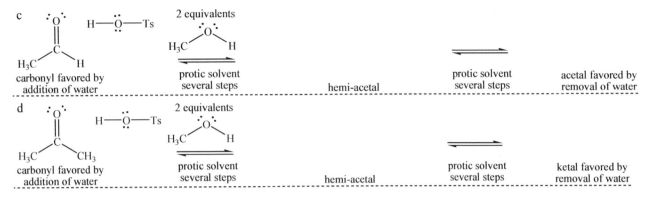

Problem 16 – Write the reverse mechanism for formation of the carbonyl compound and two alcohols from the acetal using aqueous acid solution.

In a closely related series of steps it turns out that carboxylic acids and alcohols can form esters and water using catalytic toluene sulfonic acid. This is called the Fischer ester synthesis. Using aqueous sulfuric acid will hydrolyze the ester back to the carboxylic acid and the alcohol. Again, this is an equilibrium controlled by the removal or addition of water. The steps in the forward direction are shown below.

Generic Fischer Ester Synthesis Reaction - addition of an alcohol to a carboxylic acid, followed by elimination of water from the tetrahedral intermediate, leads to overall acyl substitution. an acetal/ketal). To form the ester requires the removal of water to shift the equilibrium to the right. If water is added the equilibrium will shift to the left, reforming the carboxylic acid and alcohol groups. This is an equilibrium reaction that can be pushed in either direction.

Problem 17 – Write the reverse mechanism for formation of the carboxylic acid and alcohol from the ester using aqueous acid solution (aqueous acid hydrolysis of an ester).

Epoxides in Acid Solution

As was the case with carbonyl compounds in acid, epoxides first add a proton. The positive charge in the intermediate is distributed over the oxygen and both carbon atoms (C_a and C_b, below). The weak nucleophile is attracted to the backside of the more substituted carbon because it carries a larger partial positive charge than the less substituted carbon (just like carbocations). This is the opposite epoxide carbon that was attacked in strong base (but still backside attack because of the bridging oxygen atom on the top). If the more substituted carbon is chiral, we should see a change in the absolute configuration (from R to S or S to R).

Generic Epoxide Addition Reaction in Aqueous Acid (hydration)

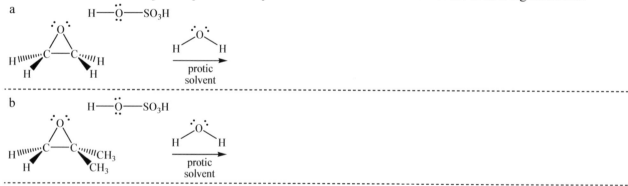

C_a has greater partial positive charge than C_b because it is more substituted, so the weak nucleophile attacks it faster. Notice that a different carbon of the unsymmetrical epoxide was attacked than was the case in strong nucleophile/base conditions. Because that carbon is chiral in this example, we can see the inversion of configuration (from S to R), even though there is an "OH" on both carbon atoms.

Generic Epoxide Addition Reaction in Alcoholic Acid

C_a has greater partial positive charge than C_b because it is more substituted, so the weak nucleophile attacks it faster. Notice that a different carbon of the unsymmetrical epoxide was attacked than was the case in strong nucleophile/base conditions. This is easy to see in this example because "methoxy = -OCH$_3$" (from the attacking weak nucleophile) is different than "hydroxy = -OH" (from the epoxide oxygen).

Problem 18 – Show the expected product and provide a mechanism for each of the following reactions.

c

d

Phosphorous ylids: good carbanion nucleophiles used to synthesize specific alkenes

A Nobel Prize winning approach to synthesizing alkenes puts double bonds exactly where you want them by connecting two carbon fragments together. When this reaction won the Nobel Prize it was one of the few reactions that could tie together two elaborate pieces in a controlled, predictable way. It is still used for that today, but other reactions have been developed that have become even more prominent: Heck (Pd), Suzuki (Pd), Stille (Pd) and Metathesis (Mo, W, Ru) reactions are all used for similar purposes. They have all won Nobel Prizes, but we don't have time to cover them. These reactions all use transition metals and don't have simple arrow-pushing mechanisms, though we could try and propose some for fun. On the other hand, simple Wittig reactions involve straight forward arrow-pushing mechanisms to make alkenes, consistent with our approach to organic chemistry, and they connect two simpler parts into a larger molecule in a predictable way. One carbon of the double bond comes from the Wittig reagent (a phosphorous ylid) and one carbon is a carbonyl carbon. This makes for interesting synthetic problems and is easily incorporated into synthesis of a target molecule. Two different carbon skeletons are stitched together, leaving an obvious clue of a double bond. As in some of our other reactions, two approaches are often possible. Wittig reactions will force you to use your organic logic in a logical manner, always looking for good nucleophile/electrophile connections. As you might expect, there are many variations of the Wittig reaction, and we will use two of them that are helpful to our goal of learning the logic of organic chemistry.

The Wittig Reaction

Step 1 – Make a triphenylalkylphosphonium halide salt (an S_N2 reaction leads to the Wittig salt). We have lots of R-Br electrophiles to choose from.

triphenylphosphine
(phenyl = Ph)

S_N2 reaction

triphenylpropylphosphonium
bromide salt (Wittig salt)

Step 2 – Remove the weakly acidic C-H proton next to the phosphonium ion (pK$_a$ ~ 31) with a very strong base (n-butyl lithium) to make the Wittig reagent (the phosphorous ylid). We will use n-butyl lithium as our base (others are possible).

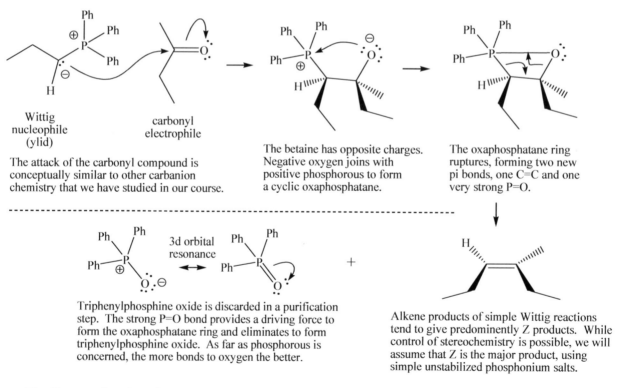

Delocalization into the empty phosphorous 3d orbitals provides a possible explanation for the stabilization of the carbanion site in the ylid. This might also be explained by the close proximity of an opposite positive charge on the phosphorous. This is the same argument used for sulfur ylids.

A possible justification for stabilization of the carbanion site next to phosphorous.

empty **full**
3d orbital **2p orbital**

The empty 3d orbital of phosphorous might accept excess electron density from the full carbanion 2p orbital. This could allow delocalization of electron density onto the positively charged phosphorous and balance the charge. Theoretical chemists argue that overlap of the 3d and 2p orbitals is not very efficient.

Step 3 – React the nucleophilic Wittig reagent (ylid) with an electrophilic carbonyl compound (aldehyde or ketone) to form an alkene. The nucleophilic ylid attacks the electrophilic carbonyl carbon and then a four atom oxyphosphatane ring forms when phosphorous forms a very strong P-O bond with oxygen. Phosphorous and oxygen form an additional bond extruding the two doubly bonded carbon atoms in a concerted elimination. Simple Wittig reactions tend to form Z alkenes when there is a choice.

Wittig
nucleophile
(ylid)

carbonyl
electrophile

The attack of the carbonyl compound is conceptually similar to other carbanion chemistry that we have studied in our course.

The betaine has opposite charges. Negative oxygen joins with positive phosphorous to form a cyclic oxaphosphatane.

The oxaphosphatane ring ruptures, forming two new pi bonds, one C=C and one very strong P=O.

3d orbital resonance

Triphenylphosphine oxide is discarded in a purification step. The strong P=O bond provides a driving force to form the oxaphosphatane ring and eliminates to form triphenylphosphine oxide. As far as phosphorous is concerned, the more bonds to oxygen the better.

Alkene products of simple Wittig reactions tend to give predominently Z products. While control of stereochemistry is possible, we will assume that Z is the major product, using simple unstabilized phosphonium salts.

The Z stereochemistry is proposed to form in the "kinetic approach" (one faster choice) of the nucleophilic and electrophilic carbon atoms. The ylid and the carbonyl compound are considered to approach in a somewhat staggered orientation. We propose that the larger groups are farther apart (lower E_a?) in this approach, it leads to Z stereochemistry in the alkene after elimination.

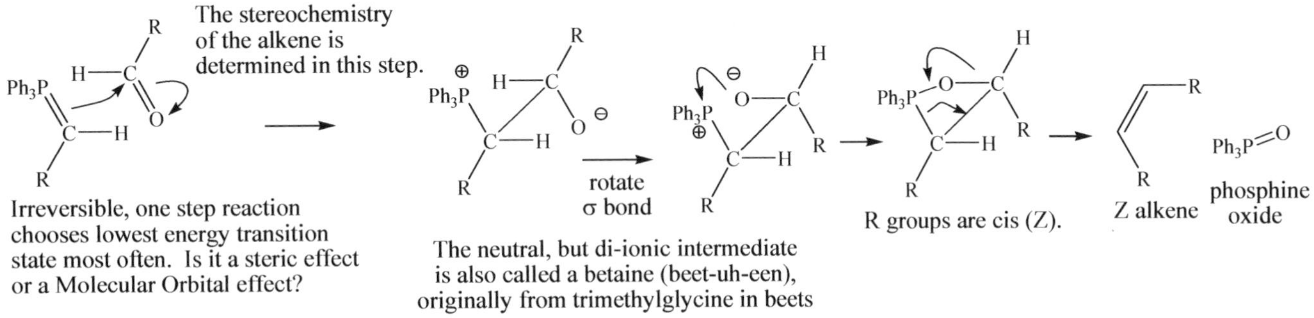
The stereochemistry of the alkene is determined in this step.
Irreversible, one step reaction chooses lowest energy transition state most often. Is it a steric effect or a Molecular Orbital effect?
rotate σ bond
The neutral, but di-ionic intermediate is also called a betaine (beet-uh-een), originally from trimethylglycine in beets
R groups are cis (Z).
Z alkene
phosphine oxide

The Wittig reaction joins an RX compound together with a carbonyl compound in a C=C double bond. Often either strategy is acceptable. This makes for a nice problem in that two different carbon skeletons can be proposed which are joined together leaving an easily recognized clue of the C=C double bond (split right down the middle). After alkenes have been studied, the problem can be made slightly more difficult by including one additional reaction beyond the actual double bond. You would have to retrosynthetically step back one step to the double bond and recognize that it could come from a Wittig reaction, but we will save that possibility until our next topic (alkenes and alkynes).

Example – Propose a synthesis for the following molecule. Notice that using our approach from simple hydrocarbons that both halves of the "Wittig" alkene were an RX compound at some point along the way to the alkene.

option 1

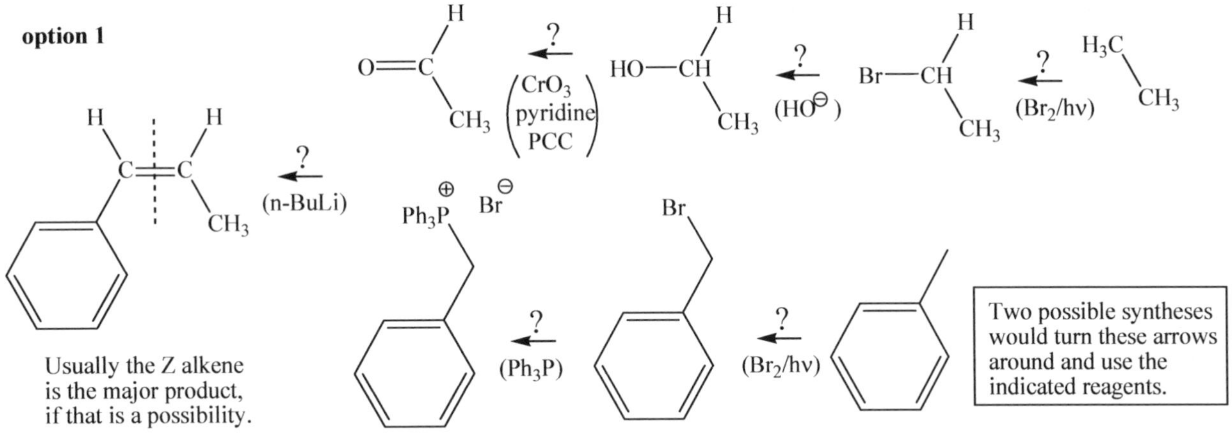
(n-BuLi)
CrO_3 pyridine PCC
$(HO^{\ominus})$
$(Br_2/h\nu)$
(Ph_3P)
$(Br_2/h\nu)$
Usually the Z alkene is the major product, if that is a possibility.
Two possible syntheses would turn these arrows around and use the indicated reagents.

or

option 2

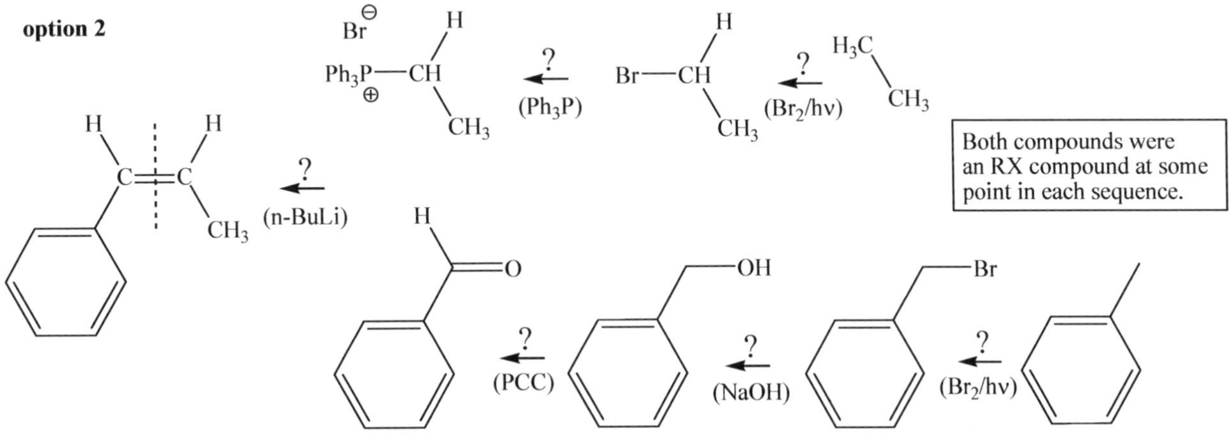
(n-BuLi)
(Ph_3P)
$(Br_2/h\nu)$
Both compounds were an RX compound at some point in each sequence.
(PCC)
(NaOH)
$(Br_2/h\nu)$

Chapter 12

Schlosser Modification of the Wittig reaction to make E alkenes

To get an E alkene, Schlosser developed an extra step in the Wittig reaction above, while at low temperature. Right after the ylid adds to the C=O bond, before the elimination, n-butyl lithium is used to take off a proton from the carbon atom attached to phosphorous. This allows quick inversion of configuration, back and forth at the carbanion site. The proton is then added back on with mild acid so that the more stable configuration forms, leading to an E alkene after elimination of the triphenylphosphine oxide.

Problem 19 – Propose a reasonable synthetic approach for the following target molecules. Remember, not every double bond we see has to come from a Wittig reaction. Wittig reactions are most useful when a "less stable" double bond has to be prepared, or if other functionality is present in molecule that cannot survive harsh elimination conditions (more typical of real life situations). When the molecule is uncomplicated by other functionality and the most stable double bond is the target, straight forward E1 and/or E2 reactions might be acceptable alternatives. You can use our usual sources of carbon and typical organic reagents.

Chapter 12

Carbonyl and epoxide electrophiles used at this point in our course (addition, elimination and substitution reactions).

weak electrophiles / strong base/nucleophiles	carbonyl compounds				epoxide compounds			
	methanal	aldehydes	ketones	esters	ethylene oxide	monosubstituted epoxide	geminal substituted epoxide	cyclohexene oxide
hydroxide	carbonyl addition	carbonyl addition	carbonyl addition	acyl substitution	epoxide addition in S_N2-like mechanism	epoxide addition in S_N2-like mechanism	epoxide addition in S_N2-like mechanism	epoxide addition in S_N2-like mechanism
alkoxides	carbonyl addition	carbonyl addition	carbonyl addition	not discussed in our course	epoxide addition in S_N2-like mechanism	epoxide addition in S_N2-like mechanism	epoxide addition in S_N2-like mechanism	epoxide addition in S_N2-like mechanism
cyanide	carbonyl addition	carbonyl addition	carbonyl addition	not discussed in our course	epoxide addition in S_N2-like mechanism	epoxide addition in S_N2-like mechanism	epoxide addition in S_N2-like mechanism	epoxide addition in S_N2-like mechanism
terminal acetylides	carbonyl addition	carbonyl addition	carbonyl addition	not discussed in our course	epoxide addition in S_N2-like mechanism	epoxide addition in S_N2-like mechanism	epoxide addition in S_N2-like mechanism	epoxide addition in S_N2-like mechanism
Grignard reagents	carbonyl addition	carbonyl addition	carbonyl addition	acyl substitution	epoxide addition in S_N2-like mechanism	epoxide addition in S_N2-like mechanism	epoxide addition in S_N2-like mechanism	epoxide addition in S_N2-like mechanism
cuprates	not discussed in our course	not discussed in our course	not discussed in our course	not discussed in our course	not discussed in our course	not discussed in our course	not discussed in our course	not discussed in our course
lithium aluminium deuteride (hydride)	carbonyl addition	carbonyl addition	carbonyl addition	acyl substitution	epoxide addition in S_N2-like mechanism	epoxide addition in S_N2-like mechanism	epoxide addition in S_N2-like mechanism	epoxide addition in S_N2-like mechanism
LDA = lithium diisopropylamide	not discussed in our course	enolate chemistry	enolate chemistry	enolate chemistry	not discussed in our course	E2-like mechanism to form allylic ROH	E2-like mechanism to form allylic ROH	E2-like mechanism to form allylic ROH
in acid = H_2SO_4	carbonyl addition (hydration)	carbonyl addition (hydration)	carbonyl addition (hydration)	acyl substitution (ester hydrolysis)	epoxide addition with nucleophilic attack at more partial positive carbon	epoxide addition with nucleophilic attack at more partial positive carbon	epoxide addition with nucleophilic attack at more partial positive carbon	epoxide addition with nucleophilic attack at more partial positive carbon
in acid = TsOH	carbonyl addition (acetal)	carbonyl addition (acetal)	carbonyl addition (ketal)	not discussed in our course	epoxide addition with nucleophilic attack at more partial positive carbon	epoxide addition with nucleophilic attack at more partial positive carbon	epoxide addition with nucleophilic attack at more partial positive carbon	epoxide addition with nucleophilic attack at more partial positive carbon

If you have worked through all of the material thus far, then you have a pretty good idea about the major concepts necessary to understand the remaining material on the other organic functional groups that we still have to cover in our course. Next we will survey a sampling of typical reactions using the common functional groups we have studied and how we can use them to make more complicated molecules. We call this *organic synthesis.*

Chapter 12

Organic Synthesis – Working Backwards

As more reactions are learned, more complicated structures can be made. This process is called synthesis. It is often called an "art", although a lot of analytical thinking goes into making a new structure. There will be a target molecule (TM) that we want to make, and it will usually have some functional group feature(s) that provides us with clues as to how it was prepared. Common synthetic steps include functional group interconversions, FGI, (making one functional group into another functional group, e.g. RBr → ROH or ROH → RBr). Additionally, forming carbon-carbon bonds will be especially important to us. A third type of reaction involves protecting functional groups while some transformation is attempted, and then deprotecting them later when the transformation is accomplished. We barely mention protection and deprotection (necessary when multiple functional groups are present). In a short course like ours, the emphasis is almost entirely on nucleophile-electrophile strategies of functional group conversions and carbon-carbon bond forming reactions. Clearly, recognizing "nucleophilic" carbons and "electrophilic" carbons is an important skill. Even though free radical approaches are known, we will not emphasize them in forming carbon-carbon bonds. There is also a lot of chemistry done with transition metals that is almost magical, and we will not emphasize this chemistry either, except for the Grignard reactions and a little copper chemistry. Our synthesis problems will typically be simple molecules (mostly with one functional group) with relatively straight forward choices, based on the reactions we have studied.

We often only have one way of making a particular functional group. Your choices are very limited, and you either know it or you don't. In any case, every single step of a synthetic approach will be a one step problem. In all organic reactions, there is a reactant (or reactants), reagents (reaction conditions) and a product formed (or products formed). As a problem, any one of these three parts could be missing, and you would need to supply the missing part. You want to catalog and summarize each reaction that we study so you can retrieve it when you need to make a specific functional group.

Reactant(s) $\xrightarrow{\text{Reagent(s)}}$ Product(s) This equation represents a known reaction (transformation) and can be arranged in a variety of ways as a problem (see below).

- -

Reactant(s) $\xrightarrow{\text{Reagent(s)}}$? Product(s) is missing.

Reactant(s) $\xrightarrow{\text{?}}$ Product(s) Reagent(s) is missing.

? $\xrightarrow{\text{Reagent(s)}}$ Product(s) Reactant(s) is missing

When chemists are given a target molecule (TM) to synthesize, they generally begin there (at the end), and work their way backwards to an acceptable starting structure that can be purchased (given). This is called retrosynthetic analysis and was developed by E.J. Corey, from Harvard, in the 1960s. He won the 1990 Nobel Prize, in part, for this work (and the fact that he was so good at synthesizing complex organic molecules and discovering new reactions). He is credited with discovering over 300 new reactions. E.J. Corey was born in 1928 and is still active in research today. We will try and use a similar approach in our synthetic planning. In each backwards step of a synthesis, we are presented with a structure that was prepared in a single step from something. This is actually a more difficult synthesis problem than the three variations presented above, because two of the three components are missing.

Product(s) $\xleftarrow{\text{?}}$? Retrosynthetic perspective (TM - 1)
(target molecules = TM)   (TM - 1) Reactant and Reagent(s) are missing.

Chapter 12

A new one step problem, similar to the example, just above, is continually created until we reach an acceptable starting point. Usually, there will be functional group clues in the product that suggest what sort(s) of reaction(s) might have been used to make the target molecule. Additionally, carbon-carbon bond formation is essential to almost every synthesis, and mostly requires a nucleophile/electrophile strategy (always in this book). Often there is more than one approach, so not everyone will generate the same sequence of reactions (or the order of the steps might be different). One approach will usually be better than the others, but we will be content with any reasonable synthetic plan.

A possible sequence of steps is shown below with a logical question that must be answered for each step. E.J. Corey showed the backward steps using a different type of arrow, pointing in the opposite direction, to show the "backwards" thinking of retrosynthetic steps. He left out the reagents, to be filled in later, when the actual synthesis is proposed. To me, E.J. Corey's backward arrows seem too confusing for beginners, so I turned around the arrows and put a question mark above them to indicate what we are probably thinking (Ask yourself: How do I do this?).

E.J. Corey's way of representing "retrosynthetic steps".

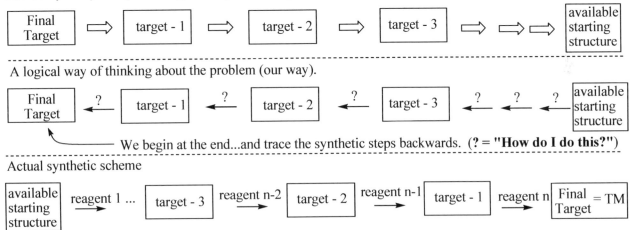

A logical way of thinking about the problem (our way).

We begin at the end...and trace the synthetic steps backwards. (? = "How do I do this?")

Actual synthetic scheme

Occasionally functional group clues are completely missing in a TM, and when that happens, it demands a higher level of sophistication to recognize what could have been there. As beginners, we won't typically attempt those sorts of problems.

Some possible synthetic strategies that are even better than making a target molecule include the following (though these won't be available to you, as a student of organic chemistry):

1. Check chemical catalogs and buy the compound, if available, and if you have enough money.
2. Pay a specialty chemical company to make it for you (can be very expensive).
3. Check the chemical literature and follow someone else's established approach (though literature approaches don't always work as well as the authors claim).

Chapter 12

Let's take a look at some of the functional group possibilities from simple RBr starting structures using the reactions that we have studied. Our actual starting points are methane, ethane, propane, cyclohexane and bromobenzene. The bromoalkane starting points in a-j can be made from these given compounds. We have learned reactions to accomplish many functional group transformations (FGI) and problem 21 will give you a chance to convince yourself that you can do so. Parts I and j use n-butane isomers, not one of our given starting materials, and are given just to complete $C_1 - C_4$ possibilities. We will propose syntheses for the C_4 and C_5 RBr compounds after problem 21.

Problem 21 – Show the necessary synthetic steps to make each functional group below. Most of them are compounds missing from our chemical catalog. The first sequence (a) shows some generic transformations possible. Answers are given for d and e in the key, since they are all pretty similar to the generic patterns and cover most of the possibilities.

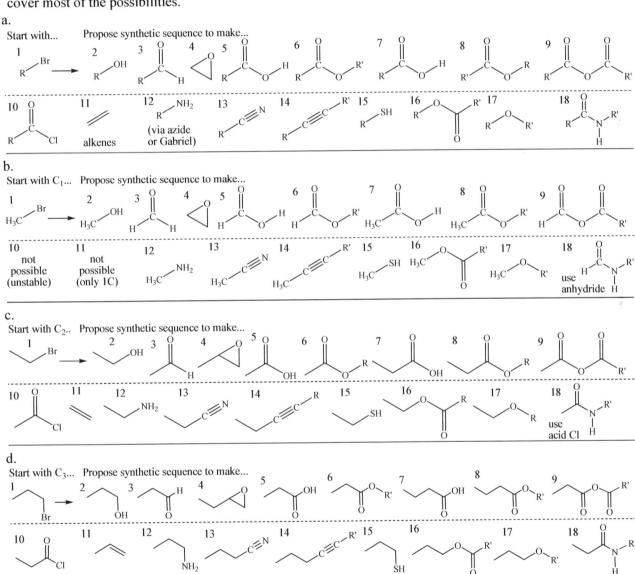

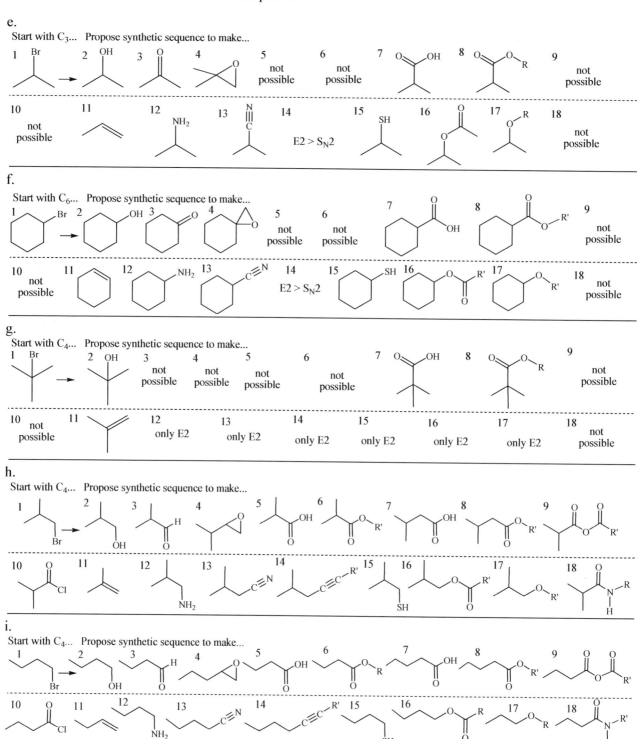

e.

Start with C₃... Propose synthetic sequence to make...

1 Br   2 OH   3 O   4   5 not possible   6 not possible   7 O OH   8 O O R   9 not possible

10 not possible   11   12 NH₂   13 N   14 E2 > S_N2   15 SH   16 O O   17 O R   18 not possible

f.

Start with C₆... Propose synthetic sequence to make...

1 Br   2 OH   3 O   4 O   5 not possible   6 not possible   7 O OH   8 O O R'   9 not possible

10 not possible   11   12 NH₂   13 C≡N   14 E2 > S_N2   15 SH   16 O O R'   17 O R'   18 not possible

g.

Start with C₄... Propose synthetic sequence to make...

1 Br   2 OH   3 not possible   4 not possible   5 not possible   6 not possible   7 O OH   8 O O R   9 not possible

10 not possible   11   12 only E2   13 only E2   14 only E2   15 only E2   16 only E2   17 only E2   18 not possible

h.

Start with C₄... Propose synthetic sequence to make...

1 Br   2 OH   3 O H   4 O   5 OH O   6 O O R'   7 OH O   8 O O R'   9 O O O R'

10 O Cl   11   12 NH₂   13 C≡N   14 C≡C R'   15 SH   16 O O R'   17 O R'   18 O N R H

i.

Start with C₄... Propose synthetic sequence to make...

1 Br   2 OH   3 O H   4 O   5 OH O   6 O O R   7 OH O   8 O O R'   9 O O O R'

10 O Cl   11   12 NH₂   13 C≡N   14 C≡C R'   15 SH   16 O O R   17 O R   18 O N R' H

j.

Start with C_4... Propose synthetic sequence to make...

1 Br   2 OH   3 O   4 O(epoxide)   5 not possible   6 not possible   7 O OH   8 O O R   9 not possible

10 not possible   11 (diene) mixture   12 NH$_2$ 13 N C   14 E2 > S$_N$2   15 SH   16 O(ester)   17 O R   18 not possible

Retrosynthetic Examples (C_4 and C_5 RBr target molecules)

A variety of retrosynthetic approaches are provided below for C_4 and C_5 bromoalkanes, which are very versatile structures. Not all possibilities are shown because we haven't covered all the functional groups. These new carbon skeletons can be used in S_N and E chemistry, organometallic compounds, carbonyl compounds and epoxides to extend the possibilities even farther (alkenes, alkynes and aromatics are still to come). The retrosynthetic examples below are only worked back to C_1-C_3 structures. It is assumed that you can convert methane, ethane or propane to the indicated starting points, as indicated in the examples above. It is clear in these examples how powerfully organometallic compounds are combined with carbonyl, epoxide and RX compounds. After working though these examples, take a look at the C_6 RX compounds in the table on page 402 of the S_N/E topic. You should be able to make most of those compounds, and X = Br can serve as a marker for many additional S_N2 and organometallics reactions that can make even more target molecules, beyond your wildest dreams.

Possible approaches for the C_4 R-Br compounds lost from the catalog. There are actually many more possibilities. In most examples below ROH precedes RBr, which is also a versatile compound for other functional groups.

e = epoxide strategy, nucleophile and OH on vicinal (neighbor) carbon atoms
c = carbonyl strategy, nucleophile and OH on the geminal (the same) carbon atoms

1-bromobutane

2-bromobutane

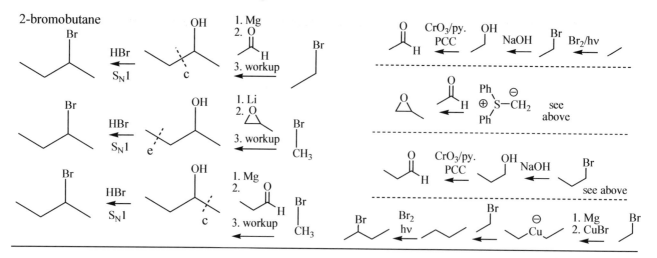

1-bromo-2-methylpropane

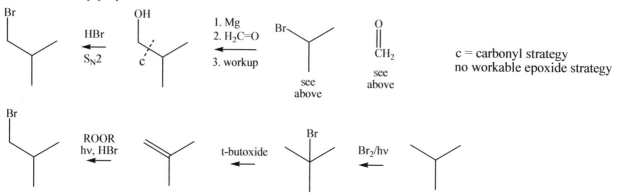

c = carbonyl strategy
no workable epoxide strategy

2-bromo-2-methylpropane

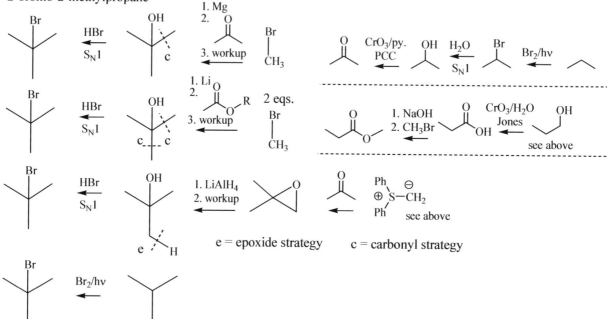

e = epoxide strategy   c = carbonyl strategy

Chapter 12

Possible approaches for the C$_5$ R-Br compounds lost from the catalog. There are actually many possibilities.

1-bromopentane

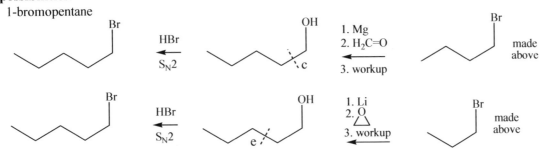

e = epoxide strategy, nucleophile and OH on vicinal (neighbor) carbon atoms
c = carbonyl strategy, nucleophile and OH on the geminal (the same) carbon atoms

2-bromopentane

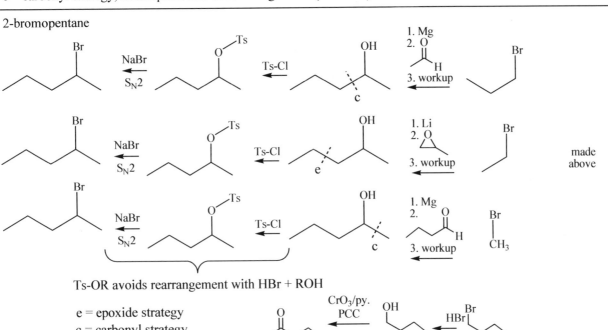

Ts-OR avoids rearrangement with HBr + ROH

e = epoxide strategy
c = carbonyl strategy

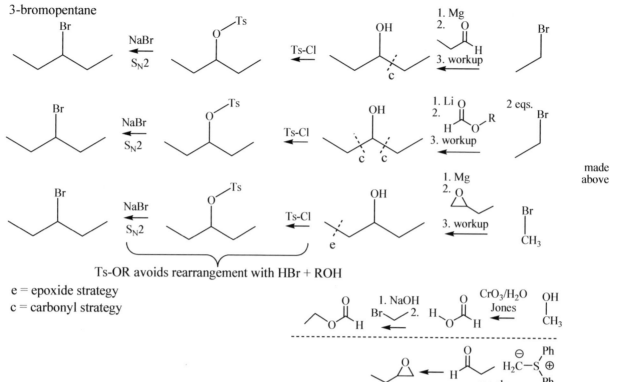

3-bromopentane

e = epoxide strategy
c = carbonyl strategy

Ts-OR avoids rearrangement with HBr + ROH

1-bromo-2-methylbutane

c = carbonyl strategy
no workable epoxide strategy

2-bromo-2-methylbutane

2-bromo-2-methylbutane

2-bromo-2-methylbutane

see above

1. Mg
2. O (acetone)
3. workup

Br (bromoethane)

HBr / S_N1

see above

1. Li
2. O—R (ester)
3. workup

Br / CH₃   2 eqs.   see above

HBr / S_N1

1. Mg
2. O (epoxide)
3. workup

Br / CH₃   see above

HBr / S_N1

e = epoxide strategy
c = carbonyl strategy

2-bromo-3-methylbutane

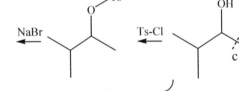

Br / NaBr ← O—Ts / Ts-Cl ← OH

1. Mg
2. O—H (acetaldehyde)
3. workup

Br (2-bromopropane)

see above

Br / NaBr ← O—Ts / Ts-Cl ← OH

1. Li
2. O—H
3. workup

Br / CH₃

Ts-OR avoids rearrangement with HBr + ROH

c = carbonyl strategy
no workable epoxide strategy

1-bromo-3-methylbutane

HBr / S_N2 ← (e) OH

1. Mg
2. O (ethylene oxide)
3. workup

Br (2-bromopropane)

see above

Br

HBr / S_N2 ← (c) OH

1. Li
2. O / H—H (formaldehyde)
3. workup

Br

Br

e = epoxide strategy
c = carbonyl strategy

1-bromo-2,2-dimethylpropane

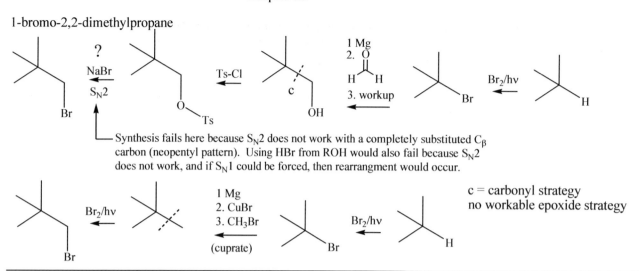

Synthesis fails here because S_N2 does not work with a completely substituted C_β carbon (neopentyl pattern). Using HBr from ROH would also fail because S_N2 does not work, and if S_N1 could be forced, then rearrangment would occur.

c = carbonyl strategy
no workable epoxide strategy

List of $C_6H_{13}X$ targets (X = Br, OH, NH_2, SH, SR, CN, CO_2H, etc.)

1   X

2   X

3   X

4   X

5   X

6   X

7   X

8   X

9   X

10   X

11   X

12   X

13   X

14   X

15   X

16   X

17   X

The following example shows how the reactions learned, thus far, can be used to propose the synthesis of a moderately complex target molecule. We will propose a retrosynthetic approach, beginning with the target molecule (at the end) and work our way backwards, step-by-step, to allowed starting materials (our 7 hydrocarbon molecules). A typical retrosynthesis might look something like the following. More than one approach is shown. Many additional examples are presented later in this chapter. Remember, it is assumed you know how to do functional group interconversions covered in this book using our allowed starting molecules.

target molecule (TM)   TM - 1   electrophile   TM - 2

In "a", one atom has to be a nucleophile and one atom is an electrophile. It seems more logical that oxygen is the nucleophile, so the carbon atom must be the electrophile.

nucleophile

We don't have any C4 straight chains so we will have to make this acid using a 3+1 or a 2+2 strategy. Either would work in this problem.

(3+1) strategy

TM - 3   TM - 4   TM - 5   TM - 6

nucleophile   TM - 7

---

target molecule (TM)   TM - 8   nucleophile   electrophile

Both of these were made above.

In "b", one atom has to be a nucleophile and one atom is an electrophile. Again, it seems more logical that oxygen is the nucleophile as a carboxylate, so the carbon atom must be the electrophile, as RBr.

The ester could also be made using Fischer ester synthesis, combining the carboxylic acid + propan-1-ol with catalytic TsOH and removal of water.

TM - 1 = make ester using acid chloride and alcohol

TM - 2 = make acid chloride using acid and PCl_3

TM - 3 = make acid using Grignard reaction, RBr + CO_2

TM - 4 = make primary RBr using alkene + HBr/ROOR/hv

TM - 5 = make alkene using E2 with RBr + t-butoxide

TM - 6 = make RBr using allowed propane + Br_2/hv

TM - 7 = make ROH using RBr +NaOH (S_N2)

TM - 8 = make ester using carboxylate (from acid + NaOH), then add RBr (S_N2)

(2+2) strategy to make the 4C carboxylic acid

TM - 11

TM - 9   c   TM - 10

TM - 12 CH_2 TM - 13 CH_3 TM - 14 CH_3 TM - 15 CH_4

H_3C—S   TM - 16 CH_3

TM - 9 = make acid using 1o alcohol + Jones (CrO_3/H_2O)

TM - 10 = make 4C alcohol uisng Grignard (epoxide + RMgBr from RBr)

TM - 12 = make epoxide using sulfur ylid chemistry, from sulfur salt + n-butyl lithium, then C=O compound

TM - 13 = aldehyde from 1° alcohol + PCC (CrO_3/pyridine)

TM - 11 = make RBr using allowed ethane + Br_2/hv

TM - 14 = make 1° ROH using RBr +NaOH (S_N2)

TM - 15 = make RBr using allowed ethane + Br_2/hv

TM - 16 = make sulfur salt from RBr + Ph_2S, (S_N2)

Chapter 12

Chapter 13 – Alkenes and Alkynes

   In our previous chapters we selectively chose our reactions to reveal typical situations that show the organic logic we are trying to understand, and we will do the same here in our studies of alkene and alkyne pi bonds. There are several variations of carbon-carbon pi systems in organic chemistry and biochemistry and the most common are shown, just below. We will split them into two groups, non-aromatic pi systems and aromatic pi systems. We will cover aromatic chemistry in our next chapter. In this chapter we will limit our study to a few representative reactions of alkenes and alkynes.

alkene pi systems   alkyne pi systems   conjugated pi systems   cumulated pi systems   aromatic pi systems

In our short course we only look at limited examples of these two (and aromatics).

We will study aromatic pi systems separately.

1. Pi bonds are concentrated regions of electron density (negative in charge) above and below (...or in front and in back) of the bonded atoms. In mechanisms and resonance we push pi electrons in a manner similar to lone pair electrons.

top and bottom   front and back

2. Sigma bond electrons are directly between the bonded atoms, underlying the pi electrons. They are always the first bond formed. If there is only one bond, it will be a sigma bond. Pi bonds are always the second and third bonds between bonded atoms, when present. Because of that exposure, CC pi bonds usually react before C-C or C-H single bonds. I suggest that you draw in two dots to show the pi electrons, just like lone pairs. It helps to focus your thinking on how these reactions begin.

3. Often a position adjacent to a pi bond (allylic, benzylic) has increased reactivity, because of resonance possibilities in an intermediate or transition state. They are an occasional exception to the greater reactivity of pi bonds over sigma bonds.

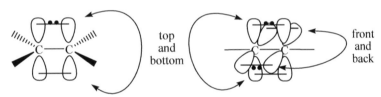

Allylic (and benzylic) positions
are stabilized by adjacent pi bonds,
whether carobcation, carbanion or
free radical.

4. Carbon-carbon pi bonds tend to be weaker and tend to be more reactive than carbon-carbon sigma bonds. They also tend to be shorter because of greater electron density between the bonded atoms.

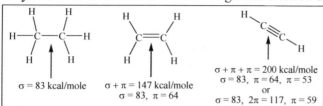

5. Atoms having pi bonds are sp² or sp hybridized since an atom's 2p orbital is required for a pi bond.

6. Lone pairs adjacent to a pi bond make the pi bond more electron rich (resonance) and carbonyl or nitrile pi bonds adjacent to a carbon-carbon pi bond make it more electron poor (resonance). A CC pi bond takes on some of the features of what it is conjugated with (lone pair or heteroatom pi bond). Inductive donation by R groups can also make a pi bond more electron rich.

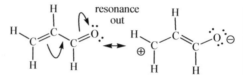

pi bond is electron rich because of "N", the N lone pair pushes electron density into the pi bond.

pi bond is electron poor because of "C=O", the C=O pulls electron density out of the C=C pi bond.

Inductive effects of "R" groups also make CC pi bonds more electron rich.

Problem 1 – Specify each C=C pi bond below as electron rich or electron poor or neither?

Almost all of the reactions in this chapter involve electron donation from a nucleophilic C=C pi bond to an electron poor reagent (electrophile, oxidizing reagent or free radical). Most of alkene and alkyne reactions that we study involve two electron transfers, and we will concentrate on those. There are reactions where single electron transfers occur and free radical intermediates are present. We already presented one of those reactions with the addition of HBr in the presence of a peroxide catalyst and light. However, the reagents in this chapter all follow the nucleophile/electrophile logic that is the primary emphasis of this book.

The logic of the first step of reactions with alkenes, alkynes, conjugated, cumulated and aromatic pi systems is the same: pi electron pair donation.

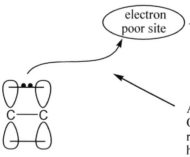

The pi bond is an electron rich region. It is able to donate its electrons to electron poor sites. Draw in the 2 electrons to focus your thniking.

electron poor site

Often this is an electrophile, but can also be an electron poor oxidizing reagent, or the LUMO (lowest unoccupied molecular orbital) of a different electron poor pi system reacting with the electron rich HOMO (highest occupied molecular orbital) of the pi bond in a concerted reaction. We avoid HOMO/LUMO reactions in our course.

Arrow pushing often shows the donation of both pi electrons. Often this is the first step of one, two or more steps. In free radical reactions one electron movement occurs and requires half headed arrows. In concerted reactions there is usually a cyclic movement of electrons (often 6), where electrons lost in the first transfer by one of the carbon atoms are returned in a final transfer back to that atom. You can study concerted reactions in a more advanced course.

Chapter 13

Besides the mechanistic steps of each reaction, there are complications of stereoselectivity and regioselectivity. Learning these ideas is definitely a challenge, but it's a challenge we cannot avoid. Because there are two carbon atoms in a pi bond, both carbon atoms must be analyzed when considering the mechanistic details of a reaction. Which carbon reacts with what is a regioselective choice (which "region" reacts?). Which relative face (top or bottom) each group adds to is a stereoselective choice (next). It would be nice if we could avoid these complications, but nature doesn't give us that option, if we really want to understand the chemistry.

Stereoselective Choices

Stereoselectivity in a reaction occurs when one stereoisomer is preferred over any of the others possible. This often occurs in reactions with alkenes or alkynes. In the structures below, wiggly lines indicate mixed stereochemistry and the dashes and wedges follow our usual conventions. We will encounter all possible outcomes in the reactions we study, depending on the reagents reacting. The terms used to describe the way reagents add to the face of a pi bond come from descriptions we used in describing conformations of rotating single bonds: "syn" and "anti".

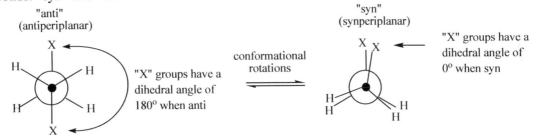

"cis" alkenes (Z alkenes) – this alkene example is symmetrical

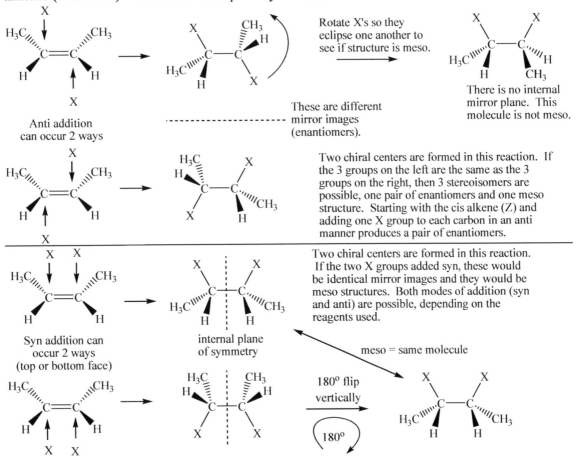

Rotate X's so they eclipse one another to see if structure is meso.

There is no internal mirror plane. This molecule is not meso.

These are different mirror images (enantiomers).

Anti addition can occur 2 ways

Two chiral centers are formed in this reaction. If the 3 groups on the left are the same as the 3 groups on the right, then 3 stereoisomers are possible, one pair of enantiomers and one meso structure. Starting with the cis alkene (Z) and adding one X group to each carbon in an anti manner produces a pair of enantiomers.

Two chiral centers are formed in this reaction. If the two X groups added syn, these would be identical mirror images and they would be meso structures. Both modes of addition (syn and anti) are possible, depending on the reagents used.

Syn addition can occur 2 ways (top or bottom face)

internal plane of symmetry

meso = same molecule

180° flip vertically

180°

"trans" alkenes (E alkenes) – this alkene example is symmetrical

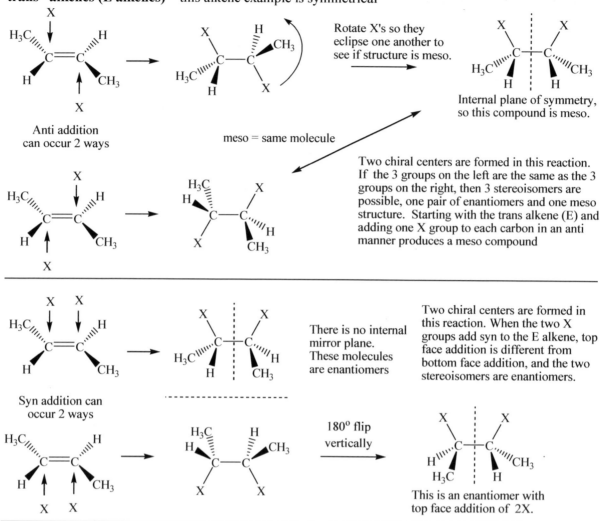

Anti addition
can occur 2 ways

meso = same molecule

Rotate X's so they
eclipse one another to
see if structure is meso.

Internal plane of symmetry,
so this compound is meso.

Two chiral centers are formed in this reaction.
If the 3 groups on the left are the same as the 3
groups on the right, then 3 stereoisomers are
possible, one pair of enantiomers and one meso
structure. Starting with the trans alkene (E) and
adding one X group to each carbon in an anti
manner produces a meso compound

There is no internal
mirror plane.
These molecules
are enantiomers

Two chiral centers are formed in
this reaction. When the two X
groups add syn to the E alkene, top
face addition is different from
bottom face addition, and the two
stereoisomers are enantiomers.

Syn addition can
occur 2 ways

180° flip
vertically

This is an enantiomer with
top face addition of 2X.

Summary of observations – If X = Y and R1 = R2 then meso stereoisomers are possible. If X ≠ Y or R1 ≠ R2
then meso stereoisomers are not possible. Additions to top face or bottom face can lead to various possible
outcomes. Remember, stereoisomers have the same connections of atoms, but have a different orientation in
space.

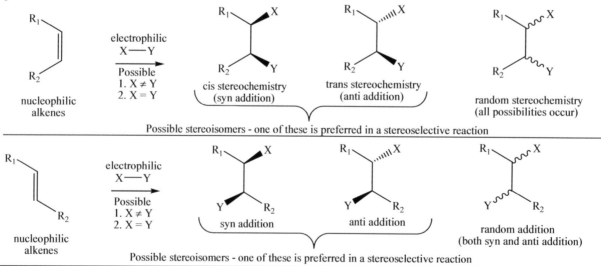

Chapter 13

Rings make the facial addition easier to see because they have a permanent top and bottom face that does not rotate.

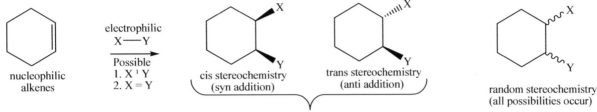

While triple bonds don't have any chiral centers, they can make E/Z stereoisomers.

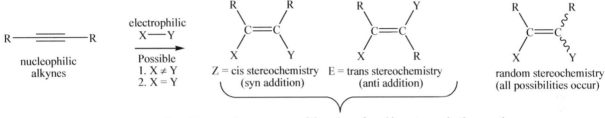

Problem 2 – What stereochemical relationships exists in the products of the following electrophilic addition reactions?

Chapter 13

Regioselective choices

Regioselectivity in a reaction occurs when one distinguishable region preferentially reacts in a certain manner over another that could have reacted in a similar way. In C=C pi bonds each carbon atom can be considered to be a different region. Unsymmetrical pi bonds often react in a regioselective way (when C1 ≠ C2). If X = Y, then we cannot distinguish different "regions" (carbons) of a pi bond. Stereoselectivity is not implied in this example.

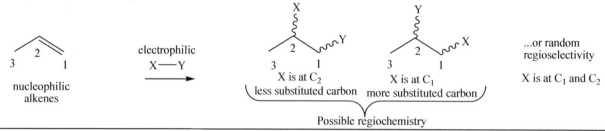

Regioselectivity is also possible in unsymmetrically substituted alkynes. Stereoselectivity is not implied in this example.

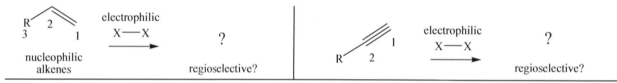

Problem 3 – Can regioselectivity be observed in an electrophilic addition to an alkene or alkyne when X = Y?

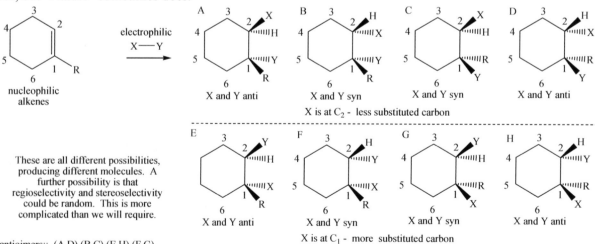

1-methylcyclohex-1-ene illustrates a range of very complicated possible outcomes. The actual products obtained from specific reagents are usually much simpler. We won't require this level of complication in our course, but "Nature" sometimes does.

These are all different possibilities, producing different molecules. A further possibility is that regioselectivity and stereoselectivity could be random. This is more complicated than we will require.

X is at C_2 - less substituted carbon

X is at C_1 - more substituted carbon

enantioimers:: (A,D) (B,C) (E,H) (F,G)
diasteromers: (A,B) (A,C) (D,B) (D,C) (E,F) (E,G) (H,F) (H,G)
(A,B,C,D) are not stereoisomers with (E,F,G,H), they have different connections of the atoms

Regioselective and Stereoselective Choices

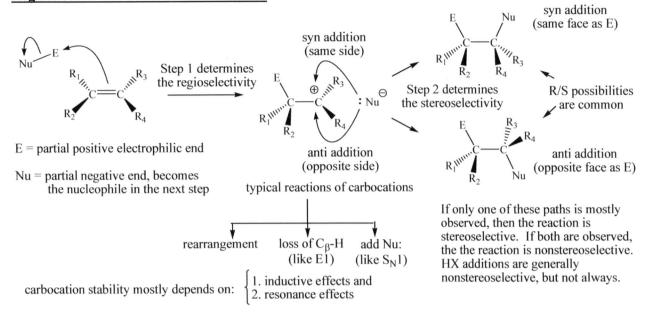

E = partial positive electrophilic end

Nu = partial negative end, becomes
the nucleophile in the next step

typical reactions of carbocations

If only one of these paths is mostly
observed, then the reaction is
stereoselective. If both are observed,
the the reaction is nonstereoselective.
HX additions are generally
nonstereoselective, but not always.

rearrangement   loss of C_β-H   add Nu:
                  (like E1)   (like S_N1)

carbocation stability mostly depends on: { 1. inductive effects and
                                             2. resonance effects

General electrophilic addition mechanisms

Because our goal is to understand the "essential" logic of organic chemistry, we will only consider 6 different reaction conditions reacting with alkenes and alkynes. Four of our reactions are considered to be electrophilic additions to the pi bonds (HBr, H_2SO_4/H_2O, Br_2 and Br_2/H_2O). We will show one oxidation reaction of alkenes (OsO_4) and one reduction reaction of alkenes (Pd/H_2). We will add two additional reductions of alkynes so that we can make E or Z alkenes.

Many electrophilic reagents are able to pull electrons from a pi bond, leaving one carbon without electrons (a carbocation site) and the other carbon sigma bonded to the electrophile. The electrophilic portion of a reagent often reacts in a way that is obvious from bond polarities (HBr and H_3O^+/H_2O). However, sometimes the electrophilic character is not obvious, but rather induced when a nonpolar reactant comes in proximity of the electrons of a pi bond (Br_2). If the solvent is also nucleophilic, it may also react with the carbocation in competition with any other nucleophiles present (Br_2/H_2O). Time to look at the details.

Addition of HBr to alkenes

We'll begin with our easiest example, electrophilic HBr addition to alkenes. (This is different than free radical addition of HBr / peroxide / light.) The hydrogen end of the H-Br bond is clearly partially positive, so it will always receive the electrons when HBr reacts in a nucleophile/electrophile manner. The carbon that can better form a carbocation will give up its share of the pi bond to the other carbon, which will make a sigma bond with the acidic proton. In general, the more substituted carbon will make a more stable carbocation ($3° > 2° >> 1° >> CH_3$), meaning the less substituted carbon will bond to the proton. This is known as Markovnikov's rule (just another way of saying the most stable carbocation forms fastest).

Remember from our $S_N1/E1$ topic, carbocations have 3 common choices: 1. Add a nucleophile, 2. Lose a beta proton to make a pi bond, or 3. rearrange to a new carbocation, which starts those 3 choices all over again. These same three possibilities are present when carbocations from electrophilic addition to alkenes.

Immediately after the first step, the carbocation intermediate assumes an opposite role of electrophile in reacting with the bromide nucleophile. We will assume that the intermediate carbocation is sp^2, trigonal planar, and can be attacked from either side by the bromide (not stereoselective = random addition).

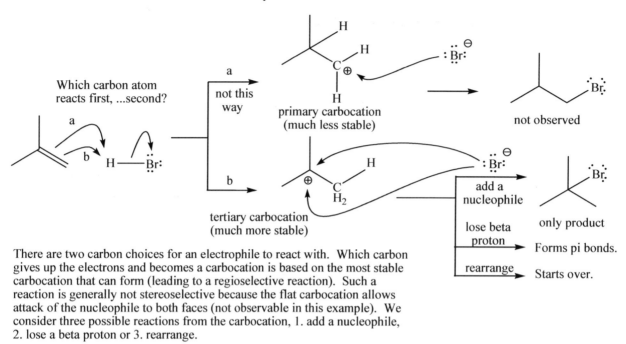

There are two carbon choices for an electrophile to react with. Which carbon gives up the electrons and becomes a carbocation is based on the most stable carbocation that can form (leading to a regioselective reaction). Such a reaction is generally not stereoselective because the flat carbocation allows attack of the nucleophile to both faces (not observable in this example). We consider three possible reactions from the carbocation, 1. add a nucleophile, 2. lose a beta proton or 3. rearrange.

We will limit ourselves to 10 different example alkenes that show different aspects of the regioselectivity, stereoselectivity and possible rearrangement.

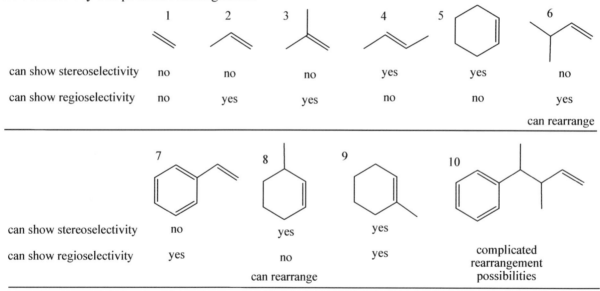

Chapter 13

Example reactions of H-Br with the representative alkenes.

1. Ethane cannot show regioselectivity or stereoselectivity with any of our reagents. The mechanistic details are provided.

nucleophile | electrophile (Lewis acid) | form most stable carbocation, 3 possible choices, 1. add nucleophile, 2. lose beta C-H, 3. rearrangement | Attack from top and bottom faces is possible here, but not observable in this example.

2. Propene can show regioselectivity, but cannot show stereoselectivity with any of our reagents. The mechanistic details are provided.

nucleophile | electrophile (Lewis acid) | Attack from top and bottom faces is possible here, but not observable in this example.

3. 2-methylprop-1-ene can show regioselectivity, but cannot show stereoselectivity with any of our reagents. Add in any missing mechanistic details.

nucleophile | electrophile (Lewis acid) | form most stable carbocation, 3 possible choices, 1. add nucleophile, 2. lose beta C-H, 3. rearrangement

4. trans-but-2-ene cannot show regioselectivity, but can show stereoselectivity with the right reagent. HBr won't show stereoselectivity, but DBr would show what happens. Write a complete mechanism.

nucleophile | electrophile (Lewis acid) | enantiomers

5. Cyclohexene cannot show regioselectivity, but can show stereoselectivity with the right reagent. Write a complete mechanism.

nucleophile | electrophile (Lewis acid)

6. 3-methylbut-1-ene can show regioselectivity, but cannot show stereoselectivity with any of our reagents. This example will also rearrange whenever carbocations form. Add in any missing mechanistic details.

nucleophile   electrophile (Lewis acid)

7. Styrene (vinylbenzene, ethenylbenzene) can show regioselectivity, but cannot show stereoselectivity with any of our reagents. Write a complete mechanism.

nucleophile   electrophile (Lewis acid)   (resonance)

8. 3-methylcyclohex-1-ene is good for showing when rearrangements are possible. Add in any missing mechanistic details.

nucleophile   electrophile (Lewis acid)   rearrangement

9. 1-methylcyclohex-1-ene can show regioselectivity and can show stereoselectivity if the reagent reacting allows it. HBr only allows us to see regioselectivity. Write a complete mechanism.

nucleophile   electrophile (Lewis acid)

10. 3-methyl-4-phenylpent-1-ene is a good example for showing complicated rearrangement possibilities. Add in any missing mechanistic details.

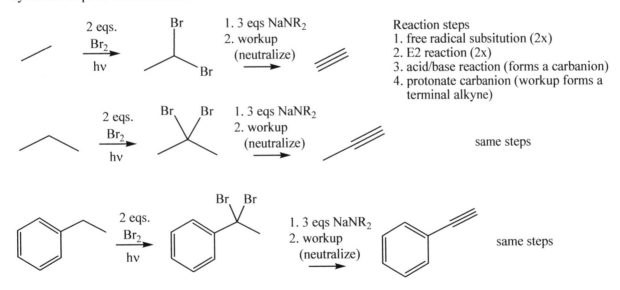

Problem 4 – Point out if any stereoisomers are formed in the following reactions. Indicate if they are enantiomers or diastereomers. If chiral centers are drawn, indicate if the absolute configuration is R or S in your drawings. Could the opposite configurations also form? Are the reactions regioselective and/or stereoselective? D (deuterium) is an isotope of hydrogen and reacts in a similar way, but is observably different.

a.

H—Br

b.

D—Br

Addition of HX to alkynes

Currently, our only method for making alkynes involves free radical substitution twice (putting on 2 leaving groups). Only three of our allowed 7 starting hydrocarbons can be used to make alkynes. The synthetic steps to do this are reshown below.

2 eqs.
Br₂
hv

Br
Br

1. 3 eqs NaNR₂
2. workup
(neutralize)

Reaction steps
1. free radical subsitution (2x)
2. E2 reaction (2x)
3. acid/base reaction (forms a carbanion)
4. protonate carbanion (workup forms a terminal alkyne)

2 eqs.
Br₂
hv

Br   Br

1. 3 eqs NaNR₂
2. workup
(neutralize)

same steps

2 eqs.
Br₂
hv

Br   Br

1. 3 eqs NaNR₂
2. workup
(neutralize)

same steps

We can extend a terminal alkyne with 3 different electrophiles (1. methyl or primary RBr, 2. aldehydes or ketones and 3. epoxides), but first make the carbanion conjugate base.

Reactions with alkynes occur in a similar manner, using similar logic (except addition is possible once or twice). It is generally more difficult to add H-X a second time, so it is sometimes possible to stop after a single addition (but not always). Similar regioselectivity is expected the second time, for similar reasons. To distinguish between adding once or twice in problems, we will use 1 equivalent for single addition and 2 equivalents for double addition. Begin by deciding which carbon atom would make the more stable carbocation and then protonate the other carbon of the triple bond.

Example reactions of H-Br with the representative alkynes.

1. Ethyne cannot show regioselectivity or stereoselectivity with any of our reagents.

2. Propyne can show regioselectivity, and can show stereoselectivity with some of our reagents. Write out a mechanism.

3. Phenylethyne can show regioselectivity, and can show stereoselectivity with some of our reagents. Add in any missing mechanistic details.

Chapter 13

4. But-2-yne cannot show regioselectivity, but can show stereoselectivity with some reagents. However, HBr is not stereoselective. Write a complete mechanism.

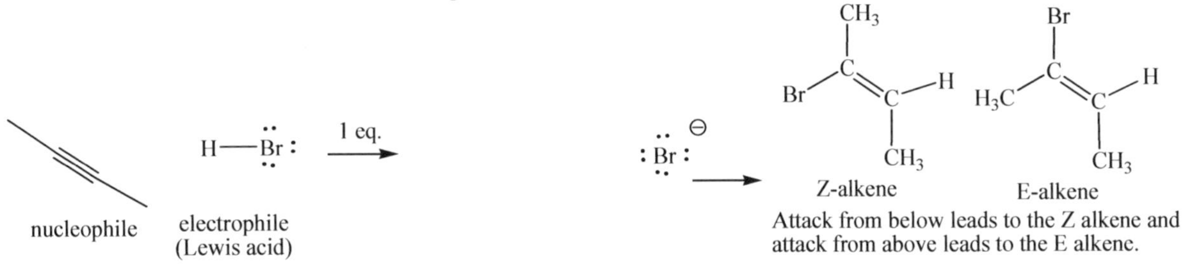

Attack from below leads to the Z alkene and attack from above leads to the E alkene.

Problem 5 – Provide a mechanistic explanation for the following reactions. Are there any stereocenters to consider? If so, point out R/S and/or E/Z possibilities.

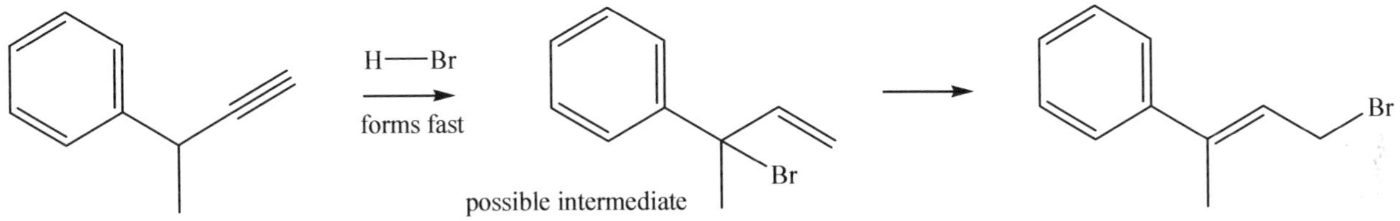

Problem 6 – Explain the regioselectivity in the following reaction. Is this reasonable considering the triple bond is substituted on both ends? Was the reaction stereoselective, as shown? If so, was the addition syn, anti or random?

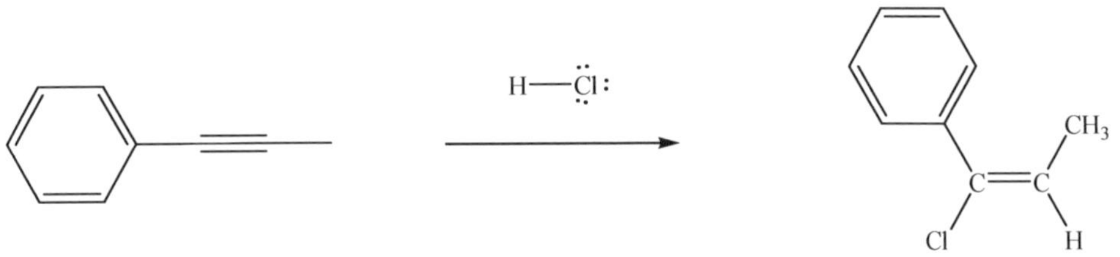

Addition of water to carbon-carbon pi bonds in acidic solution = hydration conditions (H_2SO_4/H_2O)

Strong acids protonate to donatable electron pairs. We have seen this with lone pair electrons on alcohols, ethers, carbonyl compounds, amines, nitriles and pi bond electrons, in the acid/base topic. In HX addition to alkenes (just above), the newly formed carbocation is attacked by an X^- counter ion. Almost identical chemistry occurs when aqueous sulfuric acid is mixed with alkenes. Instead of a halide ion, water is the nucleophile attacking the carbocation. This requires a subsequent proton transfer away from oxygen, leading to an alcohol product (similar to earlier S_N1 chemistry of "RX + H_2O"). Rearrangements are a possibility.

First we need to decide what our acid is. In aqueous reactions we will always show sulfuric acid, H_2SO_4. When the solvent is not water (for example, an alcohol) we will always use toluenesulfonic acid (organic sulfuric acid).

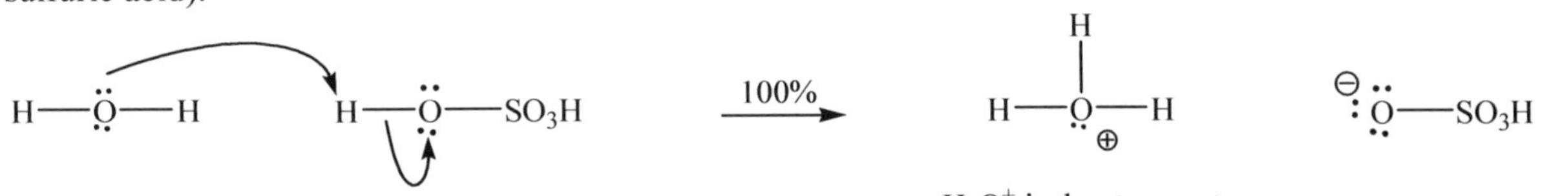

H_3O^+ is the strongest possible acid in water.

Hydration of an alkene (adding water)

10 Alkenes to practice on (This isn't much different than adding H-Br.)

1. Ethane cannot show regioselectivity or stereoselectivity with any of our reagents. A complete mechanism is provided.

nucleophile   electrophile (Lewis acid)

carbocation, 3 choices, add nucleophile, lose beta C-H, rearrangement

Attack from top and bottom faces is possible here, but not observable in this example. Water acts as a nucleophile in the first step and acts as a base in the second step.

2. Propene can show regioselectivity, but cannot show stereoselectivity with any of our reagents. Add in any necessary mechanistic details.

3. 2-methylprop-1-ene can show regioselectivity, but cannot show stereoselectivity with any of our reagents. Add in any missing mechanistic details. Add in any necessary mechanistic details.

4. trans-but-2-ene cannot show regioselectivity, but can show stereoselectivity with some of our reagents, but not with H_3O^+/H_2O. Write a complete mechanism.

R/S enantiomers

Chapter 13

5. Cyclohexene cannot show regioselectivity, but can show stereoselectivity with some of our reagents, but not with H_3O^+/H_2O. Write in a complete mechanism.

6. 3-methylbut-1-ene can show regioselectivity, but does cannot show stereoselectivity with any of our reagents. This example will also rearrange whenever carbocations form, as with H_3O^+/H_2O. Add in any missing mechanistic details.

7. Styrene (vinylbenzene, ethenylbenzene) can show regioselectivity, but cannot show stereoselectivity with any of our reagents. Add in any necessary mechanistic details.

8. 3-methylcyclohex-1-ene is good for showing when rearrangements are possible, as with H_3O^+/H_2O. Add in any missing mechanistic details.

9. 1-methylcyclohex-1-ene can show regioselectivity and can show stereoselectivity depending on the reagents reacting with it. Write a complete mechanism.

10. 3-methyl-4-phenylpent-1-ene is good for showing complicated rearrangement possibilities, as with H_3O^+/H_2O. . Add in any missing mechanistic details.

Addition of an alcohol to carbon-carbon pi bonds in acidic solution = making ethers (TsOH / ROH)

Using alcohols as 'organic' water and toluenesulfonic acid as 'organic' sulfuric acid, alkenes can make ethers using reactions almost identical to those, just above. Rearrangements are, again, a possibility. Our acid is the protonated alcohol, which gets protonated by toluenesulfonic acid, TsO-H (organic sulfuric acid).

$CH_3O^+H_2$ is the strongest possible acid in methanol.

The mechanism is essentially identical, when an alcohol is used as the solvent instead of water. However, an ether is the final product instead of an alcohol. Since carbocations are generated in both of these reactions, rearrangements to similar or more stable carbocations are possible and expected.

Chapter 13

Problem 7 – Explain the following observation. Include a complete mechanism.

H—O—CH₃ (protonated)
H—O—CH₃

major   minor

?

Maybe you noticed the similarity of mechanisms in dehydration of an alcohol (E1 mechanism) and hydration of an alkene (addition mechanism). The only difference in mechanisms appears to be the direction of the reactions. If both of these are useful reactions, there must be something that drives the reactions in one direction or the other. In fact the conditions for each of these reactions are very different.

In the dehydration of alcohols, concentrated sulfuric acid is used and almost no water is present (any water present is protonated as H_3O^+). Also, the reaction is run at a high temperature, which distills the alkene away from the mixture, continually shifting the equilibrium to alkene formation, until all of the alcohol is removed from the equilibrium mixture.

E1 mechanism – dehydration of an alcohol

R = carbon or hydrogen in this problem

Equilibrium is shifted to the right by the distillation of the alkene (Le Chatelier's principle). Watch out for rearrangements.

Water becomes completely protonated in concentrated sulfuric acid.

Alkene is removed by distillation.

When hydration of an alkene is the goal, the reaction is run at a cooler temperature and lots of water is present. Just a catalytic amount of acid is necessary and the large amount of water pushes the equilibrium of the reaction toward the alcohol product, which is also favored thermodynamically.

Hydration of an alkene (electrophilic addition in lots of water)

R = carbon or hydrogen in this problem

A large concentration of water and cooler temperature pushes equilibrium to the right.
1. Make the most stable carbocation in the first step.
2. Watch out for carbocation rearrangements.

As a student, knowing the mechanism well in one direction means you probably know the mechanism in the opposite direction too. This is especially helpful when reactions are reversible, which is true for a number of mechanisms that we learn (very common when water is involved). Learning a mechanism in one direction, often teaches two concepts at once. Also, every mechanism you learn increases your insight and intuition in newly encountered mechanisms and gives you a new opportunity to challenge your organic logic.

Problem 8 – Point out if any stereoisomers are formed in the following reactions. Indicate if they are enantiomers or diastereomers or neither. If chiral centers are drawn, indicate if the absolute configuration is R or S in your drawings. Could the opposite configurations also form? Are the reactions regioselective and/or stereoselective?

a.

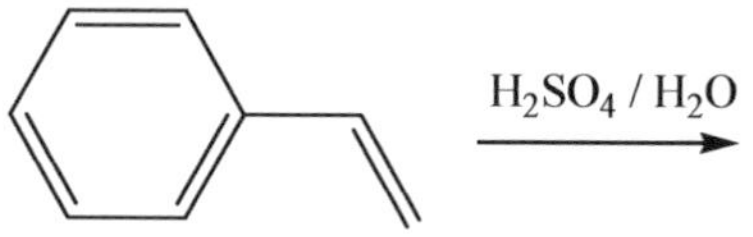

H_2SO_4 / H_2O

b.

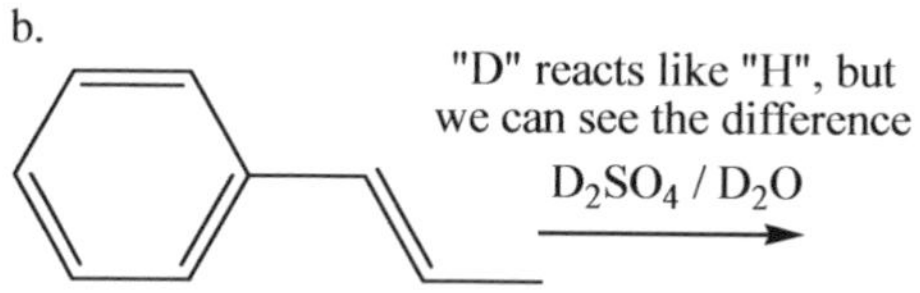

"D" reacts like "H", but we can see the difference

D_2SO_4 / D_2O

Problem 9 – Provide an arrow pushing mechanism for the following transformation. Hint: Find the extra proton to decide where the reaction began, then find the carbon-carbon bonds that formed, then where the nucleophile added. A more complicated version of this reaction occurs every time your body makes cholesterol (that's a 24/7 reaction).

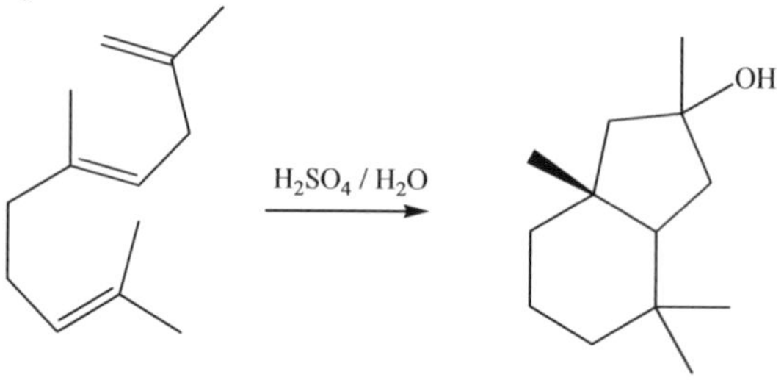

H_2SO_4 / H_2O

OH

Problem 10 – Dihydropyran (DHP) is a special alkene-ether (called an enol ether). It adds alcohols with high regioselectivity (as shown) and is, kinetically, a fast reaction. It is also used to protect alcohols (disguised as ethers), while a different reaction is run that normally would react with an alcohol group (called a protecting group). Explain the regioselectivity and speed of the reaction. Hint: An oxygen lone pair of the enol ether is important to your answer. Provide an arrow pushing mechanism.

pyran
(for reference)

dihydropyran

$R\overset{O}{\diagdown}H$

H——ŌTs

toluene sulfonic acid

$\left(= THPO\diagup R \right)$

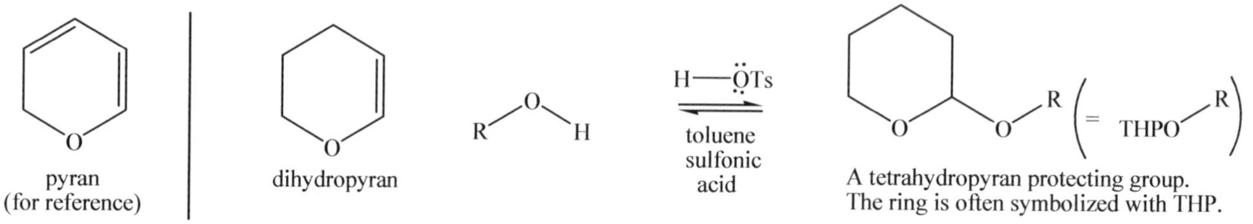

A tetrahydropyran protecting group. The ring is often symbolized with THP.

Alkynes in aqueous acid – addition of water and tautomerization – making ketones from alkynes

Adding mercuric ion (Hg^{+2}) as a catalyst in aqueous acid helps the addition of water to the first pi bond of alkynes (however we will ignore the effect of the mercury ion). The first addition forms what is called an enol ("en" = double bond and "ol" = alcohol). However, the reaction does not stop there. Instead, the reaction keeps right on going to form a "keto" tautomer from the "enol" via a fast tautomerization process. Tautomerization steps are very common mechanistic changes that pop up in all sorts of places in organic chemistry and biochemistry. You should be able to write this mechanism in acid, forwards and backwards. The same is true for tautomerism in base solution, which we show, just below.

We will use a simpler mechanism that only uses aqueous acid as an easier alternative to showing the Hg^{+2} ion.

Chapter 13

Hydration (= addition of water) and tautomerization mechanism (without mercury using only aqueous acid)

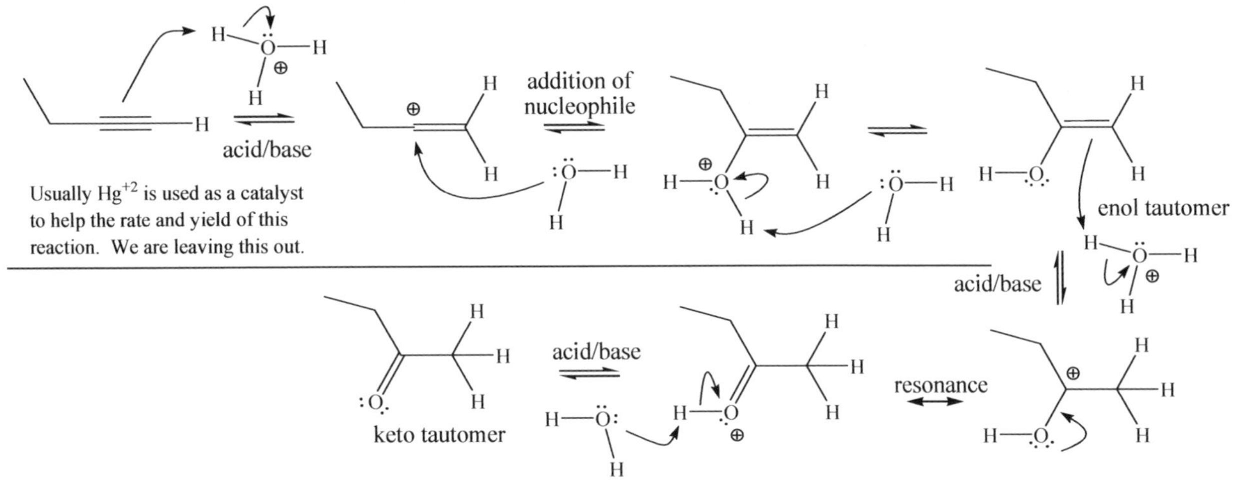

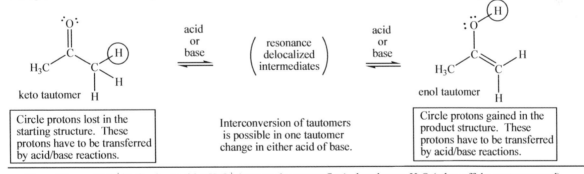

Usually Hg^{+2} is used as a catalyst to help the rate and yield of this reaction. We are leaving this out.

Keto/Enol Tautomerization

Tautomers are isomers that differ by the location of a proton and a pi bond. To be official tautomers, a heteroatom or atoms is part of the system (different than carbon, often oxygen or nitrogen or both). In the simplest case, there are at two isomers in equilibrium with one another that differ in the location of a proton and a pi bond (there may be many, many more tautomers possible in more complex systems). The tautomers are interchangeable by 1. proton transfer, 2. resonance intermediates and 3. proton transfer. The "keto" isomer, has a heteroatom in a pi bond and in the "enol" tautomer has two carbons forming a pi bond. This simple pattern can occur in an infinite number of systems, from very simple to very complex. A possible approach to figuring out what to do in keto/enol tautomer problems is shown below. We will only consider four simple tautomer interconversions (involving one tautomeric change), but you need to how to show this occurs four different ways.

A simple keto/enol tautomer system: forwards and backwards in acid and in base = four different problems

| keto tautomer | acid or base ⇌ | resonance delocalized intermediates | acid or base ⇌ | enol tautomer |

Circle protons lost in the starting structure. These protons have to be transferred by acid/base reactions.

Interconversion of tautomers is possible in one tautomer change in either acid of base.

Circle protons gained in the product structure. These protons have to be transferred by acid/base reactions.

acid conditions = H_3O^+/H_2O, best acid = H_3O^+ (puts on the proton first), best base = H_2O (takes off the proton second)

base conditions = $HO^\ominus/H_2O$, best base = $HO^\ominus$ (takes off the proton first), best acid = H_2O (puts on the proton second)

Problem 11 – The following are tautomer equilibria problems, without the mechanistic details first and with the mechanistic details next (so you can check your arrow pushing). Supply all of the mechanistic details in both directions, using H_3O^+/H_2O conditions or H_2O/HO^- conditions, as specified. There are four potential problems with each set of equilibrium arrows. (keto to enol in acid, keto to enol in base, enol to keto in acid and enol to keto in base). All of the reactions follow a similar pattern: 1. start off with proton transfer, 2. resonance intermediates, and 3. finish with proton transfer. If the first step puts on a proton (using strong acid, H_3O^+), the last step will take off a proton (using H_2O as the base). If the first step takes off a proton (using strong base, HO^-), the last step will put on a proton (always using H_2O as the acid). Water is always the weak partner (conjugate) to the stronger component (acid or base).

The four overall tautomeric changes using propanone (acetone).

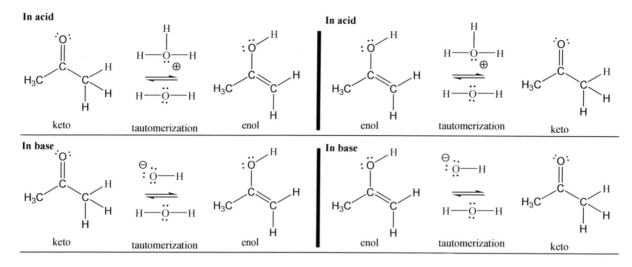

Mechanism steps without the details (you add those in).

Tautomers in acid

| keto tautomer | in acid = proton on | same resonance as b | | in acid = proton off | enol tautomer |

| enol tautomer | in acid = proton on | same resonance as a | | in acid = proton off | keto tautomer |

Tautomers in base

| keto tautomer | in base = proton off | enolate / same resonance as d | | in base = proton on | enol tautomer |

| enol tautomer | in base = proton off | enolate / same resonance as c | | in base = proton on | keto tautomer |

Chapter 13

Mechanism steps with the details (check your answers above).

Tautomers in acid

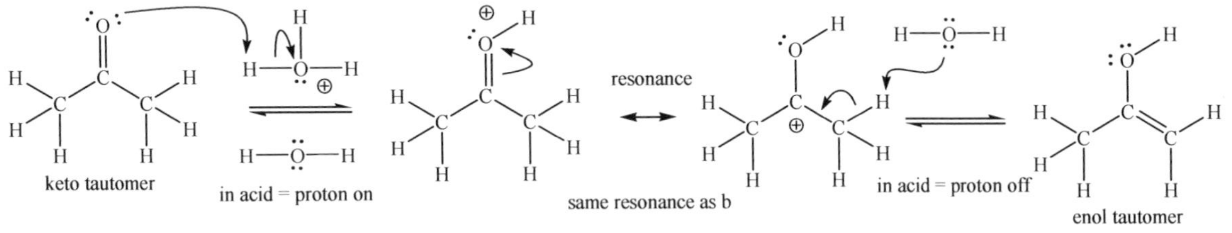

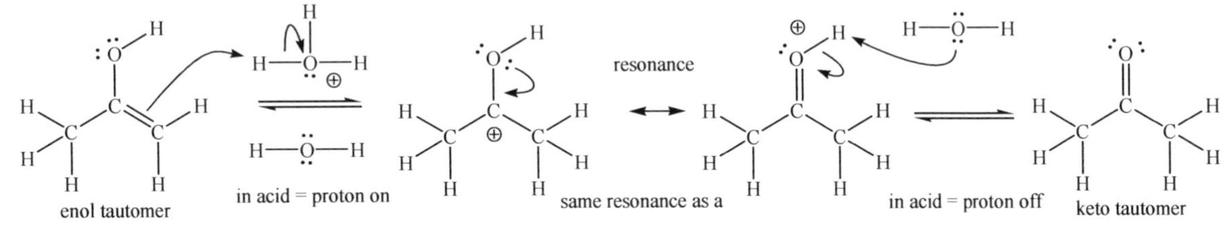

Tautomers in base

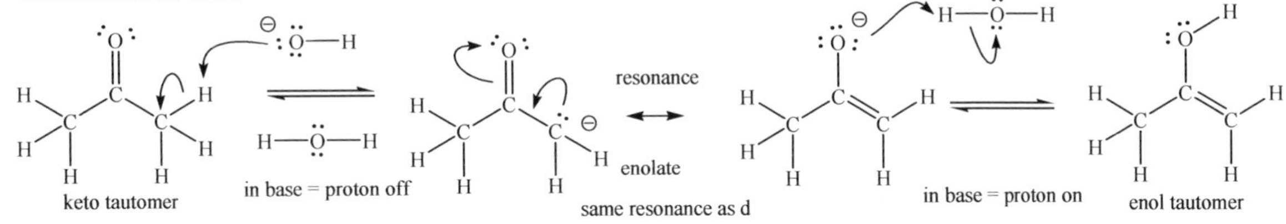

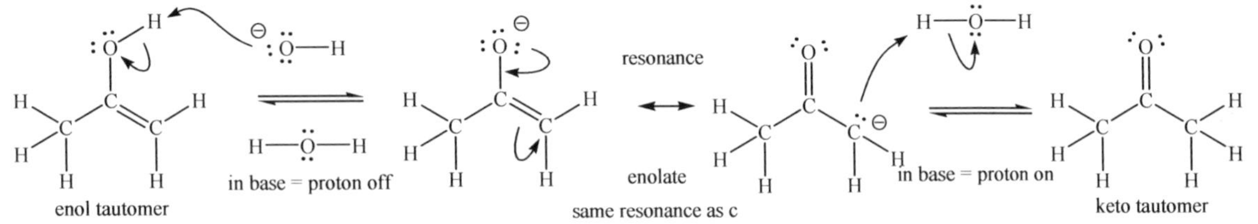

Problem 12 – Most students are familiar with glycolysis from biology classes. There are two steps in glycolysis that involve double tautomer reactions. The first is the transformation of glucose to fructose at the beginning and the second is an equilibrium situation between glyceraldehyde and dihydroxy acetone in the middle. Supply the necessary mechanistic details to show these transformations. Stereochemistry features have been omitted. Use B: as the base and B⁺-H as the acid for proton transfers.

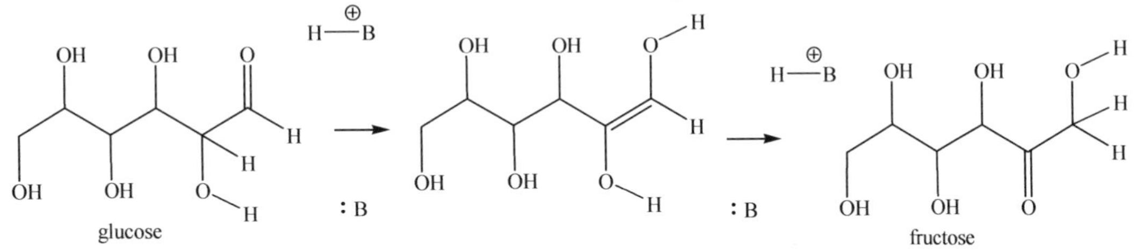

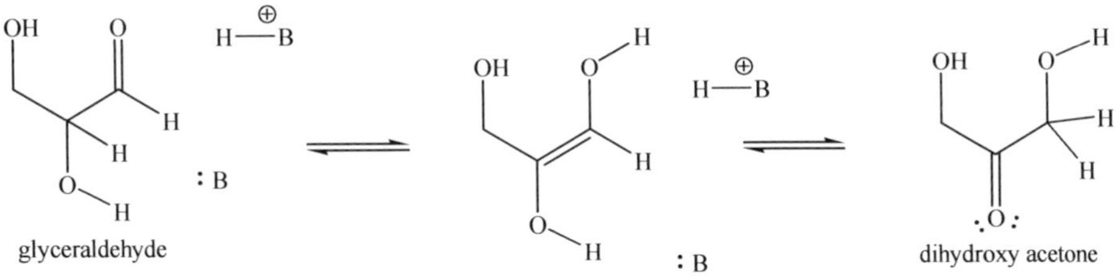

Additions of halogens to alkenes, (X_2 = Cl_2 or Br_2)

The energetics of halogen addition to alkenes span a wide range from too exothermic to endothermic (reversible). Fluorine (F_2) is too reactive and potentially explosive, while iodine (I_2) addition is not thermodynamically favored. Iodine is a good electrophile if a different nucleophile is present, particularly if it occurs intramolecularly (iodolactone formation is a common, useful application). However, bromine (Br_2) is the only halogen that we will study.

ΔH°_f (kcal/mole)	$H_2C{=}CH_2$	X—X			ΔH°_{rxn} (kcal/mole)
$X_2 = F_2$	+12	+0		-104	-116
$X_2 = Cl_2$	+12	+0		-31	-43
$X_2 = Br_2$	+12	+0		-10	-22
$X_2 = I_2$	+12	+0		+16	+4

At first glance, reactions of H-Br seem very different from reactions of Br-Br. The hydrogen bromide molecules are obviously polar, while the bromine molecules are clearly nonpolar. It is not clear what is electrophilic when the two bonded atoms are identical. However, Br_2 has a very weak bond (46 kcal/mole), and is polarizable (meaning its electron clouds are easily distortable). When electrons from an alkene pi bond approach a bromine molecule, the bromine electrons are polarized away from the pi electrons. This distorts the electron density so that the end of the bromine molecule nearest the pi bond is partially positive and the end away from the pi bond is partially negative. The weakness of the Br-Br bond that is breaking, and the strength of the C-Br bond that is forming, causes a rupture of the Br-Br bond and a bridging bromine atom leans across the two carbon atoms of the pi bond forms forming a more stable cation. The bromine bridging intermediate is called bromonium cation. A larger bromine atom with more layers of electrons and lone pairs carries the positive charge better than either carbon atom as a carbocation, having only six electrons. The bridging bromine forces the very specific "anti" stereochemistry in the subsequent addition of the bromide nucleophile, at the more partially positive carbon atom. The mechanism is a little more complicated than we are used to because there is a third arrow to show the bromine lone pair form the bridge.

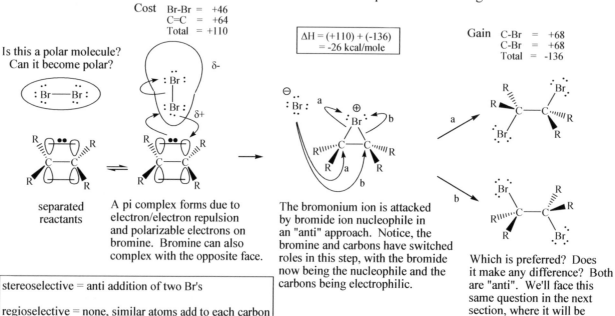

Cost   Br-Br = +46
C=C = +64
Total = +110

$\Delta H = (+110) + (-136)$
= -26 kcal/mole

Gain   C-Br = +68
C-Br = +68
Total = -136

Is this a polar molecule? Can it become polar?

separated reactants

A pi complex forms due to electron/electron repulsion and polarizable electrons on bromine. Bromine can also complex with the opposite face.

The bromonium ion is attacked by bromide ion nucleophile in an "anti" approach. Notice, the bromine and carbons have switched roles in this step, with the bromide now being the nucleophile and the carbons being electrophilic.

Which is preferred? Does it make any difference? Both are "anti". We'll face this same question in the next section, where it will be clearly seen which way the nucleophile attackes.

stereoselective = anti addition of two Br's

regioselective = none, similar atoms add to each carbon cannot tell which added first or second.

Chapter 13

We almost completely emphasize pi resonance in our course. However, sigma resonance is possible too and the three structures below show how the positive charge is spread over all three atoms (C_a, C_b and Br) in the ring. The first resonance structure is best because it has an additional bond and full octets on all of the atoms, but the other two are informative and decide which carbon atom is "more" partially positive.

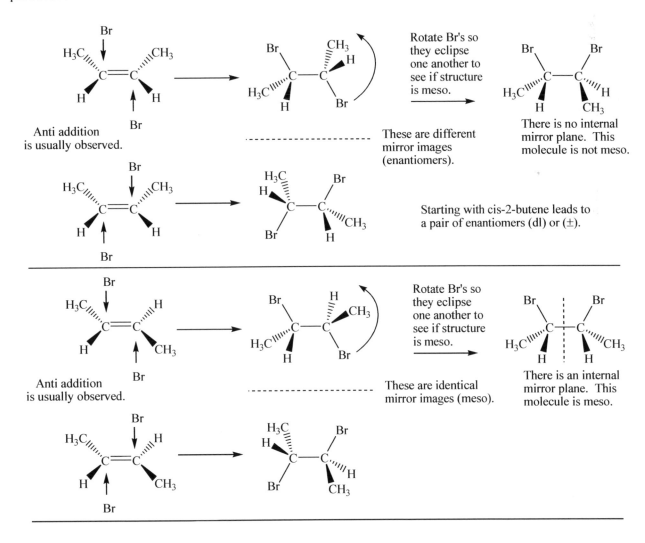

Tha amount of $\delta+$ on Ca and Cb depends on substitution patterns at those carbons ($3° > 2° > 1°$, and resonance possibilities).

The bridging bromine blocks approach of the bromide nucleophile from the same face and forces attack from the anti face. We might expect that bromide attack at the more substituted carbon would be preferred, but we can't really tell when both of the atoms adding are the same. We will be able to tell in the next section when halohydrins are formed and it does occur in the manner expected (nucleophile attack at the more partial positive carbon). However, we can tell if the addition was anti or syn if the alkene allows it. Anti addition to but-2Z-ene produces a racemic mixture of dl enantiomers (shown below), while anti addition to but-2E-ene produces the meso dibromobutane.

If the addition of Br_2 to cis-2-butene was "syn" it would have produced a meso structure. That is NOT the result in this reaction.

Syn addition is not usually observed.

If the reaction happened this way, these would be identical mirror images and they would be meso structures. But, it doesn't happen this way (usually).

internal plane of symmetry

Problem 13 – Do the following pi systems allow one to observe anti addition (as opposed to syn addition or random addition)? Write the products, clearly showing stereochemical features. Indicate if products are enantiomers, diastereomers, meso compounds or achiral structures.

10 Alkenes to practice on

1. Ethane cannot show regioselectivity or stereoselectivity with any of our reagents.

Nucleophilic bromide attack on the bromonium ion is anti. Attack at either carbon leads to the same product.

2. Propene cannot show regioselectivity or stereoselectivity with bromine addition.

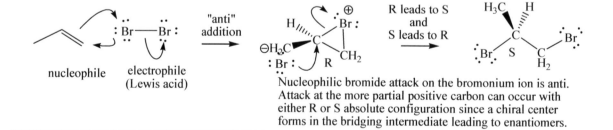

R leads to S and S leads to R

Nucleophilic bromide attack on the bromonium ion is anti. Attack at the more partial positive carbon can occur with either R or S absolute configuration since a chiral center forms in the bridging intermediate leading to enantiomers.

3. 2-methylprop-1-ene cannot show regioselectivity or stereoselectivity with bromine addition. Add in any missing mechanistic details.

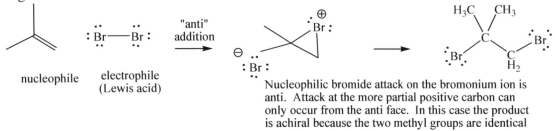

nucleophile   electrophile
                 (Lewis acid)

Nucleophilic bromide attack on the bromonium ion is anti. Attack at the more partial positive carbon can only occur from the anti face. In this case the product is achiral because the two methyl groups are identical

4. trans-but-2-ene cannot show regioselectivity, but can show stereoselectivity with bromine addition. Add in any necessary mechanistic details.

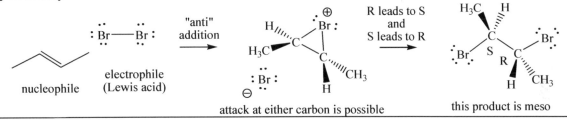

nucleophile   electrophile
                 (Lewis acid)

attack at either carbon is possible   this product is meso

5. Cyclohexene cannot show regioselectivity, but can show stereoselectivity with bromine addition. Write in a complete mechanism. Supply the necessary mechanistic details.

nucleophile   electrophile
                 (Lewis acid)   from the bromonium bridge

An enantiomeric pair of stereoisomers is obtained (SS) and (RR).

6. 3-methylbut-1-ene cannot show regioselectivity or stereoselectivity with bromine addition. Normally we would expect rearrangement with this molecule, but the bromonium bridge prevents rearrangement. Write a complete mechanism.

R leads to S
and
S leads to R

nucleophile   electrophile
                 (Lewis acid)

S and R
enantiomers

7. Styrene (vinylbenzene, ethenylbenzene) cannot show regioselectivity or stereoselectivity with bromine addition. Write a complete mechanism.

nucleophile   electrophile
                 (Lewis acid)   (resonance)

enantiomers are obtained

8. 3-methylcyclohex-1-ene does not show regioselectivity but can show stereoselectivity with bromine/water addition. Normally we would expect rearrangement with this molecule, but the bromonium bridge prevents rearrangement.

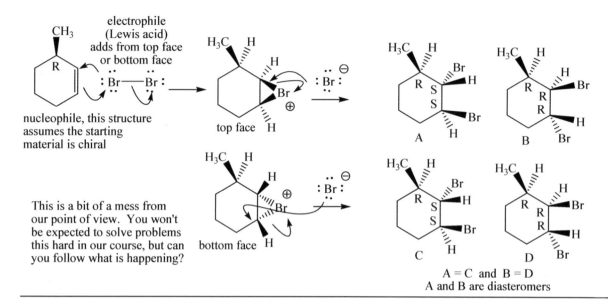

9. 1-methylcyclohex-1-ene cannot show regioselectivity but can show stereoselectivity with any of our reagents. Write a complete mechanism.

nucleophile   electrophile (Lewis acid)   from the bromonium bridge   An enantiomeric pair of stereoisomers is obtained (SS) and (RR).

10. 3-methyl-4-phenylpent-1-ene cannot show regioselectivity or stereoselectivity with bromine addition. Normally we would expect rearrangement with this molecule, but the bromonium bridge prevents rearrangement. Add in any necessary mechanistic details.

nucleophile   electrophile (Lewis acid)   "anti" addition   enantiomers

Alkyne synthesis – from dibromoalkanes

Bromine addition to an alkene allows us a specific route to dibromoalkanes. Bromination of alkenes adds exactly two bromine atoms to the carbon skeleton, which we can then use to force a double elimination reaction providing an efficient route to alkynes.

Problem 14 – Fill in the expected products for each reaction of the indicated sequence.

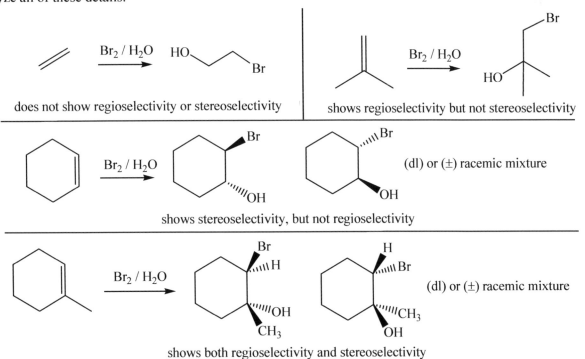

Bromohydrin formation and reaction (very similar to halogen additions, just above)

Bromine mixed with water can add in a slightly different manner. While electrophilic addition of bromine by itself cannot distinguish which carbon atom is attacked in the second step, a bromine/water mixture does make that regioselective distinction. The experimental results show that water attacks the more substituted and more partially positive carbon atom over the less hindered carbon in the second step. This reaction shows both regioselectivity and stereoselectivity. An actual alkene may show both of these, either of these or neither of these of these selectivities. This definitely requires greater sophistication on your part to analyze all of these details.

Chapter 13

Problem 15 – Bromine and water addition to methylcyclohexene produces a product having two favored chair conformations. Which chair conformation is "most" preferred and why?

Partially positive bromine accepts electrons from alkene pi bond and shares its lone pair to make a bridge.

Partially negative oxygen of water donates electrons to more partially positive tertiary carbon over secondary carbon, and attacks from the opposite side of the bromine bridge.

Proton transfer from trivalent oxygen is required to the best base available, water.

The chair starts equilibrating immediately upon reacting.

This result shows both the regioselectivity, and the "anti" stereoselectivity (trans product).

Problem 16 – Do the following pi systems allow one to observe anti addition (as opposed to syn addition or random addition)? Write the products, clearly showing stereochemical features. Indicate if products are enantiomers, diastereomers, meso compounds or achiral structures.

10 alkenes to practice on (Br$_2$ / H$_2$O addition to alkenes)

1. Ethane cannot show regioselectivity or stereoselectivity with any of our reagents.

Nucleophilic bromide attack on the bromonium ion is anti. Attack at either carbon leads to the same product.

2. Propene can show regioselectivity but not stereoselectivity with bromohydrin addition.

3. 2-methylprop-1-ene can show regioselectivity but not stereoselectivity with bromohydrin addition. Add in any missing mechanistic details.

Chapter 13

4. trans-but-2-ene cannot show regioselectivity, but can show stereoselectivity with bromohydrin addition. Add in any necessary mechanistic details.

attack at either carbon is possible

5. Cyclohexene cannot show regioselectivity, but can show stereoselectivity with bromohydrin addition. Write in a complete mechanism. Supply the necessary mechanistic details.

6. 3-methylbut-1-ene can show regioselectivity but cannot show stereoselectivity with bromohydrin addition. Normally we would expect rearrangement with this molecule, but the bromonium bridge prevents rearrangement. Write a complete mechanism.

7. Styrene (vinylbenzene, ethenylbenzene) can show regioselectivity but cannot show stereoselectivity with bromohydrin addition. Write a complete mechanism.

8. 3-methylcyclohex-1-ene cannot show regioselectivity but can show stereoselectivity with bromohydrin addition. Normally we would expect rearrangement with this molecule, but the bromonium bridge prevents rearrangement. Add in any necessary mechanistic details.

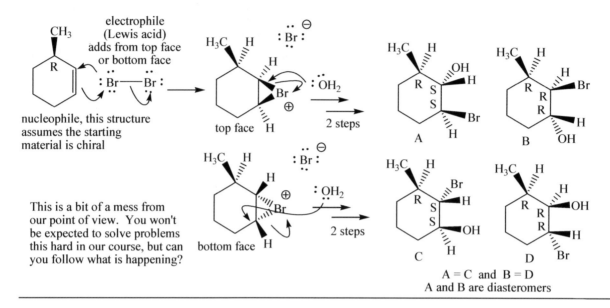

A = C and B = D
A and B are diasteromers

9. 1-methylcyclohex-1-ene can show regioselectivity and can show stereoselectivity with bromohydrin addtion. Write a complete mechanism.

: Br :⊖

: Br——Br :

nucleophile   electrophile
(Lewis acid)

⟶   ⟶   ⟶

H₂O :   H₂O :

An enantiomeric pair of
stereoisomers is
obtained (SS) and (RR).

10. 3-methyl-4-phenylpent-1-ene can show regioselectivity but not stereoselectivity with bromohydrin addition. Normally we expect rearrangement with this molecule, but the bromonium bridge prevents rearrangement. Add in any necessary mechanistic details.

: Br :⊖

: Br——Br :

CH₃
CH   H
   CH C CH₂
   CH₃   ⟶   ⟶
              electrophile   2 steps
nucleophile   (Lewis acid)   H₂O :

enantiomers

Chapter 13

Special case of an intramolecular S_N2 reaction - synthesis of oxiranes (epoxides) from bromohydrins

Using two of our previously studied reactions (acid/base and S_N2) we can make very useful epoxides electrophiles from bromohydrins with a completely different strategy than the sulfur ylids/carbonyl strategy presented earlier. A possible mechanism is shown below.

Highly strained oxiranes (epoxides) can form via intramolecular S_N2 reactions using our newly created bromohydrins above. S_N2 reactions are regiospecific and stereospecific. The carbon atom initially bonded to the leaving group, is the same carbon atom that becomes attached to the nucleophile, and attack has to occur from the backside of the carbon with the leaving group.

In the bromohydrin synthetic approach to epoxides, the nucleophilic oxygen atom and the good leaving group are on vicinal (neighbor) carbon atoms. Mildly basic conditions cause some of the alkoxide anion to form, making the oxygen a stronger nucleophile. If the correct "anti" geometry for backside attack can be attained, then an intramolecular S_N2 reaction can occur to form the epoxide.

Problem 17 – a. Develop an arrow pushing mechanism for epoxide formation in the following reaction. What is the nucleophilic atom, how is it formed and what approach does it need for a successful reaction? What conformation in a cyclohexane ring allows this approach? What conformation in an open chain allows for this approach (see part b)?

Which conformation allows backside attack? | Overall reaction.

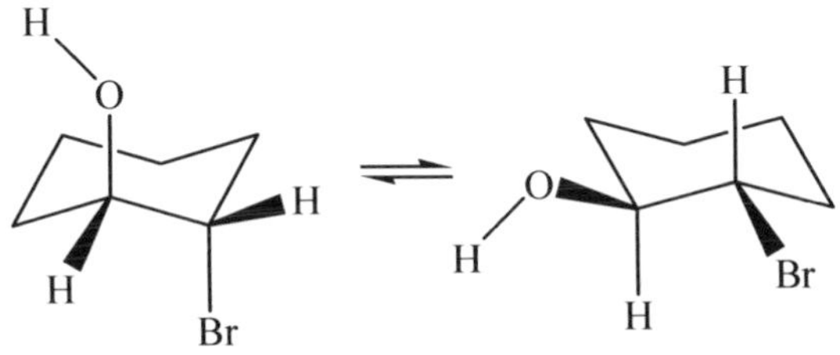

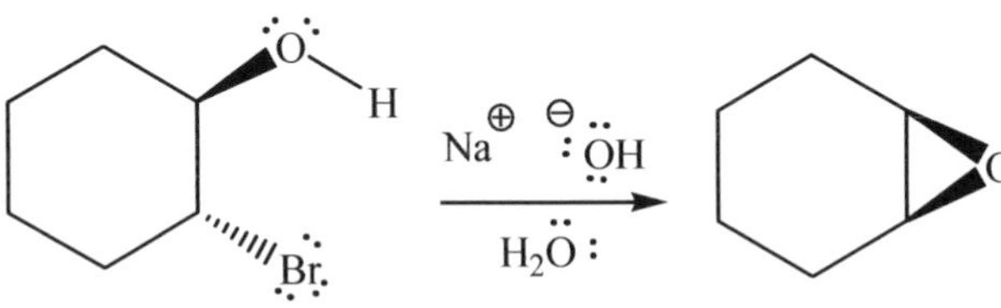

Clues: "H" and "Br" are both gone. Look at the conditions of the reaction. Is one of these steps likely to occur first? What took the place of the missing Br? What is the stereochemistry of the oxygen bridge on the cyclohexane? The mechanism occurs in two familiar steps (acid/base and S_N2).

b. Predict the product of the following reaction by showing a mechanism. Be very careful about the required conformation to react, which determines the stereochemistry of the product. You may have to rotate a bond to show the reaction.

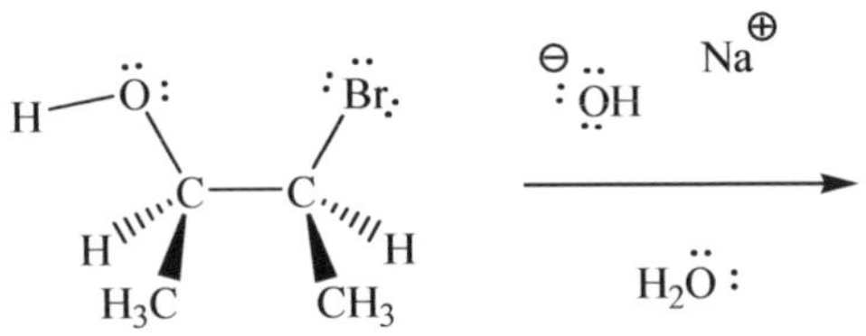

Our earlier limitation on epoxide synthesis is gone, we can now make epoxides in a ring. If we can make an alkene, we can carry it on two more steps to make an epoxide. No longer are we limited by our sulfur ylid / carbonyl group to epoxides.

Problem 18 – Propose the necessary steps for the transformations indicated below. Part d requires that you make a carbon-carbon bond.

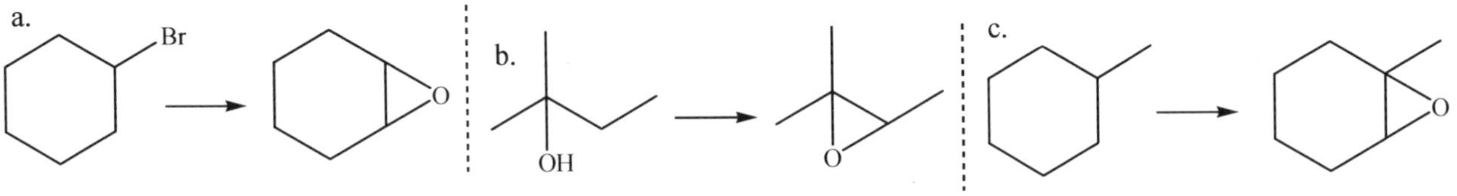

Chapter 13

Problem 19 – Show how each of the following transformations could be accomplished. Do you know the mechanistic steps for each reaction? Where possible, specify if the addition is Markovnikov or anti-Markovnikov. Regioselectivity is obvious in these problems. Is stereoselectivity also evident? Are stereocenters created? If so what sorts of stereoisomers are present, enantiomers or diastereomers or neither?

a.

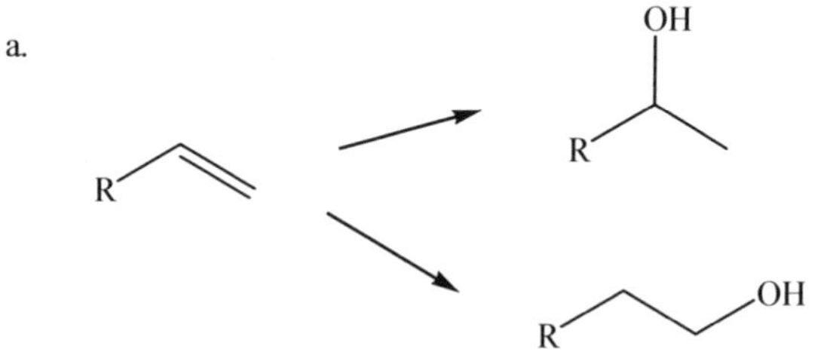

b.

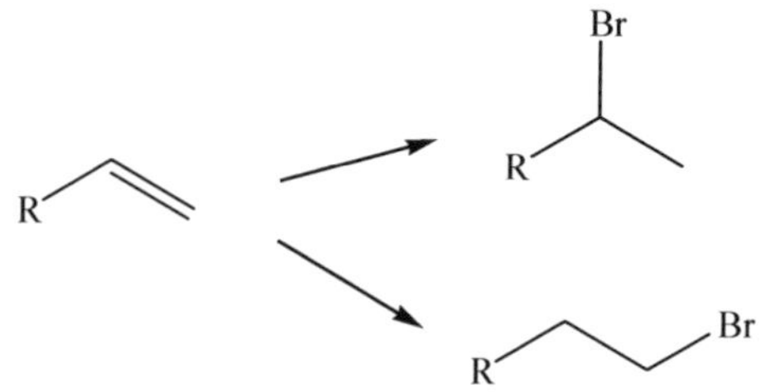

c.

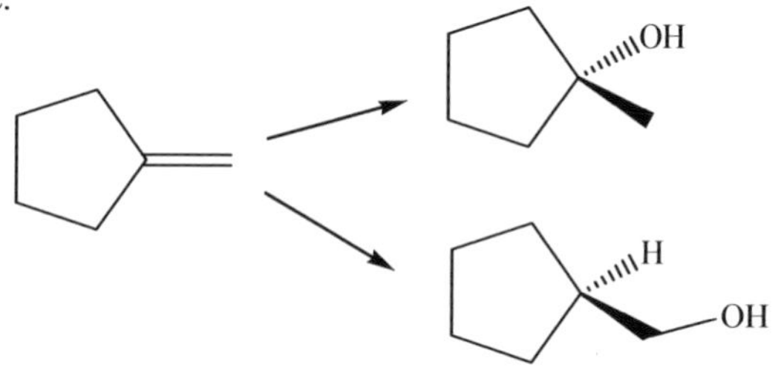

d.

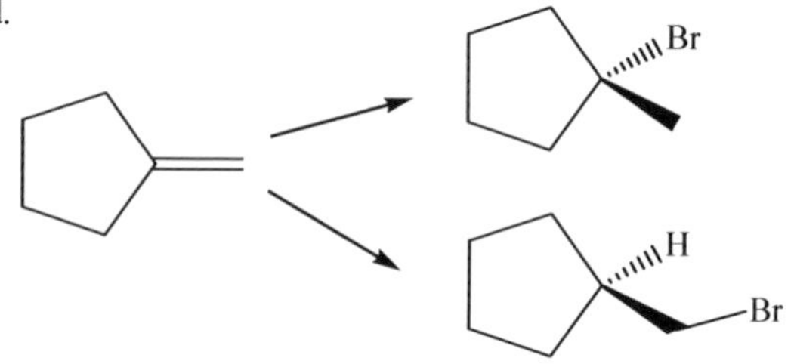

Dihydroxylation – anti versus syn diols

Dihydroxylation indicates the addition of two OH groups on vicinal carbon atoms of an alkene pi bond (vicinal = neighbors). There are several ways to do this stereoselectively (could be anti or syn), but we will only present a two strategies common to introductory organic chemistry (one anti and one syn). Not all epoxides permit us to distinguish between syn and anti addition. In those instances syn and anti addition appear to be the same. To create two hydroxyl groups anti to one another, we can react epoxides in either aqueous acid or aqueous base. Both of those conditions result in anti addition of the vicinal diols. We can easily see this mode of addition if we use cyclohexene oxide. If we include a methyl at one of the carbon atoms of the epoxide, we can also observe the regioselectivity. In strong acid, the positive charge attracts the weak nucleophile to the more substituted carbon (S_N1-like), while in strong base the nucleophile attacks the less hindered position (S_N2-like).

Epoxides in acid or base form anti diols

Epoxide reaction in aqueous acid.

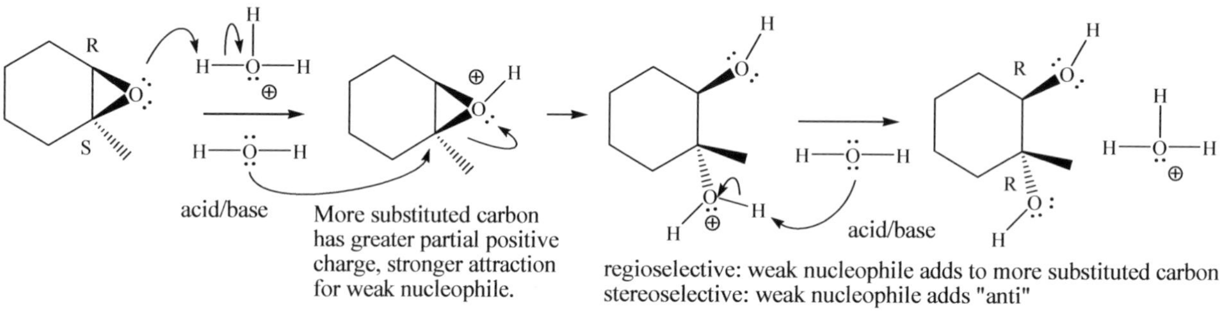

acid/base

More substituted carbon has greater partial positive charge, stronger attraction for weak nucleophile.

acid/base

regioselective: weak nucleophile adds to more substituted carbon
stereoselective: weak nucleophile adds "anti"

Chapter 13

Problem 20 – Write a mechanism for the following reaction in acidic alcohol and basic alcohol.

a. Open epoxides in alcoholic acid.

b. Open epoxides in alcoholic base (supply any missing mechanistic details).

Less substituted carbon has less steric hindrance and is attacked faster in S_N2 approach of strong nucleophile.

regioselective: strong nucleophile adds to less substituted carbon
stereoselective: strong nucleophile adds "anti" because of the bridge

Osmium Tetroxide – an example oxidation reaction of alkenes

Syn addition of two vicinal alcohol groups is possible using osmium tetroxide (OsO_4) or potassium permanganate ($K^+ MnO_4^-$). While there are some significant differences, these two reagents accomplish very similar transformations. Two big differences are cost and toxicity. Potassium permanganate is much less expensive and much less toxic. Osmium's toxicity is accentuated by the fact that it is molecular and has some volatility (that means you could breathe it). With such strong advantages, permanganate would seem to be the preferred reagent. Osmium tetroxide's advantage is that it is much more selective and provides much higher yields. Synthetic chemists who actually do the experiments use OsO_4 …and work very carefully! We will only choose osmium tetroxide, because of our limited approach in this course and our chance of exposure is pretty minimal from writing its structure on a piece of paper.

Osmium Tetroxide
oxidation state = +8
MW = 254 g/mole

Potassium Permanganate
oxidation state = +7
MW = 158 g/mole

Cost from Aldrich, 2016

Osmium tetroxide	250 mg = $236	1g = $419	or 1 mole = $106,000
Potassium Permanganate	25 g = $40	2.5 Kg = $245	or 1 mole = $15

A possible reaction mechanism for the dihydroxylation is presented below. The actual addition of the oxygen atoms is considered to be a concerted reaction. There are competing mechanisms, but we will use the one shown below, which shows a cyclic arrangement of five atoms with six electrons transferring around the cyclic transition state. This mechanism works for osmium tetroxide and potassium permanganate. Osmium is very electron poor with an oxidation state of +8 (manganese in permanganate is also very electron poor and has an oxidation state of +7). This large electron deficiency pulls the pi electrons away from one carbon in a pi bond, pushes two electrons to osmium (for itself), and transfers two electrons to the original carbon losing the electrons, all in one concerted reaction. The oxygen atoms are delivered together with "syn" stereochemistry to the alkene carbons in this single step and locked in place. The rest of the chemistry is hydrolysis of the osmium ester to the diol (addition of water). The hydrolysis steps are very similar to base hydrolysis of organic esters and organometallic reactions with esters. Often, when you learn a mechanistic theme in organic chemistry and biochemistry it repeats itself in other places.

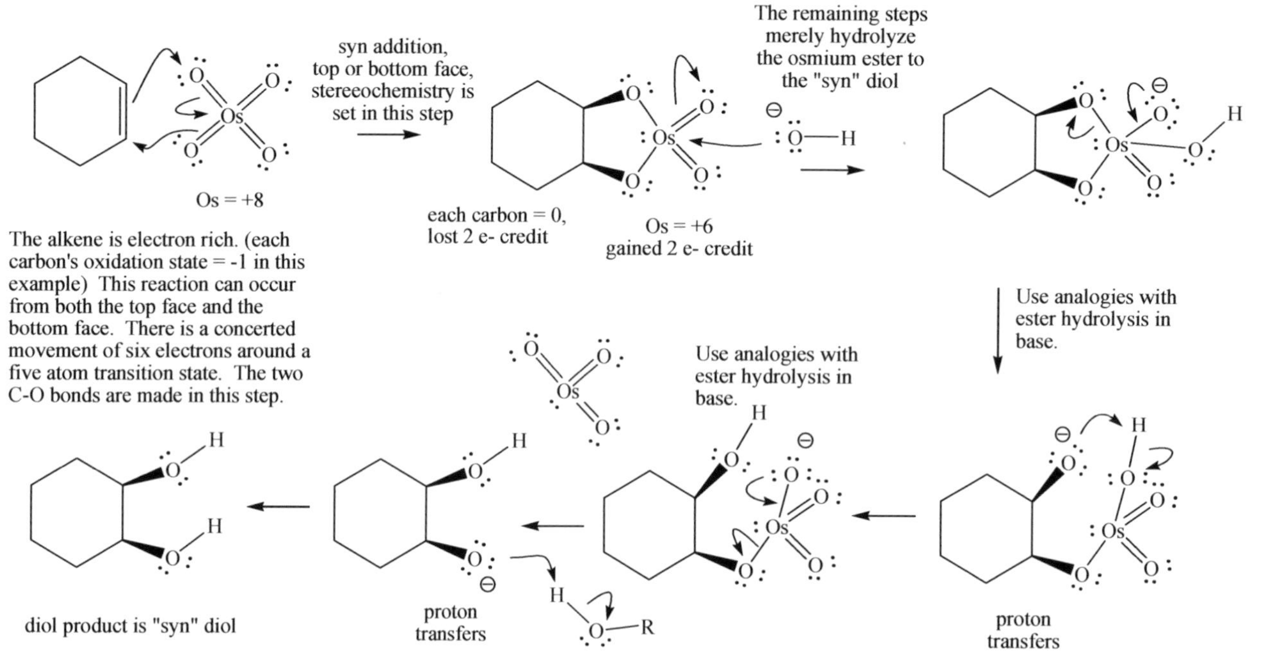

The remaining steps merely hydrolyze the osmium ester to the "syn" diol

syn addition, top or bottom face, stereeochemistry is set in this step

Os = +8

The alkene is electron rich. (each carbon's oxidation state = -1 in this example) This reaction can occur from both the top face and the bottom face. There is a concerted movement of six electrons around a five atom transition state. The two C-O bonds are made in this step.

each carbon = 0, lost 2 e- credit

Os = +6 gained 2 e- credit

Use analogies with ester hydrolysis in base.

Use analogies with ester hydrolysis in base.

Use analogies with ester hydrolysis in base.

diol product is "syn" diol

proton transfers

proton transfers

Because of the high cost of osmium tetroxide and its toxicity, only a small catalytic amount is typically used. It is regenerated by an oxidizing agent (i.e. a peroxide or an amine oxide), which is much less expensive and reduces exposure to the chemist doing the reaction.

Re-oxidation mechanism to regenerate OsO_4 (from OsO_3 and a peroxide or an amine oxide, R_3NO)

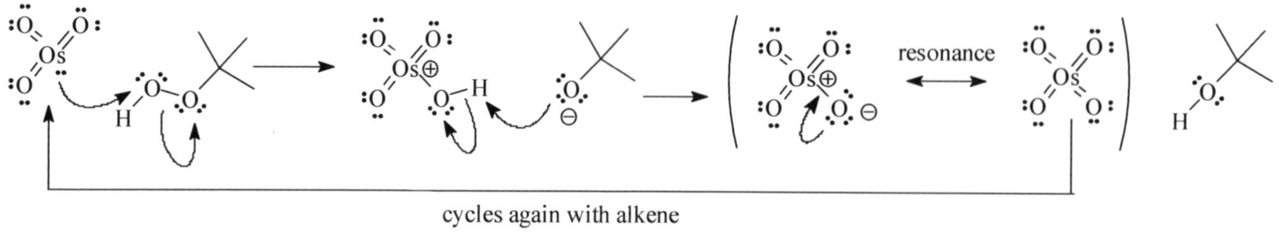

resonance

cycles again with alkene

Problem 21 – Propose a mechanism for re-oxidation of osmium trioxide to osmium tetraoxide using morpholine N-oxide.

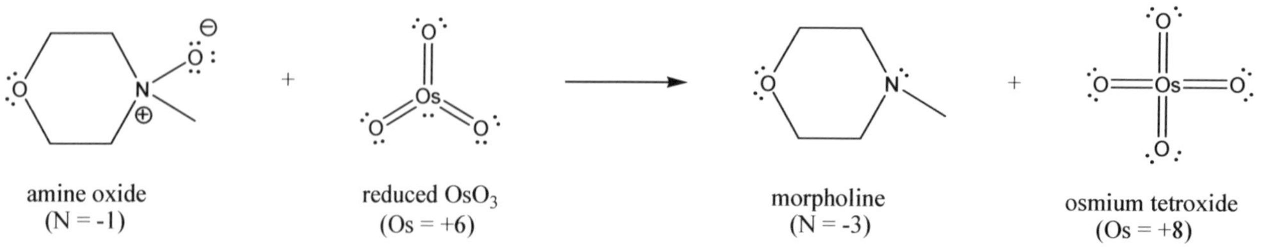

+ → +

amine oxide (N = -1)

reduced OsO_3 (Os = +6)

morpholine (N = -3)

osmium tetroxide (Os = +8)

Problem 22 – Do the following pi systems allow one to observe anti addition (as opposed to syn addition or random addition)? Write the products, clearly showing stereochemical features. Indicate if products are enantiomers, diastereomers, meso compounds or achiral structures. Because the mechanism is so long we will only show one full mechanism (above), only fill in the missing mechanistic details in one and leave the rest for you to be able to write out a mechanism for yourself.

10 Alkenes to practice on (This isn't much different than adding H-Br.)

1. Ethane cannot show regioselectivity or stereoselectivity with any of our reagents.

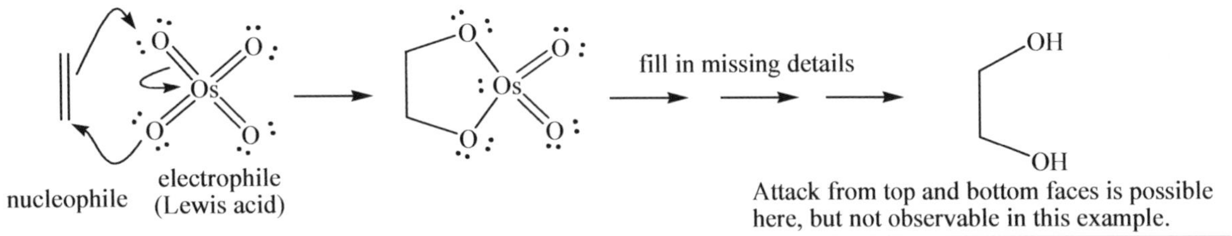

nucleophile   electrophile (Lewis acid)

fill in missing details

Attack from top and bottom faces is possible here, but not observable in this example.

2. Propene cannot show regioselectivity or stereoselectivity with OsO$_4$.

nucleophile   electrophile (Lewis acid)   R and S enantiomers are made

fill in missing details

Attack from top and bottom faces is possible here, but not observable in this example.

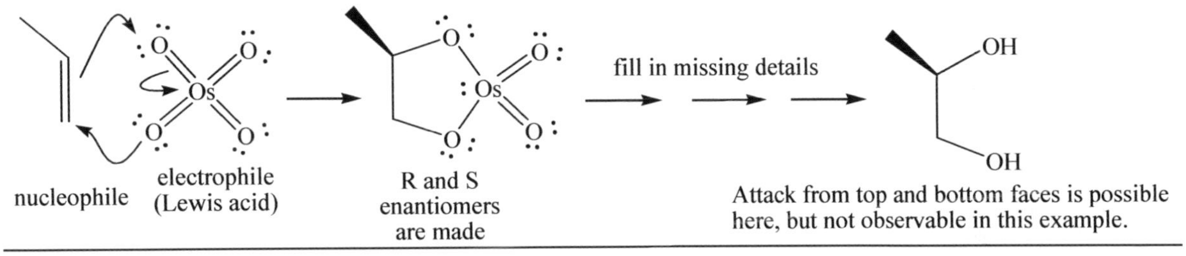

3. 2-methylprop-1-ene cannot show regioselectivity or stereoselectivity with OsO$_4$. Add in any missing mechanistic details.

nucleophile   electrophile (Lewis acid)

fill in missing details

Attack from top and bottom faces is possible here, but not observable in this example.

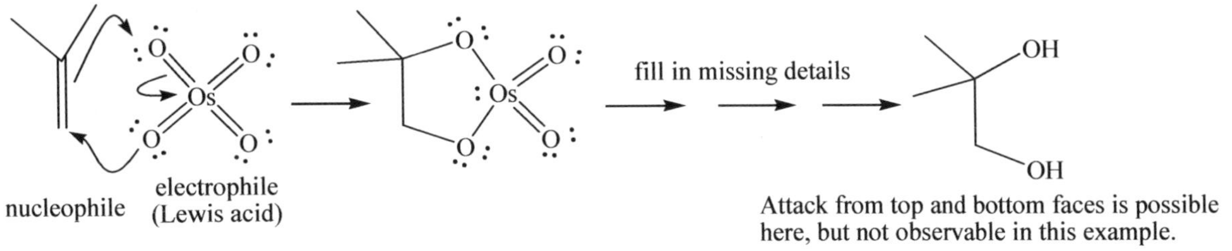

4. trans-but-2-ene cannot show regioselectivity, but can show stereoselectivity with OsO_4. Write a complete mechanism.

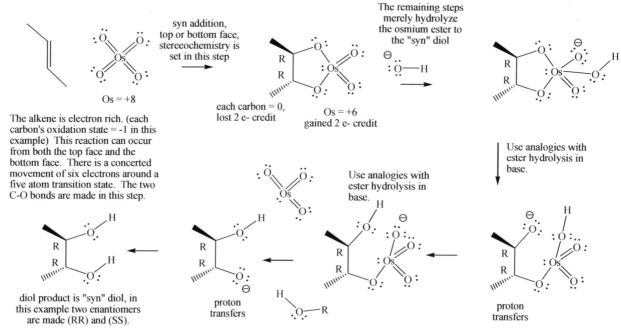

The alkene is electron rich. (each carbon's oxidation state = -1 in this example) This reaction can occur from both the top face and the bottom face. There is a concerted movement of six electrons around a five atom transition state. The two C-O bonds are made in this step.

diol product is "syn" diol, in this example two enantiomers are made (RR) and (SS).

proton transfers

Use analogies with ester hydrolysis in base.

Use analogies with ester hydrolysis in base.

proton transfers

5. Cyclohexene is shown in the original example.

nucleophile   electrophile (Lewis acid)

Attack from top and bottom faces is possible here, and leads to a "meso" compound.

6. 3-methylbut-1-ene cannot show regioselectivity or stereoselectivity with OsO_4. This example will also rearrange whenever carbocations form, but does not happen here. Add in any missing mechanistic details.

nucleophile   electrophile (Lewis acid)

Attack from top and bottom faces is possible here, and leads to enantiomers.

7. Styrene (vinylbenzene, ethenylbenzene) cannot show regioselectivity or stereoselectivity with OsO_4. Write a complete mechanism.

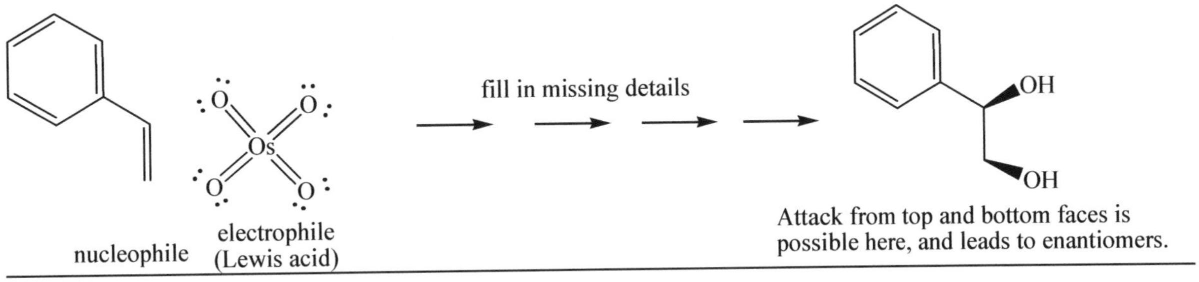
fill in missing details
OH
OH
nucleophile
electrophile
(Lewis acid)
Attack from top and bottom faces is possible here, and leads to enantiomers.

8. 3-methylcyclohex-1-ene cannot show regioselectivity but does show stereoselectivity with OsO_4. It is good for showing when rearrangements are possible, but not here. Add in any missing mechanistic details.

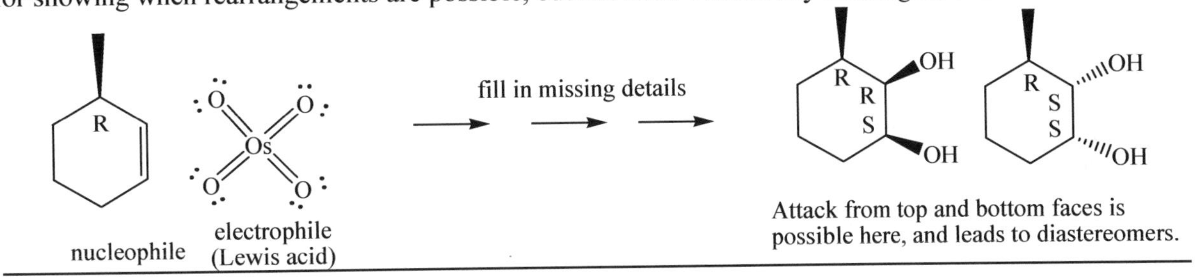
R
fill in missing details
R
R
S
OH
OH
R
S
S
OH
OH
nucleophile
electrophile
(Lewis acid)
Attack from top and bottom faces is possible here, and leads to diastereomers.

9. 1-methylcyclohex-1-ene cannot show regioselectivity but does show stereoselectivity with OsO_4. Write a complete mechanism.

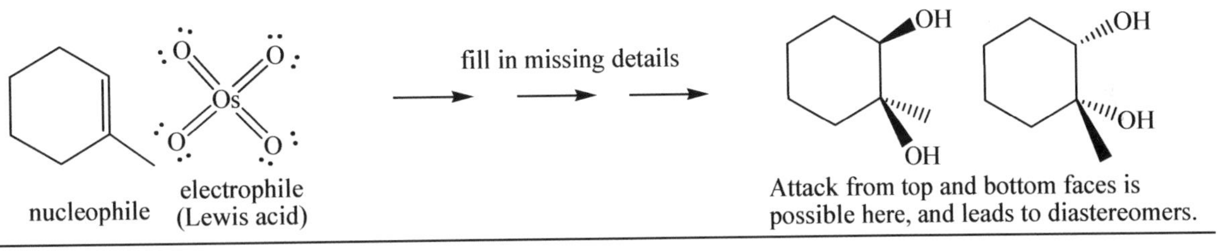
fill in missing details
OH
OH
OH
OH
nucleophile
electrophile
(Lewis acid)
Attack from top and bottom faces is possible here, and leads to diastereomers.

10. 3-methyl-4-phenylpent-1-ene cannot show regioselectivity or stereoselectivity with OsO_4. It is good for showing complicated rearrangement possibilities, but not here. Add in any missing mechanistic details.

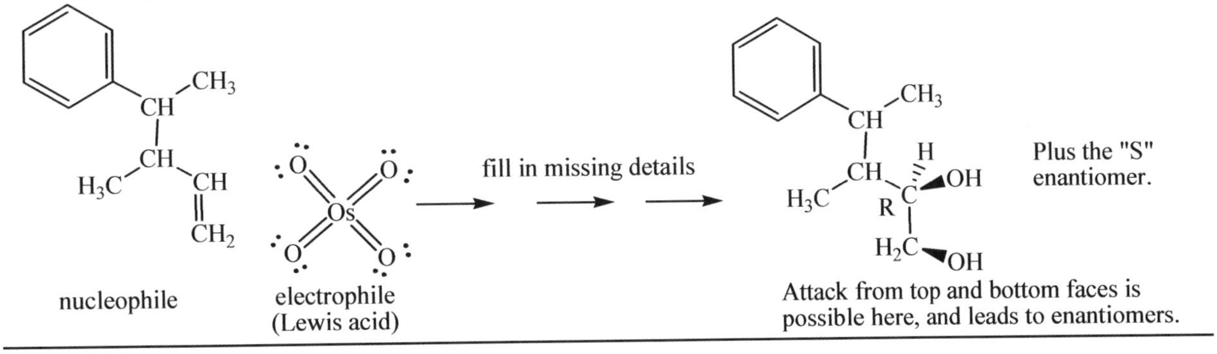
CH3
CH
H3C
CH
CH
CH2
fill in missing details
CH3
CH
H
CH
C
OH
H3C
R
H2C
OH
Plus the "S" enantiomer.
nucleophile
electrophile
(Lewis acid)
Attack from top and bottom faces is possible here, and leads to enantiomers.

Problem 23 – What are the expected products and stereochemistries in the following reactions? How do the two approaches compare?

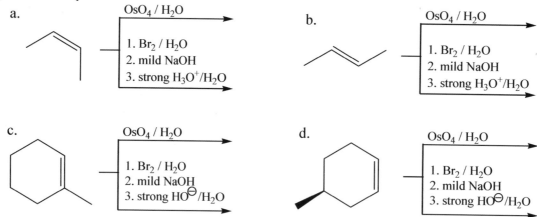

Catalytic Hydrogenation of carbon-carbon pi bonds

As with most of the reactions we study, there are many ways to accomplish the goal under consideration. This is especially true for reducing carbon-carbon pi bonds with diatomic hydrogen (adding two hydrogen atoms to the carbons of a pi bond). We will limit our approaches to some common reactions shown below. Addition of hydrogen gas to pi bonds occurs easily with the use of transition metal catalysts (one hydrogen atom adds to each carbon of the pi bond with syn stereochemistry). Hydrogenation is a very fast, exothermic reaction, "IF" conditions are induced to lower the high activation energy required to break the hydrogen-hydrogen bond (104 kcal/mole). There is no obvious polar mechanism for this reaction, but we will introduce a simplistic view that captures the essential features. If you take a more advanced course, you can consider HOMO/LUMO interactions to propose a more sophisticated mechanism. There are many metals that will work as a catalyst in this reaction, but we will only use palladium (Pd).

Common metal catalyst in hydrogenation reactions

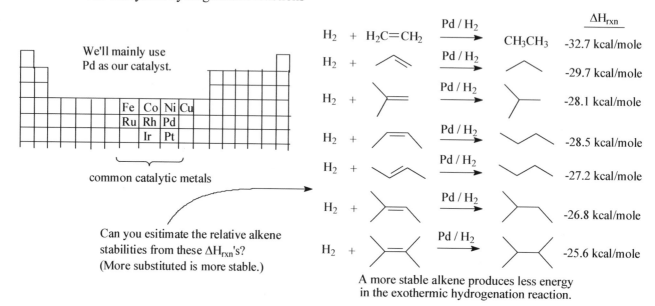

Chapter 13

Catalysts provide a lower energy pathway to products. Such a pathway usually involves more steps because the reacting substances have to come together, then react and then dissociate. Even though there are more steps, the overall energy barrier is lower and this is what allows the reaction to proceed faster.

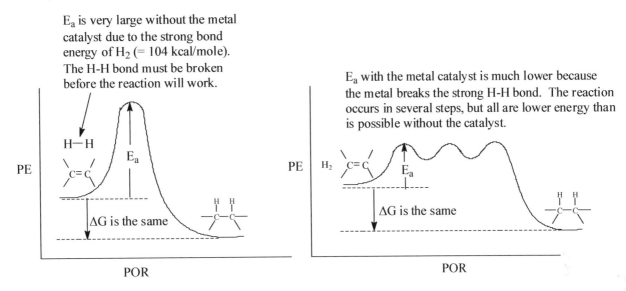

E_a is very large without the metal catalyst due to the strong bond energy of H_2 (= 104 kcal/mole). The H-H bond must be broken before the reaction will work.

E_a with the metal catalyst is much lower because the metal breaks the strong H-H bond. The reaction occurs in several steps, but all are lower energy than is possible without the catalyst.

Our strategy in this short view of organic chemistry is to look for recognizable electron pair donor sites (nucleophiles) having lone pairs or pi bonds, and direct the flow of electron density towards good electron pair acceptor sites (electrophiles), often having positive polarity or a high oxidation state. Such sites are not so obvious in hydrogenation reactions. Molecular orbital theory offers an alternative way of viewing these same reactions, but we are not using MO theory. The problem with our approach, here, is that it is somewhat "magical" and we cannot see the orbitals that are donating and those that are accepting electrons. My hope in proposing our simplistic mechanism is that it gives you an idea about how electrons are moving as bonds are made and broken. Palladium is catalytic because, not only does it help the reaction work faster, it regenerates itself in the last step. That's good because most of the transition metal catalysts are very expensive. The same face of a pi bond accepts both of the hydrides from the palladium. That means they are delivered to the same side (cis or syn addition). This assumption is not perfect, but we will look at it that way.

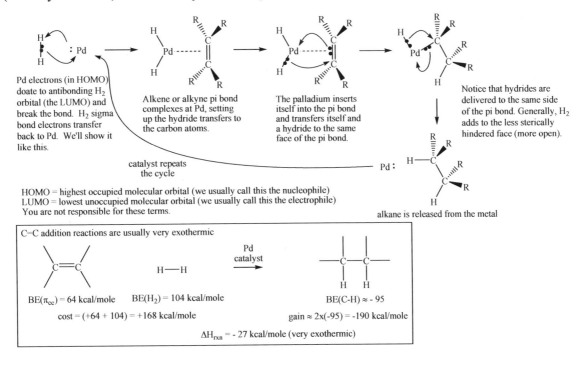

Pd electrons (in HOMO) doate to antibonding H_2 orbital (the LUMO) and break the bond. H_2 sigma bond electrons transfer back to Pd. We'll show it like this.

Alkene or alkyne pi bond complexes at Pd, setting up the hydride transfers to the carbon atoms.

The palladium inserts itself into the pi bond and transfers itself and a hydride to the same face of the pi bond.

Notice that hydrides are delivered to the same side of the pi bond. Generally, H_2 adds to the less sterically hindered face (more open).

catalyst repeats the cycle

alkane is released from the metal

HOMO = highest occupied molecular orbital (we usually call this the nucleophile)
LUMO = lowest unoccupied molecular orbital (we usually call this the electrophile)
You are not responsible for these terms.

C=C addition reactions are usually very exothermic

$BE(\pi_{cc}) = 64$ kcal/mole $BE(H_2) = 104$ kcal/mole $BE(C-H) \approx -95$

cost = (+64 + 104) = +168 kcal/mole   gain $\approx$ 2x(-95) = -190 kcal/mole

$\Delta H_{rxn} = -27$ kcal/mole (very exothermic)

Chapter 13

Because hydrogen, H_2, added in the reaction looks just like hydrogen in a molecule, we will use deuterium, D_2, in our example reactions so that the "syn" addition is observable, when possible.

Problem 24 – Do the following pi systems allow one to observe anti addition (as opposed to syn addition or random addition)? Write the products, clearly showing stereochemical features. Indicate if products are enantiomers, diastereomers, meso compounds or achiral structures. Because the mechanism is so long we will only show one full mechanism (above), only fill in the missing mechanistic details in one and leave the rest for you to be able to write out a mechanism for yourself.

10 Alkenes to practice on (This isn't much different than adding H-Br.)

1. Ethane cannot show regioselectivity or stereoselectivity with any of our reagents.

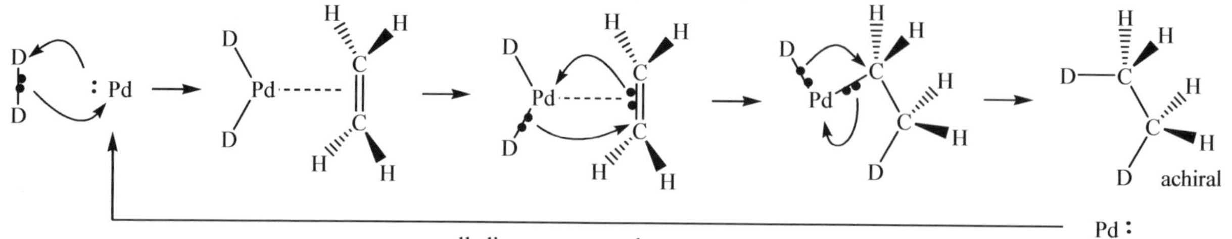

palladium repeats cycle

2. Propene cannot show regioselectivity or stereoselectivity with D_2/Pd.

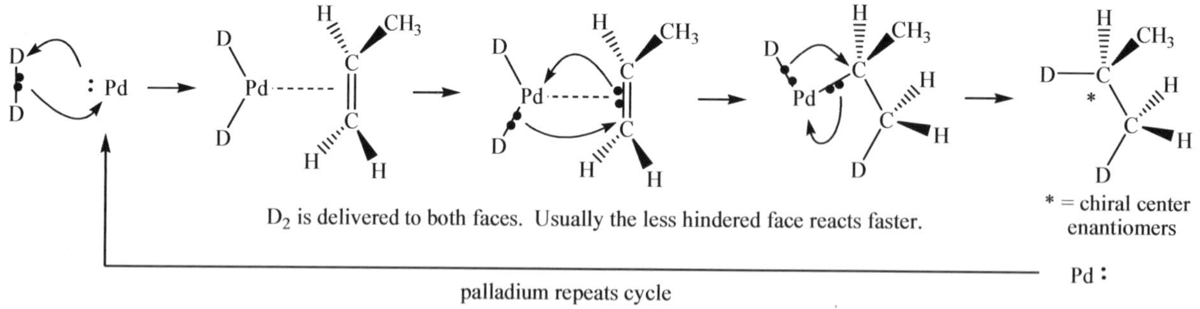

D_2 is delivered to both faces. Usually the less hindered face reacts faster.

* = chiral center
enantiomers

palladium repeats cycle

3. 2-methylprop-1-ene cannot show regioselectivity or stereoselectivity with D_2/Pd. Add in any missing mechanistic details.

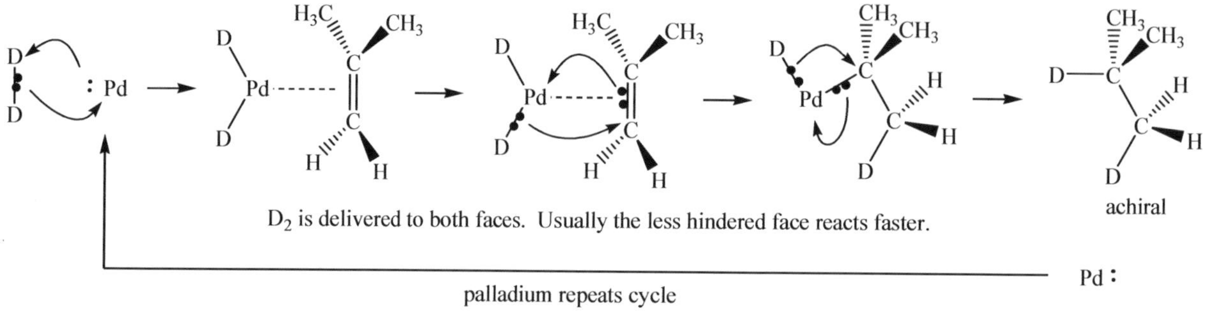

D_2 is delivered to both faces. Usually the less hindered face reacts faster.

achiral

palladium repeats cycle

4. trans-but-2-ene cannot show regioselectivity or stereoselectivity with D_2/Pd. Write a complete mechanism.

D_2 is delivered to both faces. Usually the less hindered face reacts faster.

* = chiral centers
enantiomers

palladium repeats cycle

5. Cyclohexene cannot show regioselectivity, but can show stereoselectivity with D_2/Pd. Write in a complete mechanism.

D_2 is delivered to both faces. Usually the less hindered face reacts faster.

* = chiral center
meso compound
achiral

palladium repeats cycle

6. 3-methylbut-1-ene cannot show regioselectivity or stereoselectivity with D_2/Pd. This example will also rearrange whenever carbocations form, but not here. Add in any missing mechanistic details.

D_2 is delivered to both faces. Usually the less hindered face reacts faster. No rearrangement.

* = chiral centers
enantiomers

palladium repeats cycle

7. Styrene (vinylbenzene, ethenylbenzene) cannot show regioselectivity or stereoselectivity with D_2/Pd. Write a complete mechanism.

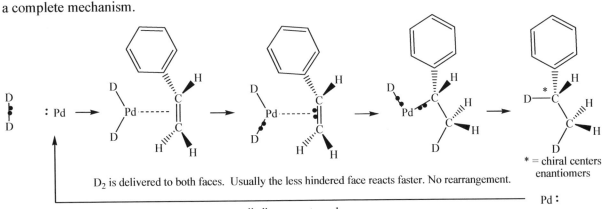

D_2 is delivered to both faces. Usually the less hindered face reacts faster. No rearrangement.

* = chiral centers
enantiomers

palladium repeats cycle

8. 3-methylcyclohex-1-ene cannot show regioselectivity but can show stereoselectivity with D_2/Pd. It is good for showing when rearrangements are possible, but not here. Add in any missing mechanistic details.

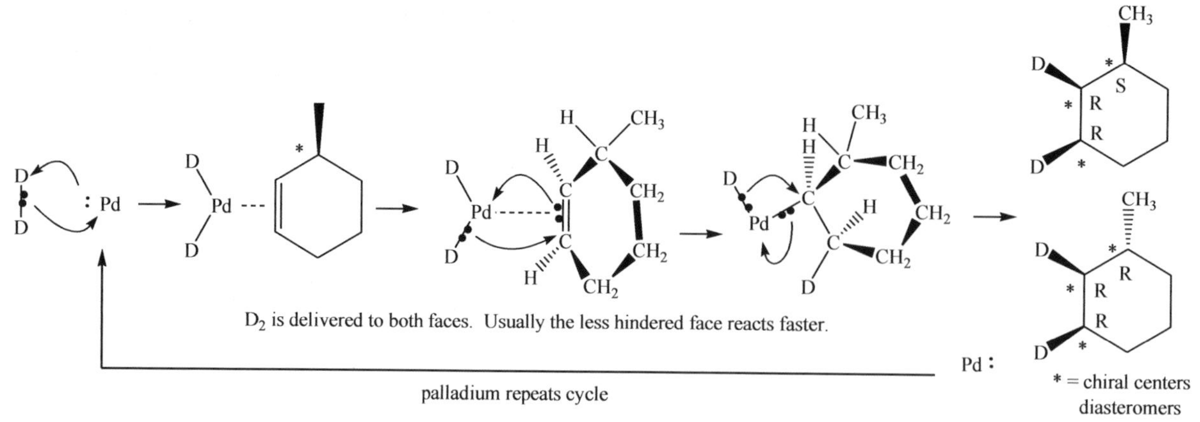
D
D
: Pd
Pd
Pd
CH3
CH2
D2 is delivered to both faces. Usually the less hindered face reacts faster.
palladium repeats cycle
Pd :
* = chiral centers
diasteromers

9. 1-methylcyclohex-1-ene cannot show regioselectivity but can show stereoselectivity with D_2/Pd. Write a complete mechanism.

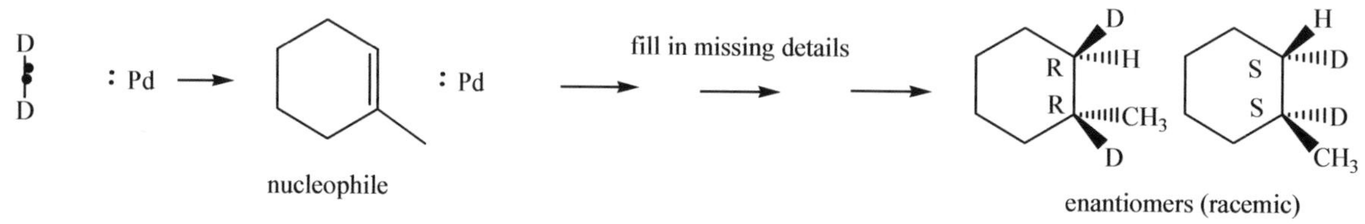
D
D
: Pd
nucleophile
: Pd
fill in missing details
enantiomers (racemic)

10. 3-methyl-4-phenylpent-1-ene cannot show regioselectivity or stereoselectivity with D_2/Pd. This example will also rearrange whenever carbocations form, but not here. Add in any missing mechanistic details.

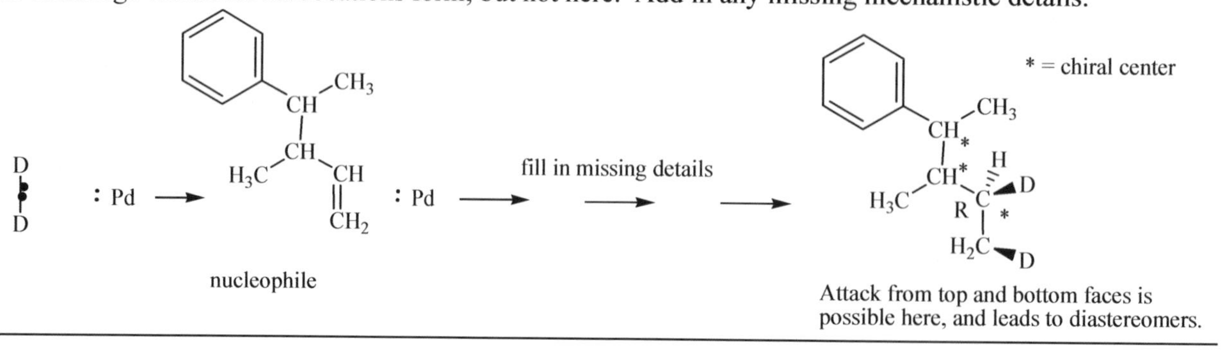
D
D
: Pd
nucleophile
: Pd
fill in missing details
* = chiral center
Attack from top and bottom faces is possible here, and leads to diastereomers.

Reduction of alkynes

Because there are two pi bonds in alkynes, reductions present varied possibilities. Using the same reducing conditions as alkenes (Pd / H$_2$), but two equivalents of hydrogen (Pd / 2 H$_2$) alkynes reduce all the way to alkanes. However, it would be nice if the triple bond could be reduced to a double bond as either cis or trans. As you might expect, all of these are possible and we will present one set of conditions for each, even though there are other ways to do the same things.

Pd / 2 moles H$_2$

alkanes

Complete reduction to alkane functionality is possible with many catalysts. We will use Pd catalyst and hydrogen gas for reducing alkenes or alkynes to alkanes.

Pd / 1 mole H$_2$ / quinoline

Lindlar's catalyst

Z alkenes

Z alkenes are produced from syn (cis) addition of two hydrogen atoms. This is accomplished with Lindlar's catalyst (Pd catalyst poisoned with quinoline), and it only adds 1 mole of H$_2$ (g) in a syn manner.

R—C≡C—R

alkynes

Na / NH$_3$ (liq) at -33°C, often some ROH is added for better protonation of the carbanions

E alkenes

E alkenes are produced from anti (trans) addition of two hydrogen atoms. This is accomplished with sodium metal in liquid ammonia, called "Birch" conditions.

To stop the catalytic reduction at the alkene stage we need to make the catalyst less efficient. This is accomplished by poisoning the catalyst. There are many contaminants that will poison a metal catalyst (not always a good thing, especially if it is another functional group in the same molecule). Quinoline is an intentional poison added to the palladium on a solid support of CaCO$_3$ or BaSO$_4$, and is called **Lindlar's catalyst**. Quinoline is a large, flat bicyclic heteroaromatic ring system. The heteroatom is nitrogen, which uses its lone pair to complex at the metal. The single bond connection likely allows the ring to sweep out a very large volume element near its complexation site. This effectively blocks any other large ligands in the vicinity of the quinoline (a steric effect). Alkynes are linear (and skinny) and can complex at the metal (HOMO/LUMO interactions) without being bumped off by the quinoline. This allows the metal to transfer its hydrogen atoms to the pi bond. However, once reduced to alkenes, they become flat and planar and require more room when bonding at the metal. Quinoline likely prevents alkenes from complexing at the metal, and therefore prevents them from receiving hydrogen atoms from the metal.

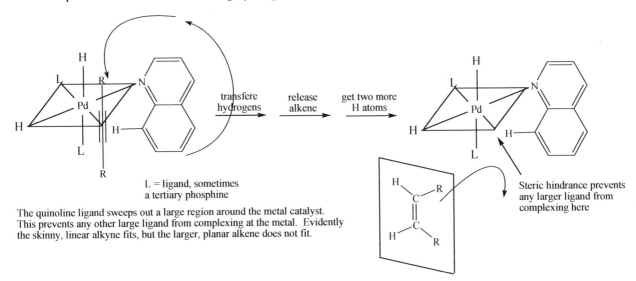

L = ligand, sometimes a tertiary phosphine

transfere hydrogens → release alkene → get two more H atoms

Steric hindrance prevents any larger ligand from complexing here

The quinoline ligand sweeps out a large region around the metal catalyst. This prevents any other large ligand from complexing at the metal. Evidently the skinny, linear alkyne fits, but the larger, planar alkene does not fit.

Alkyne reactions with sodium metal in liquid ammonia = trans alkene syntheses. (referred to as Birch reduction conditions)

Sodium metal in liquid ammonia allows two hydrogen atoms to add to opposite sides of an alkyne, forming E-alkenes. This approach compliments Lindlar's catalyst and gives us the ability to synthesize Z or E alkenes from alkynes. Highly reactive sodium metal is used to supply high potential energy electrons, which transfer to the antibonding pi orbital of the alkyne, forming a radical carbanion. The very basic radical/carbanion strips a proton from the ammonia solvent (or an alcohol proton donor, if added), becoming just a free radical. The reactive free radical accepts another high-energy electron from the metal forming a highly basic carbanion. The larger "R" group chooses a trans orientation to the other R group to minimize steric hindrance. When the carbanion adds a proton, the E configuration becomes the alkene stereochemistry. A possible mechanism is suggested below. You are not responsible for writing out this mechanism.

A single electron transfer occurs from sodium metal into the lowest unoccupied moelcular orbital (LUMO), making a radical-anion.

The very basic carbanion site picks up a proton from the ammonia solvent. (Sometimes an alcohol is added to make this easier.) We ignore that.

Notice that full headed arrows are drawn to show 2 electron movement and half headed arrows are drawn to show 1 electron movement.

The first two steps are essentially repeated one more time each, forming the E alkene, because the larger "R" groups avoid one another in the protonation step (a steric effect).

E-alkenes
(trans)

Chapter 13

Problem 25 – Propose reasonable synthetic steps to accomplish the indicated transformations. Use the starting material and our much smaller chemical catalog (on the next page) to make the TM-1 and continue on from there. No mechanisms are needed, but write out the reagents, workup and products for each synthetic step. If you have already made something in a previous step, you can start there and just refer back to the number where you made it.

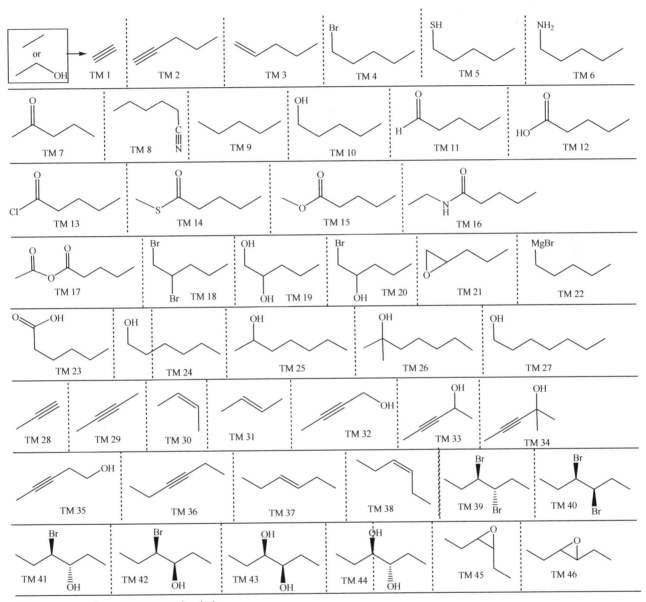

Make up some of our own target molecules!

These are our starting structures for synthesis in this book. We have seen that we can generate 13 different RBr structures and 3 RBr2 structures used to make alkynes. Bromobenzene is given until our next (and final) chapter. Our targets remain pretty simple, being C1 through C6 structures with various functional groups emphasized in this course.

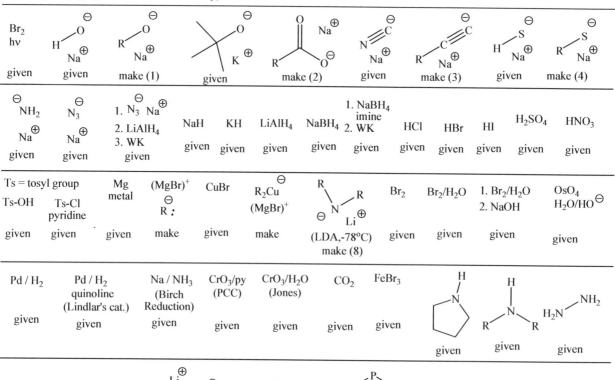

Reagents and/or Reaction Conditions - some we can buy (given whenever we want) and some we have to make (usually by acid/base chemistry).

Chapter 13

Chapter 14 – Aromatic Chemistry

The science of organic chemistry was born in the mid 19th century, in part, from solving the puzzle of the structure of aromatic compounds. Many organic compounds have aromatic parts of their structures in combination with other functional groups. Because aromatic compounds were very instrumental in the early work of organic chemistry, helping to formulate chemical structural theories, many common names were created, making their nomenclature more challenging. As is the case throughout this book, we will present just a small sample of what is known to give the flavor of the topic.

There are many, many common names. A few frequently used examples are provided. Notice that the prefix "benz" shows up again and again. You do not have to know all of these names for this course.

benzoic acid · benzoic anhydride · methyl benzoate · benzoyl chloride · benzamide · benzaldehyde

benzonitrile · benzyl alcohol · benzyl amine · benzyl bromide · aniline · anisol · phenol

nitrobenzene · toluene · 1,2-xylene · cumene · anthranilic acid · benzene sulfonic acid · Tylenol 4-hydroxyacetanilide · Aspirin

You will be responsible for two aromatic substituent patterns – phenyl and benzyl. "Ar" is a generic symbol for any aromatic system when more specificity is desired than just "R".

Ph · φ is an older represntation and not used much anymore. · Bn benzyl symbol · Ar generic aromatic symbol · R generic carbon symbol

phenyl = Ph = φ = the benzene ring as a substituent

benzyl = the benzene ring plus an extra CH₂ as a substituent

A more systematic nomenclature approach is necessary for more complicated structures

Rings with only two substituents can be named with "ortho, meta, para" terminology or using numbers from the highest priority substituent (=1). Ortho = (1,2), meta = (1,3) and para = (1,4). With more than two substituents, find a parent name (above) and use numbers to indicate where each substituent is relative to the parent substituent.

three possible disubstituted patterns · highest priority substituent (= 1) · o = ortho = 1,2-substitution (2 of these) · m = meta = 1,3-substitution (2 of these) · p = para = 1,4-substitution (1 of these)

o-bromobenzonitrile 2-bromobenzonitrile · 2-bromo-4-hydroxybenzonitrile

Chapter 14

Problem 1 – Provide an acceptable name for the following aromatic compounds. Look for a parent name in the above examples (=1) and use the lowest possible number to the next substituent. If there are only two substituents, you can use ortho, meta or para prefixes.

a. CO_2H b. OH c. NH_2 / Br d. $CONH_2$ / OCH_3 e. Br / Cl / OCH_3 f. CHO

Use numbers when more than two substituents.

Benzene looks a lot like an alkene, however, its reactions are very different. Both alkenes and aromatic compounds are weakly **nucleophilic** (electron donating) using their pi electrons, but in general aromatic compounds are less nucleophilic than alkenes. Alkenes tend to undergo addition reactions, while aromatic compounds tend to undergo substitution reactions.

alkene pi electrons are localized (between the C=C)

Most alkene reactions involve electron donation to strong electrophiles. Addition reactions are the usual end-result.

aromatic pi electrons are spread out over the entire ring (shown with resonance structures)

resonance

Most aromatic reactions involve electron donation to stronger electrophiles. Substitution reactions are the usual end-result.

aromatic compounds are less reactive & undergo substitution reactions

alkenes are more reactive and undergo addition reactions

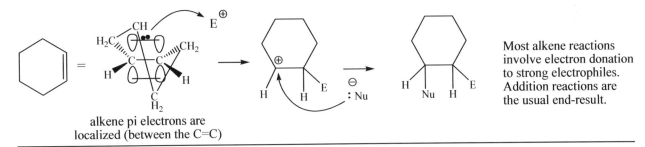

a.

Br_2 → no reaction

Br_2 / $FeBr_3$ cat. → Br + H-Br

Br_2 → Br / Br

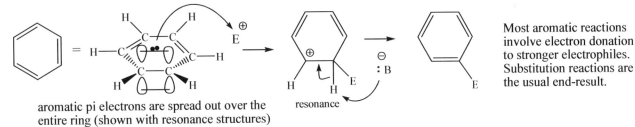

b.

D_2SO_4 / D_2O Δ harder → D

D_2SO_4 / D_2O Δ easier → OD / D

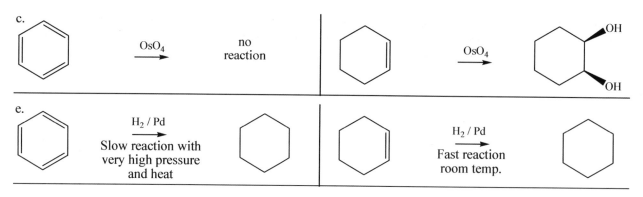

The first reaction below shows the heat of hydrogenation for a single pi bond in a six carbon ring. We might expect twice this value in the second reaction where two pi bonds are reduced. However, the reaction is less exothermic than expected by about 1.8 kcal/mole (the diene is more stable). A possible explanation for this extra stability is that the pi electrons are delocalized in the conjugated arrangement of pi bonds. We might propose that benzene should have a heat of hydrogenation of 3(-28.6) = -85.8 because it has three pi bonds. Real benzene must be much more stable than this because the "real" heat of hydrogenation is much less than expected at -49.8 kcal/mole. The energy of aromaticity could be estimated as being about 36.0 kcal/mole. Other approaches estimate even different values for aromaticity, but still very large, suggesting that benzene's cyclic structure and arrangement of pi bonds is exceptionally stabilizing. So, while hydrogenation reactions of pi bonds are normally very exothermic and alkenes react quickly when the proper catalyst is present (Pd for us), most benzene rings are much less reactive than expected and require harsher conditions to make the reactions work.

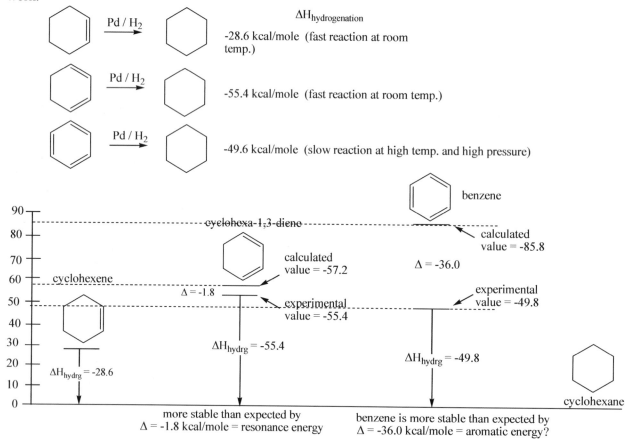

Another exceptional feature of aromatic compounds is that 1,2-disubstituted aromatics do not have isomeric structures. Early chemists tried for decades to make the apparently different isomers, but only one pattern has ever been discovered in aromatic compounds, unlike alkenes.

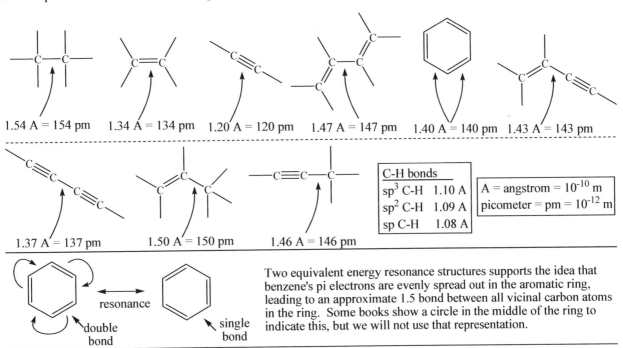

Bond lengths in benzene are all equal and intermediate between a double and single bond. The pi electrons can be written in more than one position in aromatics, which we call resonance. Another explanation states that a greater percent s character shortens the bonds because the s electrons are held tighter (closer to the nucleus), and it's possible both reasons are responsible for the shorter bonds.

1.54 A = 154 pm   1.34 A = 134 pm   1.20 A = 120 pm   1.47 A = 147 pm   1.40 A = 140 pm   1.43 A = 143 pm

- -

C-H bonds	
sp^3 C-H	1.10 A
sp^2 C-H	1.09 A
sp C-H	1.08 A

A = angstrom = 10^{-10} m
picometer = pm = 10^{-12} m

1.37 A = 137 pm   1.50 A = 150 pm   1.46 A = 146 pm

resonance
double bond   single bond

Two equivalent energy resonance structures supports the idea that benzene's pi electrons are evenly spread out in the aromatic ring, leading to an approximate 1.5 bond between all vicinal carbon atoms in the ring. Some books show a circle in the middle of the ring to indicate this, but we will not use that representation.

Problem 2 – Multiple benzene rings can be connected together and many variation are known. Collectively, these are called *polycyclic aromatic hydrocarbons* (PAH). Three famous examples are shown below. Draw the other resonance structures. Use the resonance structures to predict which bonds might be the shortest in the structures below? Hint 1 – Draw the pi bonds all the way around the perimeter of the PAH ring system, then shift all of them over one time like we do with benzene: Next find any "benzene" looking rings and draw the "benzene" resonance. If you do that for every ring, you should have drawn all of the resonance structures. Hint 2 - Some experimental x-ray data is provided just below these structures from a 1951 journal article.

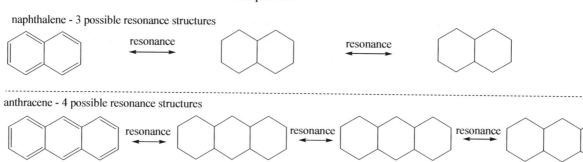

naphthalene - 3 possible resonance structures

anthracene - 4 possible resonance structures

phenanthrene - 5 possible resonance structures

Estimated resonance energies (RE) are shown in parentheses. Bond lengths are in pm = picometers = 10^{-12} m.

benzene
(RE = 36)

all bonds =140 pm

naphathelene
(RE = 61)

A = 142 pm
B = 136 pm
C = 140 pm
D = 140 pm

anthracene
(RE = 83)

A = 140 pm
B = 142 pm
C = 139 pm
D = 139 pm
E = 144 pm

phenanthrene
(RE = 91)

A = 134 pm
B = 142 pm
C = 142 pm
D = 134 pm
E = 137 pm
F = 138 pm
G = 138 pm
H = 146 pm

   The two most common allotropes of carbon are graphite and diamond (allotrope = Greek = "other form"). Graphite has almost infinite planes of fused benzene rings. It is structurally strong in the direction that would fracture those planes, but it is relatively easy to slide the planes in parallel directions, which gives it a lubricating property. Diamond, on the other hand, has an almost infinite 3 dimensional structure (sp^3 hybrid) that is very strong in all directions.

graphite (2D extended carbon network of benzene rings

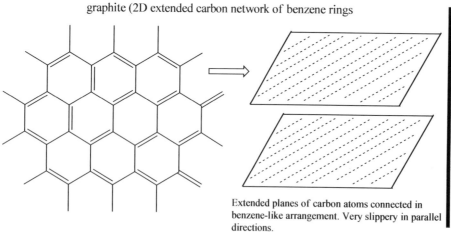

Extended planes of carbon atoms connected in benzene-like arrangement. Very slippery in parallel directions.

diamond (3D extended network of sp^3 hybridized carbon atoms)

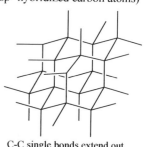

C-C single bonds extend out in all 3 dimensions. Diamond is rigid and hard, in contrast to slippery graphite.

Chapter 14

Other recently discovered allotropes of carbon are *Buckminsterfullerenes* (Bucky balls, Fullerenes) and nanotubes. The most famous of these is the C_{60} structure below. A Nobel prize was awarded to their discoverers in 1996 (Harold Kroto, Robert Curl and Richard Smalley). Uses range from microelectronic conductors to potential medicines.

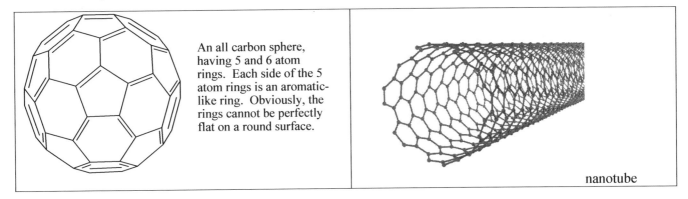

An all carbon sphere, having 5 and 6 atom rings. Each side of the 5 atom rings is an aromatic-like ring. Obviously, the rings cannot be perfectly flat on a round surface.

nanotube

Benzo[a]pyrene – a potent carcinogen

Benzo[a]pyrene is a very famous PAH compound, formed as a by-product of combustion processes (smoking, diesel exhaust, overdone cooking, forest fires,…). Another less famous PAH compound is 7,12-dimethylbenz[a]anthracene. This class of compounds was initially isolated from coal tar in 1933. It is estimated that 3000 tons are released into the environment every year. Some of these compounds are potent carcinogens. The actual carcinogens are not the hydrocarbons, but oxidized structures formed by oxidizing enzymes in the body. Benzo[a]pyrene diol-epoxide is a well studied example. It is formed by the action of cytochrome P-450 enzymes in the liver, which can oxidize carbon structures in many ways, epoxidation being a common method. One epoxide is opened to a diol by a reaction similar to that studied earlier. The diol-epoxide readily reacts with the DNA base guanine at the N-7 or O-6 positions (numbering scheme on the next page), which also methylate with CH_3-X electrophiles in epigenetics. Such reactions can inhibit base pairing and DNA transcription or replication, or might possibly turn on or turn off genes. In combination with several other errors (defective suicide gene defect = p53 and others), this can cause cancer in a cell. Benzo[a]pyrene diol-epoxide has a flat extended collection of rings that is possibly confused with a steroid and actively transported to the nucleus, where it can react with DNA. Normally, oxidizing a "foreign molecule" in the body is a good strategy. More OH groups means more water soluble and faster excretion in the urine or feces. In this case, that strategy backfires and a potent carcinogen is made and confused with the bodies other chemicals.

Epoxidation of pi bonds

Opening of epoxides to diols which increases water solubility (blood solubility allows elimination from the body).

Benzopyrene starts off the right way but takes a deadly detour to the nucleus.

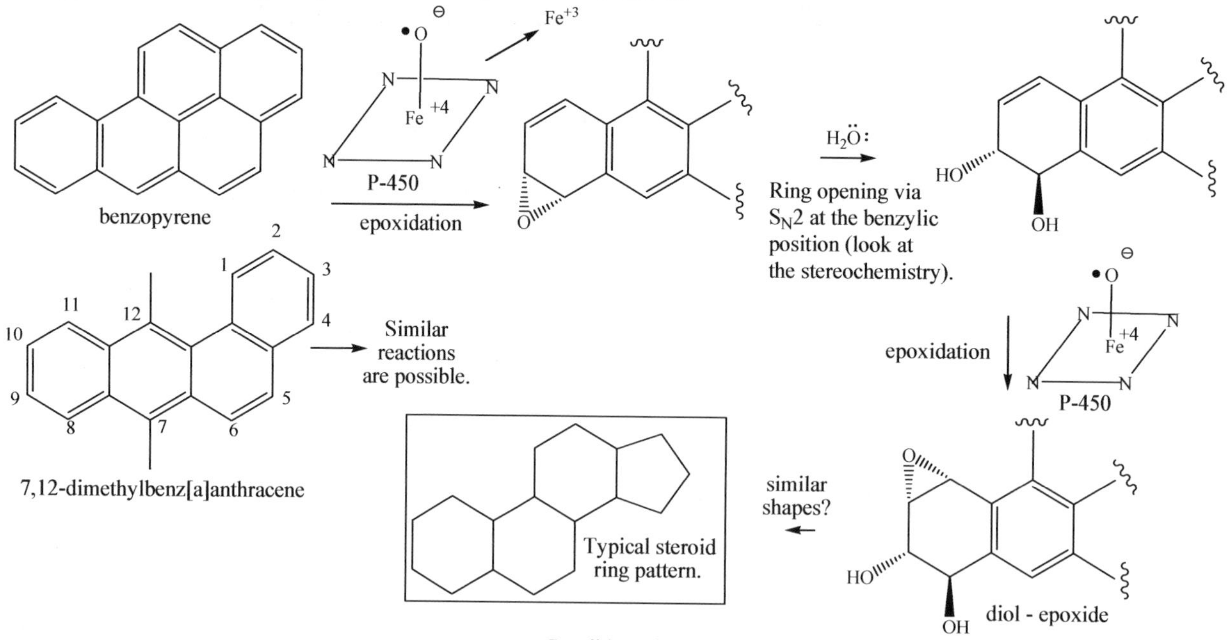
Fe+3
Fe +4
P-450
epoxidation
benzopyrene
H2Ö:
Ring opening via SN2 at the benzylic position (look at the stereochemistry).
HO
OH
epoxidation
P-450
Similar reactions are possible.
7,12-dimethylbenz[a]anthracene
Typical steroid ring pattern.
similar shapes?
diol - epoxide

Possibly actively transported to nucleus, where it reacts with DNA. The cell probably confuses it for a steroid.

Epigenetics puts methyl groups on DNA and acyl groups on lysine amino acids of histones. Random alkylation of DNA (or other biomolecules) leads to confused biochemistry, which can lead to cancer.

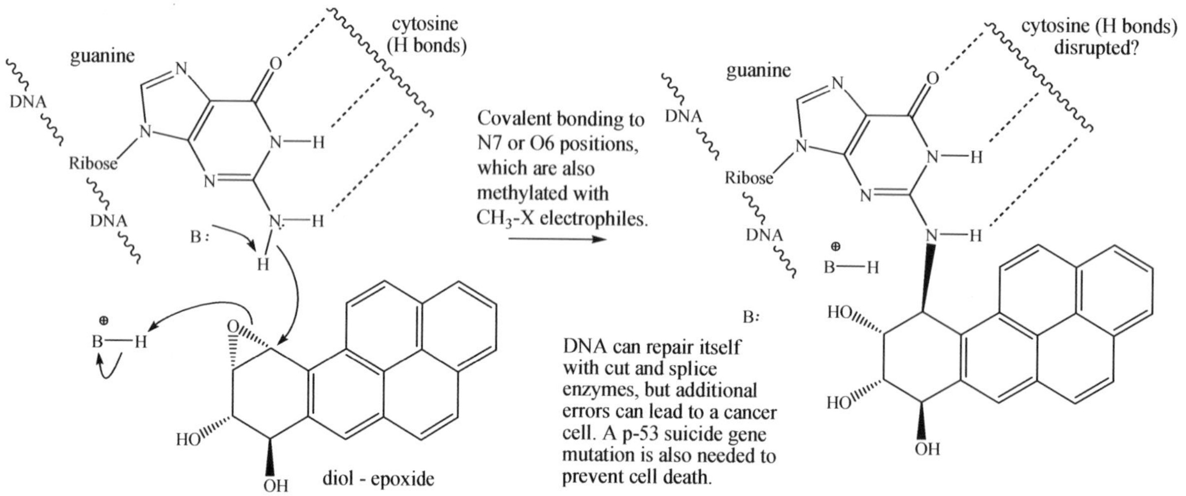
guanine
cytosine (H bonds)
DNA
Ribose
B:
diol - epoxide
Covalent bonding to N7 or O6 positions, which are also methylated with CH3-X electrophiles.
cytosine (H bonds) disrupted?
DNA can repair itself with cut and splice enzymes, but additional errors can lead to a cancer cell. A p-53 suicide gene mutation is also needed to prevent cell death.

Related chemistry: N7 or O6 alkylation - alkylation of DNA bases can be intended, as in methylation in epigenetics to turn off genes, or it can be random by external electrophiles causing serious damage (e.g. cancer).

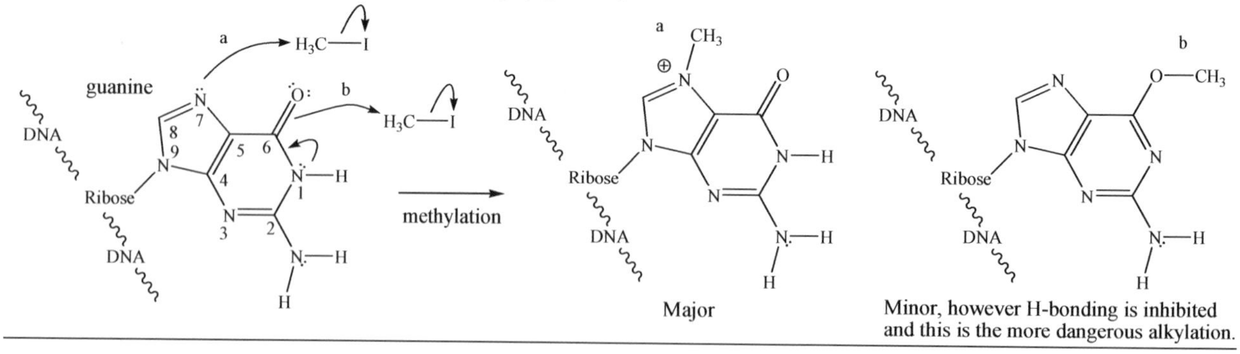
guanine
DNA
Ribose
H3C—I
methylation
Major
Minor, however H-bonding is inhibited and this is the more dangerous alkylation.

Reactions for our course – Two patterns of electrophilic aromatic substitution are presented

All of these reactions:

1. Generate a very reactive electrophile which attacks the pi electrons of the aromatic ring.

2. Our electrophiles are generated using two general approaches (use a strong Lewis acid = $FeBr_3$ or use a strong protic acid = H_2SO_4)

3. In all of our electrophilic aromatic substitution reactions (E.A.S.) the electrophile will 1. attack the ring at a position having a hydrogen, 2. form a resonance stabilized intermediate carbocation which will 3. lose the proton to a base at the position attacked by the electrophile, reforming the aromatic pi system, resulting in overall substitution

4. If a substituent (S_1) is already present on the ring when the E.A.S. reaction occurs, it will direct the new substituent (S_2) to either the ortho/para positions or the meta position. Which substitution pattern occurs depends on whether the first substituent is good at stabilizing the intermediate carbocation (= ortho/para attack), or bad at stabilizing the intermediate carbocation intermediate (= meta attack).

Three Specific Examples of Electrophilic Aromatic Substitution (Only 3 for us, because our time is short.)

1. **Nitration** reactions are probably the most important way to attach a nitrogen atom to an aromatic ring. This is often accomplished using nitric acid, HNO_3, in combination with stronger sulfuric acid, H_2SO_4, to generate the very reactive nitronium electrophile (NO_2^+). Because these are such harsh conditions there have been a number of alternative approaches developed to reach these important compounds (that we won't cover). Nitro groups can be reduced to amines (both groups have many biologically important examples). Both nitro and amine groups on aromatic rings often enhance the biological activity (in a good way, as a biological communicator or a pharmaceutical drug, or in a bad way as a carcinogen, and sometimes both can be true at once!). Such compounds should be handled carefully.

    a. Overall reaction = nitration

nitrobenzene

    b. The reactive nitronium electrophile (NO_2^+) is generated via protonation of the OH, making a water leaving group, which is then immediately protonated, forming H_3O^+.

electrophile = nitronium ion

c. The aromatic substitution part of the reaction is essentially the same for all of the electrophiles we discuss. 1. the aromatic ring donates its electrons to the electrophile, 2. forming a resonance stabilized intermediate and 3. loss of a beta proton to reform the aromatic ring (= substitution overall).

nucleophile   electrophile   pentadienyl
                                         R$^+$ resonance

Problem 3 - Propose a mechanism for the nitrosylation of benzene using nitrous acid and hydrochloric acid. Since the NO$^+$ electrophile is less reactive, it requires a more reactive aromatic compound. The methyl side chain activates the ring towards electron donation. Can you think of a reason why electron donation is easier? The relative reactivity of aromatic compounds is discussed later in this chapter.

toluene   4-nitrosotoluene

H_2O

Water would be protonated in the strongly acidic mixture.

The nitro group can be reduced to an amino group with HCl and Fe. Each metallic iron can supply two electrons (Fe → Fe^{+2} + 2e-) to reduce the nitro group (3 iron atoms need to donate 6 total electrons), and the HCl protonates the reduced nitrogen. We won't require a mechanism for this reaction (complicated and speculative), but we can speculate (if you want to). A complete speculative mechanism is written out on the next page and a blank template is on the subsequent page in case you want to try and fill in the details yourself.

nitrobenzene

Problem 4 – The amino group is often biologically important, and it has much synthetic potential, including diazonium chemistry, which we do not have time to cover. The following sequence is a very speculative mechanism for the reduction of a nitro group to an amine group, with a total of 6 electrons from 3 iron atoms. We show 2 electron transfers from the iron in the first two reduction steps and two single electron reduction steps in the last iron reduction step (to show a possible free radical variation). You don't need to know this, but speculation is fun. All of the mechanistic details are left out in a second scheme in case you want to try and fill them in yourself.

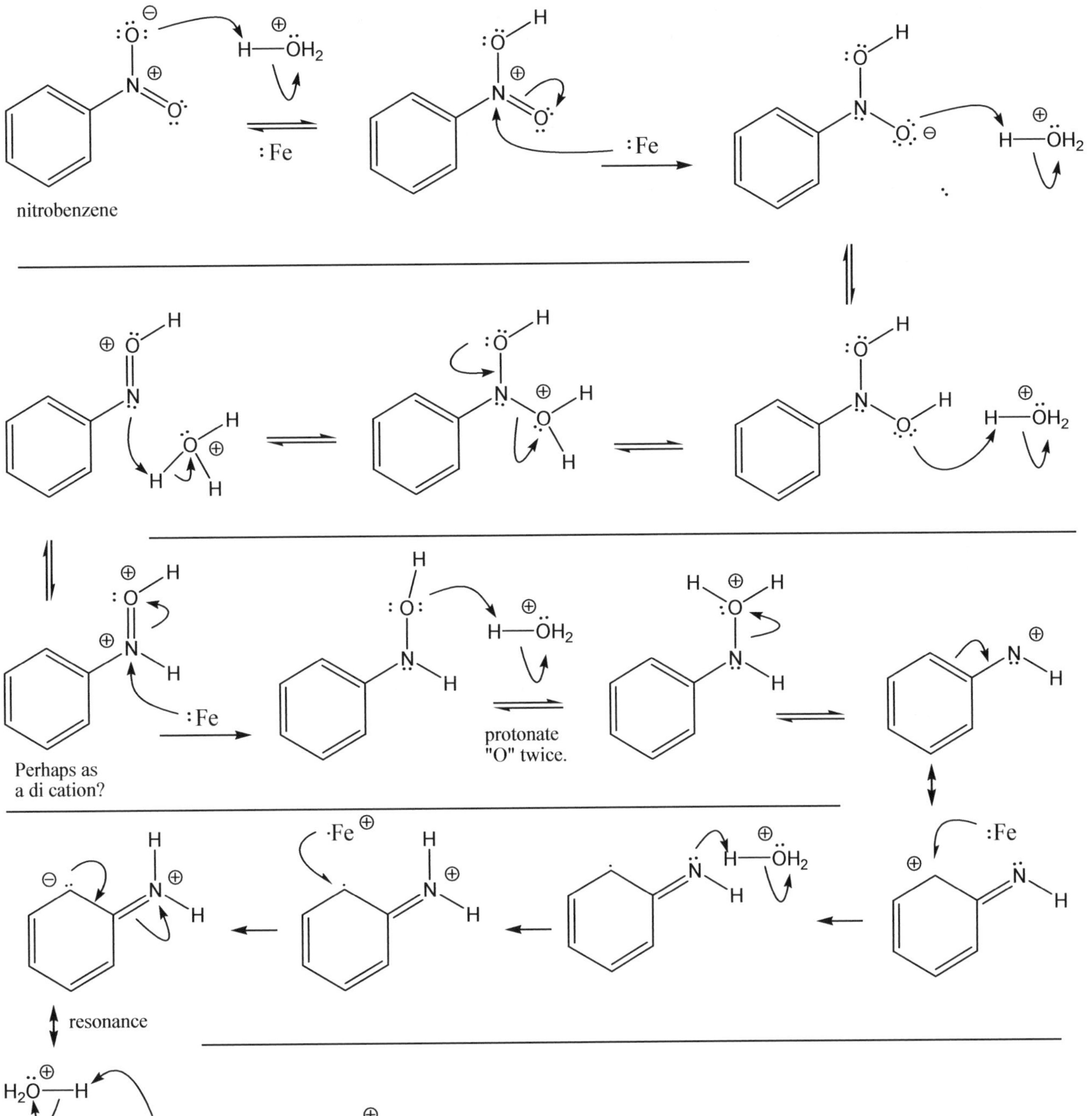
nitrobenzene
:Fe
:Fe
Perhaps as
a di cation?
:Fe
protonate
"O" twice.
·Fe
:Fe
resonance

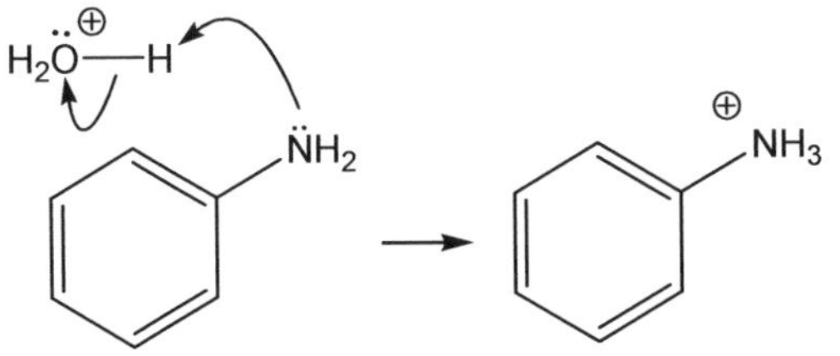

Aniline would be protonated in the acidic medium. This could be neutralized in a workup step.

Aromatic amines are important in themselves, and are the starting point for versatile diazonium chemistry, discussed later.

Problem 5 – Fill in missing mechanistic details in the proposed mechanism for reduction of an aromatic nitro group to an aromatic amine.

Perhaps as a di cation?

protonate "O" twice.

resonance

Aromatic amines are important in themselves, and are the starting point for versatile diazonium chemistry, discussed later.

Aniline would be protonated in the acidic medium. This could be neutralized in a workup step.

Examples of powerful acting amines and nitro compounds (just a few, the list is very long)

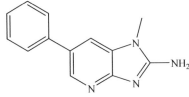

PhIP is one of the most common heterocyclic aromatic hydrocarbons formed when meat (all kinds) is cooked. More forms with higher cooking temperatures and longer cooking time. Is is strongly suspected of being a carcinogen.

PhIP (2-Amino-1-methyl-6-phenylimidazo[4,5-b]pyridine)

Nitro compounds are very rare in nature. 2-nitrophenol is an aggregation phermone in ticks.

PhIP in beef patties
fried, medium rare 0.29 ng/g
fried, well done 0.73 ng/g
fried, very well done 7.33 ng/g

Rats fed PhIp at 25 ppm developed mammary tumors. When injected with 5mg / Kg they showed genetic defects.

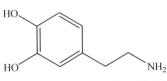

Dopamine - a neurotransmitter released by nerve cells to transmit signals to other nerve cells. Loss of dopamine secreting neurons causes Parkinson's disease.

2-Amino-1-methyl-6-phenylimidazo(4,5-b)pyridine = PhIP

Methamphetamine - powerful psychoactive drug with many undesirable side effects, including destruction of dopamine secreting neurons. Long withdrawal period (up to one year) and can cause psychosis. * = chiral center, only one enantiomer is active (dextrorotatory).

* = chiral center

2. **Sulfonation** of aromatic rings is possible with sulfuric acid, H_2SO_4 , and sulfur trioxide, SO_3, (together they are called oleum). Oleum is not only strongly acidic, it can be very oxidizing as well. These are very harsh conditions. We will show a simplistic mechanism using only H_2SO_4. Our proposed mechanism is very closely related to nitration, just above.

a. Overall reaction = sulfonation

H_2SO_4
(catalyst)

with SO_3 = oleum

This reaction can be run in reverse with lots of water.

benzenesulfonic acid

$H_2O \longrightarrow H_3O^{\oplus}$

Water would be protonated in the strongly acidic mixture.

b. The electrophile is generated via protonation of OH, making a water leaving group, which is immediately protonated, forming H_3O^+. Sulfonium ion, HSO_3^+ can be, simplistically, thought of as the electrophile (we will look at the reaction this way). This saves us two proton transfer steps in our mechanism, but SO_3 is the real electrophile.

For our purposes, this can be viewed as the electrophile. However, experimental evidence suggests that SO_3 is probably the real electrophile.

c. The aromatic substitution reaction is essentially the same for all of the electrophiles we discuss. However, a big difference in this example is the reaction can be run in reverse by adding lots of water. We won't actually propose this.

nucleophile   electrophile

pentadienyl
R^+ resonance

No water shifts equilibrium to the right.
Lots of water shifts equilibrium to the left.

benzenesulfonic acid

H_2SO_4

The sulfonic acid group can be made into a sulfonyl chloride (acid chloride) using phosphorous trichloride (PCl_3) in our course. Sulfonyl chloride can be made into a sulfonate esters using ROH, or a sulfonamides using NH_3, RNH_2 or R_2NH. Sulfonamides were some of the first antibiotics and many are still used this way today.

Possible mechanism using PCl_3 to make sulfonyl chloride.

Problem 6 – Propose a reasonable mechanism for each of the following two reactions.

benzenesulfonyl chloride   alkyl benzenesulfonates   benzenesulfonyl chloride   alkyl benzenesulfonamides

One way sulfa drugs work is by interfering with folate synthesis in bacteria. Bacteria have to make their folate, but we get ours from our diets. The key compound is sulfanilamide (in this example), which is close in shape and chemical reactivity with para-aminobenzoic acid (PABA), a part of dihydrofolate in the biochemistry of living organisms. Sulfanilamide interferes with bacteria's ability to synthesize folate, making them die (until they evolve a way around the problem). Sulfanilamide was actually part of a dye molecule that chemists originally thought was the antibiotic. Thousands of variations were made and used in the 1930s and early 1940s, until the penicillins came along. Uncountable patients were saved from a likely death from infection, including many American soldiers in World War II and the leader of England in World War II, Winston Churchill. I, myself, worked with an Army Veterinarian in the early 1970s who always sprinkled the yellow powdered sulfa drug in the wounds after all surgeries, telling me "Phil, it covers all the mistakes." About 3% of the general population is sensitive to sulfa drugs, but about 60% of HIV patients are sensitive.

One carbon metabolism
happens betwen these
two nitrogen atoms.

many repeats
of this part

sulfanilamide interfers
at this position

dihydrofolate
(Vit. B9)
Necessary to make DNA
and RNA bases.

sulfanilamide

para aminobenzoic acid = PABA

3. Halogenation of an aromatic ring is possible with Cl_2 or Br_2 and a strong Lewis acid (FeX_3, X = Cl or Br). We only propose Br_2 / $FeBr_3$ (bromination).

a. Overall reaction = bromination and chlorination

$$\text{benzene} + Br-Br \xrightarrow[\text{(cat.)}]{FeBr_3} \text{bromobenzene} + H-Br$$

b. Generation of the electrophile (E^+). $FeBr_3$ makes a Br a better leaving group from Br_2, just like and extra H^+ makes an OH into a better leaving group (water).

Lewis base   Lewis Acid

Br could be electrophilic
here with $FeBr_4^\ominus$ as a
leaving group with a push
from the aromatic ring...

...or Br^+ could be
electrophilic here.
We'll do it this way.

$FeBr_4^\ominus$ can act as a
base in a final step to
rearomatize the ring

c. The aromatic substitution reaction is essentially the same for all of the electrophiles we discuss (here Br_2, and $FeBr_3$)

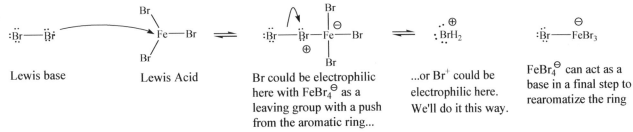

nucleophile   electrophile

pentadienyl R^+ resonance

Brominating (or chlorinating) an aromatic ring allows access to the many organometallic reactions studied previously. Grignard reagents are available from bromobenzene and other aromatic compounds. The following problem serves as a reminder of those reactions/ Putting a bromine on the aromatic ring also allows many other important reactions that we won't study: Heck, Suzuki, Stille reactions, and more. These reactions recently won the 2010 Nobel Prize.

Problem 7 – Remember, aromatic R-X compounds do not undergo S_N1 or S_N2 reactions, but they can be used with the Mg and Li organometallic reactions studied previously. Fill in the necessary reagents to accomplish the indicated transformations. Start with benzene.

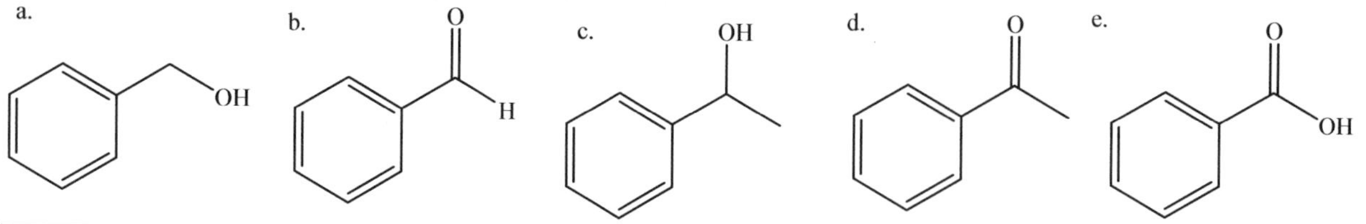

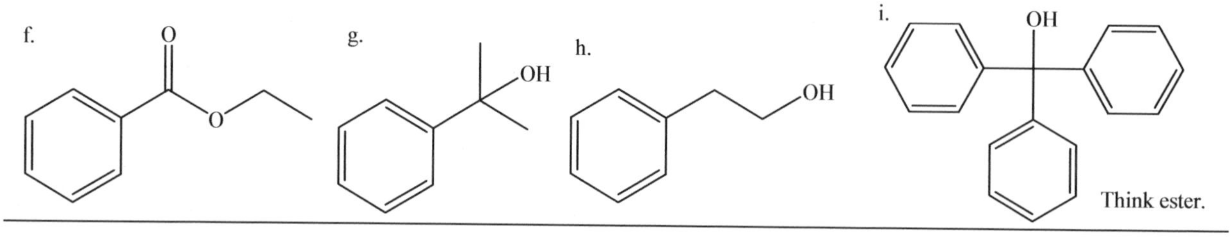

Problem 8 – Warning – "hard problem." Provide a mechanism for the following reaction (this is just a minor product of the desired reaction, bromination of t-butylbenzene). What is the electrophile (what is newly present in the product)? Which aromatic carbon must be attacked? What is the leaving group and why is it relatively stable? What happens to the leaving group in this reaction? When an aromatic ring carbon with a substituent is attacked, it is called "ipso" attack.

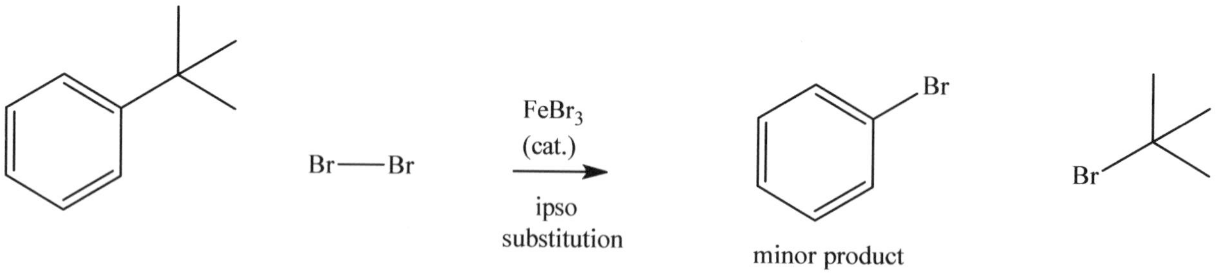

$$\text{Br} \longrightarrow \text{Br} \quad \xrightarrow[\substack{\text{ipso} \\ \text{substitution}}]{\substack{\text{FeBr}_3 \\ \text{(cat.)}}}$$

minor product

Reactions of Substituted Aromatic Rings with Electrophiles – Substituent Directing Effects

All six positions in benzene are equivalent in reactions with the above electrophiles. However, once a substituent is present, there are four different positions that can be attacked: the substituted carbon (called ipso attack), two positions immediately adjacent to the substituent (called ortho), two positions once removed from the substituted carbon (called meta) and one position directly across from the substituted carbon (called para). Which position(s) is (are) preferentially attacked by electrophiles depends primarily on what the substituent is, though the reactivity of an electrophile has some influence on the product distribution as well (more reactive = less selective). Substituents tend to fall into two different categories: ortho/para directors and meta directors (we will not emphasize "ipso" substitution). Understanding a substituent's effect on the pentadienyl carbocation intermediate is crucial for understanding the directing influence of any substituent. Those substituents that stabilize positive charge produce more ortho/para product (usually at a faster rate than benzene = activated), while substituents that destabilize positive charge produce more meta product (always at a slower rate than benzene = deactivated).

Chapter 14

<u>There are four types of possible positions to attack.</u>

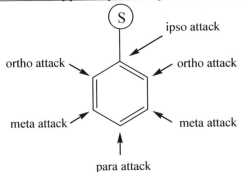

$\boxed{S}$ = substituent

You just saw these substituent possibilities:
S = -Br, -Cl, -R, -COR, -NO$_2$, -SO$_3$H

The "S" substituent tends to direct electrophiles to the

a. ortho and para positions   or
b. meta positions

   Activating substituents make the incoming electrophile react faster than they would with benzene (S = H). Such substituents are called activating groups. All activating groups are ortho/para directors. They help stabilize the pentadienyl carbocation intermediate.

   Deactivating substituents make the incoming electrophiles react slower than they would with benzene (S = H). Such substituents are called deactivating groups. Almost all deactivating groups are meta directors. They destabilize the pentadienyl carbocation intermediate. The halogens are an exception that we will ignore.

ortho / para directors and activating

Ar—N̈H$_2$ Ar—N̈R$_2$ Ar—Ö̈H   Ar—Ö̈R   Ar—N̈(H)—C(=O)R   Ar—Ö̈—C(=O)R   Ar—(phenyl)

All of these substituents can stabilize an adjacent positive charge, by resonance using a lone pair of electrons or an adjacent pi bond.

Ar—R   Alkyl groups stabilize adjacent positive charge, by an inductive effect.

Activating substituents make aromatic rings react faster with electrophiles than with benzene because they lower the activation energy, E_a.

ortho / para directors and slightly deactivating

Ar—F̈:   Ar—C̈l: Ar—B̈r: Ar—Ï:

"Ar" = aromatic ring

These halogens can stabilize an adjacent positive charge by resonance, but destabilize by their inductive effect. Because of the mismatch of p orbital size, the resonace effect is weaker than the inductive effect. (Remember, acid chlorides, RCOCl.) Surprisingly, fluorine is the least deactivating halogen. Its strong inductive withdrawal is balanced by the strongest resonance donation from a halogen (both carbon and fluorine use 2p orbitals).

---

meta directors are always deactivating

Ar—$\overset{\oplus}{N}$H$_3$ Ar—$\overset{\oplus}{N}$R$_3$ Ar—$\overset{\oplus}{N}$(=O)($\overset{\ominus}{O}$)   Ar—C≡N   Ar—CF$_3$ Ar—CCl$_3$

Ar—S(=O)(=O)—OH   Ar—C(=O)H   Ar—C(=O)R   Ar—C(=O)OH   Ar—C(=O)OR   Ar—C(=O)NR$_2$

All of these substituents have a strongly polarized positive atom attached to the aromatic ring, which destabilizes the carbocation intermediate when directly adjacent. Meta attack avoids this.

Deactivating substituents make aromatic rings react slower with electrophiles than with benzene.

All electrophiles can attack ortho, meta or para (ignoring ipso attack). In each case the positive charge is delocalized to 3 different positions. Ortho and para attack put the positive charge on the same carbon attached to the substituent. If the substituent stabilizes positive charge then ortho and para attack will be favored. If the substituent destabilizes positive charge then meta attack will be favored.

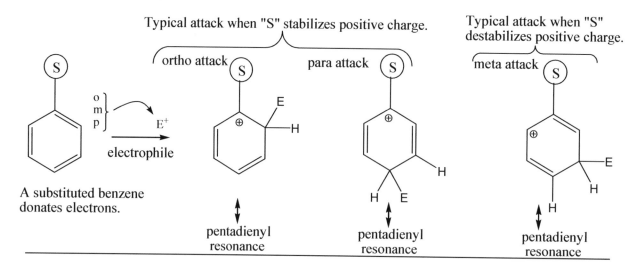

1. If "S" stabilizes the cationic intermediate Ar$^+$, then the reaction will go faster (S = activator) and resonance will be better when the electrophile is at the ortho and para positions relative to the substituent, S, since this puts the positive charge on the same carbon as the substituent.

2. If "S" destabilizes the cation in intermediate Ar$^+$, then the reaction will go slower (S = deactivator) and resonance will be poorer when the electrophile is at the ortho and para positions because the positive charge and S are on the same carbon.. The meta position looks relatively better than the ortho and para positions (i.e. meta looks less poor).

3. A large steric effects in the substituent or the electrophile can inhibit the amount of ortho substitution compared to para substitution.

Typical ortho para directors

a. Atoms with a lone pair of electrons next to the aromatic ring are good resonance donors to an empty carbocation site and will stabilize such an intermediate. This is an important clue that the directing effect of such substituents is ortho/para.

All of these substituent atoms have a lone pair of electrons next to the aromatic ring.

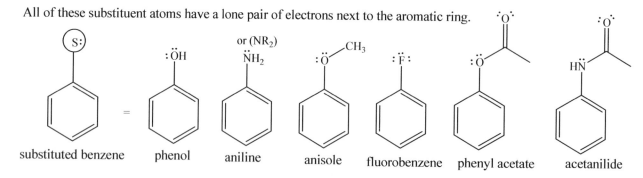

Chapter 14

The electron density profile that such substituents present to the incoming electrophiles is greater electron density (activation) than benzene (the ring of comparison), particularly so at the ortho and para positions. Positively polarized electrophiles will be more strongly attracted to these positions and react faster there. For all of the above substituents, the inductive effect is electron withdrawing from the ring (deactivating), but the resonance donating effect dominates (except for halogens, the exception we are ignoring).

Resonance effect of phenol (-OH as a substituent) _before_ attack by an electrophile.

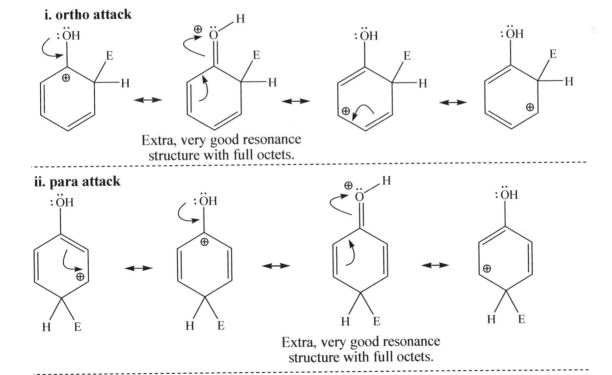

The first and last resonance structures are best, and represent the resonance of aromaticity. The three resonance structures in the middle show where the greatest electron density lies in the ring (ortho and para). These are the positions most attractive to the electrophile.

After attack by an electrophile – where is positive chare in the intermediate?

All substituents with lone pairs can share some electron density via resonance with the carbocation intermediate when the electrophile attacks at the ortho or para positions.. The extra resonance structure is especially good since it forms an extra bond and fills octets for all of the atoms. This stabilizes the transition state (lowers the E_a). Attack at the meta position does not produce this extra resonance structure. Ortho (2 positions) has a numbers advantage over para (1 position), but steric bulk in the substituent and/or the electrophile can slow the rate of ortho attack, making para attack the major product.

iii. meta attack

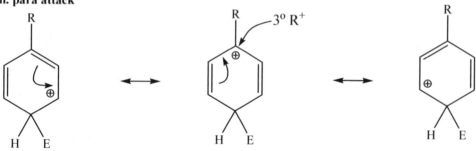

Meta attack misses the extra resonance structure from the lone pair on "S", so the lone pair is never next to the carbocation carbon.

b. Alkylbenzenes are also activating substituents and ortho and para directors. However, they don't have any lone pairs. Alkyl substituents contribute a tertiary carbocation resonance structure with ortho/para attack, not available with meta attack. This lowers the activation energy (E_a) and increases the rate of attack at these positions. The larger the alkyl group, the less ortho product formed due to a steric effect.

i. ortho attack - Large steric size in "R" or the electrophile reduces the amount of ortho product compared with para.

The 3° R⁺ resonance structure helps lower the E_a by stabilizing the transition state.

ii. para attack

The 3° R⁺ resonance structure helps lower the E_a by stabilizing the transition state.

iii. meta attack

Meta attack misses the 3° R⁺ resonance structure, as the positive charge is never next to the substituent carbon.

Chapter 14

Problem 9 – Explain the data in the table below (percent mononitration products of alkylbenzenes).

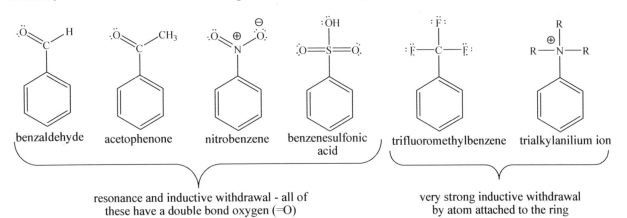

R =	% ortho	% para	% meta
-CH₃	62.1	34.6	3.3
-CH₂CH₃	50.3	46.6	3.6
-CH(CH₃)₂	44.2	53.5	2.3
-C(CH₃)₃	12.8	82.9	4.3

Typical Meta Directors

Typical meta directors are always deactivators (react slower than benzene = reference compound). Atoms with a pi bond to oxygen or having a positively polarized center next to the aromatic ring interact unfavorably with an adjacent positive charge. Such atoms are strong resonance and/or inductive electron withdrawers. If an empty carbocation site appears next to such centers it will strongly destabilize the intermediate (raising the activation energy and slowing the reaction). Such a pattern is an instant clue the substituent will be a meta director, and deactivate the aromatic ring relative to benzene by pulling electron density away from the ring.

benzaldehyde   acetophenone   nitrobenzene   benzenesulfonic acid   trifluoromethylbenzene   trialkylanilium ion

resonance and inductive withdrawal - all of these have a double bond oxygen (=O)

very strong inductive withdrawal by atom attached to the ring

Meta directing substituents make the ring particularly electron poor at the ortho and para positions. Positively polarized electrophiles attacking the ring are more likely to avoid those positions and react more often with the meta positions (though at a slower rate than they would react with benzene). For all of these substituents, the inductive effect is electron withdrawing from the ring, and when there is a "=O", an additional resonance effect reinforces the inductive effect.

Example of the charge distribution in an aromatic ring _before_ attack by an electrophile: (benzaldehyde, "S" = -CHO as a substituent) – has both inductive and resonance withdrawal.

carbonyl resonance

The first and last resonance structures are best and represent the resonance of aromaticity. The second resonance structure shows carbonyl resonance and the positively polarized carbon atom attached to the ring. The next 3 resonance structures in the middle show where the least electron density is present in the aromatic ring due to the resonance effect of the carbonyl group. These are the positions least attractive to the electrophile. The carbonyl group is also inductively electron withdrawing.

Chapter 14

Example of the charge distribution in aromatic ring _after_ attack by an electrophile: (benzaldehyde, "S" = -CHO as a substituent)

When electrophiles add ortho or para there is a face to face encounter of positive charges, which is destabilizing (higher E_a and slower rate). Attack at the meta position does not produce this conflict and is the least destabilized intermediate having the lowest activation energy (fastest rate). Attack at all positions is slower than with benzene, but meta is the preferred approach.

ortho attack

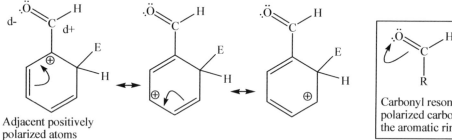

Adjacent positively polarized atoms destabilize this resonance contributor.

Carbonyl resonance puts a positively polarized carbon directly attached to the aromatic ring.

para attack

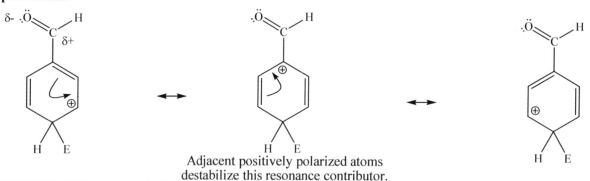

Adjacent positively polarized atoms destabilize this resonance contributor.

meta attack

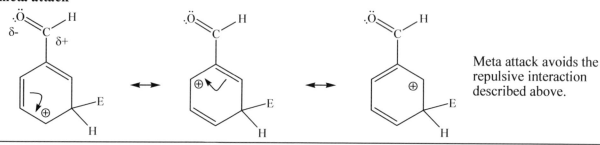

Meta attack avoids the repulsive interaction described above.

Problem 10 – Write out mechanisms for the following reactions, explaining the preferred regioselectivity.

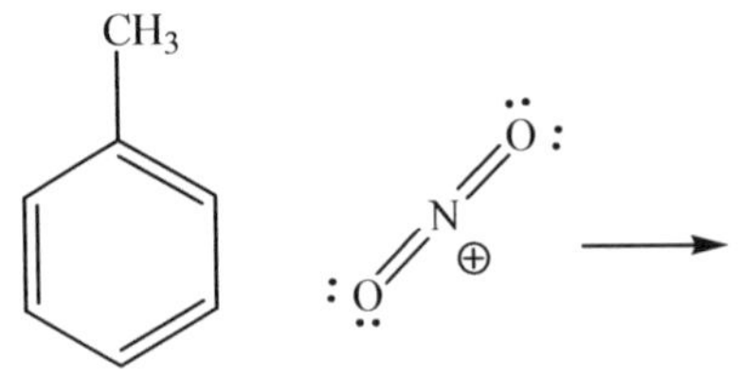

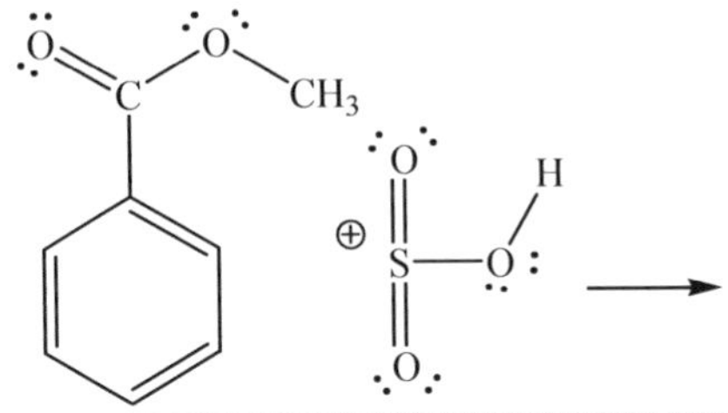

Chapter 14

Problem 11 – Predict what type of directing effect each substituent has below (o/p or m). Predict the major expected product(s) in reactions with a. HNO$_3$ / H$_2$SO$_4$ b. Br$_2$ / FeBr$_3$ c. H$_2$SO$_4$ / Δ.

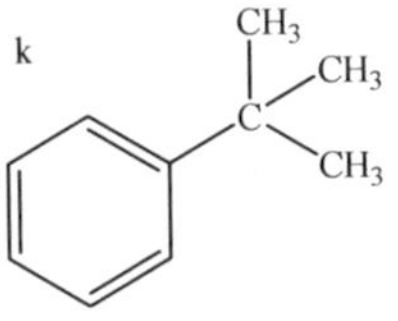

Problem 12 – Predict what the product will be and where the substitution will occur in each of the following reactions.

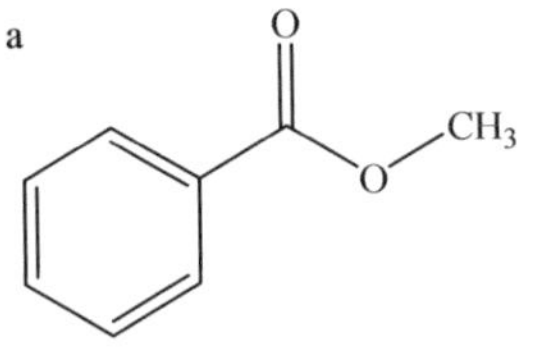

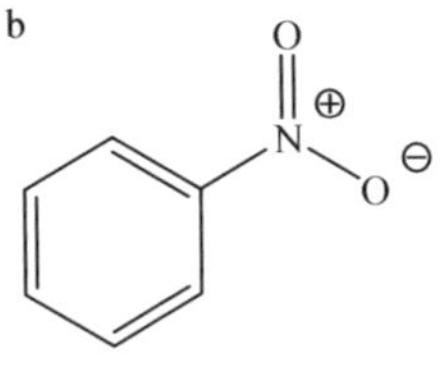

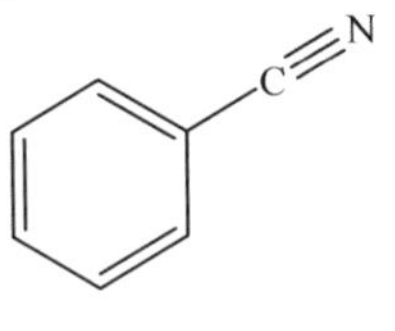

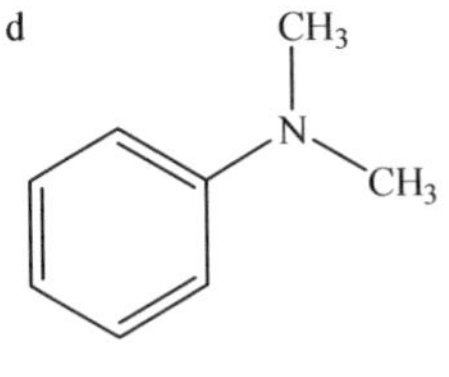

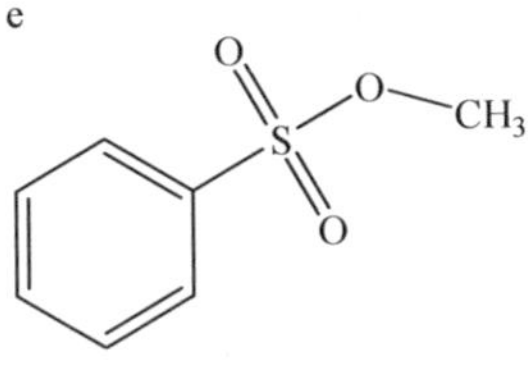

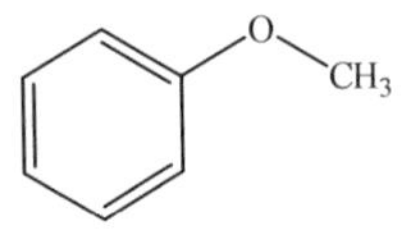

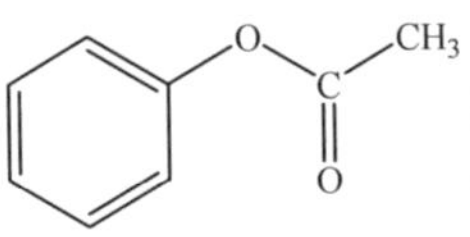

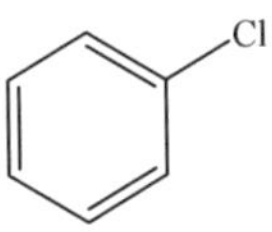

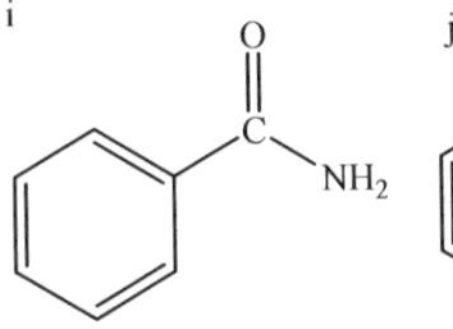

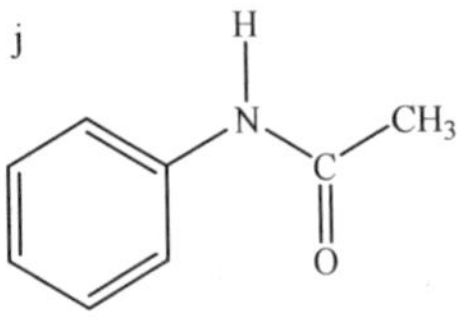

a. 4-phenylanisol

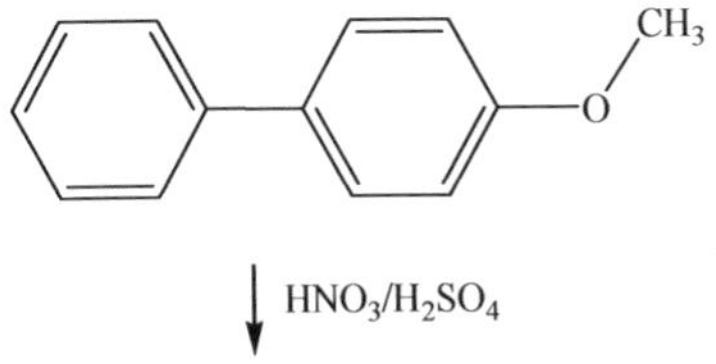

b. 4-phenylbenzoate

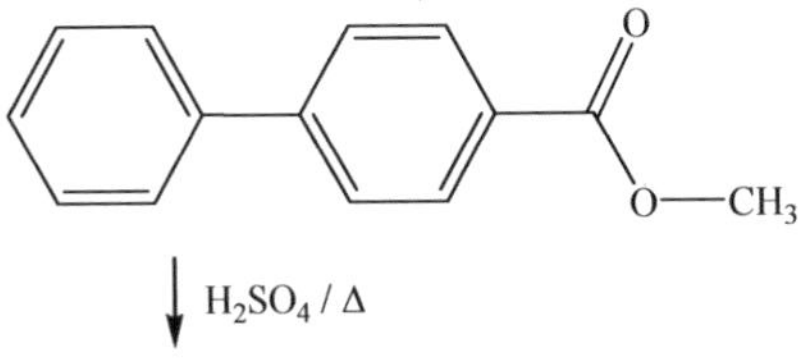

c. 4-(4-nitrophenyl)phenol

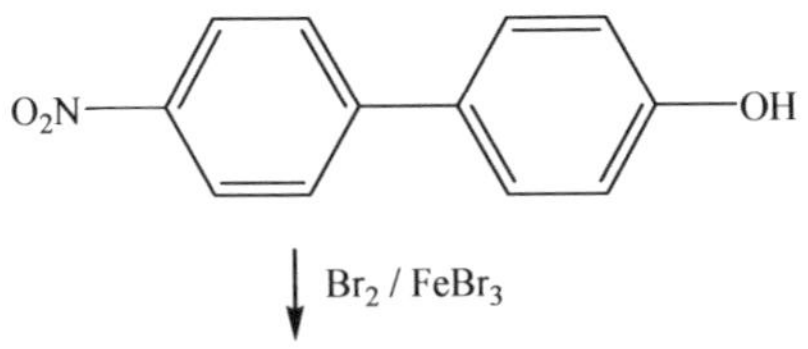

HNO$_3$/H$_2$SO$_4$ H$_2$SO$_4$ / Δ   Br$_2$ / FeBr$_3$

Topics are listed below in order for Chapters 1-14. There is no index so this is probably the best way to get an overview of where different topics can be found. While the problem count may appear low in some chapters, most problems have multiple parts, sometimes many parts. You don't necessarily have to do every part of every problem, but you have to do enough to learn and understand the material. More than most other courses, organic chemistry requires a constant, steady effort to learn the material. It continually builds on what you have learned before. You need to write out your answers using pencil and paper, lots of paper! Some students have told me that white boards worked for them, writing and erasing, over and over. Your success will depend on your foundation from other science and math courses, and your special talents at learning. It certainly won't be the same for every person in the course. You are the most qualified person to make that evaluation. Also, you probably wouldn't be taking this course if it wasn't important to your future goals. Is it worth one quarter or one semester of disciplined hard work to reach that career goal, a goal you plan to work at for 30-40 years of your life? Don't let momentary, mindless distractions divert you from succeeding in those dreams.

Most students take organic chemistry so they can understand biochemistry. Organic is a helpful stepping stone towards that goal. Attempting to learn organic chemistry the wrong way is a nightmare. Learning organic chemistry the right way is an exciting journey towards understanding the chemistry of life. It's really your choice. Whatever you do, do not underestimate the effort needed to pass this course.

List of Chapter Contents by page number

Chapter 1 – atoms, Zeff, IP, X, bonds, hybridization, 2D, 3D structures, resonance, bond line formulas

Chapter 2 – physical properties, types of bonding, solvents, solutes, solutions

Chapter 3 – functional groups, isomers, degrees of unsaturation

Chapter 4 - nomenclature

Chapter 5 – conformations, chains

Chapter 6 – conformaitons, cyclohexanes

Chapter 7 - stereochemistry

Chapter 8 – acid/base chemistry, inductive effects, resonance effects, pKa/Ka

Chapter 9 – SN/E reactions (RX compounds, carbonyl and epoxide electrophile analogies)

Chapter 10 Index – Free Radical Substitutions at sp3 C-H bonds and HBr addition to alkenes

Chapter 11 – Oxidation, Imines, Acid Derivatives, Synthesis

Chapter 14

Chapter 12 – Epoxides, Grignards, LDA, Enolates, Wittig and Synthesis

Chapter 13 – Alkene and Alkyne Chemistry

Chapter 14 – Aromatic Chemistry

Chapter 14

Made in the USA
San Bernardino, CA
21 September 2017